Microbial Fuels

Technologies and Applications

Microbial Fuels

Technologies and Applications

Edited by
Farshad Darvishi Harzevili
Serge Hiligsmann

CRC Press
Taylor & Francis Group
Boca Raton London New York

CRC Press is an imprint of the
Taylor & Francis Group, an **informa** business

CRC Press
Taylor & Francis Group
6000 Broken Sound Parkway NW, Suite 300
Boca Raton, FL 33487-2742

First issued in paperback 2019

© 2018 by Taylor & Francis Group, LLC
CRC Press is an imprint of Taylor & Francis Group, an Informa business

No claim to original U.S. Government works

ISBN-13: 978-1-4987-6379-0 (hbk)
ISBN-13: 978-0-367-87760-6 (pbk)

Library of Congress Cataloging-in-Publication Data

Names: Darvishi Harzevili, Farshad, editor. | Hiligsmann, Serge, editor.
Title: Microbial fuels : technologies and applications / edited by Farshad Darvishi Harzevili and Ir Serge Hiligsmann.
Description: Boca Raton : CRC Press, [2018] | Includes bibliographical references and index.
Identifiers: LCCN 2017009169| ISBN 9781498763790 (hardback : acid-free paper) | ISBN 9781498763806 (ebook)
Subjects: LCSH: Biomass energy. | Industrial microbiology.
Classification: LCC TP339 .M49 2018 | DDC 662/.88--dc23
LC record available at https://lccn.loc.gov/2017009169

Visit the Taylor & Francis Web site at
http://www.taylorandfrancis.com

and the CRC Press Web site at
http://www.crcpress.com

Contents

SECTION II   Liquid Biofuels

SECTION III   Prospects for Future Development of Biofuels

Preface

Microbial fuels are renewable sources of energy produced by microorganisms growing on different substrates. Microbial fuels can supplement or even replace fossil fuels and significantly reduce greenhouse gas emissions.

There are different classifications for biofuels. The microbial fuels are mainly classified based on the source (substrate) and state of the product (product). Microbial fuels are obtained from a wide range of substrates and can be divided roughly into three generations. First-generation, or conventional, biofuels are produced from sugar, starch, or vegetable oil, which are found in arable crops and can be easily extracted using conventional technology. Second-generation biofuels are obtained from nonfood products, such as lignocellulosic biomass or woody crops, agricultural residues, or waste. However, a series of physical, chemical, and biological treatments are required to convert lignocellulosic biomass to fuels. Third-generation biofuels are derived from algal biomass. Alternatively, some microorganisms can convert the CO, H_2, and CO_2 of syngas to different types of fuels. All the substrates may be converted to gaseous and liquid biofuels by microorganisms.

Successful development in the biofuel field requires major contributions in a wide range of disciplines, particularly microbiology, biochemistry, molecular biology, chemistry, biochemical engineering, and bioprocess engineering. Recently, new methods of metabolic engineering, industrial systems biology, synthetic biology, and X-omics science have been used to modify microorganisms involved in biofuel production.

Microbial Fuels: Technologies and Applications covers recent developments in technologies and applications of microbial fuels. Chapter 1 reviews microbial fuels from their historical roots to their different processes. Section I (Chapters 2 through 9) discusses the available technologies and the promising bioprocesses to produce microbial gaseous biofuels, such as biomethane and biohydrogen. Section II (Chapters 10 through 12) considers microbial liquid biofuel production from bioethanol to biodiesel and many derivatives. Section III (Chapters 13 through 15) argues about the prospects for the future development of microbial biofuels. The book is written in simple and clear text, and we also used many figures and tables to make it easier to understand. Furthermore, case studies are included at the end of some chapters. This book highlights major trends and developments in the field of microbial fuels and is written by experienced researchers in their respective fields.

Overall, this book will serve as a suitable reference for students, scientists, and researchers at universities, industries, corporations, and government agencies interested in bioenergy and clean technology, environmental and waste management, biotechnology, applied microbiology, bioprocess or fermentation technology, and all disciplines related to biofuel technologies and industries.

Farshad Darvishi Harzevili
Harzevil, Gilan, Iran

Serge Hiligsmann
Brussels, Belgium

Acknowledgments

We thank the experienced authors for their sound and enlightening contributions. We are extremely grateful to Michael Slaughter (acquiring editor) for his continued interest, critical evaluation, constructive criticism, and support.

On behalf of the authors, we would like to thank Scott Oakley (editorial assistant), the project coordinator, the production editor, the cover designer, and their respective teams at CRC Press/Taylor & Francis Group for their valuable efforts to develop our manuscript into a high-quality book.

The Editors

Editors

Farshad Darvishi Harzevili earned a BSc in biology at the University of Guilan, Iran. He earned his MSc and PhD in industrial microbiology and microbial biotechnology from the University of Isfahan, Iran. He is currently a faculty member and head of the microbial biotechnology and bioprocess engineering group at the University of Maragheh, Iran. His main interest is in the biotechnological and environmental applications of yeasts, especially the use of agro-industrial wastes and renewable low-cost substrates in the production of biotechnologically valuable products such as microbial enzymes and biofuels. He is also interested in the expression of heterologous proteins, metabolic engineering and synthetic biology of yeasts.

Serge Hiligsmann earned a master of chemical engineering and a PhD degree in engineering sciences at the University of Liège, Belgium. He is currently head of the 3BIO-BioTech unit at the Brussels Polytechnic School of Université Libre de Bruxelles. His researches are situated in the field of biotechnology and bioprocesses. More precisely, he investigates and develops biotechnological processes for the production of cell biomass, metabolites, and bioenergies based on aerobic or anaerobic fermentation and/or valorization of residual organic materials.

Contributors

Rémy Bayard
Université de Lyon
INSA Lyon
Research Group on Wastes, Water, Environment
    and Pollutions (DEEP)
Lyon, France

Philippe Bogaerts
3BIO–Biomodeling, Bioinformatics and Bioprocesses
Brussels School of Engineering
Université Libre de Bruxelles
Brussels, Belgium

Pierre Buffière
Université de Lyon
INSA Lyon
Research Group on Wastes, Water, Environment
    and Pollutions (DEEP)
Lyon, France

Ayse Dilan Celebi
Industrial Process and Energy Systems Engineering
Ecole Polytechnique Fédérale de Lausanne
EPFL Valais-Wallis
Sion, Switzerland

Wei Ning Chen
School of Chemical and Biomedical Engineering
Nanyang Technological University
Singapore

Farshad Darvishi
Microbial Biotechnology and Bioprocess Engineering Group
Division of Microbiology
Department of Biology
University of Maragheh
Maragheh, Iran

Sarina J. Ergas
Department of Civil and Environmental Engineering
University of South Florida
Tampa, Florida

Giovanni Esposito
Department of Civil and Mechanical Engineering
University of Cassino and Southern Lazio
Cassino (FR), Italy

Jean-François Flot
Laboratory of Evolutionary Biology and Ecology
Department of Organismal Biology
Université Libre de Bruxelles
Brussels, Belgium

Fabrice Franck
InBios-Phytosystems
Laboratory of Bioenergetics
University of Liège
Liège, Belgium

Luigi Frunzo
Department of Mathematics and Applications Renato
    Caccioppoli
University of Naples Federico II
Naples, Italy

Isabelle France George
Laboratory of Ecology of Aquatic Systems
School of Bioengineering
Université Libre de Bruxelles
Brussels, Belgium

Anish Ghimire
Department of Civil and Mechanical Engineering
University of Cassino and Southern Lazio
Cassino (FR), Italy
and
Department of Environmental Science and Engineering
Kathmandu University
Dhulikhel, Nepal

Bart Ghysels
InBios-Phytosystems
Laboratory of Bioenergetics
University of Liège
Liège, Belgium

Dimitri Gilis
3BIO–Biomodeling, Bioinformatics and Bioprocesses
Brussels School of Engineering
Université Libre de Bruxelles
Brussels, Belgium

Damien Godaux
InBios-Phytosystems
Laboratory of Genetics and Physiology of Microalgae
University of Liège
Liège, Belgium

Serge Hiligsmann
3BIO-Biotechnology and Bioprocess Department
Brussels School of Engineering
Université Libre de Bruxelles
Brussels, Belgium

Gregory R. Hinds
Department of Civil and Environmental Engineering
University of South Florida
Tampa, Florida

Carlo Saverio Iorio
Department of Chemical Physics
Université Libre de Bruxelles
Brussels, Belgium

Olivier Janssens
High School of Agro-Industries and Biotechnologies HELHa
Fleurus, Belgium

Amit Kumar
Environmental Biotechnology Center
University of Massachusetts
Amherst, Massachusetts

and

UNESCO-IHE Institute for Water Education
Delft, the Netherlands

Piet N.L. Lens
UNESCO-IHE Institute for Water Education
Delft, the Netherlands

Grégoire Léonard
Products, Environment, and Processes (PEPs)
Department of Chemical Engineering
University of Liège
Liège, Belgium

Vincenzo Luongo
Department of Civil, Architectural and Environmental
   Engineering
University of Naples Federico II
Naples, Italy

Hatim Machrafi
Department of Chemical Physics
Université Libre de Bruxelles
Brussels, Belgium

François Maréchal
Industrial Process and Energy Systems Engineering
Ecole Polytechnique Fédérale de Lausanne
EPFL Valais-Wallis
Sion, Switzerland

Christophe Minetti
Department of Chemical Physics
Université Libre de Bruxelles
Brussels, Belgium

Patrícia Madeira da Silva Moura
Bioenergy Unit
National Laboratory of Energy and Geology (LNEG)
Lisbon, Portugal

Joana Resende Ortigueira
Bioenergy Unit
National Laboratory of Energy and Geology (LNEG)
Lisbon, Portugal

Andreas Pfennig
Products, Environment, and Processes (PEPs)
Department of Chemical Engineering
University of Liège
Liège, Belgium

Francesco Pirozzi
Department of Civil, Architectural and Environmental
   Engineering
University of Naples Federico II
Naples, Italy

Puhulwella G. Rathnasiri
Department of Chemical and Process Engineering
University of Moratuwa
Moratuwa, Sri Lanka

Marianne Rooman
3BIO–Biomodeling, Bioinformatics and Bioprocesses
Brussels School of Engineering
Université Libre de Bruxelles
Brussels, Belgium

Ganesh Dattatray Saratale
Department of Food Science and Biotechnology
Dongguk University
Seoul, Republic of Korea

Rijuta Ganesh Saratale
Research Institute of Biotechnology and Medical Converged
   Science
Dongguk University
Seoul, Republic of Korea

Shivom Sharma
Industrial Process and Energy Systems Engineering
Ecole Polytechnique Fédérale de Lausanne
EPFL Valais-Wallis
Sion, Switzerland

Jiahua Shi
School of Chemical and Biomedical Engineering
Nanyang Technological University
Singapore

Carla Alexandra Monteiro da Silva
Department of Geographic, Geophysics and Energy
   Engineering
Faculty of Sciences
Lisbon University
Lisbon, Portugal

David C. Stuckey
Department of Chemical Engineering
Imperial College London
London, United Kingdom

Bo H. Svensson
Department of Thematic Studies
Environmental Change
Biogas Research Center
Linköping University
Linköping, Sweden

Antoine P. Trzcinski
School of Civil Engineering and Surveying
Faculty of Health, Engineering and Sciences
University of Southern Queensland
Queensland, Australia

Idania Valdez-Vazquez
Unidad Académica Juriquilla Instituto de Ingeniería
Universidad Nacional Autónoma de México
Querétaro, Mexico

Miriam H.A. van Eekert
LeAF BV
and
Subdepartment of Environmental Technology
Wageningen University and Research
Wageningen, the Netherlands

Grietje Zeeman
Subdepartment of Environmental Technology
Wageningen University and Research
and
LeAF BV
Wageningen, the Netherlands

Qiong Zhang
Department of Civil and Environmental Engineering
University of South Florida
Tampa, Florida

1

Impressive Potential of Microorganisms to Achieve the Transition from Fossil Fuels to Biofuels

Farshad Darvishi and Serge Hiligsmann

Contents

1.1 Introduction

Fuel is a substance that under chemical reaction with oxygen (burning) converts into heat energy and other types of energies. Wood combustion is the first use of fuel by humans.

Later, coal was extracted and used as a fuel from around 1000 BC in China. However, it became more common as a power source after development of the steam engine. In the nineteenth century, gas extracted from coal was used for street lighting in London. After the discovery of petroleum and natural gas, these materials replaced coal. Fossil fuels include coal, petroleum, and natural gas formed from the fossilized remains of plants and animals at high temperature and pressure in the absence of oxygen in the earth's crust over millions of years (Karim 2012). According to *International Energy Outlook (IEO) 2016*, provided by the U.S. Energy Information Administration (EIA), fossil fuels will account for 78% of the total world energy consumption up to 2040 and remain the largest source of energy (Figure 1.1).

However, the total world marketed energy consumption share of fossil fuels should decline from 33% in 2012 to 30% in 2040 (www.eia.gov/forecasts/ieo). Fossil fuels are

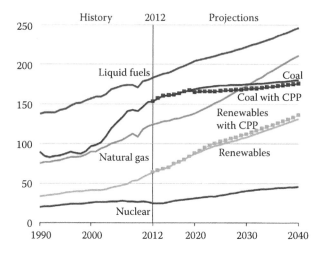

Figure 1.1 Total world energy consumption by energy source, 1990–2040 (quadrillion Btu). Dotted lines for coal and renewables show projected effects of the U.S. Clean Power Plan (CPP). (From U.S. Energy Information Administration, *International Energy Outlook 2016*, U.S. Energy Information Administration, Washington, DC, 2016, www.eia.gov/forecasts/ieo.)

nonrenewable resources since they formed over millions of years ago. This is a major drawback regarding the use of these fuels. Furthermore, the release of greenhouse gases such as carbon dioxide (CO_2) during fossil fuel combustion is another problem. Indeed, the greenhouse gases contribute to global warming and the temperature increase of the earth's surface. On the basis of IEO 2016, the world energy-related CO_2 emissions will rise from 32.2 billion metric tons in 2012 to 35.6 billion metric tons in 2020 and to 43.2 billion metric tons in 2040 (Figure 1.2).

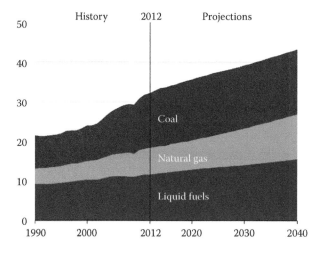

Figure 1.2 World energy-related carbon dioxide emissions by fuel type, 1990–2040 (billion metric tons). (From U.S. Energy Information Administration, *International Energy Outlook 2016*, U.S. Energy Information Administration, Washington, DC, 2016, www.eia.gov/forecasts/ieo.)

Since the replacement and conversion of energy type to another type is difficult, the use of new energy sources is necessary to avoid environmental pollution. New energy resources should be renewable and carbon-free in the future, because their consumption does not release pollutant greenhouse gases and has no destructive effects on weather and the environment. Renewable energies naturally replenish on a human timescale, for example, sunlight, wind, rain, tides, waves, geothermal heat, fuel cells, and biomass. They are expected to be the world's fastest-growing energy source, with their consumption increasing an average 2.6% per year between 2012 and 2040 (Figure 1.1).

Among the renewable energies, biomass considers any biological material derived from living or recently living organisms. Biofuels derived from biomass cover solid, liquid, and gaseous fuels. Microbial fuels are one of the biofuel types with renewable energy found from microorganisms and their derivatives. Microbial fuels can supplement or even replace fossil fuels. Their use could significantly reduce greenhouse gas emissions. All renewable energy sources are almost available periodically and not portable or storable; therefore, they cannot be used for fuel, especially in transportation. Microbial fuels, unlike other renewable energy sources, are usable without interruption and not limited to the season, time, and certain conditions. The cost of microbial fuels is competitive and cheaper than that of other types of renewable energy sources (Singh et al. 2016).

1.2 Brief History

Ethanol was the first man-made fuel. It is a liquid fuel with microbial origin. The history of ethanol is very long, that is, since humans learned to carry out the fermentation of sugar into ethanol as one of the earliest organic reactions. Dried ethanol residues found on 9000-year-old pottery in China suggest that Neolithic people may have consumed alcoholic beverages. Iranian pioneer scholar Zakariya Razi (854–930 AD), known in the West as Rhazes, discovered and purified alcohol (ethanol) and pioneered its use in medicine (Modanlou 2008) (Figure 1.3).

In the eighteenth century, alcohol-burning stoves were used for cooking and the warming of homes, and vegetable oils and fats lit up streets in Europe and America. In 1860, German inventor Nicholas Otto used ethanol as fuel in engines. Some decades later, in 1896, Henry Ford built the first automobile equipped with an engine designed to run on pure ethanol. Recently, the Energy Policy Act of 1992 in the United States defined ethanol blends with at least 85% ethanol to be an alternative fuel.

Louis Pasteur produced butanol by biological means in 1861, and Auguste Fernbach developed a bacterial fermentation process to produce butanol using potato starch in 1910. During World War I, the chemist Chaim Weizmann developed acetone-butanol-ethanol (ABE) fermentation by *Clostridium acetobutylicum*. After World War II, industrial ABE fermentation declined rapidly because of the cheaper petrochemical production of butanol. In 2006, the DuPont and British Petrol companies announced that they reconstituted the industrial-scale ABE fermentation in the United Kingdom. Now, butanol is generating interest as a renewable biofuel for scientists and companies.

Figure 1.3 Iranian pioneer scholar Zakariya Razi (854–930 AD), who discovered and identified ethanol by distillation. (The statue of Razi in the United Nations office in Vienna is part
of the Scholars Pavilion.)

Microbial gaseous fuels too have long historical roots as liquid fuels. Biogas as a
gaseous fuel typically refers to a gas produced by the biological breakdown of organic
matter by microbial fermentation or anaerobic digestion in the absence of oxygen. It
comprises primarily methane (CH_4) and carbon dioxide (Ayadi et al. 2016).

Ancient Persians understood that rotting vegetable matter gives off a flammable
gas, and Marco Polo mentioned the use of covered sewage tanks in China. In the
sixteenth century, scholar and architect Sheikh Bahai designed and constructed a
public bathroom known as Sheikh Bahai's bathroom in Isfahan, Iran (Figure 1.4).
It ran and provided hot water to the public by a single candle (a small flame) with

Figure 1.4 Sheikh Bahai's public bathroom, where biogas was used to heat water in sixteenth-century Isfahan, Iran.

an automatic flame ignition system for a long time. In fact, Sheikh Bahai used the flammable gas or biogas that was naturally produced in a nearby cesspool for heating the bathwater.

The first sewage plant was built in Bombay in 1859, and the idea for the manufacturing of gas was brought to the United Kingdom, for gas lighting in street lamps and homes, in 1895.

Louis Pasteur tried to produce biogas from horse droppings, and he explained that the rate of production was sufficient to cover the energy needs for street lighting in Paris. In the early 1900s, systems for the treatment of sewage were developed in the United Kingdom and Germany. Centralized drainage systems were further installed in many towns in Europe, and anaerobic digestion was seen as a means to reduce the volume of solid matter in sewage. However, the resulting gas was occasionally used as a source of energy.

In the 1930s, the use of farm manure to generate methane was developed, again in Bombay. Indian villagers by the Khadi and Village Industries Commission (KVIC) developed it for use in the early 1960s. China started a similar program in the 1960s and claimed that 5 million plants had been built by the early 1980s (Deublein and Steinhauser 2010).

Besides, Michael Potter, a professor at the University of Durham, used microbes to produce electricity in 1911 (Potter 1911). In fact, he established microbial fuel cells (MFCs),

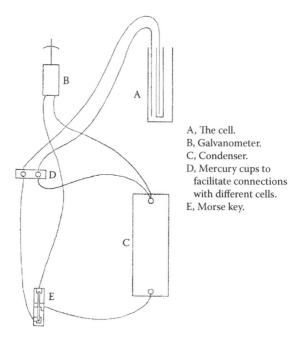

A, The cell.
B, Galvanometer.
C, Condenser.
D, Mercury cups to facilitate connections with different cells.
E, Morse key.

Figure 1.5   First plan of MFC. (From Potter, M.C., *Proc. R. Soc. B*, 84, 260–276, 1911.)

or biological fuel cells, and generated electricity from microorganisms for the first time (Figure 1.5). In the twenty-first century, researchers have been trying to use MFCs for electricity generation on a commercial scale, simultaneously with wastewater treatment.

1.3 Classification of Microbial Fuels

There are different classifications for biofuels. The microbial fuels are mainly classified based on the source (substrate) and state of the product (product) (Figure 1.6) (Hollinshead et al. 2014).

Microbial fuels are obtained from a wide range of substrates and can be divided roughly into three generations: First-generation, or conventional biofuels, are produced from sugar, starch, or vegetable oil, which are found in arable crops and can be easily extracted using conventional technology. Second-generation biofuels are obtained from nonfood products, such as lignocellulosic biomass or woody crops, agricultural residues, or waste. However, a series of physical, chemical, and biological treatments are required to convert lignocellulosic biomass to fuels. Third-generation biofuels are derived from algal biomass (Mousdale 2010).

Direct hydrolysis and fermentation of these materials to biofuels can be challenging by microorganisms. A promising approach for the production of fuel is indirect fermentation, after plant material pyrolysis, to produce synthesis gas, or syngas.

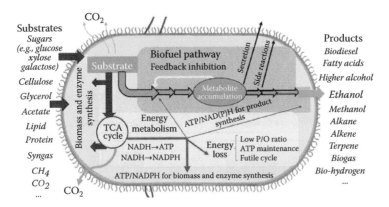

Figure 1.6 Microbial fuels are mainly classified based on the source (substrate) and state of the product (product). TCA, tricarboxylic acid. (From Hollinshead, W. et al., *Front. Microbiol.*, 5, 1–8, 2014.)

Alternatively, some microorganisms can convert the carbon monoxide (CO), hydrogen (H_2), and CO_2 of syngas to different types of fuels.

All the substrates may be converted to gaseous, liquid, or solid products by microorganisms. Therefore, gaseous, liquid, and solid biofuels are three types of microbial fuels according to the state of microbial products (Naik et al. 2010).

1.3.1 Gaseous Microbial Fuels

The gases hydrogen, methane, and carbon monoxide can be burned or oxidized with oxygen to release energy for use as clean fuels. Hydrogen is naturally produced by archaea, anaerobic and facultative aerobic bacteria, cyanobacteria, and lower eukaryotes in a single or mixed (consortium) cultures. Two main routes, photofermentation and dark fermentation, are generally considered (Figure 1.7). However, enzymatic and microbial electrolysis, or a combination of these processes, may be used for hydrogen production (Chandrasekhar et al. 2015).

Biogas usually concerns the mixture of mainly methane and carbon dioxide, but may have small amounts of hydrogen sulfide (H_2S), moisture, and siloxanes. The degradation of organic material with the production of biogas occurs in four stages, namely, hydrolysis, acidogenesis, acetogenesis, and methanogenesis (Figure 1.8).

In the hydrolysis stage, insoluble complex organic matters hydrolyze by fermentative microorganisms into soluble monomers, such as sugars, fatty acids, and amino acids. Then, the products further hydrolyze to acetate, hydrogen, carbon dioxide, and intermediate products containing propionic acid, acetic acid, formic acid, and so on, by acidogenic microorganisms. Acetogenic microorganisms convert the intermediate products from acidogenesis into methanogenic substrates. In the final stage, methanogens convert acetic acid into methane, carbon dioxide, and water (Chandrasekhar et al. 2015).

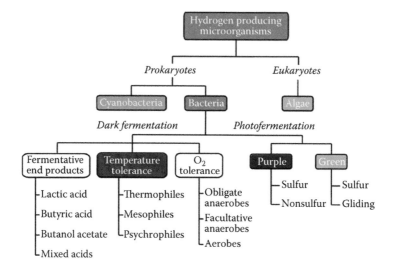

Figure 1.7   Two main routes of biohydrogen production (photofermentation and dark fermentation) and their involved microorganisms. (From Chandrasekhar, K. et al., *Int. J. Mol. Sci.*, 16(4), 8266–8293, 2015.)

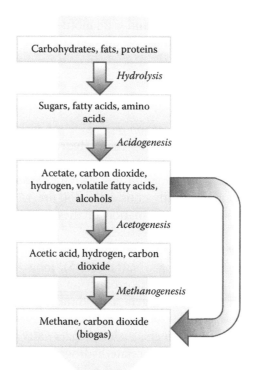

Figure 1.8   Biogas is produced from organic material in four stages: hydrolysis, acidogenesis, acetogenesis, and methanogenesis. (From http://ttrenewable.com/expertise/biogas-production/.)

1.3.2 Liquid Microbial Fuels

According to AEO2011 (Annual Energy Outlook) provided by the EIA, liquid biofuel consumption (1% in 2009) will reach 3% in 2035 (Figure 1.9).

Microbial liquid fuels can be divided into four classes, depending on the pathways: fermentative short-chain alcohols, nonfermentative short-chain alcohols, isoprenoid-derived hydrocarbons, and fatty acid-derived hydrocarbons (Figure 1.10) (Rude and Schirmer 2009).

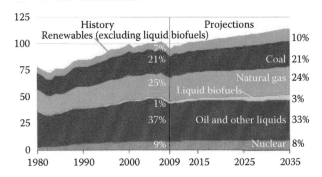

Figure 1.9 Total world energy consumption by type of fuel, 1980–2035 (quadrillion Btu). (From U.S. Energy Information Administration, AEO2011 early release overview, U.S. Energy Information Administration, Washington, DC, 2011, www.eia.gov/forecasts/ieo.)

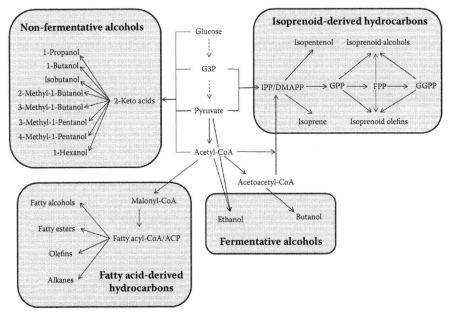

Figure 1.10 Metabolic pathways leading to microbial fuels. G3P, glyceraldehyde-3-phosphate; CoA, coenzyme A; ACP, acyl carrier protein; IPP, isopentenyl pyrophosphate, DMAPP, dimethylallyl pyrophosphate; FPP, farnesyl pyrophosphate; GPP, geranyl pyrophosphate; GGPP, geranylgeranyl pyrophosphate. (From Rude, M.A., and Schirmer, A., *Curr. Opin. Microbiol.*, 12(3), 274–281, 2009.)

All fermentative and nonfermentative short-chain alcohols can function as good gasoline replacements or blends and include ethanol, butanol, and isobutanol. Ethanol is not a new microbial fuel, and butanol has recently found renewed interest.

Sesquiterpenes like farnesol and farnesene, as isoprenoid-derived hydrocarbons, develop as precursors to diesel fuels. Monoterpenes such as pinene, sabinene, and terpinene are discussed as the potential next generation of jet fuel components. However, higher isoprenoid compounds (>C30) are explored for use as fuel feedstock.

Most fatty acid-derived hydrocarbons, such as ethyl hexadecanoate and pentadecane, are also suitable for the production of diesel fuel. The green microalga *Botryococcus braunii* has been considered a promising source of hydrocarbons up to now, with accumulation of hydrocarbons up to 75% of dry mass. Oleaginous microorganisms producing microbial oils or lipids are also known as single-cell oils (SCOs) (Rude and Schirmer 2009; Cohen and Ratledge 2010).

1.3.3 Other Types of Microbial Fuels

Higher-molecular-weight isoprenoid- or fatty acid-derived hydrocarbons, like waxes and polymers, are solid and insoluble in water. They can serve as feedstock for oil refineries.

MFCs are one of the most promising energy sources for electricity generation. Three different generations of MFCs can be distinguished by their historical development and mechanisms of electron transfer. In the first generation, or mediated MFCs, synthetic redox mediators are combined with microorganisms to transfer electrons to the anode in the cell. After the 1970s, unmediated MFCs emerged with bacteria typically having electrochemically active redox proteins, such as cytochromes, on their outer membrane and being able to transfer electrons directly to the anode. In the second generation, the natural mediating properties of sulfate-reducing microorganisms are used in the cell. In the third generation, MFCs are based on the anodophillic species, like *Geobacter sulfurreducens*, and require no soluble mediator. Each type of MFC operates under similar environmental conditions. In terms of substrate to power conversion efficiency, the second generation was most efficient. When output is expressed as power per unit of cells, the third generation is 28-fold higher than the other ones. Additionally, the second and third generations may be used advantageously in wastewater treatment and power generation from organic matter (Ieropoulos et al. 2005).

1.4 From Domestication of Microbial Processes to Biofuel Production: A Sustainable Synergy to Promote with Organic Waste Treatment

1.4.1 Biodegradation Processes of Organic Matter and Energy Recovery

Considering the available substrates to produce biofuels, three categories may be determined (Table 1.1) with specific advantages and disadvantages. The third

TABLE 1.1 Three Categories of Substrates for Biofuels according to the Carbon Content and Socioeconomical Characteristics

Category	Description	Advantages	Disadvantages
1	Carbon-rich agricultural substrates (such as starch from crops) with relatively stable and homogeneous feedstocks	Lower technological needs and higher biofuel yields due to available industrial methods of extraction; suitable for different biofuel types	Substrates usually in competition with food and feed use and at a relatively high cost
2	Cellulosic compounds and algae targeted for the so-called second- and third-generation processes	No competition with nutrition and significant contribution to CO_2 removal	Lower biofuel yields and/or higher technological needs due to more complex biomass (lignin barrier, cell resistance, etc.)
3	Residual substrates from the former categories or municipal solid waste	Low or negative cost due to large energy and manpower requirements for their current management and treatment	High process complexity; limited biofuel yields and types

category of substrates dealing with residual substrates represents a serious opportunity since their cost is generally quite lower than those of the other categories. As a consequence, although the process will be more complex, they need to be investigated in the manner of sustainable management.

Basically, any organic material of animal, vegetal, or microbial origin is naturally expected to undergo biodegradation mechanisms and, consequently, is suitable for biofuel production. Depending on the physicochemical conditions, such as temperature and moisture content, they will sooner or later be colonized by microorganisms, mostly bacteria, yeast, or fungi. Those will find nutrients to grow while producing gaseous compounds (CO_2, H_2, and CH_4 in particular) and other substances having a positive impact (particularly in compost) or negative effects (toxic and/or inhibiting compounds, such as sulfurs) on the surrounding environment. Depending on their level and how long they are maintained, these negative effects and some other initial physicochemical parameters will affect further biodegradation. Indeed, pH, temperature, moisture content, and inhibiting compounds will reduce biodegradation kinetics or completely inhibit the process when surpassing some thresholds. On the contrary, in some activities, it is of interest to take advantage of these thresholds, for instance, in biomass storage procedures aiming to maintain the initial characteristics of the substrates, whatever their category (Table 1.1). Indeed, it is crucial to control the moisture content of organic materials below about a 20% water content threshold to prevent any biodegradation of organic matter, for example, in wheat silos used in some bioethanol production processes. A second major example concerns the storage of organic materials at a low pH of about 4, as in grass and maize silage.

However, when organic matter is disposed of due to human activities, that is, when domestic or industrial wastewater and solid wastes are collected or massively rejected in the environment, they will undergo significant phenomena that should be considered and managed thoroughly depending on the deposit's size. By contrast, a certain amount of these organic matters produced every day by a human is suitable for biofuel production, depending on socioeconomic factors and origin. In fact, every day in the world from about 0.5 to 1.5 kg of fermentable waste is generated per capita, taking into account the organic matter of household waste (about 30%, not including paper and cardboard), the biodegradable material in solution and/or suspension in domestic wastewater (about 10%), and fermentable materials from industrial waste produced in the territory and reported to the size of the population (about 60%) (Walloon Region 1998). Considering that nearly 30% of these wastes are not valued for the compounds they contain (e.g., sewage sludge and refuse of nonrecoverable production in agriculture), the objectives of this introduction is to explain succinctly the phenomena that will naturally develop within and at the expense of the organic material to generate metabolites, including energy molecules such as ethanol, CH_4, and H_2. It will also discuss the parameters of influence. Basically, the following sections of this introduction will provide evidence that these materials should be managed short- and long-term to avoid nuisance caused by the biodegradation phenomena, and how to treat them or, by contrast, keep these material properties during the storage period. The aim is also to identify, progressively and on the basis of biogas processes in general, which organic matter should be considered a priority and in which physicochemical conditions. The main bibliographic sources are synthetic books (Senior 1990; Gendebien et al. 1992; Reinhart and Townsend 1998; Lens et al. 2004; Cervantes et al. 2006; De Lemos Chernicharo 2007; Townsend et al. 2015); there is no specific reference in the text in order not to burden the reading.

1.4.2 Biodegradation Mechanisms from Complex Organic Matter

The complex organic material that constitutes the residues of plants and animals, their derivative manufactured products (paper, cardboard, sugars, fats, agro-food products, etc.), or their mixtures in matrices, such as wastewater or unsorted waste, is a source of nutrients for many organisms and especially microorganisms. Irrespective of the origin of the latter (they are either already present in the initial organic matter or present in the surrounding environment), the biodegradation mechanisms are generally divided into two groups, depending on the presence or absence of oxygen. They are called aerobic and anaerobic biological metabolism, respectively. In parallel to the biodegradation process, some complex molecules, such as lignin or tannin, that present in the organic material, such as wood, will undergo polymerization mechanisms that are more commonly known as the humus formation process or humification. These natural mechanisms are relatively slow and are carried out preferably under aerobic conditions with relatively low or no negative impact on the environment.

By contrast, the real biological degradation process is relatively rapid aerobically but can also be very slow under anaerobic conditions and/or when the optimum conditions of temperature and moisture are not satisfied, whatever the biological degradation step (Rodriguez et al. 2005). Biodegradation leads to intermediate molecules, for example, acids and alcohols, whose impact on the environment can have serious consequences when the amounts are significant and their effluents are not managed. Since oxygen is particularly appreciated by most microorganisms, they will grow rapidly at the expense of the available organic matter, rejecting mostly water and carbon dioxide. By contrast, when the organic material is sufficiently concentrated (e.g., in wastewater), dense, or in solid form, and the oxygen supply is not forced, concentration gradients are easily formed between the surface of the biomass or of the matrix containing the organic materials (e.g., water in the case of wastewater) and the lower layers. Therefore, the initial aerobic conditions soon evolve to anoxia and further to anaerobic conditions in the core of the liquid or solid, or even a few centimeters from the external surface. As a consequence, four biodegradation stages will occur from complex organic material to complete mineralization. The four successive steps are called hydrolysis, acidogenesis, acetogenesis, and methanogenesis (Figure 1.11). Without any further external action (e.g., mixing, grinding, massive supply of oxygen, and biomass stack reversal, like in composting management), the anaerobic conditions and the associated biodegradation mechanisms will continue as long as the microorganisms have profuse organic matter and the minimum conditions of temperature and moisture are meant.

The hydrolysis, that is, the first step of the biodegradation process, is related to the extracellular enzymes (protein-based catalytic molecules) produced by microorganisms. These enzymes are responsible for the hydrolysis (breakage) of macromolecules, such as carbohydrates like cellulose (the main compound of wood and paper), starch, proteins, and fats, into smaller, soluble molecules of sugars, amino acids, or fatty acids. At the acidogenesis step, these molecules are converted by acidifying bacteria to small molecules (five or six carbon atoms), including alcohols and acids, some of which are called volatile fatty acids (VFAs). Very often, the acid-generating bacteria

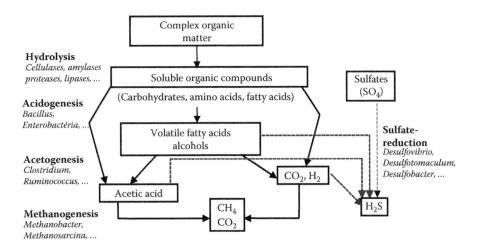

Figure 1.11 Four characteristic steps of the anaerobic biodegradation process of complex organic matter and the involvement of sulfate reduction.

of the genera *Bacillus* and *Clostridium*, or the Enterobacteriaceae group (*Escherichia*, *Enterobacter*, *Citrobacter*, etc.), are also responsible for the production of the hydrolytic enzymes involved in the "spontaneous" biodegradation of organic matter. The fundamental reason is related to the significant resistance, abundance, and growth rate of these microorganisms using many complex organic molecules. The acetogenic bacteria group that is involved in the third stage is very heterogeneous (*Clostridium*, *Ruminococcus*, *Streptococcus*, *Acetobacterium*, etc.) and produces acetic acid, hydrogen, and carbon dioxide from VFAs. These bioconversion reactions are, however, thermodynamically unfavorable unless the hydrogen partial pressure P_{H2} is lower than the range 10^{-4} to 10^{-2} atm (Kaspar and Wuhrmann 1978; Barnes et al. 1983). This may occur through syntrophic combination with hydrogen-consuming homoacetogenic microflora (acetate formation from hydrogen and carbon dioxide) and/or methanogenic archaebacteria (Lowe et al. 1993; Pohland and Kim 2000).

Methanogenic archaea form a separate microbial group (*Methanobacter*, *Methanosarcina*, etc.) and are strict anaerobes. Consequently, this step will only occur in the total absence of oxygen gas. A first group converts hydrogen and carbon dioxide to methane, and a second converts methane to acetic acid and carbon dioxide. These two mechanisms are usually the fourth and final stage of anaerobic biodegradation of complex organic material, leading to the production of a gas mixture rich in CO_2 and CH_4, called biogas, and mineralization of the original material.

This four-step biological process can be operated in bioreactors, as introduced in Chapter 2 and further described in Chapters 3 through 6 of this book. Chapter 3 deals with applications to wastewater and next chapters deal with applications to solid waste. Chapter 4 focuses on the microbiology, biochemistry, and operating conditions in general. Chapters 5 and 6 deal more specifically with municipal solid waste and agriculture waste, respectively, since both are the major current applications of anaerobic digestion at the industrial scale with relatively different operating conditions from each other. Besides, these natural mechanisms also occur spontaneously and with less control in landfills, and every time a sufficient amount of organic waste is disposed of without any control. The environmental impact is not limited and should be described.

1.4.3 Physicochemical Conditions and Environmental Impact Associated with Anaerobic Biodegradation

Overall, after the first steps that induce fast acidification of the medium down to a pH of about 4, VFA consumption and ammonia production (NH_3 and NH_4^+) from protein degradation cause a slow rise in pH, which progressively stimulates the methanogenic activity. Ultimately, the pH may reach 9 or higher values that gradually inhibit methanogenesis and the whole microflora, namely due to the important production of ammonia. The H_2 molecule is involved in most of these biological mechanisms without necessarily being detected if the sequence of the four stages is well established, that is, if the main limiting factor is the hydrolysis of the original organic matter. By contrast, the hydrogen production from simple carbohydrates or easily hydrolizable substrates may be operated in bioreactors and optimized separately from methane production. This is described in Chapter 7, focusing on bacterial H_2 production, also called dark fermentation. Chapters 8 and 9 complete the topic regarding algal and bacterial photofermentative H_2 production.

Among the microorganisms involved in the anaerobic biodegradation of organic matter, the methanogenic microflora is the most vulnerable. Figure 1.12 emphasizes the most influential parameters involved in the process of methane production. Deep anaerobic conditions, optimal moisture (55%–80% relative to the total weight) (Senior 1990), and a pH between 6.8 and 8 are essential for methanogenesis occurrence. A carbon-to-nitrogen (C/N) ratio in the optimal range of 20–40 will nevertheless affect the CH_4 production yield and rate. Regarding the temperature, the choice between a thermophilic regime (about 55°C) and a mesophilic regime (30°C–40°C) will depend on different factors, which are thoroughly presented in Chapters 3 and 4. A more acidic pH and a high concentration of salts and especially sulfate (e.g., found in soils near the sea or in some building waste, such as plaster) are conditions favoring the sulfate-reducing bacteria (Figure 1.11). These microorganisms (*Desulfovibrio*, *Desulfotomaculum*, etc.) are also strictly anaerobic and use hydrogen, acetic acid, alcohols, and VFAs as substrates to form CO_2 and hydrogen sulfide (easily detectable due to its rotten egg smell). This molecule is particularly toxic and volatile, and hence favors sulfate-reducing bacteria growth, especially from the same substrates as methanogens, leading rapidly to complete inhibition of the whole microflora. Hydrogen sulfide is also harmful for the environment and equipment used for biogas treatment. Limitation of this bioconversion pathway is therefore required.

At the scale of a large mass of complex organic matter (at least several liters or kilograms) located in favorable conditions for biodegradation and containing an active microflora, the anaerobic biodegradation phase following the short aerobic phase will be gradually extended to all biodegradable molecules based on their accessibility and nature. Three categories of organic materials are generally considered (Table 1.2) according to the biodegradation rate or half-life (time to degrade half the initial amount of material).

In anaerobic bioreactors for CH_4 and H_2 biogas production (Chapters 3 through 7), mainly the materials of the first category, or even part of the second, will undergo major biodegradation. By contrast, the other materials will be only slightly degraded.

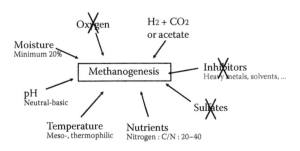

Figure 1.12 Major parameters influencing methanogenesis.

TABLE 1.2　Three Categories of Substrates for Biofuels according to the Biodegradation Rate or Half-Life

Category	Biodegradation Rate	Half-Life Time (from Favorable to Unfavorable Biodegradation Conditions)	Examples
1	Rapid biodegradation	Several hours to several months	Kitchen residues, noncellulosic plant materials, animal and meat residues
2	Moderately slow biodegradation	Some weeks to 5 years	Microbial biomass, vegetable oil and wax, transformed cellulosic materials such as paper and cardboard
3	Slow biodegradation	Few months to 15 years	Natural cellulosic materials (wood and wooden manufactured pieces)

The most accessible molecules, that is, soluble in aqueous solution or at the surface of the particles of complex materials in suspension, will rapidly undergo the process of biodegradation and produce carbon dioxide, acids, and then methane. Depending on their nature or biodegradability rate, some molecules already reach the methanogenic stage, while others, in particular cellulose, will still undergo biochemical attack by enzymes. A fortiori, when the first cellulose molecules will have passed the different successive steps to methane formation, certain proteins or low-molecular-weight carbohydrates will have already disappeared. It is therefore understandable why different phases of biodegradation may be highlighted in large biomass volumes, such as landfills or in discontinued bioreactor systems. Also, different compositions in the liquid phase (soluble molecules) and biogas will characterize each of these phases.

Figure 1.13 shows the typical profile of the concentrations of the various major components over time. This timescale is relative and not linear because it will vary

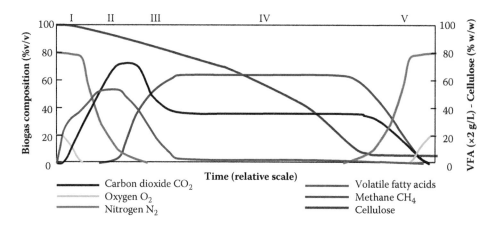

Figure 1.13　Evolution profile vs. time (nonlinear relative scale) of biogas composition and other key molecules (substrates and VFAs) during anaerobic biodegradation of organic matter.

widely, depending on the above-mentioned parameters, such as the nature and accessibility of molecules to degrade. It can last a few days for readily biodegradable materials to several years for the least favorable. In the first phase, which corresponds to the aerobic phase, biogas consists mainly of carbon dioxide and water. However, this type of biogas is produced in small quantities if the aerobic phase is short. It therefore concerns only the beginning of the accumulation of material in the digester, the landfill, or any other disposal site and, subsequently, the top layer when in direct contact with the atmosphere. Phases II and III correspond to the transition between the aerobic phase and the stable methanogenic anaerobic phase. Phase II is characterized by the hydrolytic and acidogenic phase associated with a large production of both soluble molecules, such as the VFAs, and biogas containing hydrogen in addition to carbon dioxide. Phase III shows increasing methane production with the concentration in the biogas progressively reaching about 50%–70%. Hydrogen becomes nondetectable since it is instantly consumed by methanogenic archaea. Only inhibition of these microorganisms would evidence a H_2 gaseous molecule. The duration of these first three phases is a few days, a few weeks, or up to about 1 year according to the environmental conditions already mentioned for large solid waste deposits. Phase IV is the stable methanogenic phase. Methane production is constant and consistent, with the hydrolysis and VFA production mainly from cellulose, since most of the other organic compounds should have been degraded. This phase will last for some decades, as long as biodegradable molecules are both available and accessible in the core of large-sized particles (wood pieces and derived manufactured products). Phase V is the terminal phase of the activity due to the progressive lack of biodegradable material. VFA and gas production eventually stop, and air again enters the relatively inert biomass.

Many biodegradation intermediate compounds (sugars, acids, peptides, alcohols, etc.), being soluble, constitute the major load in leachates, that is, the liquid spent from biomass during the first three biodegradation phases. In the further phases, leachates will contain progressively more humic and fulvic acids via their condensation process from nitrogen- and carbon-containing complex materials, such as lignin. Gradually increasing their molecular weight, they eventually precipitate or bind to nondegradable material. Meanwhile, physicochemical properties of the leachates may undergo huge changes. These leaching liquids contain microbial biomass and chemical compounds, that is, minerals and organic substances with up to about 50 g/L biochemical oxygen demand (BOD) and chemical oxygen demand (COD) and 5 g/L total nitrogen. They may also contain other toxic elements (such as heavy metals) that were present in the initial materials, released, and mobilized by microbiological processes. If they are not collected, these liquids can mix with surface water or groundwater and therefore cause severe pollution due to both their quantitative and qualitative aspects (ecotoxicological elements).

In addition to the major components (methane, carbon dioxide, hydrogen, and sulfide), the biogas also carries a multitude of organic substances in trace amounts (several tens to hundreds of parts per million of benzene, toluene, mercaptans, etc.). The origin of these substances also varies depending on specific processes of biological degradation (e.g., sulfate reduction) and/or initial biomass chemical composition regarding molecules, such as sulfurous proteins and chemical contaminants. The risks related to biogas can be divided into risks to humans (toxicity of trace substances, asphyxia for field operators, explosion of methane and hydrogen, and fires) and risks to the atmosphere (besides CO_2, CH_4 is one of the major greenhouse gases and smog initiators). Therefore, the biogas has to be treated to avoid further environmental impact. This is usually carried out by collecting the biogas and burning it in a controlled flare, furnace, or explosion engine of a combined heat power (CHP) unit.

1.4.4 Valorization of Organic Matter and Life Cycle Assessment

Since organic material will definitely be degraded, this is an opportunity to recover some of its energy content. Anaerobic digestion involving the four biodegradation steps leading to biogas will achieve a theoretical yield of about 300–350 L of CH_4/kg COD, that is, about 3–3.5 kWh/kg COD, whatever the initial organic substrate—carbohydrates, proteins, or fats (Table 1.3). The energy recovery through ethanol production (as a model of alcoholic liquid fuel) from simple or complex carbohydrates will achieve a slightly higher yield of 3.81 kWh/kg COD or 0.61 L C_2H_5OH/kg COD. However, this process is not able to recover energy from proteins and fats.

The bioethanol process is described at the lab and industrial scales in Chapters 10 and 11, along with other alcohol production. Chapter 12 focuses on other liquid fuels produced from algal biomass. Chapters 13 and 14 deal with current and perspective tools to investigate, control, improve, and implement microbial fuel production in industrial sectors. Finally, Chapter 15 analyzes the use and prospects for microbial fuels in developing countries. The existing industrial competitors (Figure 1.14) to energy recovery from biomass through microbial fuel production are thermochemical processes, such as incineration and gasification, or aerobic biological processes, such as composting (McKendry 2002a; Fruergaard and Astrup 2011; Sørensen 2011; Arena et al. 2015). An extensive study in Sweden (Sundqvist 2002,

TABLE 1.3 Equations for Biochemical Conversion of Organic Matter into Methane and Ethanol Fuel, Fuel Yields, and Energy Potential per Kilogram Chemical Oxygen Demand

Organic Material	Chemical Conversion Equation	Fuel Potential (L/kg COD)			Energy Potential
		Fuel	CO_2	Total	kWh/kg COD
Glucose	$C_6H_{12}O_6 \rightarrow 3\ CH_4 + 3\ CO_2$	349	349	698	3.48
Cellulose/ starch	$(C_6H_{10}O_5 + H_2O \rightarrow 3\ CH_4 + 3\ CO_2)_n$	349	349	698	3.48
Proteins	$C_4H_6ON + 2.75\ H_2O \rightarrow 2.13\ CH_4 + 1.88\ CO_2$	298	263	561	2.97
Fats	$C_{55}H_{106}O_6 + 25.5\ H_2O \rightarrow 39.25\ CH_4 + 15.75\ CO_2$	350	141	491	3.49
Glucose	$C_6H_{12}O_6 \rightarrow 2\ C_2H_5OH + 2\ CO_2$	0.61	233	234	3.81

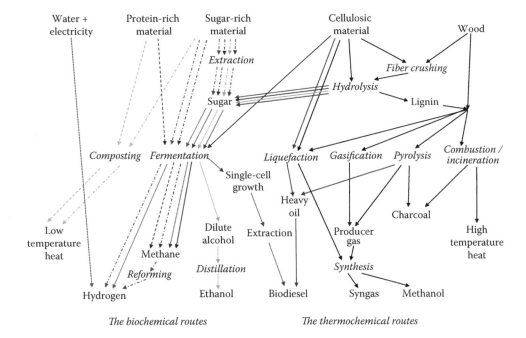

Figure 1.14 Different technological ways for energy recovery from biomass.

2004) found that for organic or mixed household refuses and similar waste, both the anaerobic digestion associated with CHP and incineration were more competitive technologies than composting or biogas valorization through waste management in landfills. These results are confirmed by other studies (Fruergaard and Astrup 2011; Naroznova et al. 2016).

Compared with microbial fuels, the main advantages of incineration are its simple and holistic engineering approach to recover energy from any combustible materials, including the nonfermentable plastics and lignin. Moreover, controlled costs are achievable for efficient treatment of incineration effluents (i.e., fumes and ashes). The major drawback of incineration is the lack of opportunity to return any compounds to the ground, that is, for further biomass growth, whereas the minimum needs for soil-structuring fertilizer to avoid gradual depletion of the soil are about 5 tons/ha/year (Hogg 2004). According to the Swedish investigations, the anaerobic digestion has the advantages, but not all of the disadvantages, of composting, that is, those related to potential acidification and eutrophication of the natural environment and also a disadvantage due to the high consumption of primary energy for providing air to the microorganisms in the core of the biomass during the composting process. When associated with biogas valorization through CHP, the impact of anaerobic digestion is lowest on both the production of greenhouse gases (10%–15% lower than incineration) and the consumption of alternative energy (3%–7% favorable). In addition, anaerobic digestion facilities, of a few megawatts, are feasible and even favorable regarding the impact of material transports while thermochemical facilities, and also microbial fuel processes needing large thermal facilities, for example, for cost-effective distillation, will benefit from economies of scale for setups from nearly a dozen to several tens of megawatts.

These benefits of anaerobic digestion lead to a total cost (financial and environmental) that is 10%–18% higher than the incineration costs for large and smaller facilities, respectively. However, these cost differences decrease depending on the material moisture content because the amount of water represents an energy cost assigned to evaporation during incineration and not biochemical digestion. The same conclusion can be drawn for other thermochemical processes, such as gasification. Therefore, the material containing less than 20%–30% moisture is preferably recovered by thermochemical processes (incineration or gasification). For substances containing more than 50% water, or more than 20%–30%, depending on specificities and opportunities, treatment by fermentation is favored (McKendry 2002a,b).

A decade ago, Kyazze et al. (2007) raised the interest in the two-stage anaerobic digestion process with the opportunity to independently produce two gaseous fuels, hydrogen and methane, with very specific and complementary properties. This opportunity is particularly suitable for installations dealing with mainly nonlignocellulosic carbohydrate-rich materials due to the fact that hydrolysis is not the limiting mechanism, and therefore the production of acids and H_2 can be significant and rapid with respect to the kinetics of methanogenesis. The supply shocks would therefore have been the most damaging for the methanogenic step. Many other authors (Antonopoulou et al. 2008; Ding and Wang 2008; Hafez et al. 2010; Park et al. 2010) confirmed the feasibility of the two stages operated in various conditions to optimize the whole anaerobic digestion. In addition, the work of Ruggeri and Tommasi (Ruggeri et al. 2010, 2011; Tommasi 2010; Tommasi et al. 2012) showed that the energy return on energy investment (EROEI) (ratio between the total useful energy generated and the amount of energy expenses for the whole setup life) ranges from 15 to 45 for the two-stage anaerobic digestion process and is relatively more sustainable than those of photovoltaic (EROEI of 3–9) and some wind (EROEI 5–80) turbines, but of the same order of magnitude as medium-sized hydropower facilities (EROEI 30–70).

In the context of transport, methane "upgrading" has been investigated for use as a fuel in vehicle engines. While it is very promising, attention has to be paid to the highly efficient control of NOx emissions having a serious impact on LCA results. By contrast, LCA carried out by Djomo and Blumberga (2011) showed a 50% reduction in greenhouse gas emissions in engines powered by biohydrogen compared with diesel or hydrogen from methane reformation (LCA considering the entire chain from production to consumption of clean gaseous fuel and compressed to 450 bar). Additionally, biogas upgrading associated with anaerobic digestion, whether in one or two stages, is competitive with bioethanol. Currently, the transport market supports higher-energy supply costs than those considered for stationary applications. Therefore, additional costs related to the pretreatment of raw materials would not be prohibitive and should not be avoided a priori. Consequently, methane and/or hydrogen production might be considered from (hemi-)cellulose materials for second-generation bioethanol. Moreover, besides its robustness, its thorough digestion capacities with or without pretreatment (only the time factor has a significant impact on the energy conversion efficiency), and its flexibility (for small and large facilities ranging from a few hundred kilowatts to several megawatts), the versatility of the anaerobic digestion is appreciable. Indeed, it can consider a large variety of substrates, such as fats and proteins, as well as carbohydrates. In addition, no limitation is suspected in relation to the nature of these materials except regarding the time required for complete biodegradation (especially for complex materials like lignocellulosic biomass). By contrast, ethanol fermentation is limited to simple carbohydrates (Chandra et al. 2012) and requires special care to control microbial ecology by promoting the development and activity of yeast. Therefore, bioethanol production processes are often profitable for large facilities where transportation costs have little impact on the overall economics of the process. In addition, residues of these processes (called stillage) still have significant value. Interesting opportunities exist in the animal feed sector, however, almost exclusively for stillage of the first-generation processes for bioethanol production. This sector is less accessible for stillage produced by second- and third-generation processes, while anaerobic digestion is an alternative technology enabling a draw of up to 10% additional energy potential from agricultural waste (Chandra et al. 2012).

1.5 Conclusion

Biofuels are a promising source of sustainable energy. Microbial fuels are currently of interest due to the increase in global energy demand by emerging economies and the recent increases in global oil prices. Furthermore, they are environmentally friendly fuels because their consumption creates far less environmental pollution than fossil fuels. Due to the benefits of biofuels, efforts should be made to increase their production. Microbial fuels develop according to the characteristics of each region in terms of available resources and climatic conditions. New methods of metabolic engineering and synthetic biology should enable us to extend a substrate's range utilization and increase the efficiency of microorganisms in order to achieve cost-effective production of microbial fuels.

References

Antonopoulou, G., Stamatelatou, K., Venetsaneas, N., Kornaros, M., and Lyberatos, G. 2008. Biohydrogen and methane production from cheese whey in a two-stage anaerobic process. *Industrial & Engineering Chemistry Research* 47:5227–5233.

Arena, U., Ardolino, F., and Di Gregorio, F. 2015. A life cycle assessment of environmental performances of two combustion- and gasification-based waste-to-energy technologies. *Waste Management* 41:60–74.

Ayadi, M., Saurabh, J.S., Vinayak, L.P., Satinder, K.B., and Cheikh, R.B. 2016. History and global policy of biofuels. In *Green Fuels Technology: Biofuels*, ed. Soccol, C.R., Brar, S.K., Faulds, C., and Ramos, L.P. Cham, Switzerland: Springer International Publishing.

Barnes, D., Bliss, P.J., Grauer, B., Kuo, E.M., Robbins, K., and McLean, G. 1983. Pretreatment of high-strength wastewaters by an anaerobic fluidized bed process. 1. Overall performance. *Environmental Technology Letters* 4(5):195–202.

Cervantes, F.J., Spyros, G., and Van Haandel, A.C. 2006. *Advanced Biological Treatment Processes for Industrial Wastewaters.* Integrated Environmental Technology Series. London: IWA Publishing, p. 345.

Chandra, R., Takeuchi, H., and Hasegawa, T. 2012. Methane production from lignocellulosic agricultural crop wastes: A review in context to second generation of biofuel production. *Renewable and Sustainable Energy Reviews* 16(3):1462–1476.

Chandrasekhar, K., Lee, Y.J., and Lee, D.W. 2015. Biohydrogen production: Strategies to improve process efficiency through microbial routes. *International Journal of Molecular Sciences* 16(4):8266–8293.

Cohen, Z., and Ratledge, C. 2010. *Single Cell Oils.* Urbana, IL: AOCS Press.

De Lemos Chernicharo, C.A. 2007. *Anaerobic Reactors.* Biological Wastewater Treatment Series. London: IWA Publishing, p. 174.

Deublein, D., and Steinhauser, A. 2010. *Biogas from Waste and Renewable Resources: An Introduction.* Weinheim, Germany: Wiley-VCH Verlag.

Ding, H.B., and Wang, J.Y. 2008. Responses of the methanogenic reactor to different effluent fractions of fermentative hydrogen production in a phase-separated anaerobic digestion system. *International Journal of Hydrogen Energy* 33:6993–7005.

Djomo, S.N., and Blumberga, D. 2011. Comparative life cycle assessment of three biohydrogen pathways. *Bioresource Technology* 102(3):2684–2694.

Fruergaard, T., and Astrup, T. 2011. Optimal utilization of waste-to-energy in an LCA perspective. *Waste Management* 31(3):572–582.

Gendebien, A., Pauwels, M., Constant, M., Ledrut-Damanet, M.-J., Nyns, E.J., Willumsen, H.-C., Butson, J., Fabry, R., and Ferrero, G.-L. 1992. *Landfill Gas. From Environment to Energy.* Brussels: Commission of the European Communities.

Hafez, H., Nakhla, G., and El Naggar, H. 2010. An integrated system for hydrogen and methane production during landfill leachate treatment. *International Journal of Hydrogen Energy* 35:5010–5014.

Hogg, D. 2004. Costs and benefits of bioprocesses in waste management. In *Resource Recovery and Reuse in Organic Solid Waste Management*, ed. Lens, P., Hamelers, B., Hoitink, H., Bidlingmaier, W. London: IWA Publishing, pp. 95–121.

Hollinshead, W., He, L., and Tang, Y. 2014. Biofuel production: An odyssey from metabolic engineering to fermentation scale-up. *Frontiers in Microbiology* 5:2–8.

Ieropoulos, I.A., Greenman, J., Melhuish, C., and Hart, J. 2005. Comparative study of three types of microbial fuel cell. *Enzyme and Microbial Technology* 37(2):238–245.

Karim, G.A. 2012. *Fuels, Energy, and the Environment.* Boca Raton, FL: CRC Press/Taylor & Francis Group.

Kaspar, H.F., and Wuhrmann, K. 1978. Kinetic parameters and relative turnovers of some important catabolic reactions in digesting sludge. *Applied and Environmental Microbiology* 36(1):1–7.

Kyazze, G., Dinsdale, R., Guwy, A.I., Hawkes, F.R., Premier, G.C., and Hawkes, D.L. 2007. Performance characteristics of a two-stage dark fermentative system producing hydrogen and methane continuously. *Biotechnology and Bioengineering* 97:759–770.

Lens, P., Hamelers, B., Hoitink, H., and Bidlingmaier, W. 2004. *Resource Recovery and Reuse in Organic Solid Waste Management.* Integrated Environmental Technology Series. London: IWA Publishing, p. 516.

Lowe, S.E., Jain, M.K., and Zeikus, J.G. 1993. Biology, ecology and biotechnological applications of anaerobic bacteria adapted to environmental stresses in temperature, pH, salinity, or substrates. *Microbiological Reviews* 57(2):451–509.

McKendry, P. 2002a. Energy production from biomass (part 2): Conversion technologies. *Bioresource Technology* 83(1):47–54.

McKendry, P. 2002b. Energy production from biomass (part 3): Gasification technologies. *Bioresource Technology* 83(1):55–63.

Modanlou, H.D. 2008. A tribute to Zakariya Razi, an Iranian pioneer scholar. *Archives of Iranian Medicine* 11(6):673–677.

Mousdale, D.M. 2010. *Introduction to Biofuels*. Boca Raton, FL: CRC Press/Taylor & Francis Group.

Naik, S.N., Goud, V.V., Rout, P.K., and Dalai, A.K. 2010. Production of first and second generation biofuels: A comprehensive review. *Renewable and Sustainable Energy Reviews* 14(2):578–597.

Naroznova, I., Møller, J., and Scheutz, C. 2016. Global warming potential of material fractions occurring in source-separated organic household waste treated by anaerobic digestion or incineration under different framework conditions. *Waste Management* 58:397–407.

Park, M.J., Jo, J.H., Park, D., Lee, D.S., and Park, J.M. 2010. Comprehensive study on a two-stage anaerobic digestion process for the sequential production of hydrogen and methane from cost-effective molasses. *International Journal of Hydrogen Energy* 35:6194–6202.

Pohland, F.G., and Kim, J.C. 2000. Microbially mediated attenuation potential of landfill bioreactor systems. *Water Science and Technology* 4(3):247–254.

Potter, M.C. 1911. Electrical effects accompanying the decomposition of organic compounds. *Proceedings of the Royal Society of London B* 84(571):260.

Reinhart, D., and Townsend, T.G. 1998. *Landfill Bioreactor Design and Operation*. Boca Raton, FL: CRC Press.

Rodriguez, C., Hiligsmann, S., Ongena, M., Charlier, R., and Thonart, P. 2005. Development of an enzymatic assay for the determination of cellulose bioavailability in municipal solid waste. *Biodegradation* 16:415–422.

Rude, M.A., and Schirmer, A. 2009. New microbial fuels: A biotech perspective. *Current Opinion in Microbiology* 12(3):274–281.

Ruggeri, B., Sanfilippo, S., Tommasi, T., and Fino, D. 2011. Process energy sustainability evaluation through a LCA approach. *Chemical Engineering Transactions* 25:629–634.

Ruggeri, B., Tommasi, T., and Sassi, G. 2010. Energy balance of dark anaerobic fermentation as a tool for sustainability analysis. *International Journal of Hydrogen Energy* 35(19):10202–10211.

Senior, E. 1990. *Microbiology of Landfill Sites*. Boca Raton, FL: CRC Press.

Singh, R.S., Pandey, A., and Gnansounou, E. 2016. *Biofuels: Production and Future Perspectives*. Boca Raton, FL: CRC Press/Taylor & Francis Group.

Sørensen, B. 2011. *The Energy Conversion Processes. Renewable Energy*. 4th ed. Boston: Academic Press, pp. 337–531.

Sundqvist, J.-O. 2002. Material and nutrient recycling and energy recovery from solid waste: Systems perspective. In *Water Recycling and Resource Recovery in Industry Analysis, Technologies and Implementation*, ed. Lens, P., Hulshoff, P.L., Wilderer, P., Asano, T. London: IWA Publishing, pp. 524–544.

Sundqvist, J.-O. 2004. System analysis of organic-waste management schemes—Experiences of the ORWARE model. In *Resource Recovery and Reuse in Organic Solid Waste Management*, ed. Lens, P., Hamelers, B., Hoitink, H., Bidlingmaier, W. London: IWA Publishing, pp. 45–70.

Tommasi, T. 2010. *Experimental Evaluation of Design Bioreactor Parameters for Dark BioH2 Production Using Organic Wastes.* Torino: Politecnico di Torino.

Tommasi, T., Ruggeri, B., and Sanfilippo, S. 2012. Energy valorisation of residues of dark anaerobic production of hydrogen. *Journal of Cleaner Production* 34:91–97.

Townsend, T.G., Powell, J., Pradeep, J., Xu Q., Tolaymat, T., and Reinhart, D. 2015. *Sustainable Practices for Landfill Design and Operation.* New York: Springer Science.

Walloon Region. 1998. *Plan wallon des déchets Horizon 2010.* Namur, Belgium.

Section I

Gaseous Biofuels

2

Bioenergy Production from Waste Substrates

Amit Kumar, Anish Ghimire, Bo H. Svensson, and Piet N.L. Lens

Contents

2.1 Biofuel Production to Address Increasing Energy Needs and Environmental Concerns

More than 1.8 billion tonnes of waste is generated each year in Europe, including waste and residues from households, industry, agriculture, energy generation, the commercial sector, construction and demolition, and mining and quarry activities. This generated waste is a significant challenge for sustainable development of the human race (Rockström et al. 2009). Waste is a material that is disposed of, even though it can in many cases be used as a resource for the recovery of value-added products and/or energy. The biofuel potential of this waste is estimated to be 295 million tonnes of oil equivalent (MtOE) by 2030, which represents ~16% of the projected primary energy requirements of the European Union in 2030 (EEA 2006). Over the last decade, significant progress has been made toward cellulosic biomass conversion to biofuels, with a prime focus on synthetic biology and metabolic engineering (Rabinovitch-Deere et al. 2013). This chapter addresses the engineering of microbial properties and their transfer to bioreactors for an optimization of their use during conversion of waste substrates, including industrial, domestic, and agricultural solids, liquids, and gases to biofuels and bioenergy.

The potential of microbial conversion processes occurring naturally in a diversity of environments, whether they live intimately within soils, guts, or the far deep earth, can be harnessed to convert the energy enclosed in wastes to energy vectors as methane (CH_4), hydrogen (H_2), or ethanol, or directly into electricity. For this, these microbial processes need to be optimized and engineered in bioreactors. Exploring the potential of these naturally occurring processes can provide new perspectives and alternate technological processes that produce biofuel from wastes in a sustainable manner. In this chapter, we discuss how microbial processes occurring in different biotopes can be used as alternative solutions to meet the societal energy demand and, at the same time, improve the health of our environmental systems.

2.2 Wastes-to-Energy Bioprocesses

The concept of *waste* originates from a human perspective. In nature, elemental (material) cycles are interlinked with varying involvement of redox reactions, allowing for energy transfer: the products and by-products of one cycle become the resources in the next cycle (McDonough and Braungart 2002; Wigginton et al. 2012). Microbes mediate many of these processes, even in the most extreme environments (Newman and Banfield 2002). During their metabolic activities, microorganisms, mainly prokaryotes (bacteria and archaea) and fungi, convert wastes into biofuels, such as alkanes (biodiesel), alcohols (bioethanol and biobutanol), volatile fatty acids (butyrate and acetate), gases (methane gas and hydrogen gas), or electrons (bioelectricity). Mankind can tap the biofuel carriers released spontaneously into the environment, that is, bioconversion end products or metabolic leaks. Capture of energy carriers that remain intracellularly or as intermediates requires targeted termination of the reaction. For example, several organisms produce a variety of microbial lipases that are used intracellularly or as a biocatalyst for the transesterification reaction step in biodiesel production (Ghaly et al. 2010).

The basics of key microbial metabolism are shown in Table 2.1. Microorganisms metabolize carbon and other nutrients in various ecosystems and biotope interfaces. Thus, by exhibiting their high diversity, they explore and grow in most environments of the earth. These include geohydrothermal sediments, such as in terrestrial hot springs and deep-sea hydrothermal vents (Sowers 2009), as well as in frozen soils (Öquist et al. 2009). Furthermore, microorganisms are found in agricultural or forest soils, as endosymbionts in ciliates and other unicellular organisms, or in the rumen of ruminants and the guts of termites (Nealson 1997; Pace 1997; Head et al. 2003; Sowers 2009). Moreover, the diversity in phenotypic expression is huge, yielding an almost infinite variety of metabolic properties expressed. There are reports of the microbial conversion of compounds previously considered nonbiodegradable under

TABLE 2.1 Biofuel Production Processes, Standard Biochemical Reactions, and Metabolic Pathways

Bioprocess Type	Standard Reaction and Products	$\Delta G^{o'}$ (kJ mol^{-1})	Metabolic Pathway	Intermediate/Key Enzymes
Anaerobic fermentation	$4H_2 + HCO_3^- + H^+ \rightarrow CH_4 + 3H_2O$	-137.4	Methanogenesis	Methyl–coenzyme M reductase
	$4H_2 + 2HCO_3^- + 2H^+ \rightarrow acetate^- + H^+ + 4H_2O$	-107.0	Wood–Ljungdahl	Acetyl coenzyme A (acetyl CoA) synthase/carbon monoxide dehydrogenase
	$C_6H_{12}O_6 + 2H_2O \rightarrow 2CO_2 + 2CH_3COOH$ (acetate) $+ 4H_2$	206.0	Glycolysis	Pyruvate and acetyl CoA/pyruvate–ferredoxin oxidoreductase and hydrogenase
Photobioreactors	$Acetate + 2H_2O + light \rightarrow 2CO_2 + 4H_2$	209.2	Nitrogenase-catalyzed reaction	Nitrogenase
	$CO_2 + H_2O + sunlight \rightarrow [CH_2O]_n + O_2$	480.0	Calvin cycle	3-Phosphoglycerate molecules/ribulose-1,5-bisphosphate carboxylase
Bioelectrochemical systems	$9H^+ + 8e^- + HCO_3^- \rightarrow CH_4 + 3H_2O$	-137.4	Methanogenesis	Methyl–coenzyme M reductase
	$Acetate^- + H^+ + 2H_2 \rightarrow ethanol + H_2O$	-9.3	Carboxylate reduction	Carboxylate reductase/acetaldehyde
	$10H^+ + 8e^- + 2HCO_3^- \rightarrow acetate^- + H^+ + 4H_2O$	-107.0	Wood–Ljungdahl	Acetyl CoA synthase/carbon monoxide dehydrogenase

Source: Amend, J. P., and E. L. Shock, *FEMS Microbiol. Rev.*, 25, 175–243, 2001; Basak, N., and D. Das, *World J. Microbiol. Biotechnol.*, 23, 31–42, 2007; Thauer, R. K. et al., *Bacteriol. Rev.*, 41, 100–180, 1977.

conditions previously thought uninhabitable. Despite this, we know relatively little about these environmental microbial processes, fundamentally, conceptually, and technologically. As these processes are diverse and complex, it is a grand challenge to trace and understand the enigma of these naturally occurring microbial processes as a means to beneficially utilize our wastes, that is, to elucidate how they interact, and find out the mechanisms regulating their operation.

2.3 Engineered Waste-to-Biofuel Bioprocesses

2.3.1 Anaerobic Digestion

Anaerobic digestion (AD) is a process in which most of the chemical energy present in organic compounds is converted into methane (Figure 2.1). The AD process is a sum of several procedural subprocesses: hydrolysis, acidogenesis, acetogenesis, and methanogenesis. Methanogenesis from organic matter in anaerobic ecosystems is catalyzed by a syntrophic association between anaerobic bacteria, protozoa, and/or anaerobic fungi, and syntrophic bacteria, acetogenic bacteria, and methanogenic archaea (Kumar et al. 2008; Appels et al. 2011; Bialek et al. 2012; Mussoline et al. 2013; Ariunbaatar et al. 2014). The anaerobic bacteria, protozoa, and fungi hydrolyze biopolymers to monomers and lipids to glycerol and long-chain fatty acids, and ferment these, together with syntrophic bacteria, to acetic acid, CO_2, and H_2. These are the sole fermentation products when polymer hydrolysis is the rate-limiting step.

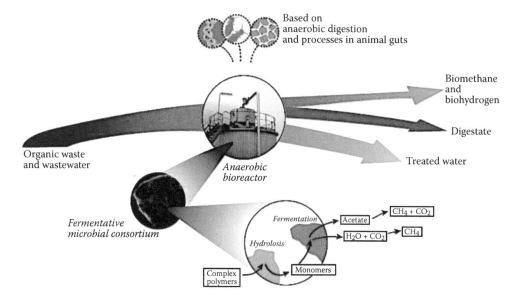

Figure 2.1 Anaerobic degradation of organic wastes to biomethane and carbon dioxide. The process is mediated by an environmental biocatalyst. A generalized pathway shows three steps for converting complex organic materials to biofuel.

This occurs in methanogenesis from biomass and when the H_2 concentration is kept below 10 Pa (Thauer et al. 2008). Acetogenic bacteria and methanogenic archaea (e.g., *Methanobacterium bryantii*, *Methanosarcina barkeri*, *Methanococcus voltae*, and *Methanosaeta concilii*) mediate this process (Sowers 2009). The latter are a phylogenetically diverse group of strictly anaerobic *Euryarchaeota* with an energy metabolism that is restricted to the formation of methane from acetic acid and H_2/CO_2, as well as formate, methanol, methanethiol, and/or methylamines (Sowers 2009).

Anaerobic microorganisms are confined to restricted, oxygen-free (anaerobic) environments, such as the soils beneath flooded rice paddies, wetlands, lake sediments, landfills, and the rumen of both domestic (e.g., cows, sheep, and camels) and wild (e.g., deer, mouse, bison, and giraffes) animals.

Anaerobic digesters have been applied for several decades for the reduction of organic pollution, accompanied by the production of CH_4, from manure; fruit, vegetable, and garden waste; sewage sludge; and other kinds of organic solid waste (Meulepas et al. 2005; Ward et al. 2008; Mao et al. 2015). Recently, industrial and domestic wastewater is also being treated by this technology. This elegantly integrates energy recovery, and thus meets environmental regulations on effluent quality and emission standards (Lettinga 1995).

Hydrogen is an intermediary product in the methanogenesis process that is, however, not available because it is rapidly taken up and converted into methane by methane-producing microorganisms. In biological hydrogen production processes, hydrogen formation and consumption are uncoupled, so that hydrogen is available as the final product (Li and Fang 2007; Ghimire et al. 2015). Hydrogen is produced by anaerobic microorganisms, such as Enterobacteriaceae and Clostridiaceae, which dispose of produced excess, reducing equivalents via hydrogen production (Hawkes et al. 2007; Ghimire et al. 2015). In recent years, the structure of the active site of the hydrogen-producing enzymes, that is, hydrogenases, has been unraveled (Kim and Kim 2011). The catalytic sites contain inorganic elements, mainly nickel, iron, and sulfur. This knowledge provides an opportunity for the production of cheap, synthetic hydrogen-producing catalysts. The first stable and active "biomimetic" hydrogen-producing catalyst was synthesized in 2001 (Gloaguen et al. 2001).

Recently, biohythane, which is a mixture of H_2 and CH_4, has attracted attention due to its versatile uses as a transportation fuel (Roy and Das 2015). The mixture (typically contains 10%–25% H_2 volume) combines the advantages of H_2 and CH_4, such as a wide flammability range, fast burning speed, and high ignition temperature. Therefore, biohythane offers an edge over stand-alone technologies producing H_2 or CH_4.

2.3.2 Photobioreactors

The photosynthetic capability of algae and cyanobacteria in converting solar energy, nutrients, and CO_2 into biomass is harnessed in photobioreactors (PBRs) (Figure 2.2). The natural diversity and distribution of photosynthetic microorganisms is very wide, and their potential remains largely unexplored. Despite this, genetic and metabolic

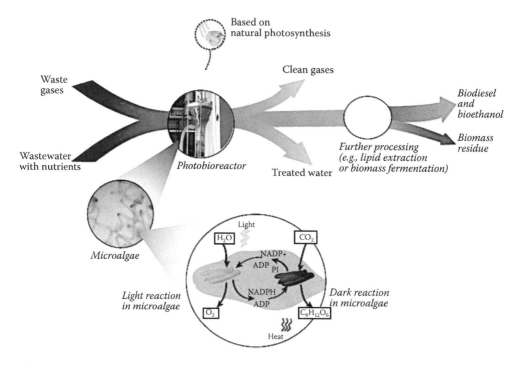

Figure 2.2 Solar energy and wastes are converted by microalgae and further processed into biodiesel and bioethanol. This process is mediated by naturally occurring photosynthesis.

engineering is also providing useful tools to harness the potential of these micro-organisms. For example, a variety of organisms synthesize alkanes via biosynthesis pathways that have remained elusive. An alkane biosynthesis pathway in cyanobacteria that converts intermediates of fatty acid metabolism to alkanes and alkenes has been described by Schirmer et al. (2010). Heterologous expression of these biosynthetic genes resulted in the production of alkanes by *Escherichia coli*. The proposed pathway of alkane biosynthesis from cyanobacteria is likely to be a valuable tool in the production of biofuel vectors.

PBRs for biofuel production (Lehr and Posten 2009), bioreactor types (Carvalho et al. 2006), and recent developments of CO_2 fixation and biofuel production (Kumar et al. 2010a) have been reviewed. The design and engineering of PBRs for the supply of nutrients and CO_2, mixing, harvesting, and processing the biomass in an energy efficient manner is still under research and development (Kumar et al. 2010b; Yuan et al. 2011). Yet, the first successful full-scale bioreactor still has to be built, and this remains an active area of development. Several challenges (e.g., reduced antenna, light dilution, light-to-dark cycle, and optical fibers) may provide higher photosynthetic efficiencies and productivities, but are yet to be explored.

2.3.3 Bioelectrochemical Systems

In bioelectrochemical systems (BESs), organics are oxidized by microorganisms that drive oxidation and reduction reactions at solid-state electrodes (Figure 2.3). BESs are not selective for a particular species or group of microorganisms, but the electrode potential has been shown to change cell surface properties (surface hydrophobicity, net surface electrostatic charge, and surface shape), shorten the doubling time, and increase the enzyme activity (Luo et al. 2005; Kumar et al. 2012, 2013a,b). A promising application of the microbial fuel cell (MFC) technology is the sediment microbial fuel cell (SMFC). SMFCs rely on the natural voltage gradient between the sediments and the overlying seawater (Bond et al. 2002; Luo et al. 2005). SMFCs for bioelectricity production have been reviewed (Lovley 2006). The performance (Rabaey et al. 2005), potentials, and limitations of BES (Girguis et al. 2010) and MFC use in sustainable wastewater treatment (Li et al. 2014) have been reviewed. Yet, the first successful full-scale MFCs still have to be built, and this is also an active area of research.

Several fundamental and applied challenges, for example, design and engineering of MFCs for least resistance, minimized losses of electrons, increased current

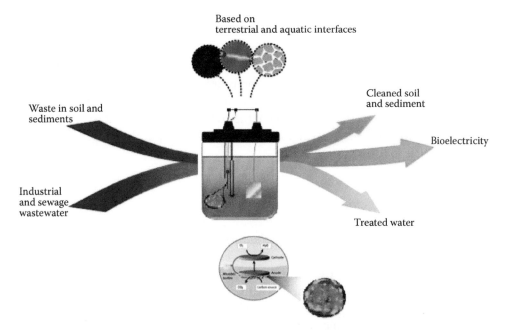

Figure 2.3 Microorganisms oxidize wastes and transfer the electrons to graphite electrodes of an MFC in sediment. These electrons flow to the cathode in the overlying aerobic water, where they react with oxygen. The process is mediated by environmental microbial processes, such as denitrification or manganese and iron reduction.

densities, and maximized conversion in a sustained manner, are still under development and may provide higher efficiencies and conversion rates that are yet to be explored.

2.4 Potential of Microbial Processes for Improved Biofuel Production from Waste

A comparison of different biofuels as products and various features of the associated production systems are presented in Table 2.2. Recent advances in synthetic biology and metabolic engineering have huge potential for biofuel production (Stephanopoulos 2007; Atsumi et al. 2008b; Lynd et al. 2008; Alper and Stephanopoulos 2009; Carroll and Somerville 2009; Peralta-Yahya et al. 2012). Other studies focusing on enzyme-based biorenewable production (Franssen et al. 2013), and conversion of CO_2 to higher alcohols in integrated microbial systems can be potentially useful in biofuel production (Li et al. 2012). In addition, waste substrate types, such as food waste to fuels and chemicals (Sharma et al. 2011; Pant et al. 2012; Lin et al. 2013), bioconversion of combined waste (rice straw and pulp and paper mill sludge) (Mussoline et al. 2013), and their potential, have been reviewed by various researchers (Esposito et al. 2012). The suitability of biomass and waste, such as the organic fraction of municipal solid waste, waste oils and animal fat, energy crops and agricultural waste, and manure and sewage sludge for bioenergy production in AD, has been reported (Appels et al. 2011). These advances will be helpful in providing solutions to the constraints in current waste control and biofuel production. However, exploring the potential of naturally occurring microbial bioconversion processes for technical use in biorecovery and reuse is still poorly understood and remains challenging. Some key emerging concepts to enhance our understanding of the complex nature of microbial processes and use them for waste processing, material recycling, and biofuel production are discussed in the following sections.

2.4.1 Microbial Respiration and Electron Exchange

Electron transfer reactions are fundamental to metabolism, and there are emerging new views on how microorganisms exchange electrons with extracellular donors or acceptors. These range from the use of electron transfer proteins located on the outer membrane (e.g., cytochromes), conductive microbial appendages (e.g., nanowires), or soluble redox shuttles (e.g., pyocyanin) (Lovley 2008; Rosenbaum et al. 2011), to the establishment of a syntrophic relationship via interspecies or intraspecies electron transfer. Direct electron transfer between microbes and minerals (such as iron and manganese [hydro-]oxides) is a widely studied interaction for extracellular respiration. For example, in aquatic–terrestrial interfaces, degradation of naturally occurring organic matter is coupled with solid-phase mineral oxide (Fe or Mn) (Nealson 1997).

Indirect electron transfer can also occur via small molecules as metal chelates by serving as electron shuttles (Gralnick and Newman 2007). In nature, various

TABLE 2.2 Comparison of Different Energy Outputs, Advantages, and Constraints of the Most Promising Wastes to Bioenergy Technologies

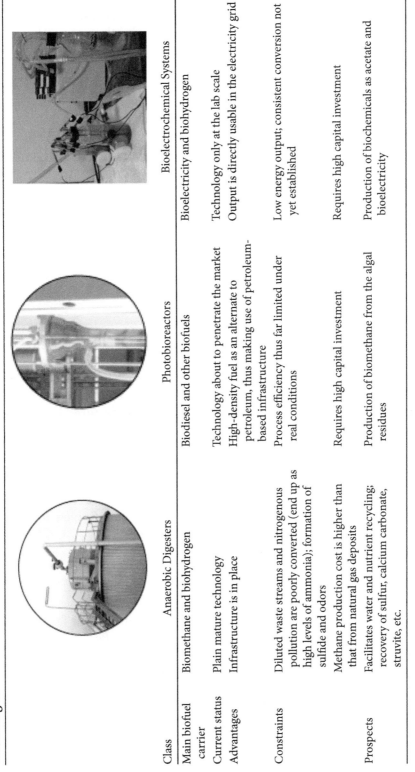

Class	Anaerobic Digesters	Photobioreactors	Bioelectrochemical Systems
Main biofuel carrier	Biomethane and biohydrogen	Biodiesel and other biofuels	Bioelectricity and biohydrogen
Current status	Plain mature technology	Technology about to penetrate the market	Technology only at the lab scale
Advantages	Infrastructure is in place	High-density fuel as an alternate to petroleum, thus making use of petroleum-based infrastructure	Output is directly usable in the electricity grid
Constraints	Diluted waste streams and nitrogenous pollution are poorly converted (end up as high levels of ammonia); formation of sulfide and odors	Process efficiency thus far limited under real conditions	Low energy output; consistent conversion not yet established
	Methane production cost is higher than that from natural gas deposits	Requires high capital investment	Requires high capital investment
Prospects	Facilitates water and nutrient recycling; recovery of sulfur, calcium carbonate, struvite, etc.	Production of biomethane from the algal residues	Production of biochemicals as acetate and bioelectricity

microorganisms use humic substances, abundantly present in soils, biochar, and natural organic matter, as electron shuttles for electron transfer (Peretyazhko and Sposito 2006). For example, pyocyanin, an electron shuttle produced by *Pseudomonas aeruginosa*, is known to have antimicrobial activity (Hassan and Fridovich 1980). Its mechanism of action is primarily the generation of toxic by-products during waste oxidation (e.g., superoxide radicals and H_2O_2), while it may also impact electron flow and energy metabolism in microbial cells.

Nielsen et al. (2010) reported that natural electric currents driven by bacterial extracellular electron transfer processes run through marine sediments and a couple of biogeochemical processes (redox reactions) in widely spatially separated layers of marine sediments, such as sulfide oxidation in the lower sediment layers, with oxygen reduction in the upper layers. The realization of these process concepts could offer opportunities for bioremediation and the production of biofuels from waste resources, including hydrophobic substances such as oil, mineral deposits such as iron and manganese oxides or elemental sulfur, and compounds whose processing yields a toxic product that might be metabolized on the cell surface.

2.4.2 Chronic Energy Stress

This is a unique ability of methanogens to adapt to environmental stresses (temperature, acidity, and salinity) and low energy availability (Valentine 2007). It allows them to outcompete bacteria for common substrates when sulfate and other oxidants are depleted and CO_2 becomes the favored electron acceptor. This can pave the way to treat waste streams having extreme salinities and temperatures, in which eubacteria and other eukaryotes cannot grow. In addition, methanogens enable polysaccharide fermentation by preventing the buildup of H_2 and other reaction products (Samuel and Gordon 2006). In particular, *Methanobrevibacter smithii* has shown a significant enrichment of gene utilization of a variety of fermentation products (H_2, CO_2, and volatile fatty acids) produced by saccharolytic bacteria (Samuel et al. 2007). This knowledge can be utilized in the symbiotic utilization of fermentative products for designing enhanced biofuel (CH_4) production systems.

2.4.3 Microbes with Extraordinary Metabolic Properties

The extraordinary metabolic diversity found in nature is still far from being fully exploited. Several microbes, including *Clostridium* and *Bacteroides*, excrete cellulosomes (complexes of cellulolytic enzymes bound together by cohesion scaffoldings) that assist in the digestion or degradation of plant cell wall materials, most notably cellulose. Therefore, these organisms can be utilized for biofuel production by enhancing saccharification of complex substances such as plant biomass. As such, *Rhodopseudomonas palustris*, a purple photosynthetic bacterium widely distributed in nature, is among the most metabolically versatile bacteria known. It degrades plant biomass and chlorinated pollutants and also generates hydrogen as a product of nitrogen fixation. *R. palustris* uses light, inorganic compounds, or organic compounds for energy and grows both aerobically and anaerobically. It is thus a model organism to probe how the web of metabolic reactions that operates within the confines of a single cell adjusts and reweaves in response to changes in light, carbon, nitrogen, and electron sources (Larimer et al. 2004).

Moreover, like the *Clostridium* gut symbionts, there might be other species that exhibit extraordinary metabolic ability in breaking down complex carbohydrates. In this regard, *Epulopiscium fishelsoni* also shows a distinct metabolic symbiosis, where it assists *Acanthurus nigrofuscus*, an herbivore surgeon, in breaking down algal biomass. *E. fishelsoni* is the largest known heterotrophic bacterium (80 × 600 μm) and is isolated from the gut of the tropical fish *Acanthurus nigrofuscus* in the Red Sea. The bacterium varies in length (30 to >600 μm) and volume (>2000-fold) both within a single host and during a daily cycle, shows a daily variability in nucleoid and cytoplasm structure, and has a peculiar mode of reproduction: daughter cells form within parental organisms and eventually emerge as mobile cells from the maternal envelope. The organism is, however, uncultured, so its actual physiology and functions within hosts are unknown, and also, its large size remains unexplained. More studies of the metabolic activity of this bacterium can open doors for its utilization in the application of algal biomass for biofuel production.

2.4.4 Extremophiles

Bioprocesses driven by organisms from extreme habitats (extremophiles), such as halophiles, barophiles, psychrophiles, thermophiles, and xerophiles, offer new opportunities for biofuel production. How these microbes drive functions in such conditions is known for only a few conditions. Recent findings of saline habitats, such as subsurface aquifer seeps, deep-sea brine pools, and ancient subglacial brines, are providing insights into the microbial metabolic activities under these conditions (DeLong 2004; Boetius and Joye 2009).

Edwards et al. (2000) report that *Ferroplasma acidarmanus*, an iron-oxidizing archaeon, not only survives, but thrives in some of the most acidic waters on earth at a seemingly impossible pH near 0.5. In Gulf of Mexico brines, which contain no sulfate, specific microbial functions—such as the formation and consumption of acetate and subsequent methane production—have been observed for the first time at high salinity and were found to vary across the chemocline (Joye et al. 2009). In an Antarctic glacier, Mikucki et al. (2009) show an active microbial assemblage cycle. The anoxic water is cryogenically concentrated, loaded with Fe^{2+}, and has no evidence of sulfide. The isotopic composition of the oxygen in the sulfate indicates that the reduction of sulfate to sulfite is occurring by microbially coupled sulfur and iron cycles driven by adenosine 5′-phosphosulfate reductase and probably originating in an organic source of sulfur. This offers a clear example of how a microbial system can survive for an extended period without photosynthesis or nutrients from an external source.

Thermophiles readily ferment pentose and/or hexose sugars from biomass and, in some cases, even structurally complex carbohydrates, a quality that is particularly important for the production of second-generation biofuels. Despite the dominance of thermophiles, other extremophiles have also been applied in this field, including methanogens (typically thermophilic, anaerobic archaea) and psychrophiles. However, the tremendous metabolic diversity found in nature, including that of extremophiles, is still far from being fully exploited.

2.5 Enhancement of Waste Processing and Biofuel Production

2.5.1 Engineering Challenges of Biofuel Production

In addition to developing engineered efficient microbes for biofuel production, mass transfer within a bioreactor system is another major technical limitation for biofuel production (Figures 2.4 and 2.5). The transfer of nutrients and metabolites and light (for photosynthetic microorganisms) determines the potential yield of the process, and thus plays a major role in the complete conversion of biomass to biofuels. Moreover, the transfer of nutrients and metabolites strongly influences biofuel production by inhibition of microorganisms due to the accumulation of metabolites or limited availability of nutrients for growth and metabolism.

Figures 2.4 and 2.5 present the parameters and processes that affect the mass and energy transfer in different biofuel production systems, which ultimately determine the production of different biofuels. In bioreactors, process conditions can be applied that support a maximum substrate conversion rate. The contacting of the waste substrate with the biocatalyst and the release and recovery of the produced biofuels impose several engineering challenges on the bioreactor development. These rely

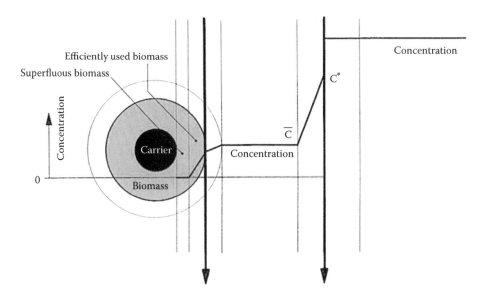

Figure 2.4 Schematic representation of mass transfers in anaerobic digesters and PBRs.

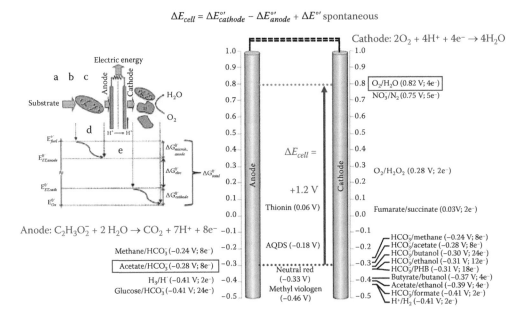

Figure 2.5 Schematic representation of mass transport and electron diffusion in BESs. (After Kumar, A. et al., *Biochem. Soc. Trans.*, 40, 1308–1314, 2012; Rabaey, K., and R. A. Rosendal, *Nat. Rev. Microbiol.*, 8, 706–716, 2010; Harnisch, F., and U. Schröder, *Chem. Soc. Rev.*, 39, 4433–4448, 2010.)

on the interplay between mass transfer and microbial kinetics. The basics of mass transfer and microbial kinetics of waste substrate bioconversion to biofuels are (1) the bulk mixing of the waste substrate entering the bioreactor, (2) the substrate transport boundary layer, (3) the diffusion layer at the biocatalysts–liquid interface, (4) diffusion through and conversion within the biocatalysts, and (5) diffusion of gaseous products via liquid boundary layer transport. An example can be a substrate availability in five zones (I–V). Figure 2.4 shows the substrate concentration gradients getting to the effective microbial biomass attached to a biofilm carrier. Moreover, in a PBR, light gradients affect the growth of microorganisms and substrate uptake.

Enhanced mass transfer increases substrate utilization, and thus the formation of products, as liquid-to-gas transfer enhances biogas production. This can be achieved by better reactor design and operation. Beckers et al. (2015) reported enhanced bio-hydrogen generation through nitrogen sparging and gentle mixing (stirring) in batch mode-operated anaerobic sequenced bioreactors. This was as a result of pronounced liquid-to-gas hydrogen transfer, which then gave rise to a reduced inhibition to the bacterial cultures present in the bioreactors. Some researchers have suggested innovative reactor designs to improve bubble formation and H_2 separation from the liquid phase (Barca et al. 2015). Show et al. (2011) have suggested partial recirculation of the effluent as a strategy to reduce high mass transfer resistance and pH gradients in anaerobic packed bed reactors (APBRs). Furthermore, mass transfer mechanisms can be studied using microfluidic techniques, which can pave the way for the scale-up development of biotechnological process (Mikaelian et al. 2015).

Improved understanding of mass or electron transfer in engineered bioreactors is required, and this can play a key role in biofuel generation from wastes (including solid waste, wastewater, and waste gas). In bioelectrical systems such as MFCs, the entire free energy of the substrate oxidation could not be converted into electricity due to thermodynamic losses that affect reactions in the anode, as well as cathode reactions (Harnisch and Schröder 2010). These losses are because of biological energy dissipation, that is, anabolic cell processes. Figure 2.5 shows oxidation of a substrate (acetate), which generates electrons that are transferred to the anode at a potential more positive than the redox potential of the substrate. Likewise, mass transfer to the electrodes and through the membranes (ion exchange) determines the production of electricity in the BES as well (Figure 2.5). Factors such as bulk mixing, electron and waste substrate boundary layer transport, their diffusion into the bacteria cell (biocatalysts), and the transfer of ions across the cell wall to the anode electrode and through the proton exchange membrane determine the mass or electron transfer in MFC systems.

2.5.2 Microbial or Biochemical Engineering

Different approaches, such as biochemical engineering (BE), genetic engineering (GE), and transcription factor engineering (TFE), are being applied to remove the technical hurdles related to biofuel production (Courchesne et al. 2009). Some examples of different strategies that can be applied for enhanced biofuel production are presented in Table 2.3. The BE approach includes strategies to enhance the production

TABLE 2.3   Some Examples of Different Approaches for Enhancing Biofuel Production

Biofuels	Strategy	Reference
Microalgal lipids	Nutrient starvation (nitrogen, phosphorous, or iron) to direct metabolic fluxes to lipid biosynthesis	Courchesne et al. 2009
	Modification of ammonium assimilation to increase xylose utilization by deleting glutamate dehydrogenase (GDH1) and overexpressing GDH2 in *Saccharomyces cerevisiae* increased ethanol yield	Roca et al. 2003
Ethanol in fermentation	Deletion of genes in *Thermoanaerobacterium saccharolyticum* involved in organic acid formation (acetate kinase, phosphate acetyltransferase, and L-lactate dehydrogenase) increased ethanol production	Shaw et al. 2008
Hydrogen in photofermentation	Deletion of the polyhydroxbutyrate-producing gene from *Rhodobacter sphaeroides* KD131 for enhancing the H_2 production rate	Kim et al. 2011
Butanol	Expression of different gene combinations for butanol production in *E. coli* modeled after the *Clostridium acetobutylicum* pathway; deletion of competing pathways; increased NADH availability	Atsumi et al. 2008a

of targeted products, that is, biofuels, by controlling the nutritional or cultivation conditions (e.g., temperature, pH, and salinity) to channel the metabolic flux generated to the desired product. The GE approach manipulates the understanding of the rate-limiting enzymes in order to enhance the channeling of particular metabolites to targeted products (biofuels) by overexpressing one or more key enzymes in recombinant microbial strains. The most recent TFE strategy is more focused on enhancing the production of a particular metabolite by means of overexpressing transcription factors that regulate the metabolic pathways involved in the production of desired metabolites.

A better understanding of interactions between environmental conditions and biochemical processes would allow new breakthroughs to be made. Furthermore, genome scale modeling to characterize the interplay between transcriptional regulation and regulatory networks has not yet been studied. Metabolic modeling could help in predicting the composition of various intermediate as well as by-products that could aid in gaining knowledge for optimizing the production of particular intended high-value products, for example, butanol, ethanediol, and succinic acid.

2.6 Conclusions

Waste or residues generally enclose a variety of energy carriers, thus providing opportunities to couple anthropogenic pollution removal to the production of biofuels. Advances in our understanding of microbial biochemical processes are providing important insights into how waste mitigation can be coupled to biofuel production paths as methane, hydrogen, or bioelectricity.

Acknowledgments

Acknowledgment to the FP 7 People Programme of the European Commission for a Marie Curie Intra European (A/6342-PIEF-GA-2009-237181) and a Marie Curie International Outgoing (PIOF-GA-2011-302964) Fellowship for Career Development, and the Biogas Research Center, Linköping University, Sweden.

References

Alper, H., and G. Stephanopoulos. 2009. Engineering for biofuels: Exploiting innate microbial capacity or importing biosynthetic potential? *Nature Reviews Microbiology* 7: 715–723.

Amend, J. P., and E. L. Shock. 2001. Energetics of overall metabolic reactions of thermophilic and hyperthermophilic archaea and bacteria. *FEMS Microbiology Reviews* 25: 175–243.

Appels, L., J. Lauwers, J. Degreve, L. Helsen, B. Lievens, K. Willems, J. Van Impe, and F. Dewil. 2011. Anaerobic digestion in global bio-energy production: Potential and research challenges. *Renewable and Sustainable Energy Reviews* 15: 4295–4301.

Ariunbaatar, J., A. Panico, G. Esposito, F. Pirozzi, and P. N. L. Lens. 2014. Pretreatment methods to enhance anaerobic digestion of organic solid waste. *Applied Energy* 123: 143–156.

Atsumi, S., A. F. Cann, M. R. Connor, C. R. Shen, K. M. Smith, M. P. Brynildsen, K. J. Y. Chou, T. Hanai, and J. C. Liao. 2008a. Metabolic engineering of *Escherichia coli* for 1-butanol production. *Metabolic Engineering* 10 (6): 305–311.

Atsumi, S., T. Hanai, and J. C. Liao. 2008b. Non-fermentative pathways for synthesis of branched-chain higher alcohols as biofuels. *Nature* 451: 86–89.

Barca, C., A. Soric, D. Ranava, M.-T. Giudici-Orticoni, and J.-H. Ferrasse. 2015. Anaerobic biofilm reactors for dark fermentative hydrogen production from wastewater: A review. *Bioresource Technology* 185: 386–398.

Basak, N., and D. Das. 2007. The prospect of purple non-sulfur (PNS) photosynthetic bacteria for hydrogen production: The present state of the art. *World Journal of Microbiology and Biotechnology* 23: 31–42.

Beckers, L., J. Masset, C. Hamilton et al. 2015. Investigation of the links between mass transfer conditions, dissolved hydrogen concentration and biohydrogen production by the pure strain *Clostridium butyricum* CWBI1009. *Biochemical Engineering Journal* 98: 18–28.

Bialek, K., A. Kumar, T. Mahony, P. Lens, and O. Flaherty. 2012. Microbial community structure and dynamics in anaerobic fluidized-bed and granular sludge-bed reactors: Influence of operational temperature and reactor configuration. *Microbial Biotechnology* 5: 738–752.

Boetius, A., and S. Joye. 2009. Thriving in salt. *Science* 324: 1523–1525.

Bond, R., D. E. Holmes, L. M. Tender, and D. R. Lovley. 2002. Electrode-reducing microorganisms that harvest energy from marine sediments. *Science* 295: 483–485.

Carroll, A., and C. R. Somerville. 2009. Cellulosic biofuels. *Annual Review of Plant Biology* 160: 165–182.

Carvalho, P., L. A. Meireless, and F. X. Malcata. 2006. Microalgal reactors: A review of enclosed system designs and performances. *Biotechnology Progress* 22: 1490–1506.

Courchesne, N. M. D., A. Parisien, B. Wang, and C. Q. Lan. 2009. Enhancement of lipid production using biochemical, genetic and transcription factor engineering approaches. *Journal of Biotechnology* 141 (1–2): 31–41.

DeLong, E. F. 2004. Microbial life breathes deep. *Science* 306: 2198–2200.

Edwards, K. J., P. L. Bond, T. M. Gihring, and J. F. Banfield. 2000. An archaeal iron-oxidizing extreme acidophile important in acid mine drainage. *Science* 278: 1796–1799.

EEA (European Environment Agency). 2006. How much biofuel can Europe produce without harming the environment? Copenhagen: EEA. http://www.eea.europa.eu/publications /eea_report_2006_7 (accessed May 5, 2014).

Esposito, G., L. Frunzo, A. Giordano, F. Liotta, A. Panico, and F. Pirozzi. 2012. Anaerobic co-digestion of organic wastes. *Reviews in Environmental Science and Bio/Technology* 1: 325–341.

Franssen, M. C., P. Steunenberg, E. L. Scott, H. Zuilhof, and J. P. M. Sanders. 2013. Immobilised enzymes in biorenewables production. *Chemical Society Reviews* 42: 6491–6533.

Ghaly E., D. Dave, M. S. Brooks, and S. Budge. 2010. Production of biodiesel by enzymatic transesterification: Review. *American Journal of Biochemistry and Biotechnology* 6: 54–76.

Ghimire, A., L. Frunzo, F. Pirozzi et al. 2015. A review on dark fermentative biohydrogen production from organic biomass: Process parameters and use of by-products. *Applied Energy* 144: 73–95.

Girguis, P. R., M. E. Nielsen, and I. Figueroa. 2010. Harnessing energy from marine productivity using bioelectrochemical systems. *Current Opinion in Biotechnology* 21: 252–258.

Gloaguen, F., J. D. Lawrence, and T. B. Rauchfuss. 2001. Biomimetic hydrogen evolution catalyzed by an iron carbonyl thiolate. *Journal of the American Chemical Society* 123: 9476–9477.

Gralnick, J. A., and D. Newman. 2007. Extracellular respiration. *Molecular* 65: 1–11.

Harnisch, F., and U. Schröder. 2010. From MFC to MXC: Chemical and biological cathodes and their potential for microbial bioelectrochemical systems. *Chemical Society Reviews* 39: 4433–4448.

Hassan, H. M., and I. Fridovich. 1980. Mechanism of the antibiotic action pyocyanine. *Journal of Bacteriology* 141: 156–163.

Hawkes, R., I. Hussy, G. Kyazze, R. Dinsdale, and D. L. Hawkes. 2007. Continuous dark fermentative hydrogen production by mesophilic microflora: Principles and progress. *International Journal of Hydrogen Energy* 32: 172–184.

Head, M., D. M. Jones, and S. R. Larter. 2003. Biological activity in the deep subsurface and the origin of heavy oil. *Nature* 426: 344–352.

Joye, S. B., V. A. Samarkin, B. N. Orcutt, I. R. MacDonald, K.-U. Hinrichs, M. Elvert, A. P. Teske, K. G. Lloyd, M. A. Lever, J. P. Montoya, and C. D. Meile. 2009. Metabolic variability in seafloor brines revealed by carbon and sulphur dynamics. *Nature Geoscience* 2: 349–354.

Kim, D.-H., and M.-S. Kim. 2011. Hydrogenases for biological hydrogen production. *Bioresource Technology* 102 (18): 8423–8431.

Kim, M.-S., D.-H. Kim, H.-N. Son, L. N. Ten, and J. K. Lee. 2011. Enhancing photo-fermentative hydrogen production by *Rhodobacter sphaeroides* KD131 and its PHB synthase deleted-mutant from acetate and butyrate. *International Journal of Hydrogen Energy* 36 (21): 13964–13971.

Kumar, A., S. Ergas, X. Yuan, A. Sahu, Q. Zhang, J. Dewulf, F. X. Malcata, and H. Van Langenhove. 2010a. Enhanced CO_2 fixation and biofuel production via microalgae: Recent developments and future directions. *Trends in Biotechnology* 28: 371–380.

Kumar, A., X. Yuan, A. K. Sahu, J. Dewulf, S. J. Ergas, and H. Van Langenhove. 2010b. A hollow fiber membrane photo-bioreactor for CO_2 sequestration from combustion gas coupled with wastewater treatment: A process engineering approach. *Journal of Chemical Technology and Biotechnology* 85: 387–394.

Kumar, A., K. Katuri, P. N. L. Lens, and D. Leech. 2012. Does bioelectrochemical cell configuration and anode potential affect biofilm response? *Biochemical Society Transactions* 40: 1308–1314.

Kumar, A., P. Ó'Conghaile, K. Katuri, P. N. L. Lens, and D. Leech. 2013a. Amine-modification of graphite using aryldiazonium salt electro-reduction improves microbial biofilm acetate oxidation current. *RSC Advances* 3: 18759–18761.

Kumar, A., A. Siggins, K. Katuri et al. 2013b. Catalytic response of microbial biofilms grown under fixed anode potentials depends on electrochemical cell configuration. *Chemical Engineering Journal* 230: 532–536.

Kumar, A., K. Yadav, T. R. Sreekrishnan, S. Santosh, and C. P. Kaushik. 2008. Treatment of low strength industrial cluster wastewater by anaerobic hybrid reactor. *Bioresource Technology* 99: 3123–3129.

Larimer, F. W., P. Chain, L. Hauser et al. 2004. Complete genome sequence of the metabolically versatile photosynthetic bacterium *Rhodopseudomonas palustris*. *Nature Biotechnology* 22: 55–61.

Lehr, F., and C. Posten 2009. Closed photo-bioreactors as tools for biofuel production. *Current Opinion in Biotechnology* 20 (3): 280–285.

Lettinga, G. 1995. Anaerobic digestion and wastewater treatment systems. *Antonie van Leeuwenhoek* 67: 3–28.

Li, H., and H. P. Fang. 2007. Fermentative hydrogen production from wastewater and solid wastes by mixed cultures. *Critical Reviews in Environmental Science and Technology* 37: 1–39.

Li, H., P. H. Opgenorth, D. G. Wernick et al. 2012. Integrated electromicrobial conversion of CO_2 to higher alcohols. *Science* 335 (6076): 1596.

Li, W. W., H. Q. Yu, and Z. He. 2014. Towards sustainable wastewater treatment by using microbial fuel cells-centered technologies. *Energy & Environmental Science* 7: 911–924.

Lin, C. S. K., L. A. Pfaltzgraff, L. Herrero-Davila et al. 2013. Food waste as a valuable resource for the production of chemicals, materials and fuels. Current situation and global perspective. *Energy & Environmental Science* 6: 426–464.

Lovley, D. R. 2006. Bug juice: Harvesting electricity with microorganisms. *Nature Reviews Microbiology* 4: 497–508.

Lovley, D. R. 2008. Extracellular electron transfer: Wires, capacitors, iron lungs, and more. *Geobiology* 6: 225–231.

Luo, Q., H. Wang, X. Zhang, and Y. Qian. 2005. Effect of direct electric current on the cell surface properties of phenol-degrading bacteria. *Applied and Environmental Microbiology* 71: 423–427.

Lynd, L. R., M. S. Laser, D. Bransby, B. E. Dale, B. Davison, R. Hamilton, M. Himmel, M. Keller, J. D. McMillan, J. Sheehan, and C. E. Wyman. 2008. How biotech can transform biofuels. *Nature Biotechnology* 26: 169–172.

Mao, C., Y. Feng, X. Wang, and G. Ren. 2015. Review on research achievements of biogas from anaerobic digestion. *Renewable and Sustainable Energy Reviews* 45: 540–555.

McDonough, W., and M. Braungart, eds. 2002. *Cradle to Cradle: Remaking the Way We Make Things*. New York: North Point Press.

Meulepas, R. J. W., A. Nordberg, J. M. Alvarez, and P. N. L. Lens. 2005. Methane production from wastewater, solid waste and biomass (Chapter 7). In *Biofuels for Fuel Cells: Renewable Energy from Biomass Fermentation*, eds. P. N. L. Lens, P. Westerman, M. Haberbauer, and A. Moreno, 121–135. London: IWA Publishing.

Mikaelian, D., B. Haut, and B. Scheid. 2015. Bubbly flow and gas-liquid mass transfer in square and circular microchannels for stress-free and rigid interfaces: Dissolution model. *Microfluidics and Nanofluidics* 19 (4): 899–911.

Mikucki, J. A., A. Pearson, D. T. Johnston et al. 2009. A contemporary microbially maintained subglacial ferrous "ocean." *Science* 324: 397–400.

Mussoline, W., G. Esposito, A. Giordano, and P. N. L. Lens. 2013. The anaerobic digestion of rice straw: A review. *Critical Reviews in Environmental Science and Technology* 43: 895–915.

Nealson, H. 1997. Sediment bacteria: Who's there, what are they doing, and what's new? *Annual Review of Earth and Planetary Sciences* 25: 403–434.

Newman, K., and J. F. Banfield. 2002. Geomicrobiology: How molecular-scale interactions underpin biogeochemical systems. *Science* 296: 1071–1077.

Nielsen, L. P., N. R. Petersen, H. Fossing, P. B. Christensen, and M. Sayama. 2010. Electric currents couple spatially separated biogeochemical processes in marine sediment. *Nature* 463: 1071–1074.

Öquist, M. G., T. Sparrman, L. Klemedtsson, S. H. Drotz, H. Grip, J. Schleucher, and M. Nilsson. 2009. Water availability controls microbial temperature responses in frozen soil CO_2 production. *Global Change Biology* 15: 2715–2722.

Pace, N. M. 1997. A molecular view of microbial diversity and the biosphere. *Science* 276: 734–740.

Pant, D., A. Singh, G. Van Bogaert, S. I. Olsen, P. S. Nigam, L. Diels, and K. Vanbroekhoven. 2012. Bioelectrochemical systems (BES) for sustainable energy production and product recovery from organic wastes and industrial wastewaters. *RSC Advances* 2: 1248–1263.

Peralta-Yahya, P. P., F. Zhang, S. B. del Cardayre, and J. D. Keasling. 2012. Microbial engineering for the production of advanced biofuels. *Nature* 488: 320–328.

Peretyazhko, T., and G. Sposito. 2006. Reducing capacity of terrestrial humic acids. *Geoderma* 137: 140–146.

Rabaey, K., G. Lissenns, and W. Verstraete. 2005. Microbial fuel cells: Performances and perspectives. In *Biofuels for Fuel Cells: Renewable Energy from Biomass Fermentation*, ed. P. N. L. Lens, P. Westerman, M. Haberbauer, and A. Moreno, 377–396. London: IWA Publishing.

Rabaey, K., and R. A. Rosendal. 2010. Microbial electrosynthesis—Revisiting the electrical route for microbial production. *Nature Reviews Microbiology* 8: 706–716.

Rabinovitch-Deere, C. A., J. W. K. Oliver, G. M. Rodriguez, and S. Atsumi. 2013. Synthetic biology and metabolic engineering approaches to produce biofuels. *Chemical Reviews* 113 (7): 4611–4632.

Roca C., J. Nielsen, and L. Olsson. 2003. Metabolic engineering of ammonium assimilation in xylose-fermenting *Saccharomyces cerevisiae* improves ethanol production. *Applied and Environmental Microbiology* 69: 4732–4736.

Rockström, J., W. Steffen, K. Noone et al. 2009. A safe operating space for humanity. *Nature* 461: 472–475.

Rosenbaum, M., F. Aulenta, M. Villano, and L. T. Angenent. 2011. Cathodes as electron donors for microbial metabolism: Which extracellular electron transfer mechanisms are involved? *Bioresource Technology* 120: 324–333.

Roy, S., and D. Das. 2015. Biohythane production from organic wastes: Present state of art. *Environmental Science and Pollution Research* 23 (10): 9391–9410.

Samuel, B. S., E. E. Hansen, J. K. Manchester et al. 2007. Genomic and metabolic adaptations of *Methanobrevibacter smithii* to the human gut. *Proceedings of the National Academy of Sciences of the United States of America* 104: 10643–10648.

Samuel, S., and J. I. Gordon. 2006. A humanized gnotobiotic mouse model of host–archaeal–bacterial mutualism. *Proceedings of the National Academy of Sciences of the United States of America* 103: 10011–10016.

Schirmer, M., A. Rude, X. Li, E. Popova, and S. B. del Cardayre. 2010. Microbial biosynthesis of alkanes. *Science* 329: 559–562.

Sharma, Y. C., B. Singh, and J. Korstad. 2011. A critical review on recent methods used for economically viable and eco-friendly development of microalgae as a potential feedstock for synthesis of biodiesel. *Green Chemistry* 13: 2993–3006.

Shaw, A. J., K. K. Podkaminer, S. G. Desai, J. S. Bardsley, S. R. Rogers, P. G. Thorne, D. A. Hogsett, and L. R. Lynd. 2008. Metabolic engineering of a *Thermophilic bacterium* to produce ethanol at high yield. *Proceedings of the National Academy of Sciences of the United States of America* 105 (37): 13769–13774.

Show, K.-Y., D.-J. Lee, and J.-S. Chang. 2011. Bioreactor and process design for biohydrogen production. *Bioresource Technology* 102 (18): 8524–8533.

Sowers, K. R. 2009. Methanogenesis. In *Encyclopedia of Microbiology*, ed. M. Schaechter, 265–285. 3rd ed. Amsterdam: Elsevier.

Stephanopoulos, G. 2007. Challenges in engineering microbes for biofuels production. *Science* 315: 801–804.

Thauer, R. K., K. Jungermann, and K. Decker. 1977. Energy conservation in chemotrophic anaerobic bacteria. *Bacteriological Reviews* 41: 100–180.

Thauer, R. K., A.-K. Kaster, H. Seedorf, W. Buckel, and R. Hedderich. 2008. Methanogenic archaea: Ecologically relevant differences in energy conservation. *Nature Reviews Microbiology* 6 (8): 579–591.

Valentine, L. 2007. Adaptations to energy stress dictate the ecology and evolution of the archaea. *Nature Reviews Microbiology* 5: 316–323.

Ward, A. J., P. J. Hobbs, P. J. Holliman, and D. L. Jones. 2008. Optimization of the anaerobic digestion of agricultural resources. *Bioresource Technology* 99 (17): 7928–7940.

Wigginton, N., J. Yeston, and D. Malakoff. 2012. More treasure than trash. *Science* 337: 662–663.

Yuan, X., A. Kumar, A. K. Sahu, and S. J. Ergas. 2011. Impact of ammonia concentration on *Spirulina platensis* growth in an airlift photobioreactor. *Bioresource Technology* 102: 3234–3239.

3

Biomethane from Industrial and Municipal Wastewater

Miriam H.A. van Eekert and Grietje Zeeman

Contents

3.1 Introduction

Remains of drainage systems to remove waste and latrines have been found in houses from the Mesopotamian Empire (3500–2500 BC); ancient Rome had its Cloaca Maxima, and there still exists a working 4000-year-old sewer system in Greece. Nevertheless, it was not until the late nineteenth century, and after a fourteenth-century long dark age, that it was recognized that municipal waste water needs to be removed from its origin and treated to prevent the outbreak of diseases (Lofrano and Brown 2010). Until then, wastewater had been discharged in surface water or so-called "night soil" (toilet waste) and collected and used for fertilization. Later, in the early twentieth century, biological oxygen demand (BOD) was introduced as a measure of pollution and the first wastewater treatment systems were installed. Recently, the recovery of nutrients, reuse of water, production of intermediates, and generation of energy have become important incentives for the treatment of wastewater from both industrial and municipal origins. This may be more feasible through separation at the source and improved design of water usage and treatment systems (Guest et al. 2009; Larsen et al. 2009). Aerobic treatment was and still is the main technology used for the treatment of municipal wastewater in the north and cold climate areas. In the twentieth century, the possible application of anaerobic

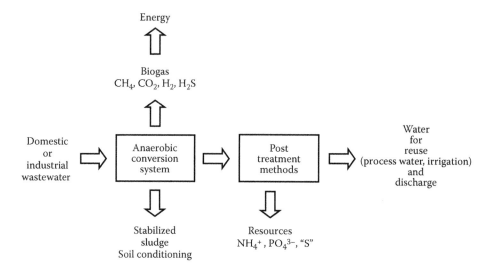

Figure 3.1 Central role of anaerobic conversion in the processing of waste streams. (Adapted from van Eekert, M.H.A et al., in *Methane and Climate Change*, ed. D. Reay, P. Smith, and A. van Amster, Earthscan, London, 2010, pp. 151–174.)

systems for the treatment of industrial wastewater and municipal wastewater in warmer climates was recognized after the development of the upflow anaerobic sludge bed (UASB) system in Wageningen in the 1970s (Lettinga 2014; van Lier et al. 2015). Nowadays, with new treatment designs and the paradigm shift toward seeing wastewater as a source of valuable resources, the application of anaerobic technology may be expanded toward treating municipal sewage in cold climates as well. Anaerobic treatment has its advantages, for example, lower excess sludge production, high applicable loadings, and lower energy demands, combined with biogas production. Initially, those were the reasons for the application of anaerobic treatment. Nowadays, the fact that nutrients (N and P) are not destroyed (e.g., emitted as N_2) but released as recoverable ions is considered an additional advantage, since this facilitates recovery. Therefore, anaerobic technology has a central role in existing and newly developed waste treatment systems (Figure 3.1). It is, however, important to consider that in most cases, posttreatment of anaerobic effluent is warranted to guarantee that limits for safe discharge of the effluent are met (von Sperling and de Lemos Chrenicharo 2002).

3.2 Basics of Anaerobic Conversion of Organic Matter to Methane

The microbiological aspects of the anaerobic degradation of organic matter to methane have been studied extensively in the past (e.g., Zehnder 1988; O'Flaherty et al. 2006). In short, organic matter is converted to biogas, which consists of methane, CO_2, and other constituents (depending on the composition of the wastewater), and biomass. In contrast with aerobic processes, the main electron acceptor in the process is not oxygen but organic products (in the case of fermentation) or inorganics, like CO_2 or sulfate or metalloids (Fe), leading to products that are relatively energy rich. This is why biomass yields are usually lower under anaerobic conditions than under aerobic conditions when CO_2 is formed (Tauseef et al. 2013). The amount of biogas produced depends largely on the nature and composition of the organic matter and the kinetics of the microorganisms involved in the process, but also on process conditions like hydraulic retention time (HRT), sludge retention time (SRT), and temperature (Braun 2007).

The conversion of organic matter to methane is a complex process that is carried out in four sequential steps, namely, hydrolysis, acidogenesis, acetogenesis, and methanogenesis, by roughly three groups of microorganisms. The microorganisms often operate syntrophically; that is, the microorganisms need to cooperate to catabolize a substrate that otherwise cannot be catabolized by either of them alone (Stams and Plugge 2009) (Figure 3.2). The microbial community in a well-functioning digester is composed in such a way that intermediates do not

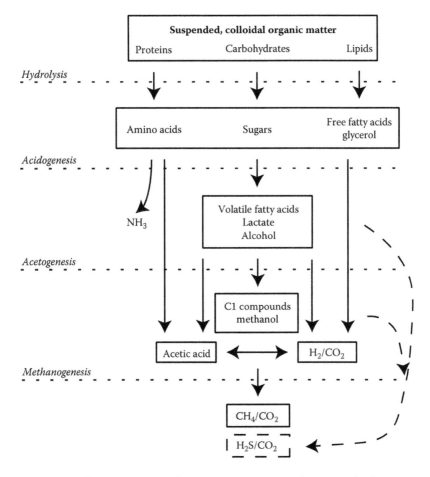

Figure 3.2 Anaerobic conversion of organic matter to methane. Dashed arrows indicate steps that may involve sulfate as the electron acceptor leading to the formation of H_2S.

accumulate (discussed and/or reviewed by, among others, Stams and Plugge [2010]; Parawira [2012]; Azman et al. [2015]). Fermentative bacteria are responsible for the hydrolysis and acidogenesis, leading to the formation of oligo- and monomers from high-molecular-weight organic compounds (fat, carbohydrates, and proteins), which are then fermented to fatty acids, alcohols, acetate, carbon dioxide, and hydrogen. Hydrolysis is catalyzed by exoenzymes that are excreted by the bacteria, and is considered to be the rate-limiting step in the conversion of organic matter to methane. Acetogenic bacteria convert fatty acids, alcohols, and other end products of the acidogenesis to acetic acid, carbon dioxide, and hydrogen. Lastly, methanogenic archaea transform a limited number of substrates, that is, acetic acid, H_2 or CO_2, methylamines, methanol, or other C1 compounds, to methane and carbon dioxide. Methanogens are often the slowest-growing organisms in a digester. Therefore, the methanogenesis is considered to be the most sensitive step in the anaerobic digestion process.

The performance of the microbial consortium (the activity) in any reactor will be influenced by the composition of the feed, environmental conditions (e.g., pH, temperature, salt, redox, and oxygen level), and the presence of toxic compounds. The composition of the feed obviously directly reflects the biological methane potential (BMP), as does the presence of other components, like nitrogen, phosphorous, and sulfur, which are essential elements that make up biomass ($CH_7NO_2S_{0.1}$). Lists with BMPs from municipal and industrial wastes have been compiled by others (e.g., Braun 2007). In theory, the conversion of 1 kg of organic matter (chemical oxygen demand [COD]) yields 0.35 Nm^3 of CH_4,* when biomass yield is neglected. In practice, the microbial characteristics (e.g., biomass yield), the presence of nutrients, the environmental conditions, and the method of operation of the anaerobic reactor determine whether this yield is actually achieved. The presence of other noncarbon constituents (N and S) in the wastewater may decrease the amount of CH_4 produced either because organic matter (COD) is used for favorable conversions with alternative acceptors (sulfate or nitrate), leading to the formation of CO_2 and H_2S or N_2, or because compounds are liberated (e.g., NH_3/NH_4) that may inhibit methane formation. Also, the presence of other microelements (Fe, Ni, Co, etc.) has an effect on the activity depending on the concentration (e.g., Zandvoort et al. 2006; Demirel and Scherer 2011; Thanh et al. 2016). Anaerobic reaction rates are not limited by the concentrations of reactants (compare with aerobic processes and the presence of O_2), but more by the microbial activity (if all other conditions needed for biodegradation, e.g., the presence of nutrients, are met) (Hartmann and Ahring 2006). In general, anaerobic digestion is considered to be more sensitive toward changes in environmental conditions (i.e., pH and temperature) than aerobic microbial degradation because of the sensitivity of methanogenic bacteria and the lower growth rate of this particular group of microorganisms. Therefore, the pH range for (the one-stage) anaerobic conversion of organic matter to methane is usually limited to pH 6.5–8.5, and alkalinity should be maintained at a level that will ensure that acidification does not occur (Ketheesan and Stuckey 2015).

---

* This is based on the following: 1 kmol methane (at 273 K and standard pressure; thus a molar volume of 22.414 m^3) represents 16 and 64 kg of COD.

The temperature should not fluctuate too much in order to maintain stability of the process. In practice, these limitations can be easily overcome provided that the alkalinity of the wastewater is monitored (Braun 2007). Anaerobic reactors should always be designed for the lowest temperature. In principle, three different temperature ranges for anaerobic digestion are distinguished: psychrophilic digestion up to 20°C, mesophilic digestion up to 40°C, and thermophilic digestion at higher temperatures (but usually around 50°C–60°C) (van Lier et al. 1997), with clearly different microbial populations in each temperature range (Chapleur et al. 2016). Wastewaters generated at temperatures below 40°C are treated in mesophilic conditions. For wastewaters that are produced below 20°C, heating may be needed, but the biogas produced in anaerobic treatment may be used for that (Sipma et al. 2010). Overall, organic $(CH_4-)COD$ (or BOD) has an energy content of 14.5 MJ/kg (McCarty 1965). Ideally, 3.45 g COD/L contains enough energy to heat the liquid by 1°C (disregarding any heat transfer efficiencies and other possible energy losses that would occur). For high-temperature wastewaters, thermophilic wastewater treatment may be the preferred option to avoid problems related to the heating and cooling of waters (precipitation). Moreover, the reactor volumes required for thermophilic treatment of the wastewater are smaller, and when water reuse is considered, the higher pathogen reduction at thermophilic conditions is a clear advantage. Still, the general idea is that thermophilic treatment is prone to more instability issues (Braun 2007).

Process instability is often due to abrupt or extensive changes in environmental parameters (pH, temperature, etc.), the presence of toxic compounds, or organic overloading of the system. As a result, the reactor system will show a sequence of events: an increase of the volatile fatty acid (VFA) concentration, lowering of the pH, decreased COD removal and methane recovery efficiencies, and overall deterioration of effluent quality. In the earlier stages of overloading, the problems may be overcome by lowering the COD load and/or increasing the pH and alkalinity. However, due to lower growth rates of the methanogens, the recovery of the process may take some time once methane formation has been affected. Hydraulic overloading may cause similar effects combined with washout of anaerobic biomass. The overall effects of these events are strongly correlated to the extent and duration (Ketheesan and Stuckey 2015).

3.3 Anaerobic Wastewater Treatment: Technology

Typically, there are two categories of systems that can be distinguished based on the applied conditions. Low-rate systems (e.g., lagoons and septic tanks) (Table 3.1) are used for the treatment of waste(water) streams with high organic matter contents. COD removal in most cases does not exceed 50%. A next-generation low-rate system is the completely stirred tank reactor (CSTR), which is completely mixed, and the anaerobic contact process usually has an external settler, with a sludge return. HRTs and SRTs are uncoupled in high-rate systems (Table 3.2), and contact between sludge and water is usually more intense, enabling higher loadings and higher COD removal efficiencies (compare Tables 3.1 and 3.2). The uncoupling of the HRTs and SRTs is

TABLE 3.1   Overview of Available Anaerobic Wastewater Treatment Systems and Characteristics

Process	General Description	Sludge	Mixing	OLR (kg COD/ m³/day)	HRT	SRT (days)
Low loaded anaerobic lagoon		F/Su/Se	−	<2	20–50 days	50–100
Anaerobic contact process	Completely mixed, degassing chamber and separate settler	Su	+	2–5	<SRT	
Anaerobic migrating blanket reactor (AMBR)	Sludge maintained by reversing flow direction		+	1–3	4–12 hours	≫HRT
Anaerobic sequencing batch reactor (AnSBR)	One reactor; feed, reaction, settling, and decanting are separated in time	Su, GS	+	1–3	6–24 hours	50–200
Anaerobic CSTR	Semisolid wastes	Su		<4	15–30 days	15–30
Plug flow	Semisolids, sometimes recirculation for inoculation	10%–18% SS	+/−	<4	20–30 days	20–30

Source: Data taken from Tauseef, S.M. et al., *Renewable Sustainable Energy Rev*, 19, 704–741, 2013; Tchobanoglous, G. et al., *Wastewater Engineering: Treatment and Resource Recovery*, 2 vols, McGraw-Hill, New York, 2014.

Sludge Note: F = flocculent; GS = granular sludge; OLR = organic loading rate; Se = settled; Su = suspended.

TABLE 3.2 High-Rate Systems

Process	General Description	Sludge	Mixing	U/D	Gas	V_{up} (m/h)	OLR (kg COD/m³/days)	HRT	SRT (days)
UASB	Suitable for suspended solids	GS	–, V_{up}, biogas	U	GLS	1–6	5–20	4–8 hours	>30
EGSB	UASB with high V_{up}, greater H/D, recirculation of effluent	GS	–, V_{up}, biogas	U	GLS	4–10	35		
IC-UASB	Two stacked UASBs with internal circulation	GS	–, V_{up}, biogas	U	GLS	8–20 × UASB			
FB	Particles in suspension and mixed by high V_{up}	BiC	–, V_{up}, biogas	U	S	10–20	20–40		
AF	Filter media: stones, pumice, wood chips, polypropylene	FF-BiC	–	U/D	S		5–20	1–3 days	≫HRT
AHP	UASB-AF	GS/FF-BiC		U					
AnMBR	Membrane solids–liquid separator	Su/F	+				5–15	≪SRT	
ABR	Series of UASBs	2%–10% VS	–	U			5–10	6–24 hours	≫HRT

Source: Data taken from Tauseef, S.M. et al., *Renewable Sustainable Energy Rev*, 19, 704–741, 2013; Tchobanoglous, G. et al., *Wastewater Engineering: Treatment and Resource Recovery*, 2 vols., McGraw-Hill, New York, 2014.

Note: ABR = anaerobic baffled reactor; AF = anaerobic filter; AHP = anaerobic hybrid process; BiC =biomass on inert carrier; F = flocculent; FB = fluidized bed; FF = fixed film; GS = granular sludge; GLS = gas-liquid-solid separator; S = other separator; Se = settled; Su = suspended; U/D = upflow/downflow.

attained by fixation of the biomass on a stagnant or suspended (inert) carrier or by the formation of granules often coinciding with improved sedimentation of the active biomass in the reactor. One of the more widely applied systems is the UASB reactor, which features a gas–solid–liquid separator. The mixing in this system is achieved via the production of biogas and maintaining an upflow velocity (V_{up}) of 1–6 m/h. The UASB was further "upgraded" to an EGSB system for the treatment of especially "difficult," mainly dissolved, industrial effluents, that is, very diluted or concentrated wastewaters, cold industrial effluents, or process streams containing potential toxicants. The most important difference between UASB and EGSB is the height–diameter ratio, which is much higher in EGSBs, leading to a higher upflow velocity of the liquid in the reactor. This, combined with the recirculation of effluent, leads to a better contact between water and sludge, which is especially important for cold and diluted wastewaters. On the other hand, the EGSB is also applied for the treatment of very concentrated waste streams or wastewaters that do not have a constant concentration because effluent recirculation establishes equalization of the flows, and thus results in less fluctuations in the COD loading.

Since the 1980s, the different systems based on either fixed biofilms or granular sludge technology have been developed and commercialized by different suppliers (Tables 3.1 and 3.2) (van Lier et al. 2015). The formation of granular sludge has been studied for several decades (Hulshoff Pol et al. 2004). Granulation is an important feature in anaerobic high-rate reactors that enables the operation of reactors at higher loadings. Also, the uncoupling of HRT and SRT in anaerobic reactors is facilitated. Nevertheless, the formation of granular sludge is not essential to establish a well-functioning anaerobic reactor. This is especially the case for UASB reactors treating municipal waste. Often, the sludge in those reactors is flocculent and not granular. However, there is no need for granulation as long as the methanogenic activity of the sludge is high enough (up to 0.5 g CH_4-COD/g volatile suspended solids [VSS]/day, for municipal sewage [van Haandel and Lettinga 1994]) and the settleability of the sludge such that the active biomass remains in the reactor. Suspended solids in the influent may also lead to problems with granulation. For most industrial wastewaters, granulation is less of a problem except for specific cases (e.g., high COD wastewaters like vinasses and distillery slops [van Lier et al. 2015]) when specific groups of microorganisms (e.g., the acetogens) are overgrowing the other trophic groups in the systems, leading to an imbalanced microbial consortium. Good-quality granular sludge is characterized by a methanogenic activity of 0.3–1.0 CH_4-COD/g VSS/day for industrial wastewater and settling velocities of 30–75 m/h (van Lier et al. 2015).

One of the latest developments in anaerobic wastewater treatment is the application of anaerobic membrane processes. These processes are especially suitable for industrial wastewaters with high salinity, high temperature, or wastewaters containing toxic, but biodegradable compounds. The sludge in these reactor systems is present in flocculent form of even as single cells. The costs of this technology are still relatively high compared with conventional anaerobic treatment, but full-scale anaerobic membrane reactors (AnMBRs) have been put into operation in Europe, Japan, and the United States for food, diary, and bioethanol industry wastewaters (van Lier et al. 2015).

Depending on the composition of the feed of an anaerobic reactor, biogas may contain CH_4, CO_2, H_2, NH_3, H_2S, volatile hydrocarbons, water, and other compounds. Depending on the further use, the biogas has to be cleaned up. In any case, uncontrolled release of biogas needs to be prevented since methane is a 25 times more potent greenhouse gas than CO_2 (over a 100-year time horizon) (IPCC 2007). Biogas may be used directly in a gas burner or combined heat and power plants to produce electricity and heat. Usually, the efficiency of combined heat power (CHP) is rather low (34%–40%). In both cases, H_2S will have to be removed with chemical or microbiological methods and the biogas needs to be dried. In some cases, trace contaminants will have to be removed as well. Upgrading to natural gas quality requires the removal of CO_2 to increase the Wobbe index of the gas to natural gas quality (Braun 2007).

3.4 Industrial Wastewater to Methane

The composition of industrial wastewater varies enormously among industries. The continuous or batch production of products may affect the composition of the produced wastewater, but also, operational factors within a plant (choice of cleaning agent) may affect wastewater characteristics. Table 3.3 lists examples of BMPs of different industrial waste streams as determined in laboratory tests. The data show that there is a large variety in methane production to be expected. Besides the origin of the wastewater and the composition, there are also other factors that determine the methane production efficiency from industrial wastewaters. In some cases (e.g., in certain agro-industries), feed to a reactor may vary seasonally. Thus, the biological treatment of these industrial wastewaters, either anaerobic or aerobic, is often

TABLE 3.3   Examples from BMPs from Different Industrial Wastewaters

Industrial Waste Stream	BMP (Nm³/kg VS)
Vegetable fat and oil wastewater	0.68–0.97
Animal fat	0.88
Fish for human consumption	2–4.31
Vegetable and fruit processing	0.26–0.78
Slaughterhouse	0.86
Primary industrial sewage sludge	0.27
Pharmaceutical waste (egg residues, blood)	0.86–1.2
Fermentation and distillery slops	0.35–0.75
Potato industry	0.30–0.79
Pulp and paper mill wastewater	0.14–0.40
Brewery	0.26–0.35
Chipboard manufacturing wastewater	0.78

Source: Data taken and adapted from Braun, R., in *Improvement of Crop Plants for Industrial End Uses*, ed. P. Ranalli, Springer, Berlin, 2007, pp. 335–416; Maya-Altamira, L. et al., *Water Res.*, 42(8–9), 2195–2203, 2008; Meyer, T., and Edwards, E.A., *Water Research*, 65, 321–349, 2014.

complicated because of the high organic strength of the wastewaters, low concentrations or even absence of certain nutrients, presence of toxic compounds, and physicochemical parameters of the wastewater, such as pH, temperature, and salinity (Sipma et al. 2010). In most of these cases, equalization tanks may facilitate the application of anaerobic treatment, but all the factors mentioned above may affect the methane production efficiency.

Extreme environmental conditions (such as pH, temperature, or salinity) occur in a variety of occasions. This is even more the case since most industries start to reuse their process water (van Lier et al. 2015). Countermeasures have to be taken to avoid problems in these "extreme" conditions. In some cases, industrial effluents need to be neutralized prior to feeding to the reactor. Sudden changes in the (industrial) activity may lead to a pH drop and subsequent overloading of the reactor and lowering of the methane production (Sipma et al. 2010). These changes in pH may have an effect on the microbial population directly (pH shock and toxicity), but also indirectly via the availability of, for example, trace elements that are essential for microbial activity. In extreme salt conditions, many anaerobic (and aerobic) microorganisms may be inhibited. Examples of industries that have high-salinity wastewaters are the petroleum refining, textile processing, leather processing, and fish and other food (e.g., cheese) processing industries (Ismail 2013). One of the difficulties with saline wastewater is the effect of the salt on granulation in high-rate anaerobic systems. In some cases, wastewaters with a high salinity are also alkali, which often necessitates other measures or technologies (like the application of an EGSB).

There are a large variety of organic or inorganic compounds that may be inhibiting anaerobic processes to a certain extent. These may be present in the industrial effluent or formed as a result of (bio-)transformation processes occurring during the treatment of the wastewater. These compounds could be heavy metals or mono- and (multi-)valent cations, but also natural or anthropogenic compounds like aromatic compounds, phenols, halogenated (organic) compounds, volatile organics, surfactants, and biocides (Sipma et al. 2010). The release of these compounds in the industrial wastewater may be temporary or permanent, and concentration levels may also fluctuate. Whether the compound actually affects the anaerobic processes is determined by the concentration and (environmental) factors, like pH, temperature, and other factors that determine bioavailability. For example, ammonium, H_2S, and VFAs are inhibiting anaerobic processes only in the undissociated form (Chen et al. 2014). Adaptation of the microbial population is possible and may or may not occur, depending on the conditions. Usually, some time is required before a microbial population is adapted. Also, in some cases bioaugmentation with microorganisms able to degrade the toxicant may be applicable (Christiansen and Ahring 1996; Schauer-Gimenez et al. 2010). Anaerobic microorganisms may respond very differently to the presence of toxic compounds compared with aerobic microorganisms. It is beyond the scope of this chapter to review the effects of the different groups of compounds that may occur in industrial wastewaters. Several other authors have compiled lists with toxicity data (e.g., Blum and Speece 1991).

Certain compounds are more easily biodegradable under anaerobic conditions (polyhalogenated hydrocarbons), whereas others are only biodegradable under aerobic conditions (monohalogenated hydrocarbons). As a result, in some cases combined anaerobic–aerobic biodegradation is needed for the complete removal of inhibitory compounds. Potential pre- or posttreatment for the removal or conversion of toxicants may be physicochemical (e.g., chemical oxidation) or biological (e.g., aerobic).

Overall, a large enough stable pool of active biomass seems to be the key solution to problems concerning overloading, the presence of toxicants, or extreme environmental conditions. Systems with longer SRTs are better capable of dealing with inhibition problems. This is partially due to the fact that often microorganisms that degrade toxicants are slow growing, but also because of concentration gradients that occur in biofilms that prevent the exposure of part of the microorganism to the toxicant (Sipma et al. 2010).

Industrial wastewaters of different origin give different methane production yields depending on production routes and technologies used. Table 3.4 gives ranges of wastewater production, COD, and theoretical methane production for different wastewaters. Possible bottlenecks, like salinity, toxicity, and other problems, are not taken into account. The wastewater from the pulp and paper industry stands out for the amount

TABLE 3.4   Wastewater Production, COD, and Theoretical Methane Production per Ton of Product

| Industry Type | Wastewater Generation | | COD | | Methane Production |
| | Typical | Range | Typical | Range | |
	(m³/ton)	(m³/ton)	(kg/m³)	(kg/m³)	(Nm³/ton Product)
Alcohol refining	24	16–32	11	5–22	74
Beer and malt	6.3	5.0–9.0	2.9	2–7	5
Dairy products	7	3–10	2.7	1.5–5.2	5
Fish processing	NA	8–18	2.5		6–13
Meat and poultry	13	8–18	4	1 2–7	15
Organic chemicals	67	0–400	3	0.8–5	56
Petroleum refineries	0.6	0.3–1.2	1	0.4–1.6	0
Plastics and resins	0.6	0.3–1.2	3.7	0.8–5	1
Pulp and paper[a]	162	85–240	9	1–15	408
Soap and detergents	NA	1.0–5.0	NA	0.5–1.2	0–2
Starch production	9	4–18	10	1.5–42	25
Sugar refining	NA	4–18	3	2 1–6	2–30
Vegetable oils	3.1	1.0–5.0	NA	0.5–1.2	0–2
Vegetables, fruits, juices	20	7–35	5	0 2–10	28
Wine and vinegar	23	11–46	1.5	0.7–3.0	10

Note: Methane production is calculated using data taken from IPCC (2006) with a methanogenic conversion factor of 0.8.

[a]Nowadays, the pulp and paper industry often uses less water and the wastewater contains lower COD concentrations due to the internal closing of water cycles (Pokhrel and Viraraghavan 2004; Meyer and Edwards 2014).

of methane produced (although the data used are very general and care should be taken with interpretation). The benefits of methane production from wastewater from the pulp and paper industry are illustrated by the case described in the next section.

3.5 Case Study in Industrial Wastewater: Treatment of Wastewater from a Paper Mill

The main constituents of paper are cellulose fibers, filler material (i.e., clay or chalk), water, and starch. The production of paper requires substantial amounts of water. This water needs to be as clean as possible to avoid problems in the production process. The process water flow rate and the composition of the process water (low N and P) are maintained at such levels that undesired bacterial growth during the papermaking process is minimized. Recycling of process water is quite common in the paper industry. Paper mill effluent, in most cases, needs to be treated before reuse is possible to comply with the demands for use as process water. Another equally important reason for the treatment of paper mill effluent is the obligation to meet the discharge limit for surface water.

This case roughly describes the treatment of wastewater that is produced by three paper mills in the Netherlands. The treatment of the wastewater (paper mill effluent = influent of the wastewater treatment system) is carried out by a private company that was especially funded for this purpose. Data used are taken from public sources (Gast 2009; Industriewater Eerbeek 2013). This case shows that anaerobic treatment of industrial wastewater may provide the necessary energy for complete purification of the wastewater to reach surface water discharge limits while producing useful resources. In fact, there is an excess of more than 60%, which is currently converted to electricity that is fed to the electricity grid. In the near future, however, the excess biogas after washing will be returned directly to the paper mill for further use as a replacement for natural gas in a gas burner. This is more energy-efficient because the generator used at the moment has a conversion efficiency of 36%. The treatment plant has been running successfully for many years. The influent characteristics, the discharge limit to be met, and the effluent characteristics of the treated wastewater in 2013 are given in Table 3.5.

A scheme of the wastewater treatment plant (WWTP) in 2013 and subsequent production of useful products is given in Figure 3.3. The influent (3.8 million m^3/year) of the WWTP (i.e., the effluent of the paper mills) contains suspended solids (fibers, calcium carbonate, and clay), as well as solubilized matter. Solids (organic and inorganic) are removed in the primary settler. Around 70% of the soluble organic matter (cellulose, VFAs, and other dissolved COD) is biodegraded in the anaerobic reactor and converted to raw biogas. The HRT of the UASB reactor is 4–5 hours, and the reactor is operated at 25°C–35°C. The effluent of the anaerobic reactor still contains soluble organic matter (VFAs, acetate, and propionate) and sulfide. These compounds, 10% of the readily biodegradable material, are removed in the aeration tank (also intended to remove odor). Remaining (not readily) biodegradable organic matter and nutrients are removed in the activated sludge plant. The aerobic plant is operated at a low loading and with an HRT of 8 hours. The overall BOD removal efficiency is 99.8%. The BOD

TABLE 3.5 Influent and Effluent Characteristics and Discharge Permit for the Entire Plant

		Permit	Influent 2013	Effluent 2013
Flow rate	×1000 m³/day		3820	
Dry matter	mg/L	<30	1670	11
COD	mg/L	<200	3188	84
BOD	mg/L	<20	1605	3
N_{total}	mg N/L	<10		7.5
N_{kj}	mg N/L			3.7
Phosphate	mg P/L	<2		0.5
Sulfate	mg SO_4/L		221	159
Chloride	mg Cl/L		99	120

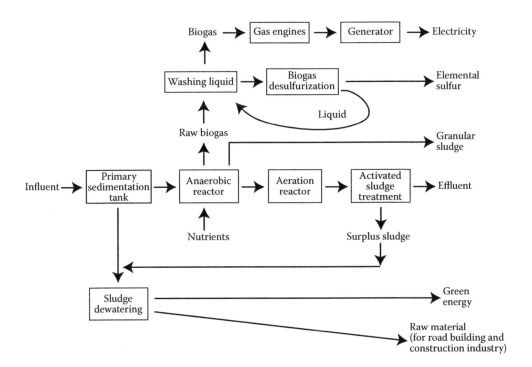

Figure 3.3 Treatment of the paper mill wastewater: from influent to effluent and products formed in the process. (Scheme adapted from Industriewater Eerbeek, Sustainability report, Industriewater Eerbeek, Arnhem, Netherlands, 2013.)

in the effluent is 4 mg/L, which is low enough to discharge the effluent in the surface water of the neighboring Ijssel River (Table 3.5). Sulfur compounds are removed from the biogas with a washing liquid, and the washing liquid is then biologically treated to produce elemental sulfur while the washing liquid is reused. The desulfurized biogas is currently used in engines and a generator for electricity production (efficiency 36%).

TABLE 3.6   Energy Balance in 2013

Parameter	Unit	Value	Remarks
Flow in 2013	m³	3,820	
COD	g/m³	3,188	
COD removed in anaerobic reactor	g/m³	2,231	70% COD reduction in anaerobic reactor
COD used for sulfate reduction	g/m³	42	Sulfate (influent–effluent)*0.67 g COD/g sulfate
COD converted to CH_4	g/m³	2,190	
CH_4 produced	Nm³/m³ influent	0.77	Methane has a COD of 64 g/mol; molar volume = 22.414 L/mol; biomass formation is neglected
CH_4 production in 2013	m³	2,931,790	CH_4 produced*yearly flow
	GJ	105,955	Heating value 36.14 MJ/m³ (Braun 2007)
	MWh	29,430	1 MWh = 3600 MJ
Electricity production	MWh	10,595	36% efficiency in CHP
Electricity needed	MWh	3,655	0.3 kWh/kg COD (influent)
Electricity excess to grid	MWh	6,940	

Source: Data taken from Industriewater Eerbeek, Sustainability report, Industriewater Eerbeek, Arnhem, Netherlands, 2013.

Sludge and settled matter from the primary settler (9000 tons/year) are dewatered and processed into green energy and raw materials for road building and the construction industry. The excess electricity production is currently supplied to the national grid, but as mentioned above, the excess biogas will be used by the paper mill in the near future. Another product of the WWTP is 1790 m³ of excess granular sludge, which is sold for the start-up of other anaerobic reactors or experimentation with pilot plants.

Yearly, the WWTP produces around 4 million m³ of biogas with approximately 80% CH_4. This amount of biogas is more than enough to supply the amount of energy needed for the whole plant, as is shown by the calculations in Table 3.6.

3.6 Municipal Wastewater: Specific Compounds and Resources

Domestic wastewater generally contains feces and urine, and toilet flush water (together called black water [BW]); shower and bath water, kitchen water, and laundry water (together called grey water [GW]); rainwater; and infiltration water. When industrial streams are also added, this is referred to as municipal wastewater.

In some cases, kitchen waste (KW) is added. In the United States, for example, kitchen grinders are often applied and connected to the sewer. In some countries, like the Netherlands, kitchen grinders are forbidden, to prevent extra load to the aerobic WWTP. When source separation systems are applied, the addition of KW to the BW can be useful to increase the recovery of energy via anaerobic treatment (Kujawa-Roeleveld et al. 2005). The per capita per day production of different components, like COD, N, P, and K, in domestic waste and wastewater streams, is shown in Table 3.7.

TABLE 3.7 Mean Resource Load in Black Water, Kitchen Waste, and Grey Water

Parameter	Unit	Black Water	Kitchen Waste[a]	Grey Water[a]
Volume	L/cap/day	1.5	0.2	79
COD	g/cap/day	61	59	52
N_{total}	g/cap/day	9.8	1.7	1.2
P_{total}	g/cap/day	1.3	0.2	0.4
PO_4^{3-}-P	g/cap/day	0.5	–	0.1
K	g/cap/day	3.7	0.2	0.8

Source: Tervahauta, T. et al., *Water*, 6, 2436–2448, 2014.

[a]For industrialized countries.

3.6.1 Combined Domestic Wastewater

Typically, GW and BW are collected within one sewer pipe, often combined with rainwater and sometimes with industrial water, and transported to a centralized municipal WWTP. The concentration of the municipal wastewater will vary with the household water consumption, rainwater collection, additional industrial water, and infiltration of water in the sewer. Typical concentrations of different components are summarized in Table 3.8.

The results in Table 3.8 show a high fraction of COD_{ss} within municipal wastewater, which explains the formation of flocculent instead of granular sludge when

TABLE 3.8 Characteristics of Municipal Wastewater in Different Regions of the World

		Bennekom, Netherlands[a]	Cali, Colombia[a]	Pedregal, Brazil[a]	Ghent, Belgium[b]	Ramallah, Palestine[c]
COD_{total}	mg/L	520	267	727	268 ± 44	1518–3812
COD_{ss}[d]	mg/L				139 ± 34	545–1925
BOD	mg/L	231	95	368		
Nitrogen	mg N/L	45	24	44	37 ± 5	54–119
Ammonia-N	mg/L		17	34	26 ± 3	47–72
Phosphorous	mg P/L	18	1.3	11	7 ± 3	10–15
PO_4-P	mg/L	14		8	4 ± 2	6–17
Sufate	mg/L	15		18		474–2060
Chloride	mg/L			110		
Alkalinity	mg/L as $CaCO_3$	350	120	388	361 ± 44	
Temperature	°C	8–20	24–27	24–26		26–40
TSS	mg/L		215	429	128 ± 27	510–1096
VSS	mg/L		107	252	100 ± 24	255–892

[a]From Seghezzo, L. et al., *Bioresour. Technol.*, 65(3), 175–190, 1998.

[b]From Kalogo, Y., and Verstraete, W., *Environ. Technol.*, 21(1), 55–65, 2000.

[c]From Mahmoud, N. et al., *Environ. Pollut.*, 126, 115–122, 2003.

[d]$COD_{suspended solids}$.

applying a UASB reactor for municipal wastewater treatment. Van Haandel and Lettinga (1994) report the presence of only flocculent sludge in full-scale UASB reactors, treating municipal sewage (van Haandel and Lettinga 1994). The high fraction of suspended solids in municipal wastewater implies that the hydrolysis is the rate-limiting step in the treatment of municipal wastewater (Zeeman and Sanders 2001). When treating municipal wastewater in a UASB reactor, a major part of the influent suspended solids is removed via settling and entrapment in the flocculent sludge bed. The removal of solids is visualized by Mahmoud et al. (2004a) (Figure 3.4). The upflow velocity, which can be calculated from the reactor height and the HRT, influences the SS removal. At a fixed influent concentration, the HRT determines loading rate, which, together with the temperature, influences the SRT. The SRT affects the sludge characteristics, which, in turn, influence the SS removal. The SS removal again affects the SRT.

The percent hydrolysis is determined by the biodegradability and the first-order hydrolysis rate constant of the particulate COD, in combination with the SRT and temperature. The amount of methane produced plus the produced biodegradable dissolved COD per volume wastewater equals the hydrolyzed COD. Anaerobic treatment of municipal wastewater is so far mainly applied in tropical and subtropical areas. The latter can be explained by the low hydrolysis rate of municipal sewage COD_{ss} when applying low temperatures. A long SRT and related long HRT are therefore required for sufficient hydrolysis and methanogenesis at temperatures of 15°C or lower. The very low COD concentration and related biogas yield (liters of biogas per liter of wastewater) of municipal sewage in many regions (Table 3.8) render energy-efficient heating of domestic or municipal wastewater impossible. For wastewaters containing a large fraction of COD_{ss}, Zeeman and Lettinga (1999) developed a formula (Equation 3.1) to calculate the HRT for a required SRT, enabling treatment of municipal wastewater with a large fraction of suspended

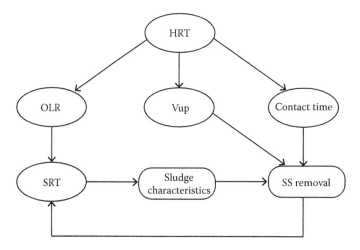

Figure 3.4 Factors influencing the removal of suspended solids in a UASB. (Adapted from Mahmoud, N. et al., *Water Res.*, 38(4), 983–991, 2004a.)

solids (Zeeman and Lettinga 1999). In addition, the upflow velocity needs to be checked.

$$HRT = (C * SS/X) * R * (1 - H) * SRT \qquad (3.1)$$

with C = the influent COD concentration (g/L), X = the sludge concentration in the reactor (g COD/L), SS = the fraction COD_{SS} of the COD_{inf}, R = the fraction of the COD_{inflSS} that is removed, H = the fraction of the removed COD_{SS} that is hydrolyzed, and HRT and SRT are in days.

The fraction H can be calculated for different SRTs based on the biodegradability and hydrolysis rate constant of the COD_{ss}. An overview of the biodegradability and first-order hydrolysis rate constants of different municipal wastewater streams at different temperatures, as reported in the literature, is given in Table 3.9.

Reported biodegradability varies between 55% and 79%, while the reported first-order hydrolysis rate constant at 15°C varies between 0.005 and 0.03 per day. Based on a first-order hydrolysis rate constant at 15°C of 0.005 per day (lowest reported), an SRT (assuming the UASB sludge bed behaves like a CSTR) of 180 days can be calculated, assuming that 90% conversion of the biodegradable suspended COD is

TABLE 3.9   Biodegradability and First-Order Hydrolysis Rate Constants for Municipal Sewage, Black Water, Grey Water, and Primary Sludge at Different Temperatures, as Reported in Literature

Type of ww	Biodegradability (%)	Hydrolysis Rate Constant (per day)	T (°C)	Reference
BW	55	ND	37	de Graaff et al. 2010a
BW	71	ND	30	Elmitwalli et al. 2001
DS	74	ND	30	Elmitwalli et al. 2001
DS		0.05 ± 0.02	30	Teixeira et al. 2008
COD_{ss} in DS	77	ND	30	Elmitwalli et al. 2001
GW	70	0.02	35	Hernandez-Leal et al. 2010
MWW	79	0.08 (carbohydrates) 0.025 (proteins) 0.06 (lipids)	25	Halalsheh et al. 2005
MWW		0.005 (carbohydrates) 0.012 (proteins) 0 (lipids)	15	Halalsheh et al. 2005
PS		0.005	15	Halalsheh et al. 2005
PS		0.021	25	Halalsheh et al. 2005
PS	0.59 (0.002)	0.0289 (0.0006)	15	Halalsheh et al. 2005
PS	0.63 (0.005)	0.07 (0.002)	25	Halalsheh et al. 2005
PS	0.57 (0.006)	0.113 (0.002)	35	Halalsheh et al. 2005

Note: DS = domestic sewage, Bennekom, except for Teixeira et al. (2008); MWW = municipal wastewater, Amman; ND = not determined; PS = primary sludge; ww = wastewater.

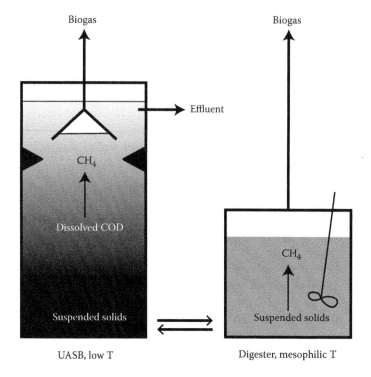

Figure 3.5 Schematic representation of a UASB–digester system for treatment of municipal wastes at low temperatures. (Based on Mahmoud, N. et al., *Water Res.*, 38(9), 2348–2358, 2004b.)

achieved. The latter would result in an HRT of 20 hours at 15°C (calculated with Equation 3.1*). Sewage temperatures in the Netherlands can, in winter, become lower than 10°C. Halalsheh et al. (2005) report an average hydrolysis and methanogenesis of 76% and 71%, respectively, during summer, and 46% and 42%, respectively, during winter when operating a UASB of 60 m³ at an HRT between 23 and 27 hours with strong municipal wastewater of the Khirbit As-Samra WWTP, Amman, Jordan (Halalsheh et al. 2005). The SRT was calculated to be 137 and 186 days for winter (mean temperature 18°C) and summer (mean temperature 25°C), respectively. The methane recovery at summer and winter conditions equals, respectively, 96% and 58% of the biodegradability (Table 3.9).

To overcome the difficulties with low-temperature anaerobic treatment of municipal wastewater, the UASB–sludge digester system is being developed (Mahmoud et al. 2004b; Zhang et al. 2013). A high loading of the low-temperature UASB is possible, due to the transfer of the nonstabilized sludge to a mesophilic sludge digester, while the mesophilic digestate is returned to the UASB for removing the dissolved COD at low-temperature conditions. A schematic representation is shown in Figure 3.5. Mahmoud et al. (2004b) show a stable performance of the UASB–sludge digester at an HRT and temperature in the UASB of 6 hours and 15°C, respectively, and a

---

* Biodegradability is 79%, C = 0.6 g COD/L, X = 20 g COD/L, R = 0.75, and SS = 0.7.

temperature of 35°C in the sludge digester, while applying a sludge recirculation rate of 22%. The recent results of Zhang (2016) showed a stable performance of a UASB–sludge digester, when the UASB is operated at 6 hours' HRT at temperatures as low as 10°C, at a sludge recirculation rate of 16%.

3.6.2 Source-Separated Municipal Wastewater

Instead of a combined collection of municipal wastewater streams, source separation can be applied (Zeeman 2012). When separately collecting BW with water-saving toilets (i.e., vacuum toilets), COD concentrations are sufficiently high for applying mesophilic conditions during anaerobic treatments of the BW. de Graaff et al. (2010a) show the long-term anaerobic treatment of BW, collected with vacuum toilets, in a UASB (HRT = 8.7 days, 25°C). Influent concentrations and a mass balance over the entire operational period of 951 days are shown in Figure 3.6.

After anaerobic treatment, the liquid effluent can be treated in a nitritation–anammox system, either a two-step (de Graaff et al. 2010b, 2011b) or a one-step (Vlaeminck et al. 2009) system, followed by struvite precipitation (de Graaff et al. 2011a). Such a BW treatment chain is applied at full scale for 250 houses in Sneek (de Graaf and van Hell 2014), an office building in Venlo, and a school in the Ukraine. A fourth application at the Institute of Ecological Research (NIOO) in Wageningen includes anaerobic treatment of BW in a UASB, followed by a photobioreactor for nutrient recovery by the growth of algae (Fernandes et al. 2015). At the new housing estate in Sneek, not only BW but also GW is treated, applying a bioflocculation system (Hernandez-Leal et al. 2010), followed by aerobic posttreatment. The bioflocculation sludge is added to the BW UASB reactor for increasing the energy production (Hernandez-Leal et al. 2010; Tervahauta et al. 2014). Additionally, KW is added to the BW, applying kitchen grinders connected to the vacuum system.

3.7 Case Studies in New Sanitation

3.7.1 Waterschoon, Sneek, the Netherlands

The Sneek case is the most monitored and reported of the four above-mentioned case studies for new sanitation. A flow scheme is presented in Figure 3.7.

A summary of the results of the new sanitation system, Waterschoon, in Sneek, is presented in Table 3.10.

Unfortunately, a relatively large amount of water is used for the kitchen grinders, resulting in a total flow of 13.7 L/cap/day (Table 3.10). The latter negatively influences the energy balance (de Graaf and van Hell 2014). For the second phase of Waterschoon, where an additional 290 houses are built, adjustments are made to reduce the water consumption of the kitchen grinders. An extrapolated energy balance for an optimized new sanitation concept, similar to Waterschoon, but for 1200 persons, is presented in Table 3.11 in comparison with that of a conventional WWTP of 100,000 person equivalent.

	Unit	Days 1–518	Days 519–915
pH	–	8.8 ± 0.22	8.6 ± 0.53
COD_{total}	g/L	9.8 ± 2.6	7.7 ± 2.5
COD_{ss}	g/L	5.1 ± 2.7	4.9 ± 2.0
$COD_{colloidal}$	g/L	1.3 ± 0.42	0.5 ± 0.22
$COD_{soluble}$	g/L	3.4 ± 0.47	2.3 ± 0.81
VFA-COD	g/L	1.5 ± 0.48	1.2 ± 0.89
N_{total}	g/L	1.9 ± 0.19	1.2 ± 0.18
NH_4-N	g/L	1.4 ± 0.15	0.85 ± 0.15
P_{total}	g/L	0.22 ± 0.067	0.15 ± 0.064
$P_{total, soluble}$	g/L	0.090 ± 0.0087	0.057 ± 0.018
PO_4-P	g/L	0.079 ± 0.0085	0.054 ± 0.027

(a)

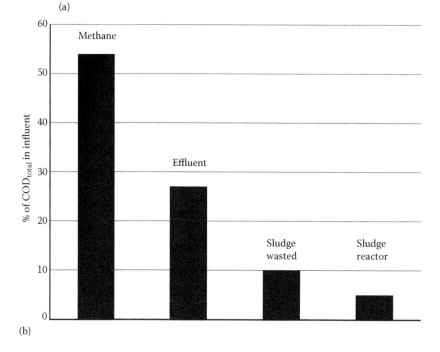

(b)

Figure 3.6 (a) Composition of BW in the study of de Graaff et al. (2010a). Two experimental periods were distinguished (days 1–518 and days 519–915), because a noise reducer was installed in the vacuum toilets for BW collection. (b) COD balance calculated over the total of 915 days of reactor run time. The total amount of COD fed to the reactor was 49 kg (of which 0.1% was inoculum sludge). (Data taken and adapted from de Graaff, M.S. et al., *Water*, 2(1), 101–11, 2010.)

Table 3.11 illustrates a positive energy balance for the optimized new sanitation concept, Waterschoon, while conventional sanitation consumes energy. In addition, calculations show that the cost of a new sanitation system for 1200 persons is 72 €/person/year, while the total cost for conventional sanitation (sewerage and treatment) is 65 €/person/year for a system of 100,000 person equivalent. Taking price volatility and uncertainty factors into account, a scale of application between 1000 and 1500

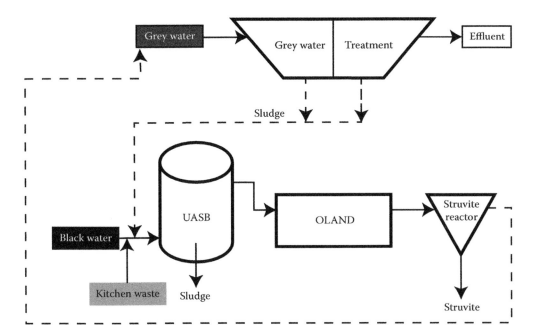

Figure 3.7   Flow scheme of the new sanitation system, Waterschoon, Sneek, the Netherlands.

persons is estimated to result in equal costs of new sanitation and conventional sanitation (de Graaf and van Hell 2014).

3.7.2 Case Study in Municipal Wastewater Treatment at a Tropical Condition, Latin America

Anaerobic treatment of municipal wastewater applying UASB technology is used in several tropical areas, especially in India (Sato et al. 2006) and Latin America (Chernicharo et al. 2015). Chernicharo et al. (2015) present an overview of the most important technologies applied for municipal wastewater treatment in Latin America. The mostly applied system in that region is the stabilization pond (38%), followed by the activated sludge process (26%). In 17% of cases, a UASB reactor is applied, and these UASBs combined treat a flow of 14.2 m³/s municipal wastewater. In Latin America, UASB reactors are a recognized technology for the treatment of municipal wastewater. Large full-scale plants have been in operation for more than 10 years (Chernicharo et al. 2015).

A general flow scheme of a municipal WWTP, including UASB technology, comprises the preliminary treatment of the wastewater prior to anaerobic treatment in a UASB, which may be equipped with a biogas flare and a sludge dewatering unit. When UASB technology is applied for the treatment of municipal wastewater, in general, a posttreatment system is implemented to fulfill discharge standards or standards for agricultural reuse. Chernicharo (2006) summarizes the average effluent concentrations and typical removal efficiencies of the main pollutants in sewage, applying

TABLE 3.10 Results of a Long-Term Monitoring Program[a] for the New Sanitation System, Waterschoon, Sneek, the Netherlands

		Measured, 79 Persons								Extrapolated, 1200 Persons
	Unit	BW + KW[a]	UASB Effluent and Products[a]	Effluent OLAND[a]	Struvite Reactor Effluent and Product	GW	Effluent GW System[d]	Combined Influent (GW + Effluent BW)[e]	Effluent Combined System[e]	Effluent Combined System[e]
COD_{total}	mg/L	9500	720	991[b]	254	636	62	528	57 (19)[c]	
COD_{ss}	mg/L	5500	120	756[b]	34	100	0	194	0	
N_{total}	mg/L	1000	960	285[b]	276	16.2	4.2	19.7	6.7 (2.6)[c]	
NH_4^+-N	mg/L	640	820	29[b]	22	6.5	1.0	2.3	0.8 (0.8)[c]	
NO_3-N	mg/L			181[b]	206		0		2.3 (1.7)[c]	
NO_2-N	mg/L			12[b]	8		0		0.1 (0.1)[c]	
P_{total}	mg/L	110	86	83[b]	20	13.2	8.7	17.1	13.5 (2.5)[c]	
PO_4-P	mg/L	80	76	65[b]	9	11.2	7.9	10.4	12.6 (2.1)[c]	
K	mg/L			420[b]						
Mg	mg/L			59[b]	89					
Flow	L/day	1084	–	–				6779		
	L/cap/day	13.7	–	–				86		82

(*Continued*)

TABLE 3.10 (CONTINUED) Results of a Long-Term Monitoring Program[a] for the New Sanitation System, Waterschoon, Sneek, the Netherlands

		Measured, 79 Persons							Extrapolated, 1200 Persons
		BW + KW[a]							
	Unit	UASB Effluent and Products[a]	Effluent OLAND[a]	Struvite Reactor Effluent and Product	GW	Effluent GW System[d]	Combined Influent (GW + Effluent BW)[e]	Effluent Combined System[e]	Effluent Combined System[e]
CH₄ production	L/kg COD$_{inf}$	271							
	m³/cap/year	13.8							12.2
Sludge production	kg COD/cap/year	3.3[f]							
	kg TS/cap/year	4.2[f]							9.2
Struvite	kgP/cap/year			0.3					

Source: Adapted from de Graaf, R., and van Hell, A.J., Nieuwe sanitatie (new sanitation), Noorderhoek, Sneek, Deelonderzoeken, STOWA, Amersfoort, Netherlands, 2014.

Note: HRT = 34 days; temperature = 35°C. The system is designed for 550 persons and operated for 79 persons. Results of extrapolation for 1200 persons are included.

[a] Monitoring period 15 months; sampled weekly.
[b] Sampled before the settler.
[c] Standard deviation in brackets.
[d] Monitoring period 3 months; sampled weekly.
[e] Monitoring period 6.5 months; sampled weekly.
[f] Due to the low applied loading, hardly any GW sludge is produced.

TABLE 3.11 Energy Balance of the New Sanitation System, Waterschoon, Sneek, the Netherlands

	Unit	Waterschoon (1200 persons)[a]	Deventer (100,000 persons)[b]
Production and delivery of drinking water	kWh$_p$/cap/year	−35	−58
Energy consumption treatment	kWh$_p$/cap/year	−50	−6
Diesel consumption WKK	kWh$_p$/cap/year	0	−3
Heat production heat pump	kWh$_p$/cap/year	477	0
Electricity consumption heat pump	kWh$_p$/cap/year	−264	0
Heat production biogas	kWh$_p$/cap/year	148	6
Electricity consumption treatment	kWh$_p$/cap/year	−52	−75
Electricity production WKK	kWh$_p$/cap/year	0	61
Electricity consumption transport ww	kWh$_p$/cap/year	−42	−13
Total energy consumption/production	kWh$_p$/cap/year	184	−88

Source: Based on de Graaf, R., and van Hell, A.J., Nieuwe sanitatie (new sanitation), Noorderhoek, Sneek, Deelonderzoeken, STOWA, Amersfoort, Netherlands, 2014.

Note: Extrapolated to 1200 persons, compared with the conventional municipal wastewater treatment system for 100,000 persons, as operated in Deventer, the Netherlands (measured). ww = wastewater; WKK.

[a]Extrapolated based on measurements and expert judgment.

[b]Measured, conventional municipal wastewater treatment system in Deventer, the Netherlands.

UASB technology in combination with different posttreatments (Chernicharo 2006). Table 3.12 presents influent and effluent BOD and COD values for these different combinations. At BOD and COD influent values of 500–700 and 250–310 mg/L, respectively, effluent values after posttreatment vary between 20 and 80 mg/L for BOD and 60 and 200 mg/L for COD. The UASB plus dissolved air flotation (UASB-DAF) achieves the best effluent values for both COD and BOD. Phosphorus is best removed

TABLE 3.12 Average Influent and Effluent BOD and COD Concentrations in Municipal Sewage Treated with UASB Technology in Combination with Different Posttreatments

	COD$_{inf}$[a]		BOD$_{inf}$[a]		COD$_{eff}$		BOD$_{eff}$	
	Min.	Max.	Min.	Max.	Min.	Max.	Min.	Max.
UASB	600	600	250	280	180	270	70	100
UASB + AS	500	600	286	294	60	150	20	50
UASB + SBAF	500	600	286	294	60	150	20	50
UASB + HTF	583	667	286	300	70	180	20	60
UASB + AF	500	667	307	320	100	200	40	80
UASB + DAF	588	600	286	294	60	100	20	50
UASB + PP	588	600	304	308	100	180	40	70
UASB + OF	600	600	300	304	90	180	30	70

Note: AF = anaerobic filter; AS = activated sludge; HTF = high-rate trickling filter; SBAF = submerged aerated biofilter.

[a]Calculated based on percent removal as presented by Chernicharo, C.A.L., *Rev. Environ. Sci. Bio/Technol.*, 5(1), 73–92, 2006.

in the UASB-DAF, NH_4^+-N and total nitrogen removal is highest in the UASB plus overland flow (UASB-OF), while fecal coliforms (FCs) and helminth eggs are best removed in UASB plus polishing ponds (UASB-PP) (Chernicharo 2006). Effluents of UASB-PP can comply with FC standards of $1*10^3$/100 mL, while helminth numbers are ≤1 egg/L. When effluents are used for irrigation fertilization, N and P will be reused for plant growth. For discharge, new technologies, like nitritation–anammox in the water line, could be implemented (Hendrickx et al. 2012). Souza et al. (2011) quantify the total methane production and the fraction dissolved methane in UASB reactors treating municipal wastewater at pilot, demo, and full scale, at temperatures between 24°C and 25°C. Considerable quantities of dissolved methane via the anaerobic effluents are reported, varying from 36% to 41% of total methane generated in the reactor (Souza et al. 2011). The dissolved methane concentrations were considerably higher than the saturation concentrations determined based on Henry's law. A total methane yield of 0.22–0.24 L/g $COD_{removed}$ was achieved, while only 0.14–0.15 L CH_4/g$COD_{removed}$ was collected via the biogas. Applying these values to a UASB treating municipal wastewater with an influent COD concentration of 600 g/m^3 and an influent COD concentration between 180 and 270 g/m^3, 48–61 L CH_4/m^3 will be collected, while 28–36 L/m^3 is expected to be emitted to the atmosphere. At lower temperatures, the emitted amount will be higher. The dissolved CH_4 not only represents a considerable energy loss, but also contributes to the greenhouse effect, as CH_4 is a 25 times stronger greenhouse gas than CO_2. An interesting development is the simultaneous nitrogen and methane removal, established by the denitrifying methanotrophic bacterium *Candidatus* Methylomirabilis oxyfera, solving two problems at once (Luesken et al. 2011). Kampman et al. (2014) achieved a maximum volumetric nitrite consumption rate of 36 mg NO_2-N/L/day applying a membrane bioreactor, operated at a temperature of 20°C. Volumetric consumption should, however, be significantly increased before full-scale application takes place.

References

Azman, S., A.F. Khadem, J.B. van Lier, G. Zeeman, and C.M. Plugge. 2015. Presence and role of anaerobic hydrolytic microbes in conversion of lignocellulosic biomass for biogas production. *Critical Reviews in Environmental Science and Technology* 45 (23):2523–2564.

Blum, D.J.W., and R.E. Speece. 1991. A database of chemical toxicity to environmental bacteria and its use in interspecies comparisons and correlations. *Research Journal of the Water Pollution Control Federation* 63 (3):198–207.

Braun, R. 2007. Anaerobic digestion: A multi-faceted process for energy, environmental management and rural development. In *Improvement of Crop Plants for Industrial End Uses*, ed. P. Ranalli, 335–416. Berlin: Springer.

Chapleur, O., L. Mazeas, J.J. Godon, and T. Bouchez. 2016. Asymmetrical response of anaerobic digestion microbiota to temperature changes. *Applied Microbiology and Biotechnology* 100 (3):1445–1457.

Chen, J.L., R. Ortiz, T.W.J. Steele, and D.C. Stuckey. 2014. Toxicants inhibiting anaerobic digestion: A review. *Biotechnology Advances* 32 (8):1523–1534.

Chernicharo, C.A.L. 2006. Post-treatment options for the anaerobic treatment of domestic wastewater. *Reviews in Environmental Science and Bio/Technology* 5 (1):73–92.

Chernicharo, C.A.L., J.B. van Lier, A. Noyola, and T. Bressani Ribeiro. 2015. Anaerobic sewage treatment: State of the art, constraints and challenges. *Reviews in Environmental Science and Bio/Technology* 14 (4):649–679.

Christiansen, N., and B.K. Ahring. 1996. Introduction of a de novo bioremediation activity into anaerobic granular sludge using the dechlorinating bacterium DCB-2. *Antonie van Leeuwenhoek* 69 (1):61–66.

de Graaff, M.S., H. Temmink, G. Zeeman, and C.J.N. Buisman. 2010a. Anaerobic treatment of concentrated black water in a UASB reactor at a short HRT. *Water* 2 (1):101–119.

de Graaff, M.S., H. Temmink, G. Zeeman, and C.J.N. Buisman. 2011a. Energy and phophorus recovery from black water. *Water Science and Technology* 63 (11):2759–2765.

de Graaff, M.S., H. Temmink, G. Zeeman, M.C.M. van Loosdrecht, and C.J.N. Buisman. 2011b. Autotrophic nitrogen removal from black water: Calcium addition as a requirement for settleability. *Water Research* 45 (1):63–74.

de Graaf, R., and A.J. van Hell. 2014. Nieuwe sanitatie (new sanitation), Noorderhoek, Sneek. Deelonderzoeken. Amersfoort, Netherlands: STOWA.

de Graaff, M.S., G. Zeeman, H. Temmink, M.C.M. van Loosdrecht, and C.J.N. Buisman. 2010b. Long term partial nitriation of anaerobically treated black water and the emission of nitrous oxide. *Water Research* 44:2171–2179.

Demirel, B., and P. Scherer. 2011. Trace element requirements of agricultural biogas digesters during biological conversion of renewable biomass to methane. *Biomass and Bioenergy* 35 (3):992–998.

Elmitwalli, T.A., J. Soellner, A. De Keizer, H. Bruning, G. Zeeman, and G. Lettinga. 2001. Biodegradability and change of physical characteristics of particles during anaerobic digestion of domestic sewage. *Water Research* 35 (5):1311–1317.

Fernandes, T.V., R. Shrestha, Y.X. Sui, G. Papini, G. Zeeman, L.E.M. Vet, R.H. Wijffels, and P. Lamers. 2015. Closing domestic nutrient cycles using microalgae. *Environmental Science & Technology* 49 (20):12450–12456.

Gast, M. 2009. Grens in besparing op water bereikt. *H2O* 24:14–15.

Guest, J.S., S.J. Skerlos, J.L. Barnard, M.B. Beck, G.T. Daigger, H. Hilger, S.J. Jackson et al. 2009. A new planning and design paradigm to achieve sustainable resource recovery from wastewater. *Environmental Science & Technology* 43 (16):6126–6130.

Halalsheh, M., Z. Sawajneh, M. Zu'bi, G. Zeeman, J.B. van Lier, M. Fayyad, and G. Lettinga. 2005. Treatment of strong domestic sewage in a 96 m3 UASB reactor operated at ambient temperatures: Two-stage versus single-stage reactor. *Bioresource Technology* 96:577–585.

Hartmann, H., and B.K. Ahring. 2006. Strategies for the anaerobic digestion of the organic fraction of municipal solid waste: An overview. *Water Science and Technology* 53 (8):7–22.

Hendrickx, T.L.G., Y. Wang, C. Kampman, G. Zeeman, H. Temmink, and C.J.N. Buisman. 2012. Autotrophic nitrogen removal from low strength waste water at low temperature. *Water Research* 46:2187–2193.

Hernandez-Leal, L., H. Temmink, G. Zeeman, and C.J.N. Buisman. 2010. Bioflocculation of grey water for improved energy recovery within decentralized sanitation concepts. *Bioresource Technology* 101 (23):9065–9070.

Hulshoff Pol, L.W., S.I. de Castro Lopes, G. Lettinga, and P.N.L. Lens. 2004. Anaerobic sludge granulation. *Water Research* 38:1376–1389.

Industriewater Eerbeek. 2013. Sustainability report. Arnhem, Netherlands: Industriewater Eerbeek.

IPCC (Intergovernmental Panel on Climate Change). 2006. Wastewater treatment and discharge. In *2006 IPCC Guidelines for National Greenhouse Gas Inventories*, ed. M.R.J. Doorn, S. Towprayoon, S.M. Manso, W. Irving, C. Palmer, R. Pipatti, and C. Wang, 6.1–6.28. Japan: IPCC, Institute for Global Environmental Strategies.

IPCC (Intergovernmental Panel on Climate Change). 2007. Climate change 2007: The physical science basis, contribution of Working Group I to the fourth assessment report of the Intergovernmental Panel on Climate Change, ed. S. Solomon, D. Qin, M. Manning, Chen Z., M. Marquis, K.B. Averyt, M. Tignor, and H.L. Miller. Cambridge: IPCC.

Ismail, S.B. 2013. Anaerobic wastewater treatment of high salinity wastewaters: Impact on bioactivity and biomass retention. PhD, Wageningen Univeristy.

Kalogo, Y., and W. Verstraete. 2000. Technical feasibility of the treatment of domestic wastewater by a CEPS-UASB system. *Environmental Technology* 21 (1):55–65.

Kampman, C., H. Temmink, T.L.G. Hendrickx, G. Zeeman, and C.J.N. Buisman. 2014. Enrichment of denitrifying methanotrophic bacteria from municipal wastewater sludge in a membrane bioreactor at 20°C. *Journal of Hazardous Materials* 274:428–435.

Ketheesan, B., and D.C. Stuckey. 2015. Effects of hydraulic/organic shock/transient loads in anaerobic wastewater treatment: A review. *Critical Reviews in Environmental Science and Technology* 45 (24):2693–2727.

Kujawa-Roeleveld, K., T. Fernandes, Y. Wiryawan, A. Tawfik, M. Visser, and G. Zeeman. 2005. Performance of UASB septic tank for treatment of concentrated black water within DESAR concept. *Water Science and Technology* 51 (1–2):307–313.

Larsen, T.A., A.C. Alder, R.I.L. Eggen, M. Maurer, and J. Lienert. 2009. Source separation: Will we see a paradigm shift in wastewater handling? *Environmental Science & Technology* 43 (16):6121–6125.

Lettinga, G. 2014. *My Anaerobic Sustainability Story*. Wageningen: LeAF.

Lofrano, G., and J. Brown. 2010. Wastewater management through the ages: A history of mankind. *Science of the Total Environment* 408:5254–5264.

Luesken, F., T. van Alen, E. van der Biezen, C. Frijters, G. Toonen, C. Kampman, T. Hendrickx, et al. 2011. Diversity and enrichment of nitrite-dependent anaerobic methane oxidizing bacteria from wastewater sludge. *Applied Microbiology and Biotechnology* 92:845–854.

Mahmoud, N., M.N. Amarneh, R. Al-Sa'ed, G. Zeeman, H. Gijzen, and G. Lettinga. 2003. Sewage characterisation as a tool for the application of anaerobic treatment in Palestine. *Environmental Pollution* 126:115–122.

Mahmoud, N., G. Zeeman, H. Gijzen, and G. Lettinga. 2004a. Anaerobic stabilisation and conversion of biopolymers in primary sludge—Effect of temperature and sludge retention time. *Water Research* 38 (4):983–991.

Mahmoud, N., G. Zeeman, H. Gijzen, and G. Lettinga. 2004b. Anaerobic sewage treatment in a one-stage UASB reactor and a combined UASB-digester system. *Water Research* 38 (9):2348–2358.

Maya-Altamira, L., A. Baun, I. Angelidaki, and J.E. Schmidt. 2008. Influence of wastewater characteristics on methane potential in food-processing industry wastewaters. *Water Research* 42 (8–9):2195–2203.

McCarty, P.L. 1965. Thermodynamics of biological synthesis and growth. *Air and Water Pollution* 9 (10):621–39.

Meyer, T., and E.A. Edwards. 2014. Anaerobic digestion of pulp and paper mill wastewater and sludge. *Water Research* 65:321–349.

O'Flaherty, V., G. Collins, and T. Mahony. 2006. The microbiology and biochemistry of anaerobic bioreactors with relevance to domestic sewage treatment. *Reviews in Environmental Science and Biotechnology* 5 (1):39–55.

Parawira, W. 2012. Enzyme research and applications in biotechnological intensification of biogas production. *Critical Reviews in Biotechnology* 32 (2):172–186.

Pokhrel, D., and T. Viraraghavan. 2004. Treatment of pulp and paper mill wastewater—A review. *Science of the Total Environment* 333 (1–3):37–58.

Sato, N., T. Okubo, T. Onodera, A. Ohashi, and H. Harada. 2006. Prospects for a self-sustainable sewage treatment system: A case study on full-scale UASB system in India's Yamuna River Basin. *Journal of Environmental Management* 80:198–207.

Schauer-Gimenez, A.E., D.H. Zitomer, J.S. Maki, and C.A. Struble. 2010. Bioaugmentation for improved recovery of anaerobic digesters after toxicant exposure. *Water Research* 44 (12):3555–3564.

Seghezzo, L., G. Zeeman, J.B. van Lier, H.V.M. Hamelers, and G. Lettinga. 1998. A review: The anaerobic treatment of sewage in UASB and EGSB reactors. *Bioresource Technology* 65 (3):175–190.

Sipma, J., M.B. Osuna, M.A.E. Emanuelsson, and P.M.L. Castro. 2010. Biotreatment of industrial wastewaters under transient-state conditions: Process stability with fluctuations of organic load, substrates, toxicants, and environmental parameters. *Critical Reviews in Environmental Science and Technology* 40 (2):147–197.

Souza, C.L., C.A.L. Chernicharo, and S.F. Aquino. 2011. Quantification of dissolved methane in UASB reactors treating domestic wastewater under different operating conditions. *Water Science and Technology* 64 (11):2259–2264.

Stams, A.J.M., and C.M. Plugge. 2009. Electron transfer in syntrophic communities of anaerobic bacteria an archaea. *Nature Reviews Microbiology* 7:568–577.

Stams, A.J.M., and C.M. Plugge. 2010. The microbiology of methanogenesis. In *Methane and Climate Change*, ed. D. Reay, P. Smith, and A. van Amster, 14–26. London: Earthscan.

Tauseef, S.M., T. Abbasi, and S.A. Abbasi. 2013. Energy recovery from wastewater with high-rate anaerobic digesters. *Renewable and Sustainable Energy Reviews* 19:704–741.

Tchobanoglous, G., H.D. Stensel, R. Tsuchihashi, F. Burton, M. Abu-Orf, G. Bowden, and W. Pfrang. 2014. *Wastewater Engineering: Treatment and Resource Recovery*. 2 vols. New York: McGraw-Hill.

Teixeira, A.R., C.A. de Lemos Chernicharo, and S.F. Aquino. 2008. Influence of particle size reduction on the hydrolysis rate of domestic raw sewage. *Engenharia Sanitaria e Ambiental* 13 (4):405–415.

Tervahauta, T., I.M. Bryant, L. Hernandez-Leal, C.J.N. Buisman, and G. Zeeman. 2014. Improved energy recovery by anaerobic grey water sludge treatment with black water. *Water* 6:2436–2448.

Thanh, P.M., B. Ketheesan, Z. Yan, and D. Stuckey. 2016. Trace metal speciation and bioavailability in anaerobic digestion: A review. *Biotechnology Advances* 34 (2):122–136.

van Eekert, M.H.A., H.J. van Dooren, M. Lexmond, and G. Zeeman. 2010. Wastewater and manure. In *Methane and Climate Change*, ed. D. Reay, P. Smith, and A. van Amster, 151–174. London: Earthscan.

van Haandel, A.C., and G. Lettinga. 1994. *Anaerobic Sewage Treatment: A Practical Guide for Regions with a Hot Climate*. Hoboken, NJ: John Wiley & Sons.

van Lier, J.B., S. Rebac, and G. Lettinga. 1997. High-rate anaerobic wastewater treatment under psychrophilic and thermophilic conditions. *Water Science and Technology* 35 (10):199–206.

van Lier, J.B., F.P. van der Zee, C.T.M.J. Frijters, and M.E. Ersahin. 2015. Celebrating 40 years anaerobic sludge bed reactors for industrial wastewater treatment. *Reviews in Environmental Science and Bio/Technology* 14 (4):681–702.

Vlaeminck, S.E., A. Terada, B.F. Smets, D. van der Linden, N. Boon, W. Verstraete, and M. Carballa. 2009. Nitrogen removal from digested black water by one-stage partial nitritation and anammox. *Environmental Science & Technology* 43 (13):5035–5041.

von Sperling, M., and C.A. de Lemos Chrenicharo. 2002. Urban wastewater treatment technologies and the implementation of discharge standards in developing countries. *Urban Water* 4:105–114.

Zandvoort, M.H., E.D. van Hullebusch, J. Gieteling, and P.N.L. Lens. 2006. Granular sludge in full-scale anaerobic bioreactors: Trace element content and deficiencies. *Enzyme and Microbial Technology* 39 (2):337–346.

Zeeman, G., and G. Lettinga. 1999. The role of anaerobic digestion of domestic sewage in closing the water and nutrient cycle at community level. *Water Science and Technology* 39 (5):187–194.

Zeeman, G., and W. Sanders. 2001. Potential of anaerobic digestion of complex waste(water). *Water Science and Technology* 44:115–122.

Zehnder, A.J.B. 1988. *Biology of Anaerobic Microorganisms*. Wiley Series in Ecological and Applied Microbiology. Hoboken, NJ: Wiley-Interscience.

Zhang, L. 2016. Anaerobic treatment of municipal wastewater in a UASB-Digester system. Temperature effect on system performance, hydrolysis and methanogenesis. PhD Thesis. Wageningen University, the Netherlands.

Zhang, L., T.L.G. Hendrickx, C. Kampman, H. Temmink, and G. Zeeman. 2013. Codigestion to support low temperature anaerobic pretreatment of municipal sewage in a UASB-digester. *Bioresource Technology* 148:560–566.

4

Microbial Biomethane from Solid Wastes:
Principles and Biotechnogical Processes

Antoine P. Trzcinski and David C. Stuckey

Contents

4.1 Introduction

Chapter 3 introduced anaerobic digestion (AD) processes dealing with liquid effluents from domestic or industrial sources. Due to mostly soluble compounds in
wastewater, there are fewer parameters impacting AD. This chapter introduces the
parameters affecting AD of solid waste and how technology has been adapted to treat
municipal or agricultural waste.

AD of solid waste is a microbial process whereby organic waste is converted to
methane gas and a stabilized compost or soil conditioner (depending on the quality of the waste treated). Besides producing two valuable by-products, it has the
advantage of diverting solid waste from landfills. The capital costs of AD plants
are high due to the pretreatment units, digester, biogas treatment, and wastewater
treatment, but nevertheless, AD has been proven to be economically viable and is an
energy-positive process. In addition to financial benefits, AD results in considerable
carbon dioxide (CO_2) emissions savings from fossil fuel substitution, in contrast
with other solid waste technologies, such as landfilling, incineration, or composting.
Codigestion of municipal solid waste (MSW) with agro-industrial residues, animal
manures, sewage sludge, and slaughterhouse waste is the trend because it allows
for higher methane production, boosting the plant economics. In Europe, the first
full-scale AD plant treating the organic fraction of municipal solid waste (OFMSW)
was commissioned in the late 1980s and is still in operation, demonstrating that it
can be considered a mature technology, and more than 168 industrial plants have
been commissioned in the last 30 years (Monson et al., 2007). This chapter reviews
the main parameters affecting the AD of solid waste: physicochemical properties
of feedstock, parameters investigated during laboratory tests, feeding and design
parameters, improvement parameters, and types of inhibition. The next sections

deal with the processes treating MSW alone and the processes applying codigestion of organic waste.

4.2 Parameters Affecting the Anaerobic Digestion of Organic Solid Waste

4.2.1 Physicochemical Properties of Feedstock

4.2.1.1 Lignin Content

Lignocellulosic substrate is a general term for material containing cellulose, hemicellulose, and lignin, and it includes the OFMSW, wood, grass, leaves, paper, wheat straw, and other agricultural residues. Cellulose is a linear polymer of glucose units that can form intra- and interchain bonds, leading to a crystalline macromolecule. It is relatively rigid with a high degree of dimensional stability in the direction of the cellulose fibers. Hemicelluloses have a more irregular structure with side groups, substituent groups, and sugars present along the length of the chain. Lignin is a randomized condensed polymer with many aromatic groups and is much more hydrophobic than cellulose or hemicelluloses (Popescu et al., 2011).

Most of the cellulose is located in highly ordered crystalline regions, in which cellulose chains or fibrils are so tightly packed that even water molecules can scarcely penetrate; cellulose is accordingly water insoluble. Less ordered portions of the assembly, called amorphous regions, typically comprise about 5% of the cellulose microstructure (Bailey and Ollis, 1986). These amorphous regions are easily hydrolyzed by, for example, acids; the crystalline regions, on the other hand, are much more difficult to decompose, and it depends on the crystallinity, crystallite size, degree of polymerization, surface area, particle size, lignin content, wood density, and so forth (Hendriks and Zeeman, 2009; Popescu et al., 2011).

However, much more recalcitrant is the lignin casing, which encloses the polysaccharide components of the biomass. Lignin is a complex and heterogeneous material, and its random arrangement makes it very resistant to chemical and enzymatic attack. The lignocellulosic complex prevents enzymes from accessing the degradable cellulose, and some studies have suggested that enzymes may adsorb preferentially onto lignin, which results in unproductive enzyme binding (Kristensen et al., 2007). As both accessibility of enzymes to the solid matter, and hydrolysis of complex compounds, constitute the rate-limiting step during anaerobic degradation (Eastman and Ferguson, 1981), pretreatment of the substrate is beneficial for the rate and extent of the anaerobic process, the biogas yield, and mass reduction of anaerobic sludge (Xiao and Clarkson, 1997). On the other hand, owing to the refractory structure of the lignocellulosic biomass, the efficiency of AD to treat agriculture residues is limited. Typical lignin percentages are listed in Table 4.1 for organic waste.

Although cellulose and hemicellulose can be degraded under anaerobic conditions, lignin is not and prevents enzyme accessibility to cellulose. Hemicellulose serves as a connection between the lignin and the cellulose fibers and gives the whole cellulose–hemicellulose–lignin network more rigidity. Therefore, only a low fraction of lignocellulosic biomass can be converted into biogas. Hence, the pretreatment

TABLE 4.1   Composition of Lignocellulosic Materials

Biomass	Cellulose (%)	Hemicellulose (%)	Lignin (%)	C/N Ratio
Corn stover	37.5	22.4	17.6	63
Wheat straw	38.2	21.2	23.4	60
Switch grass	31–45	20–31	12–18	90
Bagasse	38.2	27.1	20.2	118
Sugarcane	25	17	12	–
Rice straw	32	24	13	47
Eucalyptus	38–45	12–13	25–37	–
Giant reed stalk	33.1	18.5	24.5	–
Giant reed leaves	20.9	17.7	25.4	–
Sunflower stalk	31	15.6	29.2	–
Biomass sorghum	22.2	19.4	21.4	–
Barley straw	37.5	25.3	26.1	–
Rye straw	38	36.9	17.6	20
Napier grass	45.7	33.7	20.6	26
Newspaper	–	–	20%–23% of VS	–
Office paper	–	–	0.1%–1% of VS	–
Yard waste	–	–	4%–10% of VS	–
Food waste	–	–	0.1%–0.7% of VS	–

Source: Adapted from Kayhanian, M., *Waste Manag. Res.*, 13(2), 123–136, 1995; Owens, J.M., and Chynoweth, D.P., *Water Sci. Technol.*, 27(2), 1–14, 1993; Sawatdeenarunat, C. et al., *Bioresour. Technol.*, 178, 178–186, 2015.

of the lignocellulosic biomass is crucial to remove lignin and hemicellulose and make cellulose more accessible to the enzymes that convert carbohydrate polymers into fermentable sugars (Mosier et al., 2005), and therefore to increase the biogas potential.

4.2.1.2 Particle size distribution

The biodegradability of solid waste depends on the lignin content and the cellulose crystallinity, as well as the surface available for hydrolysis (Zhu et al., 2007). Some authors argued that the increase in the hydrolysis rate at increasing biodegradability suggests that the rate of hydrolysis of particulate organic matter is determined by the adsorption of hydrolytic enzymes to the biodegradable surface sites (Tong et al., 1990; Veeken and Hamelers, 1999). This means that a high surface area is required for an efficient enzymatic adsorption to take place. Chynoweth et al. (1993) have shown that particle size did not influence the rate of methane production in a biochemical methane potential (BMP) test of herbaceous and woody feedstocks and municipal waste in the range of 1–8 mm. Similarly, with a feedstock of MSW, Nopharatana et al. (2006) and Angelidaki and Ahring (1999) did not observe any difference in the rate of digestion and methane yield with particles in the 2–50 mm and 5–20 mm ranges, respectively. In contrast, Sharma et al. (1989) obtained 98% more biogas with 0.4 mm particles of plant stem than they did with the 6.0 mm particles. Hills and Nakano (1984) showed that the hydrolysis rate of tomato waste increased with decreasing particle size. In a more recent study (Clarkson and Xiao, 2000), different sizes of paper (ground paper, shredded strips, and whole pieces) did not show any differences, statistically, in bioconversion rate and extent. Hartmann et al. (1999) found that biogas potential did not correlate with a smaller size of fibers. Results from the maceration indicate that the biodegradability of the fibers is rather enhanced by shearing, which is not necessarily reflected by a change in size distribution. These results showed that surface area is a more important parameter than particle size.

4.2.1.3 Biodegradability of the Organic Fraction of Municipal Solid Waste

MSW is a heterogeneous lignocellulosic waste, which may be divided into a number of subfractions:

- Digestible organic fraction. Organic matter that is readily degradable, for example, kitchen waste, grass cuttings, and paper
- Combustible fraction. Slowly digestible and indigestible organic matter, for example, wood, cardboard, plastics, and other synthetics
- Inert fraction. Stones, sand, glass, metals, bones, and so forth

In general, not taking into account the recent strategies (e.g., developed in some European countries in the beginning of the twenty-first century) to separate different fractions of MSW at the domestic source, about 50%–60% of MSW consists of organic matter (Gellens et al., 1995; Veeken and Hamelers, 1999), and the OFMSW typically contains 40%–50% cellulose, 12% hemicellulose, and 10%–15% lignin by weight (Wang et al., 1994). Due to its heterogeneous character, the MSW stream requires various pretreatments before anaerobic degradation.

The composition of the OFMSW is important in determining which treatment method is most appropriate; yard waste contains lignocellulosic material, which does not readily degrade under anaerobic conditions (Braber, 1995; Veeken and Hamelers, 1999). Furthermore, wood is highly refractory to anaerobic decomposition (Chynoweth and Jerger, 1985). On the other hand, material such as kitchen waste is too wet and lacking in structure for aerobic composting. In general, if the material collected is predominantly kitchen waste, AD is the most appropriate treatment, and if yard waste dominates, aerobic composting is preferred.

MSW composition may vary considerably with various factors:

1. The period of the year. There is more fruit in the blend during the summer months, leading to a higher ratio of volatile solids to total solids (VS/TS). In contrast, the OFMSW may comprise less biodegradable items, such as fir trees after the Christmas period.
2. The region.
3. The climate.
4. The extent of recycling.
5. The collection frequency.
6. The sociocultural practices.
7. Changes in technology.

Furthermore, garden waste fractions of MSW can be expected to exhibit a high variability in yields due to the variety of plant materials, which can be included (Owens and Chynoweth, 1993).

Paper waste can represent a significant fraction of the MSW. The type of virgin fiber, pulping process, lignin content, coating, and printing may all affect the methane yield and rates. Fine printing papers and office papers are generally made from a high-grade bleached kraft pulp, which has none of the lignin present in the virgin fibers. The corrugated boxboard and brown bag paperboard are normally made from an unbleached kraft pulp, which may still have a 5% lignin content. Groundwood pulp is used in newsprint and magazine production, and contains almost all the virgin lignin, comprising up to 20% of virgin fibers. It is worth highlighting the fact that inks used in newsprint do not lower the methane yields over unprinted newsprint (Owens and Chynoweth, 1993). Also, in the same study the authors found that finely ground paper used in food packaging was highly biodegradable, and its degradation was not affected by polymer coatings. However, information concerning the influence of inks and coatings on AD is not widely available in the literature.

The ultimate biodegradability (usually expressed as a percentage of TS of VS) of a feedstock can be characterized using the BMP method (Owen et al., 1979). This standard technique allows for the determination of the methane yield (usually expressed in mL CH_4 g^{-1} VS_{fed}) by incubating a substrate under ideal conditions for AD, that is, broad-spectrum inoculum, excess inoculum, excess nutrients, substrate concentration below inhibitory levels, excess buffering capacity, mesophilic temperature, and strict anaerobic conditions (Chynoweth et al., 2002). An ultimate methane yield greater than 0.2 L g^{-1} VS is considered favorable for any feedstock (O'Keefe et al., 1996).

The following empirical relationship was developed to estimate the biodegradable fraction of an organic substrate from lignin test results (Kayhanian, 1995): biodegradable fraction (BF) = 0.83 − 0.028LC, where the BF is expressed as a fraction of the VS and LC is the lignin content, expressed as a percentage of the VS. Carbohydrates, the major component of plant residues, have a theoretical methane yield of 0.36 L g^{-1} VS. Some components, such as lignin, are not degradable under anaerobic conditions and may hinder degradation of other components (Tong et al., 1990; Xiao and Clarkson, 1997). A substantial fraction of the lignin solubilized may be in the form of medium-molecular-weight oligoaromatic compounds (McFarlane and Pfeffer, 1981).

According to McFarlane and Pfeffer (1981), these lignocellulosic materials are resistant to biodegradation, and unless some form of pretreatment is employed, the methane generated is unlikely to be economic, as large digesters operating at long retention time will be required. Originally, the readily hydrolyzable carbohydrates were named hemicellulose to differentiate them from cellulose. Three well-defined groups of hemicellulose, the xylans, mannans, and galactans, are recognized as components in all lignified plants. With organic substrates, it is usual to assume that the biodegradable VS consist entirely of cellulose. Thus, each kilogram of VS removed will produce 0.415 m^3 of methane at standard temperature and pressure (STP) or 0.453 at 25°C. This assumes that the biodegradability of lignin is negligible and the gas yield per gram of hemicellulose is similar to that per gram of cellulose metabolized.

An important aspect related to the successful application of AD, and probably the most important, is the quality of biowaste—thus the collection strategy for its management. Waste collection can therefore be considered the first treatment step for MSWs: different collection strategies lead to organic streams that can differ widely in terms of characteristics. According to the applied collection system, the OFMSW can be called mechanically sorted (MS-OFMSW), separately collected (SC-OFMSW) (e.g., fruit and vegetable waste collected from markets), or source sorted (SS-OFMSW). The latter often refers to manually sorting food scraps at the household level. The experience gained in recent years has shown that separate collection of biowaste leads to material of good quality and high biomethane potential, whereas the MS-OFMSW requires very high-energy demanding pretreatments (e.g., magnetic separator, crusher, macerator, and hydropulper) and still results in poor yields. Moreover, the possible recycling of digestate is clearly very limited because of the high presence of pollutants and inert material. From an exhaustive review of Monson et al. (2007), it was concluded that the success of source separation is crucial to the realization of a quality compost that can be used on land, and good continuous public education is key to achieve this. In other words, if one of the objectives is to apply compost on land or sell the compost to private users, a good sorting strategy must be in place at source.

In the past (1980s–1990s), the unsorted MSWs were mechanically treated and the organic fraction was separated and sent to biological treatment. However, several years of experience taught that this strategy led to very poor yields (biogas production of some 60 m^3 ton^{-1} (Bolzonella et al., 2006a) and several mechanical problems with the reactors. Edelman and coworkers (Edelman and Engeli, 2005; Edelman, 2008) reported the failure of some dry reactors in Switzerland, Mata-Alvarez and coworkers (Mace et al., 2005) demonstrated the failure of the dry reactors at the Ecoparc in Barcelona and at other Spanish sites, and the same problems were reported for the plants in Verona and Bassano in Italy (Cecchi and Bolzonella, 2005), all treating MS-OFMSW.

On the other hand, the yields recently obtained in the AD of SC-OFMSW and SS-OFMSW showed that these collection strategies are the most promising way to obtain valuable substrates, also considering the final quality of the composted digestate. The wastes clearly show different characteristics and compositions according to the adopted collection strategy. Table 4.2 reports data dealing with the composition and characteristics of differently collected biowaste. SC-OFMSW and SS-OFMSW sometimes show a dry matter content of 10% or less when fruit and vegetable wastes are prevalent, while the solids content reaches 25% for kitchen waste. Some studies refer to the waste as OFMSW, although it contains some impurities (plastic, metal, and glass), depending on the efficiency of the waste collection strategy (Hartmann and Ahring, 2005).

TABLE 4.2　Typical Characteristics of MS-OFMSW, SS-OFMSW, and SC-OFMSW

	TS (%)	VS (% TS)	N (% TS)	P (% TS)	COD/VS
MS-OFMSW	51–95	29–57	1.2–3.4	0.05–0.22	0.8–1.6
	48–72	35–91	–	–	–
SS-OFMSW	16	90	2.1	2.6	1.0
	20	88	3.2	0.4	1.0
	17–25	70–90	–	–	–
	9–26	85–94	–	–	–
	31	23	–	–	–
	12–24	91–92	2.7–3.3	0.3–0.4	0.9–1.1
SC-OFMSW	7–15	80–90	1.5–3	1–3	–
	29	63	2.2–3.4	0.4–0.6	–
	31	70	–	–	–
	13	83	–	–	–
	5–13	78–92	1.4–3.3	1.3–3.3	0.7–1.5

Source: Cecchi, F. et al., in *Comprehensive Biotechnology*, ed. M. Moo-Young, 2nd ed., Academic Press, Burlington, MA, 2011, pp. 463–472.

Note: Rows refer to different sources.

4.2.1.4　Methane Potential of Organic Solid Waste

Lignocellulosic biomass is an abundantly available bioresource with an annual (global) yield of more than 200 billion dry metric tons per year (Kumar et al., 2008). For example, the United States alone produces about 1.37 billion dry tons of such biomass per year for biofuel production (Limayem and Ricke, 2012). Common examples of these renewable resources include agri- and forest residues, and dedicated energy crops (Cherubini, 2010). Typical values of ultimate biodegradability are given in Table 4.3. It should be borne in mind that in the best possible conditions, not more than 80% of the expected BMP can be recovered in a real reactor (Peu et al., 2011).

In some cases, the data vary widely, depending on the literature source. Furthermore, the quality and composition of industrial waste depends on the raw matter origin, the chemical process used to obtain the main product, and if existent, the waste refining treatment. For instance, the use of glycerol as cosubstrate can have a negative effect due to the presence of methanol and cations (sodium or potassium from the catalyst) (Siles et al., 2009; Robra et al., 2010; Castrillón et al., 2011). Desugared molasses can also be inhibitory because of the high concentrations of sodium and potassium (Fang et al., 2011b). The inhibitory effect of olive mill waste (OMW) and wine distillery wastewater is related to the presence of phenolic compounds (Fountoulakis et al., 2002; Akassou et al., 2010), whereas limonene is the inhibitory compound present in citric waste (Mizuki et al., 1990). Other substrates, like seaweed, can lead to a biogas with high levels of H_2S, making it unsuitable for energy recovery without treatment (Peu et al., 2011). Therefore, before the addition of an unknown or insufficiently studied cosubstrate, it is highly recommended to perform laboratory experiments to detect the presence of inhibitory compounds, which could lead to a process breakdown or decrease the methane production. Regarding

TABLE 4.3   Biomass Yield and Methane Potential from Biomass

Biomass	Biomass Yield (metric ton wet weight ha^{-1})	Methane Potential (Nm^3 CH$_4$ metric ton^{-1} VS)	References
Sugar beet	40–70	387–408	Weiland, 2010
Fodder beet	80–120	398–424	Weiland, 2010
Maize	40–60	291–338	Weiland, 2010
Wheat	30–50	351–378	Weiland, 2010
Triticale	28–33	319–335	Weiland, 2010
Sorghum	40–80	286–319	Weiland, 2010
Grass	22–31	286–324	Weiland, 2010
Red clover	17–25	297–347	Weiland, 2010
Sunflower	31–42	231–297	Weiland, 2010
Wheat grain	6–10	371–398	Weiland, 2010
Microalgae		200–315	Golueke et al., 1957; Nair et al., 1983
		180[a]	González-Fernández et al., 2011
Seaweed (*Ulva* sp.)		148	Peu et al., 2011
Primary sludge	–	470	Razaviarani et al., 2013b
Secondary activated sludge	–	255 180	Girault et al., 2012; Razaviarani et al., 2013b
Greasy sludge		871	Girault et al., 2012
Lawn cuttings		288–388	Triolo et al., 2012
Wood cuttings		142–240	Triolo et al., 2012
Cow manure		148–223	Møller et al., 2004
Cattle manure		148–166	Castrillón et al., 2013
Sugar beet leaves		490	Fang et al., 2011a
Sugar beet top		500	Fang et al., 2011a
Sugar beet pulp		240	Fang et al., 2011a
Pig manure		358	Møller et al., 2004
		188	Astals et al., 2011
		306[a]	González-Fernández et al., 2011
		310	Peu et al., 2011
Glycerol		430	Astals et al., 2011, 2012; Castrillón et al., 2013
Palm oil mill effluent		367	Nurliyana et al., 2015
Newspaper		100	Kayhanian, 1995; Owens and Chynoweth, 1993
Office paper		369	Kayhanian, 1995; Owens and Chynoweth, 1993
Mixed paper waste		147	Trzcinski and Stuckey, 2012b
Yard waste		143	Kayhanian, 1995; Owens and Chynoweth, 1993
Garden waste		114	Trzcinski and Stuckey, 2012b
Food waste		472	Cho et al., 1995; Kayhanian, 1995
Food waste		357	Trzcinski and Stuckey, 2012b
MS-OFMSW		160–380	Cecchi et al., 2011
SC-OFMSW		400–600	Cecchi et al., 2011
SS-OFMSW		400–800	Cecchi et al., 2011

[a]Calculated from COD/VS ratio.

OFMSW, it is clear that the more "selected" the biowaste, the higher the biogas potential (Table 4.3).

4.2.2 Laboratory Testing Parameters

Besides the physicochemical properties listed above, it is important to characterize further the feedstock (or the mixtures of feedstocks) prior to applying the technology at a larger scale. The carbon-to-nitrogen ratio (C/N) and temperature are important parameters to be tested at the laboratory scale to determine the optimum conditions for a particular feedstock. The BMP test at thermophilic temperature can determine whether it is justified to operate at higher temperatures.

4.2.2.1 Carbon-to-Nitrogen Ratio

The C/N ratio is a measure of the relative amounts of organic carbon and nitrogen present in the feedstock. The C/N ratio of the collected waste is determined by its composition. If the C/N ratio of OFMSW is very high, the waste used as a single substrate will be deficient in nitrogen, which is needed for the buildup of bacterial communities. As a result, the gas production will be low. If the C/N ratio is very low, nitrogen will be liberated and accumulate in the form of ammonia. This will increase the pH value of the material, and a pH value higher than 8.5 will start to show a toxic effect on the methanogenic bacterial communities (Hartmann and Ahring, 2005).

The presence of paper in MSW is responsible for one of the most frequent problems associated with OFMSW digestion, that is, the high C/N ratio of these residues (Gomez et al., 2005). The C/N ratio has been estimated at 150/1 for a substrate of paper waste and wood, which is nitrogen deficient (Banks and Humphreys, 1998). Pfeffer and Liebman (1974) also determined that the typical MSW was deficient in both nitrogen and phosphorus with respect to microbial growth. A feedstock with a C/N ratio (biodegradable C) greater than 30 is considered to be deficient in nitrogen for a biological treatment process. Lignocellulosic biomass generally has a C/N ratio higher than 50:1, while some wastes (e.g., wastes from meat industries, slaughterhouses, and farms) have low (<20) C/N ratios due to their high protein contents. To circumvent this problem, several authors have proposed codigestion of the OFMSW, with either sewage sludge or residues from livestock farms, or both (Kayhanian and Rich, 1995; Gomez et al., 2005). The benefits of codigestion include the dilution of potential toxic compounds, an improved balance of nutrients, synergistic effects of microorganisms, an increased load of biodegradable organic matter, and better biogas yields. Laboratory tests (BMP) should be carried out to ensure that blending ratios result in a more balanced C/N ratio and higher methane yields.

4.2.2.2 Inoculum-to-Substrate Ratio

This ratio is used for the BMP test, which is carried out in batch to determine the anaerobic biodegradability of a waste and the residence time required for a high conversion to methane. It refers to how much anaerobic bacteria (on a volatile suspended solids [VSS] basis; volatile refers to the weight fraction that is burnt off in a test at

550°C) is introduced in the fermentation bottle relative to the amount of waste (on a VS basis). The inoculum-to-substrate (I/S) ratio proposed by Owen et al. (1979) was approximately 1, but Chynoweth et al. (1993) stated that increasing the ratio may be necessary for some types of substrates and suggested a ratio of 2. Neves et al. (2004) investigated the BMP of kitchen waste with an I/S ratio of 2, 1, 0.74, and 0.43 and found that the highest ratio prevented acidification. Lopes et al. (2004) investigated the AD of MSW using bovine rumen inoculum at an I/S ratio of 0.17, 0.11, and 0.05. They found better performance at a ratio of 0.17 compared with no inoculum in terms of methane produced per gram VS (+112%), a first-order kinetic constant (+105%), VS removal (+23%), and the average methane amount in the biogas (+39%). They speculated that this was associated with the quantity of native microorganisms present in rumen bovine fluid favoring the conversion of chemical oxygen demand (COD) to biogas. It could also be related to the amount and activity of enzymes, which is greater when more inoculum is introduced, and this is reflected by a better hydrolysis of the waste.

Raposo et al. (2006) used an I/S ratio of 1, 1.5, 2, and 3 to investigate the BMP of maize and found similar methane contents in the biogas. However, the cumulative biogas was inversely proportional to the I/S ratio. The highest methane yield was obtained at a ratio of 1, and this was associated with a greater accumulation of volatile fatty acids (VFAs). Nevertheless, the methane yields were in the range of 196–233 mL CH_4 (STP) g^{-1} VS, and may, therefore, not be significantly different from each other, as no standard deviation was reported. Nallathambi (1997) reports that methane yields from straw were very low at an I/S ratio below 0.25, and that yields increased at a decreasing rate up to an I/S ratio of 2, after which it remained relatively constant.

These considerations have direct applications to the field of AD: in the laboratory, researchers will inoculate their digester using a high proportion of anaerobic inoculum, which is preferably active (freshly collected from a running plant) and acclimatized to the waste to be treated. On a full scale, operators will try to inoculate the digester with an active population of anaerobic microorganisms using sludge from another plant, as this will significantly shorten the start-up phase, but this is not always possible due to availability, transportation costs, or price of the inoculum itself. In practice, a minimum I/S of 0.1 will be targeted for the initial amount of substrate, and after efficient methanogenesis has been established, progressive filling of the bioreactor will be carried out.

4.2.2.3 Effect of Temperature

Several two-stage mesophilic and thermophilic studies have been described in the literature that combined the benefits of the thermophilic process, that is, higher organic loading rate (OLR), higher pathogen, and VS removal (Oleszkiewicz and PoggiVaraldo, 1997; Schober et al., 1999), with the advantages of mesophilic conditions (higher process stability and less VFA accumulation) (Rintala and Ahring, 1994; Gallert and Winter, 1997). Borja et al. (1995b) have shown that thermophilic AD of olive wastewater resulted in 28% more methane than mesophilic AD. In another study (Griffin et al., 1998), the thermophilic start-up period was shorter and more stable; however, in several papers, thermophilic conditions did not improve the yields

(Gallert and Winter, 1997; Pavan et al., 2000b). The gas production was higher at mesophilic temperatures because the solubility of carbon dioxide decreases at higher temperatures, giving rise to a lower methane percentage in the biogas at thermophilic temperatures. Another explanation could be the higher production of hydrogen in the biogas at thermophilic temperatures, but the authors did not quantify this.

Pavan et al. (2000b) studied the effect of temperature on the performance of a two-stage process treating SS-OFMSW: increasing the temperature in the hydrolytic reactor up to thermophilic levels did not improve the yields or kinetics. Gallert and Winter (1997) also found that the performance was not improved when comparing two continuously stirred tank reactors (CSTRs) treating SS-OFMSW at both temperatures.

Another risk when operating at thermophilic temperatures is the accumulation of VFAs. Schober et al. (1999) found concentrations of 0.6 g L^{-1} (at OLR = 7.7 g VS L^{-1} day^{-1}) and 2.2 g L^{-1} (at OLR = 6 g VS L^{-1} day^{-1}), respectively, at mesophilic and thermophilic conditions in a reactor treating kitchen refuse. Even at higher OLRs, VFAs did not accumulate in the mesophilic digester. On the other hand, thermophilic conditions enabled higher VS removals, compared with mesophilic conditions in one stage (Oleszkiewicz and PoggiVaraldo, 1997; Schober et al., 1999). In contrast, Gallert and Winter (1997) obtained similar VS removal percentages at both temperatures.

Both microbial growth rates and VFA accumulation are affected by temperature change. Increased process temperature in general has a positive effect on metabolic rates, but also results in a higher VFA concentration. However, thermophilic reactors can better tolerate temperature variations when immobilized biomass is used. Several authors have found that anaerobic fermentation of wastes with a high concentration of ammonia was more easily inhibited and less stable at thermophilic temperatures than at mesophilic temperatures (Braun et al., 1981). A decrease in operating temperature from 60°C to 37°C in anaerobic digesters with high ammonia concentration provided relief from inhibition caused by VFAs, as indicated by an increase in biogas yield (Angelidaki and Ahring, 1994; Hansen et al., 1999). Contrary to these findings, Gallert and Winter (1997) had reported that methane production was inhibited by 0.22 g L^{-1} VFAs at 37°C and by 0.69 g L^{-1} VFAs at 55°C, indicating that thermophilic flora tolerated at least twice as much VFAs as mesophilic flora. Recent studies have reported good microbial flocculation under thermophilic conditions. The isolation of thermophilic floc-forming microorganisms can help overcome sludge flotation and washout.

Monson et al. (2007) visited 20 industrial-scale AD systems treating OFMSW and concluded that there was no significant advantage of thermophilic systems in terms of biogas production from OFMSW, but there will be benefits in terms of pathogen reduction and processing times. The authors also remarked that given that there is often no end use for excess heat, heat requirements for both thermophilic and mesophilic processes can be easily covered by excess heat produced on site.

Methods to improve digester performance that allow reduction of the reactor volume and a better process energy balance are needed to reduce the high capital costs associated with the application of AD processes. One approach to improving process efficiency is to operate the anaerobic reactor under thermophilic conditions.

Cecchi et al. (1991) used a pilot-scale (3 m³) semidry (16%–23% solids) thermophilic process to treat MS-OFMSW and obtained a gas production rate (GPR) two to three times higher than in the mesophilic process. The process was operating at hydraulic retention times (HRTs) decreasing from 15 to 8 days, and at OLRs increasing from 6 to 14 kg VS m⁻³ day⁻¹. An economic evaluation favored the thermophilic over the mesophilic process.

4.2.3 Feeding and Design Parameters

4.2.3.1 Hydraulic and Solid Retention Time

The HRT is a measure of the rate of substrate flow into and out of a reactor. The HRT is determined by the average time it takes for organic material to digest, as measured by the COD and biochemical oxygen demand (BOD) of the effluent. In a completely mixed digester that employs continuous mixing, all the content of the system have the same biomass residence or retention time. In such a system, the detention time is governed by the replication time of the slowest-growing organism of the microbial community. Below this value, the system fails due to washout of the slowest-growing organism that is necessary to the anaerobic process (Droste, 1997). The HRTs for most dry (influent solids content of above 20%) anaerobic processes range between 14 and 30 days, and for wet (influent solids content of below 20%) anaerobic processes can be as low as 3 days in classical mixed digesters, or only some hours in bioreactors with biomass retention (e.g., membrane) treating soluble waste. The optimal value varies according to the specific technology in place, the process temperature, and the solid waste characteristics. Reducing HRT reduces the size of the digester, resulting in cost savings therefore, there is an incentive to design systems that can achieve a complete digestion in shorter HRT. A shorter HRT will lead to a higher production rate per reactor volume unit, but a lower overall degradation. These two effects have to be balanced in the design of the full-scale anaerobic digester. Good mixing in the digester will ensure a good contact between microorganisms and substrate, which contributes to a reduction in HRT.

The solid retention time (SRT) refers to the average time that the solid substrate (e.g., food waste and OFMSW) spends inside the digester. It is an important parameter, as it influences the process economics. A large SRT means that the solid will remain longer inside the reactor, which means that a large digester is required to treat large OLRs. For a given size, a large SRT means that a low OLR is applied. Its increase will translate to a greater VS removal and a lower food-to-microorganism (F/M) ratio, which produces a more stabilized organic matter. However, it depends greatly on the type of solids. Lignocellulosic materials contain lignin and therefore require longer SRTs to obtain a given VS removal compared with, for instance, food waste. Kayhanian and Tchobanoglous (1992) report a better stability when operating at an SRT of 30 days than 15 days; biogas production and VS removal rates were higher at increased SRT. This latter observation was also made by Oleszkiewicz and Poggi-Varaldo (1997). The SRT substantially influences the hydrolysis of particulate substrates; the latter is often described by first-order kinetics, meaning that the hydrolysis rate decreases with increasing SRT, which indicates a nondegradable fraction (Eastman and Ferguson, 1981; Gujer and Zehnder, 1983). Fernández-Rodríguez et al. (2014) tested the SRT from 15 days to 3 days (OLRs from 4 to 20 kg VS m^{-3} day^{-1}) during the dry thermophilic digestion of OFMSW (20% TS) in a CSTR and showed that 80% VS removal can be obtained at 15 days' SRT. They reported the optimum SRT to be 5–8 days. At 4 days' SRT (16.5 kg VS m^{-3} day^{-1}), methane production decreased and VFAs accumulated. This illustrates the direct relationship between SRT and OLR; decreasing the SRT will increase the OLR and biogas production will increase until a limit is reached.

4.2.3.2 Total Solids Concentration

Water is essential for methane fermentation, as the nutrients for the microorganisms must dissolve in water before they can be assimilated. The moisture content may not only aid in bacterial movement, but also is known to influence mass transport limitations in a high-solids bed, and the balance between VFA production and the conversion of acids to methane. As the TS level is increased, there is less water available for dilution and salt concentrations increase. Most MSWs are produced at solids concentrations greater than 10%, while food waste typically has less than 10% solids, and activated sludge is usually in the range of 1%–4% TS.

Batch reaction rates were reported to have been severely limited at initial TS over 30% TS (Wujcik and Jewell, 1979) with agricultural residues as feedstock. Methane production decreases considerably between 30% and 35% TS, while the major problem encountered is the buildup of VFAs (up to 33,000 mg L^{-1} at 55% TS) (Wujcik and Jewell, 1979; D'Addario et al., 1992). Acidification can be avoided using a mature inoculum, such as in the sequential batch anaerobic composting (SEBAC) process, or the digested effluent, such as in the Dranco process. This phenomenon was also observed by D'Addario et al. (1992) with MSW as a feedstock. At 5% initial TS, the VFAs reached a plateau at 6 g VFAs L^{-1} after 8 days and methane was produced after 16 days. In contrast, at 20% initial TS, the VFAs reached 24 g VFAs L^{-1}, which resulted in high levels of nonionized acids that inhibited the acidogenesis as well as the methanogenesis. This linear relationship was also observed between VFAs and initial VS by Eastman and Ferguson (1981) with domestic primary sludge (PS) as substrate. According to Vandevivere et al. (2003), the large majority (about 90%) of industrial applications use one-stage processes, and these are evenly split between "dry" systems, where the waste is digested as received, and "wet" systems, where the waste is slurried with water to circa 12% TS. The advantages of both systems are listed in Table 4.4.

In terms of capital costs, dry and wet systems are comparable in as much as dry processes require much smaller reactor volumes but more expensive equipment. In terms of biological performance, dry systems can reach high biomass concentrations and have proven reliable. From a technical viewpoint, the dry systems appear more robust because frequent technical failures due to sand, stones, plastics, and wood were reported in wet systems (Vandevivere et al., 2003).

Monson et al. (2007) visited 20 industrial-scale AD systems treating OFMSW and concluded that, on average, dry AD systems yielded 375 m^{3} biogas ton^{-1} TS, whereas

TABLE 4.4 Advantages and Disadvantages of High Solids Content

Advantages	Disadvantages
• Much smaller reactor volumes—requires less land • Less or no liquid effluent and less water consumption (ca. 0.1 m^3 ton^{-1} waste); thus, the salt concentration can reach toxic concentrations • Smaller energy consumption for heating large volume • Smaller dewatering equipment • The plug flow ensures complete hygienization of the waste (at thermophilic temperatures)	• Slower anaerobic fermentation • More robust, thus more expensive pumps and auxiliary equipment • Less water available for diluting the salts present • Lower methane yields

Source: Monson, K.D. et al., *Anaerobic Digestion of Biodegradable Municipal Solid Waste: A Review,* University of Glamorgan, Trefforest, Wales, 2007; Vandevivere, P. et al., in *Biomethanization of the Organic Fraction of Municipal Solid Wastes,* ed. J. Mata-Alvarez, IWA Publishing, London, 2003, pp. 111–140; Wujcik, W.J., and Jewell, W.J., Dry anaerobic fermentation, presented at 2nd Symposium on Biotechnology in Energy Production and Conservation, Gatlinburg, TN, 1979.

wet AD systems yielded 663 m^3 ton^{-1} TS. Dry AD requires less land, but it should be born in mind that most of the land is not occupied by the digester but by the pretreatment units, receiving halls, and posttreatment facilities.

4.2.3.3 Organic Loading Rate

The OLR gives a measure of how much organics can be fed daily to the digester, and therefore high OLRs are desirable, but they should not be too high or they could result in the acidification of the digester. The methane production increases together with the OLR until a threshold is reached, which correlates with instabilities, and this can lead to process failure. For instance, Comino et al. (2010) obtained 70% VS removal during the degradation of cow manure and crop silage at 5.2 kg VS m^{-3} day^{-1}. However, the process failed when the proportion of the cosubstrate was increased. Nayono et al. (2010) treated food waste and OFMSW as cosubstrates and observed that the biogas production increased linearly with a OLR from 12 to 20 kg COD m^{-3} day^{-1}. The latter OLR was critical as VFAs started to accumulate, especially propionate.

Pavan et al. (2000b) have loaded up to 68.5 kg VS m^{-3} day^{-1} with fruit and vegetables in a hydrolytic reactor. The optimum operating retention times were 2–3 days in the mesophilic hydrolytic reactor, and 8–9 days in the thermophilic methanogenic reactor. Pavan et al. (2000a) investigated the treatment of MSW and biowastes in a thermophilic one-stage wet system. They found a maximum sustainable OLR for mechanically sorted MSW equal to 9.7 kg VS m^{-3} day^{-1}. The same OLR, however, was unsustainable when the feed was switched to source-separated biowaste, for which the maximum OLR was 6 kg VS m^{-3} day^{-1}. Weiland (1992) obtained a similar OLR value for highly biodegradable agro-industrial wastes under mesophilic conditions. The authors suggested this was due to VFA inhibition. Raynal et al. (1998) investigated the liquefaction of different vegetable solid wastes in different reactors and noticed that the hydrolysis yield increased with the OLR and then plateaued at

6 g COD L^{-1} day^{-1}, but the hydrolysis rate was linear with the loading rate. Oleszkiewicz and Poggi-Varaldo (1997) have obtained a sustainable OLR of 9 kg VS m^{-3} day^{-1} with a 30% TS mixture of paper, food waste, sewage sludge, and potato industry wastes. This feedstock was characterized by a high paper content (45% dry solids basis) but demonstrated a methane yield of 0.27 L g^{-1} VS fed.

Besides these laboratory-scale experiments, two plants were started in 1999 for the biomethanization of mechanically sorted MSW with a one-stage wet process; the one in Verona was designed with an OLR equal to 8 kg VS m^{-3} day^{-1}, while the one in the Netherlands has a design capacity of 5 kg VS m^{-3} day^{-1}. Possible limiting factors affecting the critical OLR are biomass concentration, mass transfer rate of substrate to bacteria, and accumulation of inhibitory substances (Vandevivere et al., 2003).

4.2.4 Parameters Improving Performance

4.2.4.1 pH
One way to optimize digester performance is to ensure that it operates at a proper pH. This is a capital parameter, as it is a sign of well-being of the bacterial population. With this aim, the two-phase configuration of an AD process enables each reactor to be optimized separately. Lagerkvist and Chen (1992) have studied reactors simulating landfill, and reported a pH of 5.3 and 7.4, respectively, in the acidogenic and methanogenic phase. It is, however, generally accepted that hydrolysis is optimum at pH 6–6.5 (Vieitez and Ghosh, 1999), and acidification at pH 5.8–6.2 (Zoetemeyer et al., 1982b), although the acid-forming bacteria can also grow well at pH 5 (Eastman and Ferguson, 1981).

In contrast, it is generally accepted that methane production from VFAs and lipids is optimal for a pH in the range of 6.3–7.8 (Lay et al., 1999). The effect of a drastic change in pH in the influent depends on the available alkalinity in the reactor, the extent and duration of the imposed change, effluent recirculation, and the concentration of VFAs. Methanogens are the most affected by changes in pH, while VFAs can still be produced at a pH below 6.5. The sharp drop in the rate below 6.3 is related to the fact that methane formation proceeds at a slower rate than the production of organic acids. The sharp drop in the rate above 7.8 may be related to a shift in NH_4^+ to the toxic, un-ionized form NH_3. The pH is known to influence enzymatic activity, because each enzyme is active only within a specific and narrow pH range, and displays maximum activity at an optimum pH (Lay et al., 1997).

Maintaining pH is especially delicate in the start-up phase because organic waste must undergo acid-forming stages before any methane forming can begin, which will lower the pH. To raise the pH during the early stages, operators must add a buffer to the system. The same procedure is followed when the pH drops during operation, for example, due to increased loading rate. It is the responsibility of an operator, therefore, to keep bicarbonate alkalinity as high as possible in order for the pH to remain high enough for methanogens to survive. The common materials used to increase the alkalinity are lime, soda ash, ammonia, ammonium bicarbonate, sodium hydroxide, and sodium bicarbonate. Generally, lime, sodium hydroxide, and ammonia are the least expensive of these chemicals (Anderson and Yang, 1992).

4.2.4.2 Alkalinity

The alkalinity is a measure of the capacity of the solution to neutralize acids. Optimal anaerobic biotechnology is characterized by nearly neutral conditions. Process imbalance can be due to low pH, which can be caused by two sources of acidity, H_2CO_3 and VFAs. The major requirement for a well-operating digester is the neutralization of the high carbonic acid concentration, which results from the high partial pressure of carbon dioxide in the reactor. The concentration of CO_2 in the liquid phase is set by Henry's law. Speece (1996) calculated that a minimum bicarbonate alkalinity of about 4000 mg L^{-1} was required to neutralize the carbonic acid in solution at a carbon dioxide partial pressure of 0.3 atm and an ionic strength of 0.2 in an anaerobic digester operating at pH 7 and 35°C. The extra alkalinity on top of that can then be used to neutralize VFAs produced by the AD. Many factors will determine the available reserve alkalinity, such as the effluent recycle, the alkalinity present in the waste stream, the alkalinity generated by the degradation of proteins, the chronic elevated VFA concentrations, and supplementation by the operator. On the other hand, if the alkalinity is excessively high, for instance, due to a high protein content of the waste stream, the pH will rise, which can shift the equilibrium concentration of ammonia toward its toxic form and inhibit the methanogens.

4.2.4.3 Phase Separation

Several authors report two-phase treatment as more rapid and stable than single stage (Massey and Pohland, 1978; Zoetemeyer et al., 1982a; O'Keefe et al., 1996; Vieitez and Ghosh, 1999). However, two-stage systems are the most complex and expensive of all systems (Vandevivere et al., 2003). Another benefit of the two-stage process, highlighted by Held et al. (2002), is the successful elimination of the odor in the overall output of the process, which could not be achieved in a CSTR alone. Their greatest advantage lies in the buffering of the OLR taking place in the first stage, allowing a more constant feeding rate to the methanogenic second stage. This is a substantial advantage in the case of substrates whose degradation is limited by methanogenesis rather than by hydrolysis, for example, cellulose-poor kitchen waste (Pavan et al., 2000a, 2000b; Vandevivere et al., 2003). These wastes, being very rapidly acidified, tend to inhibit methanogenesis in one-stage reactors when the feedstock is not adequately mixed, buffered, and dosed. This phenomenon has been experimentally observed by Pavan et al. (2000a): the one-stage CSTR, fed with a rapidly hydrolyzable fruit and vegetable waste, failed at an organic load of 3.3 kg VS m^{-3} day^{-1}, whereas the two-stage process remained stable at 7 kg VS m^{-3} day^{-1}. Banks and Humphreys (1998) also recommend a two-stage operation for substrates offering low natural buffering capacity, such as a mixture of paper and wood, as they could not maintain stable operation in a single stage because VFAs could not be buffered, leading eventually to souring. This process and others treating MSW only or through codigestion have been gathered in the tables in the following pages.

Many studies have focused on the two-stage configuration, as it allows a separate optimization in both reactors, leading to better performance. The first stage usually takes place at a pH between 4.5 and 6, while the second stage operates in a pH range of 6.5 to 8.2 (Speece, 1996). However, a strict separation of acidogenesis and methanogenesis is almost impossible without addition of chemicals for pH control. Initially, Chanakya et al. (1993) maintained a good separation when the VFA concentration was above 6 g L^{-1} and the pH below 5, but strictly separated acidogenic and methanogenic fermentation could not be maintained at the final stage when treating leaf biomass. Lettinga and Hulshoff Pol (1991) have recommended the selection of an HRT in the range of 6–24 h for the acidogenic stage. Under these conditions, there will be an accumulation of VFAs in the acidogenic reactor. The minimum solid residence time for methane production at mesophilic temperatures is approximatively 2.5–4 days. Sometimes some methane is produced at much shorter residence times because of the hydrogen-consuming bacteria, which, according to Shea et al. (1968), have a minimum cell residence time of 19 hours. Sometimes, biofilm or granule formation can shelter methanogens from low pH or high VFA concentration. This can explain why methane is produced in the first reactor before biogas production starts in the second one.

Sosnowski et al. (2003) compared the treatment of sewage sludge and OFMSW in a one-stage thermophilic reactor and a two-stage process (thermophilic CSTR followed by a mesophilic upflow anaerobic sludge blanket [UASB]) and showed that the two-phase system was more effective. In the single stage, a specific gas production (SGP) of 0.427 L g^{-1} VS was obtained with VS removal of 49.3% at 38 days' HRT, while in the two-stage process, the SGP and removal were 0.532 L g^{-1} VS and 62.7%, respectively, at 28 days' HRT (11 + 17 days), which proves the twofold advantages of the two-stage process: greater methane production and solid degradation at lower HRT.

Similarly, Bertin et al. (2013) reported a 40% methane production increase when a two-stage system was compared with a one-stage system for the treatment of cow manure and cheese whey. By contrast, Monson et al. (2007) visited 20 industrial-scale AD plants treating OFMSW and concluded that for source-sorted OFMSW, single-stage processes are preferable to multistage ones in most cases. In addition, single-stage AD is simpler and usually cheaper.

4.2.4.4 Leachate Recirculation

Liquor recycle features in a number of processes, and is important because it allows valuable buffering capacity to be retained in the system, maintaining heat, nutrients, and microorganisms within the system (O'Keefe et al., 1992; Banks and Humphreys, 1998; Nordberg et al., 2006). It also allows the toxic components in the feed and the VFAs in the reactor (Babel et al., 2004) to be diluted and to adjust the organic loading (Vieitez and Ghosh, 1999). Recirculation of the effluent was used in anaerobic processes treating various kinds of wastewaters to provide alkalinity in order to avoid failure of the acidogenic reactor (Borja et al., 1996; Banks and Humphreys, 1998). Leachate recirculation has been especially employed in reactors simulating landfills, or directly in landfills, because it increases methane yields and lowers the treatment costs (Vieitez and Ghosh, 1999; He et al., 2005), but also provides alkalinity and moisture in the leach bed (O'Keefe et al., 1992; Nordberg et al., 2006). In full-scale processes, it is used not only to dilute the feed, but also to reduce the freshwater consumption and storage capacity requirement (Nordberg et al., 2006). On the other hand, liquid recycle could lead to a buildup of toxic or recalcitrant compounds, such as ammonia (He et al., 2005), leading to an inhibition of hydrolysis, acidogenesis, and methanogenesis (Nordberg et al., 2006). Nordberg et al. (2006) therefore concluded that it is necessary to adjust the degree of liquid recirculation to reach an optimal process. However, apart from CSTR and leach bed systems, the behavior of other reactor designs with recirculation of the effluent is not well documented. Trzcinski and Stuckey (2009) operated a two-stage anaerobic process with liquid recycle in order to minimize the use of freshwater and did not observe inhibition. This was due to the fact that some effluent was purged from the process together with the digestate.

The recycling of effluent to the first stage of a two-stage process has been reported to reduce the costs of neutralizing additives, increase the pH, improve the buffering capacity, and increase the biogas production and methane content of the biogas (Jarvis et al., 1995). However, accumulation of inorganic and organic compounds may occur after initiating recirculation of the effluent liquid. This will cause a change in the digester environment and will continue until steady-state concentrations of the different compounds are obtained. Jarvis et al. (1995) obtained a 10-fold increase and a 3-fold increase in the degradation rate of mannitol and lactate, respectively, after initiating recirculation, and mannitol degradation was closely related to hydrogen partial pressure. They also observed a 10-fold increase in the number of hydrogenotrophic methanogens in the first reactor due to effluent recycling. This led to shortened periods of high hydrogen partial pressure following feeding, and a 20% increase in the total SGP. A similar observation was made by Trzcinski and Stuckey (2009), who speculated that high methane yields were possible thanks to the daily recycle of digested slurry from the SAMBR (3L) (recycle ratio as low as 0.15%), which reinoculated the hydrolytic reactor (20L) on a daily basis with methanogens. The closer syntrophic distances in the granules from the SAMBR may have sheltered the methanogens from the lower pH in the hydrolytic reactor (HR).

4.2.5 Types of Inhibition

4.2.5.1 Ammonia Inhibition

Ammonia is the result of the degradation of proteins found in food waste and garden waste in particular (Richards et al., 1991; Kayhanian and Tchobanoglous, 1992; Oleszkiewicz and PoggiVaraldo, 1997), and this can lead to an accumulation in processes treating MSW (Kayhanian and Tchobanoglous, 1992; Weiland, 1992; Vandevivere et al., 2003). Ammonia is produced by the biological degradation of nitrogenous matter, such as proteins and urea. Several mechanisms for ammonia inhibition have been proposed, such as a change in the intracellular pH, increase of maintenance energy requirement, and inhibition of a specific enzyme reaction (Wittmann et al., 1995). Ammonium ion $\left(NH_4^+\right)$ and free ammonia (NH_3) are the two principal forms of inorganic ammonia nitrogen in aqueous solution. Free ammonia has been suggested to be the main cause of inhibition since it is freely membrane permeable (De Baere, 1999). The hydrophobic ammonia molecule may diffuse passively into the cell, causing proton imbalance and/or potassium deficiency (Gallert and Winter, 1997).

Among anaerobic microorganisms, methanogens are the least tolerant and the most likely to cease growth due to ammonia inhibition (Kayhanian, 1994). There is conflicting information in the literature on the sensitivity of aceticlastic and hydrogenotrophic methanogens. Comparison of methane production with growth rates has indicated that the inhibitory effect was in general stronger for the aceticlastic than for the hydrogenotrophic methanogens (Angelidaki and Ahring, 1993), while others had observed the relatively high resistance of acetate-consuming methanogens to high total ammonia nitrogen (TAN) (involving both nitrogen forms, i.e., NH_4^+-N and NH_3-N) levels compared with hydrogen-utilizing methanogens (Zeeman et al., 1985).

A wide range of inhibiting ammonia concentrations have been reported in the literature, with the inhibitory TAN concentration that caused a 50% reduction in methane production ranging from 1.7 to 14 g L^{-1}. The significant differences in inhibiting ammonia concentration can be attributed to the differences in substrates and inocula, environmental conditions (temperature and pH), and acclimation periods (Angelidaki and Ahring, 1994). Kayhanian and Tchobanoglous (1992) report that their digester was operational at total ammonia concentrations up to about 1 g L^{-1}, but was less stable. The normal range for the three SRTs studied was 300–600 mg L^{-1}. It has been reported that the Valorga and Dranco processes remain stable at ammonia concentrations equal to 3 and 2.5 g L^{-1}, respectively. It has been speculated that the sturdiness of the dry systems toward ammonia inhibition is due to the fact that microorganisms are better shielded against toxicants; only restricted zones are affected by temporary shock loads, leaving other "pockets" little exposed to transient high levels of inhibitors (Vandevivere et al., 2003). Inhibition studies have been carried out by Wujcik and Jewell (1979) by adding ammonia as NH_4Cl to digesters treating a mixture of newsprint paper and dairy manure at different TS percentages (10%, 25%, and 30% TS). It turned out that reactor performance, quantified by biogas production rate, was not as seriously affected by increasing ammonia levels as it was by increased TS levels. For instance, at 10% TS, the reactor with ammonia levels of 6000 mg L^{-1} produced gas at a rate faster than the reactor with 600 mg L^{-1}. It is generally accepted that TAN concentrations of more than 2–3 g kg^{-1} are inhibitory if not toxic to nonadapted systems, especially where the pH is high enough to enable a significant fraction to be in the un-ionized NH_3 form, which is the most toxic. The following equation describes the equilibrium:

$$NH_4^+ + OH^- \rightleftharpoons NH_3 + H_2O \tag{4.1}$$

When the pH is sufficiently high (above 7.5), the equilibrium is shifted to the right and the free ammonia form predominates. There is a temperature- and pH-dependent relationship between free ammonia and ammonia (Anthonisen et al., 1976):

$$\text{Free ammonia [mg/L]} = \frac{[NH_4 - N] \cdot 10^{pH}}{\exp\dfrac{6334}{273 + T} + 10^{pH}} \tag{4.2}$$

The ionized form of ammonia is beneficial because the hydroxide produced in Reaction 4.1 reacts with CO_2, produced during AD, to form bicarbonate that will increase the buffering capacity:

$$CO_2 + HOH \rightleftharpoons H_2CO_3 \rightleftharpoons H^+ + HCO_3^-$$
(4.3)

This increase in buffer capacity was reported by Zhang et al. (2013) when codigesting manure and food waste. The addition of cattle manure enhanced the buffer capacity, allowing a high organic load with pH control.

There are conflicting reports on the effect of TAN concentrations. It was reported that concentrations between 1.5 and 3 g L^{-1} were inhibitory at pH levels above 7.4, and those in excess of 3 were toxic regardless of pH (Han and Shin, 2004). Lay et al. (1997) found that in the pH range of 6.5–8.5, the methanogenic activity decreases with an increase in NH_4^+-N concentration, and drops by 10% at a concentration of 1.67–3.72 g NH_4^+-N L^{-1}, 50% at 4–5.55 g, and to zero at 5.88–6.6 g. However, the lag phase time was dependent on the NH_3 level, but not on NH_4^+, and when NH_3-N was higher than 500 mg L^{-1}, a notable shock was observed; this confirms the known fact that the NH_3 level is a more sensitive factor than the NH_4^+ level for an unacclimatized bacterial system.

Some authors report different inhibitions for un-ionized ammonia values, depending on the temperature, pH, and bacterial adaptation capacity, but the inhibitory level should be close to 100–150 mg L^{-1} (Speece, 1996; Gallert and Winter, 1997; Han and Shin, 2004). Many researchers consider that the toxicity is associated with NH_3-N, depending on pH, and that concentrations in excess of ~100 mg L^{-1} may cause severe toxicity (Han and Shin, 2004).

Several researchers found a possible inhibition of the microbial activity due to ammonia accumulation in processes treating MSW (Kayhanian and Tchobanoglous, 1992; Weiland, 1992; Vandevivere et al., 2003).

To remove ammonia from the feed stream, air stripping and chemical precipitation have been used. These are technically feasible at high ammonia concentrations and in a complex wastewater matrix, although costs may be an issue. A common solution to ammonia inhibition is dilution. However, the resulting increase in volume makes this method economically unattractive (Callaghan et al., 1999) unless dilution is achieved using another substrate that is low in ammonia. Various types of inhibition can be counteracted by increasing the biomass retention in the reactor. It was found that methane yield in a CSTR could be increased by switching off the stirrer half an hour before and after substrate addition. This operation increased biomass retention due to improved sedimentation, resulting in an effluent with a reduced concentration of biomass solids (Hansen et al., 1998). Immobilizing the microorganisms with different types of inert material (clay, activated carbon, and zeolite) has been demonstrated to reduce inhibition of the biogas process and make the process more stable (Hansen et al., 1998). Ionic exchangers or adsorbents that can remove inhibitors have also been used to mitigate ammonia inhibition. Natural zeolite and glauconite show high selectivity for the ammonium ion and can be used as an ionic exchanger for ammonia

(Hansen et al., 1998). The addition of antagonistic cations such as Mg^{2+} or Ca^{2+} stabilizes the process. The positive effect of zeolite on the anaerobic process could partially be attributed to the presence of cations such as Ca^{2+} and Na^+, which have been shown to counteract the inhibitory effect of ammonia.

4.2.5.2 Volatile Fatty Acid Inhibition

Researchers have found that VFA concentrations are one of the most important parameters in AD (Hill et al., 1987). The toxic effects of high VFA concentrations have been studied and reported by several authors, and the resulting drop in pH is generally considered to be the main cause of inhibition. Several authors believe that VFA inhibition is pH related, and that the undissociated fraction is the inhibitor (Anderson et al., 1982; Baronofsky et al., 1984). Wang and Wang (1984) reported that undissociated acetic acid was much more inhibitory than the dissociated acetate ion; they mentioned that complete growth inhibition resulted when the undissociated acetic acid concentration was between 0.04 and 0.05 M (2.4–3 g L^{-1}). Babel et al. (2004) found that the inhibitory level of undissociated acids for acidogenesis was about 2300, 650, and 120 mg L^{-1} at pH 5, 6, and 7, respectively. Thus, undissociated acetic acid is responsible for growth inhibition at pH values lower than 6, while dissociated acetate ion plays a major role at pH values higher than 7 (Wang and Wang, 1984; Babel et al., 2004). Baronofsky et al. (1984) studied the effect of undissociated acetic acid on acidogenesis and found that the acid production ceased at pH 5. They postulated that acetic acid produced by the cell was responsible for acidification of the cytoplasm, resulting in an inhibition of cellular functions. Schwartz and Keller (1982) studied the production of acetic acid with a controlled pH. The highest acetic acid concentration was about 15 and 20 g L^{-1} at pH 6 and 7, respectively. Moreover, according to Yu and Fang (2003), operation at a pH of 4–5 favors the production of propionate and hydrogen, whereas operation at pH 6–7 encourages the production of acetate and butyrate. Thus, the authors recommend operation at a neutral pH range (6–7) because it will naturally minimize the concentration of undissociated acids (Babel et al., 2004), and because it will optimize the production of acetate (Yu and Fang, 2003). Propionate is a common intermediate in the metabolism of complex organics. It is normally converted to acetate and hydrogen, and this reaction is only possible if the hydrogen partial pressure is below 100 ppm (Smith, 1986). Furthermore, the propionate conversion rate will proceed much faster if there is a close proximity in granules or biofilm between the H_2 producers and the H_2 utilizers. Excessive agitation can disrupt the granule structure, reducing the rate of oxidation of fatty acids, which can lead to digester instabilities. Elevated levels of propionate are not necessarily due to excessive hydrogen because both acetate and hydrogen are inhibitory to the propionate conversion process (Van Lier et al., 1996). Elevated concentrations of propionate can be inhibitory to the acidogens, as well as the methanogens, but acclimatization of the bacteria to levels up to 8000 mg L^{-1} has been reported (McCarty and Brousseau, 1963). In another study (Hajarnis and Ranade, 1994), however, propionate was found to inhibit methane production at 5000 mg L^{-1} at neutral pH. Furthermore, when the pH was reduced the situation became worse because un-ionized propionic is known to be more toxic.

4.2.5.3 Long-Chain Fatty Acid Inhibition

Besides their high methane potential, lipids and lipid-rich wastes are not commonly used as a sole substrate in anaerobic digesters due to their inhibitory effect on anaerobic biocenosis, along with the development of other operational problems, like clogging, foaming, and biomass flotation. The inhibitory problems of lipids are mainly related to long-chain fatty acids (LCFAs). The main mechanisms of LCFA toxicity is through their adsorption onto the cell wall of microorganisms, thus inhibiting transport phenomena (Noutsopoulos et al., 2013).

Some industries produce wastewater containing significant amounts of fats, oils, and grease (FOG), for instance, the palm oil mill effluent. LCFAs or glycerol in the wastewater can be difficult to degrade and can result in long residence times or, worse, an accumulation of VFAs in the digester, resulting in a pH drop leading to digester failure. This can be due to an imbalance in methanogenic populations in the digester.

Methanogens can be inhibited by LCFAs due to their cell wall, which is similar to that of gram-positive bacteria. Indeed, LCFAs would inhibit methanogenic microorganisms by adsorption onto the cell wall or membrane, and interference with the transport or protective functions. In addition, sorption of a light layer of LCFAs on to biomass leads to sludge flotation and consequent washout (Rinzema et al., 1989). LCFA toxicity varied with the type of anaerobic sludge, and was more correlated on to the sludge physical characteristics (specific surface area and size distribution) than to their biological characteristics. Flocculent sludges, which have a higher specific surface area, suffered greater inhibition than did granular sludge. Thermophiles have been reported to be more sensitive to LCFAs than mesophiles, possibly due to different cell composition. However, biodegradation of LCFAs has been reported in both mesophilic and thermophilic environments (Hanaki et al., 1981a). It has also been suggested that LCFAs exerted a bactericidal effect and no adaptation of methanogens occurred (Hanaki et al., 1981a). However, recent studies based on the degradation of oleic acid in an anaerobic fixed bed reactor showed that acclimation improved biofilm tolerance to oleic acid. Addition of calcium has been shown to reduce LCFA inhibition, probably because of the formation of insoluble salts. However, calcium addition has not solved the problem of sludge flotation.

In Mediterranean countries, olive oil is composed of triglycerides that consist of a mixture of LCFAs, mainly oleic acid bonded to glycerol. Microbial oxidation of LCFAs has been intensively studied (Pereira et al., 2004). When exposed to LCFAs, the sludge quality may deteriorate due to a serious drop in methanogenic activity due to inhibition. Sludge flotation can arise at a lauric acid concentration exceeding 100 mg L^{-1}. Flotation results from the poor release of gas bubbles by the granules, due to the adsorption of LCFAs at the surface of the granular sludge. Moreover, the adhered LCFA film may hamper the supply of substrate to the bacteria present in the grains. Another harmful effect is the disintegration of sludge aggregates that can occur when lipids are present. This is because at a neutral pH, LCFAs act as a surfactant, lowering the surface tension, and consequently decreasing the aggregation of hydrophobic bacteria. Accordingly, this surfactant can also cause the disintegration of acetogens, examples of hydrophobic bacteria, that can degrade the LCFAs.

In addition to its high polluting power, FOG wastes, such as OMW, usually possess a high antibacterial activity exerted by various phenolic compounds. Although AD of this type of residue is feasible, and quite appealing from an energetic point of view, the presence of phenolic inhibitors decelerates the process, hinders removal of part of the COD, and detracts from its economic viability (Borja et al., 1995a). OMW is a complex effluent obtained from the traditional press and the continuous three-phase mills of olive oil production with a variable amount of lipidic compounds (4–25 g L^{-1}) (Hamdi, 1992) and phenolic compounds (3–12 g L^{-1}) (Beccari et al., 1999). The AD of raw OMW has been reported as a difficult process (Morillo et al., 2009). Boari et al. (1984) achieved instability and poor reproducibility on the start-up of UASB reactors treating OMW. They reported foam formation, accidental organic overloads, and biomass washout. Later, Zouari and Ellouz (1996) removed the oil layer from OMW to avoid inhibition caused by lipids. Nevertheless, it was revealed to be insufficient since the accumulation of recalcitrant colored compounds caused an irreversible inhibition of bacterial growth. More recently, Khoufi et al. (2006) operated an anaerobic filter packed with polyurethane foam cubes, at an OLR of 4 g COD L^{-1} day^{-1} of raw OMW. They observed a decrease of pH, biogas production, and yield, and an accumulation of VFAs.

Different strategies have been applied to enhance OMW biodegradation. Pretreatments have been proposed to detoxify the effluent (Khoufi et al., 2006). Other researchers decrease OMW toxicity by codigestion with agro-industrial residues (Dareioti et al., 2010b). Instead of removing or degrading OMW toxic compounds before the anaerobic treatment, Hamdi (1992) performed batch experiments to get insights to the problem of OMW biodegradation and toxicity. They concluded that among the OMW inhibitory substances, the darkly colored polyphenols induced the problem of OMW biodegradation, whereas the LCFAs, tannins, and simple phenolic compounds were responsible for its toxicity to methanogenic bacteria.

In a previous work on synthetic oily wastewater, Pereira et al. (2004) found that biomass encapsulated with 1 g COD-LCFAs g VS^{-1} was able to convert these biomass-associated LCFAs (mainly palmitic acid) into methane at a maximum rate of 500 mg COD-CH$_4$ g VSS^{-1} day^{-1}. In Gonçalves et al.'s work (2012), the maximum values of biomass-associated LCFAs were lower (0.18 and 0.41 g COD-LCFAs g VS^{-1}) and were mainly due to oleate, followed by palmitic and stearate. However, due to the complex nature of OMW, involving a matrix of organics, including phenolic compounds, it is not surprising that a different pattern of LCFA behavior was obtained. Neves et al. (2009) also found a lower threshold value for LCFA accumulation onto the solid matrix of about 0.18–0.22 g COD-LCFAs g TS^{-1} in the codigestion of cow manure, food waste, and oily waste. The main accumulated LCFAs shifted according to the reactor conditions. Oleate was initially the main accumulated LCFA. Then, palmitate started to accumulate at higher concentrations than oleate, while stearate was slightly increasing and palmitoleate was not detected. The LCFA degradation mechanism is still unclear (Sousa et al., 2007). Palmitic acid has been indicated as the primary detected product from oleic degradation. Oleate accumulation was observed as well by Cirne et al. (2007) in the treatment of a lipid-rich (triolein) model waste in batch tests. The experimental results suggest that the metabolic pathways or the limiting step may change during the treatment of OMW, stimulated by the increase of higher lipids or LCFA concentrations.

The intermittent feeding used by Gonçalves et al. (2012) also improved the removal of resilient phenolic compounds. The phenolic compounds that remained inside the reactor could have been degraded, adsorbed, or precipitated with proteins, according to the literature. Field and Lettinga (1987) reported that tannin polymers can be effectively adsorbed or precipitated with proteins, which lead to its toxic effect. Zouari and Ellouz (1996) mentioned that the accumulation of recalcitrant colored compounds in the sludge induced an irreversible inhibition of bacterial growth, but this was not observed by Gonçalves et al. (2012). Hanaki et al. (1981b) examined the inhibitory effect of these LCFAs on the AD process in batch experiments using synthetic substrates. Adding LCFAs to the anaerobic digesters caused the appearance of a lag period in the production of methane from acetate, and in the degradation of the LCFAs and n-butyrate. Anaerobic oxidation of LCFAs and alcohols to acetic acid, hydrogen, and carbon dioxide is carried out by the obligate hydrogen-producing acetogenic (OHPA) bacteria. Novak and Carlson (1970) observed that LCFA degradation was inhibited by hydrogen. This can be due to the accumulation of hydrogen gas that can inhibit methanogens even at low partial pressures (Trzcinski and Stuckey, 2012a). As a result, it is recommended to keep the hydrogen concentration low in anaerobic digesters, and this is achieved by maintaining a small distance between syntrophic bacteria (Kim et al., 2002; de Bok et al., 2004), that is, between the hydrogen producers and the hydrogen consumers. However, Dolfing and Bloemen (1985) found that methane formation from acetate in *Methanosarcina* spp. was inhibited by high levels of hydrogen. It is therefore advisable to control the hydrogen concentration in the digester. The addition of calcium chloride reduced the inhibitory effect of LCFAs, but it did not do so after the culture had been exposed to these acids for several hours (Siles et al., 2010). Because of the potential inhibition by LCFAs from palm oil, a two-stage process is recommended.

4.3 Large Diversity of Solid Waste Anaerobic Processes

4.3.1 Anaerobic Treatment of Municipal Solid Waste

To evaluate the performance of the AD process for the treatment of organic solid waste, three performance indexes are generally used: (1) the OLR (kg VS m^{-3} day^{-1}), which serves as an index of the processing speed or potential reduction of the reactor volume; (2) the methane yield (mL g VS^{-1}), which serves as an indicator of the efficiency of energy conversion; and (3) VS reduction (%), which provides a measure of the waste reduction potential of the reactor. In conventional single-stage wet AD, OLRs of 1–4 kg VS m^{-3} day^{-1} have been commonly reported in the treatment of food waste or the organic fractions of MSW (Mata-Alvarez et al., 1992; Heo et al., 2004; Zhang et al., 2007). Even in semidry and dry AD, OLR seldom exceeds 10 kg VS m^{-3} day^{-1} (Cecchi et al., 1991; Bolzonella et al., 2006b; Forster-Carneiro et al., 2008).

Some one-stage and two-stage laboratory- and bench-scale processes are listed in Table 4.5, and pilot- and full-scale processes are listed in Table 4.6. For each process,

TABLE 4.5 Laboratory- and Bench-Scale Anaerobic Digestion Processes for Municipal Solid Waste

Feedstock	Process Configuration: Reactors	HRT (Days)	TS (%)	OLR (g VS or COD L^{-1} Day^{-1})	GPR (L$_{biogas}$ L^{-1} Day^{-1})	SGP (L$_{biogas \, or \, methane}$ g^{-1} VS$_{fed}$)	CH$_4$ (%)	VS or COD Removal (%)	References
OFMSW	2 LB (M) UASB (M) System	2	F: 30	11 (COD)		0.15 (CH$_4$) 0.12 (CH$_4$) 0.27		95 (COD) 70	Anderson and Saw, 1992
SS-OFMSW	CSTR (M) FF (M)	2–3 7		20–25 10	4		19–74 66–80	70	Kubler and Wild, 1992
SS-OFMSW	CSTR (T) UASB (T)	7.5 16	F: 4.3	5.9 9 (COD)		0.04 (CH$_4$) 0.25–0.31 (COD)	9–29 79.5	68–81 (COD)	Rintala and Ahring, 1994
SS-OFMSW	CSTR (M)	19	F: 18.4	9.4	5.3	0.88 (VS$_{degraded}$)	67	64	Gallert and Winter, 1997
PW + wood	CSTR (M) CSTR (T) System	2.5 2.5 5	F: 3	0.92 (BVS) 0.46 (BVS)		0.33		53	Banks and Humphreys, 1998
MSW	LB (M) PB (M) System	295	16	0.31	0.026 0.075 0.046	0.27	48 48 46	30	Vieitez and Ghosh, 1999
MSW	LB (M) PB (M) System	524	16	0.27 0.13	0.06 0.03–0.027 0.045	0.023 0.34	38 70–80 50	54	Vieitez et al., 2000
MSW	CSTR (M) AF (M)	24 6	17.1	6 12.2 (COD)	4 1.8	0.52 0.31	47 61	64 44	Held et al., 2002
OFMSW	CSTR (M)	18	18	8.8	4.77	0.54	55	67.4	Castillo et al., 2005
OFMSW	CSTR (T)	15	1.5	4.3		0.32 (CH$_4$)			Angelidaki et al., 2006
SS-OFMSW	CSTR (T)	15	5	2.8		0.3–0.4 (CH$_4$)	62	80	Davidsson et al., 2007

(Continued)

TABLE 4.5 (CONTINUED)   Laboratory- and Bench-Scale Anaerobic Digestion Processes for Municipal Solid Waste

Feedstock	Process Configuration — Reactors	HRT (Days)	TS (%)	OLR (g VS or COD L^{-1} Day^{-1})	GPR (L$_{biogas}$ L^{-1} Day^{-1})	SGP (L$_{biogas\ or\ methane}$ g^{-1} VS$_{fed}$)	CH$_4$ (%)	VS or COD Removal (%)	References
OFMSW	CSTR (T)	8.9	F: 7	2.76	0.5–3	0.419	70		Sosnowski et al., 2003
	UASB (M)	20.9					75		
HS-OFMSW	CSTR (M)	20	F: 3.2–3.8	0.85	0.266	0.183	58	46	Borowski, 2015
HS-OFMSW	CSTR (T)	20	F: 3.4–3.6	0.91	0.328	0.190	58	48	Borowski, 2015
OFMSW + FW (64:36, COD)	CSTR (M)	7		19	4.51	0.37	66	50	Nayono et al., 2010
OFMSW	CSTR (M)	15	~20	4.8	1.45 (CH$_4$)	0.34		86.2	Fernández-Rodriguez et al., 2014
		8		8.8	2.37 (CH$_4$)	0.21			
		5		13	4.19 (CH$_4$)				
		4 (unstable)		16.5	3.29 (CH$_4$)				
OFMSW	Batch (T)	90	20		0.05 (CH$_4$)	0.08	65.6	79.5	Forster-Carneiro et al., 2008
OFMSW	CSTR (M)[a]	20	F: 10	4	1.35	0.148	55	75	Trzcinski and Stuckey, 2009; Trzcinski et al., 2010
	SAMBR (M)	1.6–2.3		20 (COD)	0.81	0.06	62	90 (COD)	
FW + GW	CSTR (M)[a]	2		10	4.5	0.25	63–69	81	Trzcinski and Stuckey, 2010
	SAMBR (M)	1		11.7 (COD)				96 (COD)	
FW	CSTR (M)[a]	16	7.2	9.2	6.6	0.455		92	Nagao et al., 2012

Abbreviations: AF = anaerobic filter; ASBR = anaerobic sequencing batch reactor; BVS = biodegradable volatile solids; F = feed; FF = fixed film reactor; FW = food waste; GPR = gas production rate; GW = garden waste; HS-OFMSW = hydromechanically sorted organic fraction of municipal solid waste; LB = leach bed; M = mesophilic; PB = packed bed reactor; PW = press water; SAMBR = submerged anaerobic membrane bioreactor; T = thermophilic.

[a] With biomass retention mechanism.

TABLE 4.6 Pilot- and Full-Scale Anaerobic Digestion Processes for Municipal Solid Waste

Feedstock	Process Configuration Reactors	HRT (Days)	TS (%)	OLR (g VS or COD L⁻¹ Day⁻¹)	GPR (L_{biogas} L⁻¹ Day⁻¹)	SGP ($L_{biogas\ or\ methane}$ g⁻¹ VS_{fed})	CH₄ (%)	VS or COD Removal (%)	References
					Operational Parameters		**Process Performance**		
MSW	CSTR (T)	7.4	14.3	12.9	5.2	0.37	58	76.2	Mata-Alvarez et al., 1992
OFMSW	CSTR (T) CSTR (T)	30	23–30 25	6–7 (BVS)	5.3	0.8–0.9 (BVS)	50	98 (BVS)	Kayhanian and Tchobanoglous, 1992
FYV	Kompogas (T)	17	F: 40	11.1	3.6	0.325	62		Wellinger et al., 1992
A + MS-OFMSW	CSTR (T)	8.5	F: 17.7	13.4	4.3	0.32	58.8		Cecchi et al., 1992
MSW	SEBAC (T)	42	F: 56.5–80.2	3.2	1.02–1.11	0.32–0.35	55–60	48.9–52.4	O'Keefe et al., 1992
FV	CSTR (M) CSTR (T) System	6.6 12.5 19.1	F: 8.2	12.6 5.7	4.4	0.7		82	Pavan et al., 2000b
MS-OFMSW	Valorga (M)	15–20	30	5–13.7	2.6	0.21–0.3 (CH₄)		50–70	Braber, 1995; Nallathambi, 1997; Vandevivere et al., 2003
SS-OFMSW	Kompogas (T)	13–25	30			0.21–0.3 (CH₄)		50–70	Vandevivere et al., 2003
MSW	Dranco (T)	14–21	31–57	12–15	2.8	0.187 (CH₄)	55	65	Nallathambi, 1997; Vandevivere et al., 2003
MSW	Biocell LB (M)	14–21	40	3.6–5.1					Braber, 1995; Vandevivere et al., 2003
MSW leachate	ICR (M)	1	–	23–40.5 (COD)	11 (CH₄)	0.25	72–80	80 (COD)	Luo et al., 2014

Abbreviations: A = algae; AF = anaerobic filter; ASBR = anaerobic sequencing batch reactor; BVS = biodegradable volatile solids; DW = dewatered sludge; F = feed; FF = fixed film reactor; FV = fruits and vegetables; FW = food waste; FYV = fruit, yard, and vegetable waste; GW = garden waste; GPR = gas production rate; HS-OFMSW = hydromechanically sorted organic fraction of municipal solid waste; ICR = internal circulation reactor; LB = leach bed; M = mesophilic; PB = packed bed reactor; PW = press water; SAMBR = submerged anaerobic membrane bioreactor; T = thermophilic.

the configuration is given, along with the process temperature. It is not straightforward to compare processes treating OFMSW, as performance depends heavily on the composition of the feed used for the study, and furthermore, many authors do not specify the composition or the ultimate biodegradability of the MSW feedstock. In most papers, authors only specify whether the MSW feedstock is source sorted or mechanically sorted. The former comprises hand-sorted waste, such as kitchen waste or waste fruit and vegetables from a market, with a high organic content (VS/TS ≈ 90%). The latter represents the OFMSW after a sorting plant where the plastics and glass can be removed in a hydropulper, for instance. The SS-OFMSW has a much lower solid content than MS-OFMSW. Furthermore, the MS-OFMSW presents lower biodegradability (VS/TS ≈ 45%) due to the presence of more inerts and lignin or plastics, which may have passed the sorting line together with the degradable fraction (Pavan et al., 2000a).

In the multiple-stage reactors, dilution water is generally added into the process to enhance solubilization of the substrate before loading it into the methanogenic reactor. This dilution of the substrate increases wastewater volume and decreases COD in the wastewater, resulting in decreased HRT in the methanogenic reactor. In contrast with multiple-stage digestion, semidry (10%–20% TS) and dry (20%–35% TS) digestion, which use high-TS sludge without adding water to the substrate, have achieved a high OLR and stable digestion. This high performance in operational plants can paradoxically be explained by the low VS reduction and low methane yield of these reactors. Typical VS reductions of 50%–80% are achieved in well-mixed reactors, while leach bed reactors have lower VS destruction, in the range of 30%–50%. If the proportion of food waste is high, then VS destruction of 80%–90% is common (Table 4.5). Cecchi et al. (1991) reported that the VS reduction decreased with increasing OLR in semidry AD, with VS reductions of 48%, 43%, 34%, 31%, and 37% at OLRs of 5.9, 6.9, 9.2, 10.7, and 13.5 kg VS m^{-3} day^{-1}, respectively. Interestingly, although decreased VS reductions and methane yields were observed, VFA accumulation and inhibition of methanogenesis were not observed, even at a high OLR of 13.5 kg VS m^{-3} day^{-1}. This low VS reduction indicates that undegraded organic matter remained in the digested sludge and was not transformed into short-chain fatty acids, since the mass transfer efficiency decreases at high TS contents in the anaerobic sludge, resulting in high stability without VFA inhibition of methanogenic activity. Veeken and Hamelers (1999) noted that the limitations in mass transport at high TS levels will have positive effects by preventing the accumulation of VFAs and the resulting inhibition of methanogenesis. They also proposed that improved performance in dry AD could be achieved by varying the rate of leachate recirculation, which can regulate the hydrolysis of the biowastes without VFA accumulation.

The majority of processes at laboratory scales use a conventional CSTR. However, biomass retention devices are sometimes used. For instance, Trzcinski and Stuckey (2009) used a hydrolytic reactor containing a perforated mesh that allowed them to separate in situ the partially degraded solids from the leachate. This integrated solid–liquid separation allowed for the production of a high-strength leachate to be produced and fed to the second stage, while partially degraded solids remained in the first reactor to be degraded further. The SRT and HRT were about 65 and 20 days in the first reactor, respectively. The VS removal was 75% at an OLR of 4 kg VS m^{-3} day^{-1} (Trzcinski et al., 2010). The authors used a microfiltration membrane with a 0.4 μm pore size as the biomass retention mechanism in the second stage of the process. This proved to be very effective, as more than 90% COD in the leachate was removed at an HRT of around 1–2 days. The membrane was able to retain the slow-growing methanogens, which allowed for a rapid start-up and stable operation due to a high SRT of about 300 days. However, colloids in the leachate caused early membrane fouling, which limited the flux, and therefore the lowest HRT that can be achieved in this system. Using the same process to treat a mixture of highly degradable food waste and garden waste, the authors achieved 81% VS removal in the first reactor operating at 2 days' HRT and 96% COD removal from the leachate in the second reactor at 1 day HRT (Trzcinski and Stuckey, 2010). This demonstrated the low HRT (3 days compared with 30 days in conventional AD systems) that can be achieved with good performance when a biomass retention mechanism is implemented.

The operating loads are generally lower in bench-scale than in full-scale plants, where loading rates as high as 10–18 kg VS m^{-3} day^{-1} are cited in the literature. A direct comparison is difficult, as commercial literature cites the best case studies after a long operation period, and often does not specify the amount of recycled solids. Comparison is not straightforward, as, in some studies, no nutrients and/or no seed inoculum was added to the system. This is often the case with the leach bed process, where a batch of MSW is loaded and the reactor closed, which results in longer operating time, and thus lower methane yields. In batch systems, the dry waste (30%–40%) is allowed to go through all degradation steps sequentially in the digester. Very often, the leachate is recirculated in order to disperse the inoculant, nutrients, and acids. Batch reactors are simple and cheap, and require small quantities of water; however, the OLR is generally low. Also, the high VS removal percentages reported in laboratory- and bench-scale processes should be interpreted with caution due to the better contact between microorganisms and the more homogeneous substrate.

The VS removal efficiencies are in the range of 50%–70%, which indicates that 50%–70% of the VS is actually biodegradable VS. These high VS removals have been reported in both single-stage and two-stage processes, mesophilic and thermophilic. The performance of these systems is highly dependent on the type of pretreatment applied and whether the feed is MSW or SS-OFMSW.

Nonetheless, some authors report only the COD removal from the leachate fed to the second reactor. As a result, COD removals of greater than 90% have been reported in two-phase systems (Anderson and Saw, 1992; He et al., 2005), and COD removals of 75% have been reported at 24°C (Kettunen et al., 1996). Most studies have been carried out with a one-stage mesophilic CSTR. However, a few thermophilic CSTRs have demonstrated the improvement of anaerobic codigestion, to allow higher OLRs.

AD of OFMSW is technically feasible and economically viable, and the optimum configuration (dry or wet, mesophilic or thermophilic, pretreatment or posttreatment) is site and case specific and dependent on the amount and characteristics of the waste, the aims of the project, local legislation, and circumstances such as market and land available for the digestate. Codigestion should be maximized where possible, and very often the TS content of the cosubstrate will be decisive to choose the dry or wet type of configuration. A common size of industrial AD systems for OFMSW is in the range of 11,000–20,000 tons per annum (tpa); however, plants with a capacity of 50,000–100,000 tpa have also been commissioned in the year 2000. Very few plants have a capacity greater than 100,000 tpa (Monson et al., 2007).

The plant economics is very dependent on the close proximity of the wastewater treatment plant (WWTP), waste generation site (e.g., farm, food industry, and slaughterhouse), landfill site, thermal treatment plant, and potential users for the excess heat produced. All these existing infrastructures may reduce capital and operating costs of the AD plant. For instance, if the biowaste reception and pretreatment plant are located near an existing incineration plant for municipal waste, several synergies could be realized: infrastructure savings, a grid connection, a district heating connection, a road connection, the use of a weighbridge, one security and fencing expense, planning benefits, the engineering and building of two interconnected plants on one site, savings in planning and environmental impact assessment costs, and use of the heat produced by the incinerator to pasteurize the biowaste.

Regarding the number of stages, De Baere (1999) reported in 2002 that 90% of AD systems were single stage. In 2007, the trend remained similar, and this is particularly true for SS-OFMSW, which is more biodegradable than MS-OFMSW. In the year 2000, there was interest in a two-stage AD for OFMSW, where the first stage is an aerobic percolation stage where the waste is mixed with hot water and aerated, and then sent to a high-rate AD (Monson et al., 2007).

4.3.2 Codigestion of Sewage Sludge and Municipal Solid Waste Components

Sewage sludge is characterized by a low C/N ratio and higher buffer capacity (Astals et al., 2013), and its digestion is carried out in a WWTP at low OLRs (usually about 1 g COD L^{-1} day^{-1}) at long residence times (typically 30 days) and low efficiencies around 20%–40% VS removal (Xie et al., 2007), meaning that these are often not used at their full capacity. The digesters are often overdesigned to allow for increased amounts of sludge due to an increasing population. Zupančič et al. (2008) reported that the digester is operating at 0.8 kg VS m^{-3} day^{-1} and 20 days' HRT to treat 60% PS and 40% waste activated sludge (WAS). Sewage can therefore be codigested with easily biodegradable organic matter with low alkalinity. Besides, OFMSW and food waste are produced in every municipality, which can then reduce the transportation costs. FOG from the grit chamber at the WWTP and agro-industrial waste produced near the WWTP are also ideal cosubstrates.

The anaerobic codigestion of OFMSW with sewage sludge, using an existing anaerobic digester in a municipal WWTP, seems to be especially attractive and potentially could reduce capital and operating costs. The benefits of codigestion include dilution of potential toxic compounds coming from the cosubstrate, synergistic effects of microorganisms, better methane yield per unit of digester volume by increased OLR of biodegradable matter, and improved balance of nutrients. Food waste contains highly soluble organic materials that would be rapidly converted to VFAs in the early stage of digestion. One of the interesting alternatives for the treatment of food waste is anaerobic codigestion, together with sewage sludge in the existing digester of the municipal WWTP. The codigestion has been successfully applied in laboratory and full scale for the treatment of OFMSW (various types and qualities) and sewage sludge (Tables 4.7 and 4.8). Krupp et al. (2005) reported that codigestion was a better option than OFMSW monodigestion or OFMSW composting. In practice, the codigestion is not so easy due to sorting of the OFMSW; composition variability and seasonality; and accumulation of VFAs, LCFAs, light metals, or NH_3. For instance, Pahl et al. (2008) reported an accumulation of Zn, Pb, and Ni in the WWTP digestate when MS-OFMSW was codigested. This problem, related to OFMSW sorting, does not appear in pilot- and lab-scale studies, where most of the studies have been carried out.

Table 4.7 lists the successful case studies at laboratory scale, and a wide range of OLRs were reported due to the different processes and qualities of the substrate. It can be seen that the codigestion typically results in specific methane production greater than 0.3 m^3 CH_4. kg VS^{-1}, which indicates a good biodegradability, and VS removals in the range of 40%–80% were reported. This is remarkable considering that VS removal of WAS alone is normally in the range of 20%–40%. In some instances, a slight increase in the OLR caused an accumulation of VFAs and a decrease in the GPR, indicating process instability. For instance, Liu et al. (2013) treated WAS and OFMSW at 8 kg VS m^{-3} day^{-1} and obtained the maximum GPR, but there was a VFA accumulation and a slight decrease in specific methane production) (0.73 to 0.62 m^3 kg VS^{-1}), suggesting that 8 kg VS m^{-3} day^{-1} was the critical OLR. Thermophilic processes have also been studied to codigest WAS and OFMSW, and Cavinato et al. (2013) reported a 45%–50% increase in biogas yield with the thermophilic process compared with the mesophilic one. The additional biogas could provide the higher heat requirements. There is an equal number of mesophilic and thermophilic processes treating OFMSW at full scale (Baere, 2006). An increase of 0.2 m^3 CH_4. kg VS^{-1} was reported in a two-stage thermophilic temperature-phased process compared with the mesophilic counterpart. Although higher methane yields were obtained (0.36 vs. 0.31 m^3 CH_4 kg VS^{-1}), Borowski (2015) reported that VFAs and free ammonia accumulated during the thermophilic operation to 5.7 and 1 g L^{-1}, respectively.

Heo et al. (2004) examined the biodegradability of mixtures of food waste (FW) and WAS at ratios of 10:90, 30:70, 50:50, 70:30, and 90:10 (FW/WAS) on a VS basis. The C/N ratio improved from 6.16 to 14.14, and the biodegradability of the mixtures increased from 36.6% to 82.6% as the food waste proportion of the mixture increased from 10% to 90%. The stability and performance of the single-stage anaerobic digester were investigated at the HRTs of 10, 13, 16, and 20 days with five mixtures at 35°C. During all the experiments, the pH remained in the range of 7.2–7.5 and alkalinity remained in the range of 3.1–4.7 g $CaCO_3$ L^{-1}. There was no sign of accumulation of VFAs in any of the digesters, and the buffer capacity was the highest in the digester fed with a feed mixture of 50:50. The VFAs, as a ratio of acetic acid to alkalinity (V/A), are often reported as a good tool for the observation of the stability of the anaerobic process; it should be below 0.3 for sufficient stability of the digester. The free acids and ammonium ion concentration were maximum at 37 mg L^{-1} and 1.17 g L^{-1} when the digester operated at an HRT of 20 days with a feed mixture of 50:50, which are well below the inhibitory levels reported

TABLE 4.7 Laboratory- and Bench-Scale Anaerobic Codigestion Processes of Sewage Sludge and Municipal Solid Waste

Feedstock (VS Basis Unless Stated Otherwise)	Process Configuration	Operational Parameters			Process Performance				References
	Reactors	HRT (Days)	TS (%)	OLR (g VS or COD L^{-1} Day^{-1})	GPR (L$_{biogas}$ L^{-1} Day^{-1})	SGP (L$_{biogas\ [or\ methane]}$ g^{-1} VS$_{fed}$)	CH$_4$ (%)	VS (or COD) Removal (%)	
WAS + MSW	CSTR (M)	25	1	2.9		0.36	68–72	88	Elango et al., 2007
FW + WAS (50:50)	CSTR (M)	13	0.3–0.6	2.43	1.24	0.321 (CH$_4$)	56.8	50–60	Heo et al., 2004
SS + OFMSW (50:50)	CSTR (T)	12	6.2	4		0.36	50	64	Del Borghi et al., 1999
SS + OFMSW (75:25)	CSTR (T)	11.1	F: 10	3.084		0.532	70	62.7	Sosnowski et al., 2003
	CSTR (M)	17.3					75		
SS + FW (60:40)	ASBR (T)	3.5	5.9	6.1	1.2 (CH$_4$)	0.2 (CH$_4$)	57–68	44.8	Kim et al., 2011
	ASBR (M)	3.5							
	System	7							
SS + FW (60:40)	ASBR (M)	4	3.7	3.5	0.7 (CH$_4$)	0.18 (CH$_4$)		41.6	Kim et al., 2011
	ASBR (M)	4							
	System	8							
PS + OFMSW	CSTR (T)	15	4.4	2.4	1 (CH$_4$)	0.3 (CH$_4$)		69.8	Schmit and Ellis, 2001
	CSTR (M)								
DS + FW (0.9:1)	CSTR (M)	30	12.9	5.1	1.79 (CH$_4$)	0.35 (CH$_4$)	57 ± 2	62.2 ± 1.1	Dai et al., 2013
		20		7.2	2.39 (CH$_4$)	0.332 (CH$_4$)	58 ± 1	58.1 ± 0.8	
		16		8.5	2.58 (CH$_4$)	0.303 (CH$_4$)	56 ± 1	56 ± 0.9	
		12		11.5	3.52 (CH$_4$)	0.306 (CH$_4$)	57 ± 2	55 ± 1	
		8		17.8	5.13 (CH$_4$)	0.288 (CH$_4$)	56 ± 1	52.2 ± 1	
SS + HS-OFMSW (1:1)	CSTR (M)	20	F: 5.4–6.1	2.11	1.043	0.316	64	41.5	Borowski, 2015
SS + HS-OFMSW (1:1)	CSTR (T)	20	F: 4.7–5.9	1.7	0.622	0.224	61	41	Borowski, 2015
OFMSW + PS + WAS	CSTR (M)	20	F: 7	3.1	0.96	0.29	54 ± 4	53	Griffin et al., 1998

Abbreviations: ASBR = anaerobic sequencing batch reactor; DS = dewatered sludge; F = feed; FW = food waste; GPR = gas production rate; HS-OFMSW = hydromechanically sorted organic fraction of municipal solid waste; M = mesophilic; PS = primary sludge; PW = press water; SS = sewage sludge; T = thermophilic.

TABLE 4.8 Pilot- and Full-Scale Anaerobic Codigestion Processes of Sewage Sludge and Municipal Solid Waste

Feedstock (VS Basis Unless Stated Otherwise)	Process Configuration Reactors	Operational Parameters			Process Performance				References
		HRT (Days)	TS (%)	OLR (g VS or COD L⁻¹ Day⁻¹)	GPR (L_{biogas} L⁻¹ Day⁻¹)	SGP ($L_{biogas\,[or\,methane]}$ g⁻¹ VS_{fed})	CH_4 (%)	VS (or COD) Removal (%)	
PW + FW + WAS + IPW	CSTR (T)	21	30–35	9.1	6	0.2–0.27	55–60	44–53	Oleszkiewicz and PoggiVaraldo, 1997
SS + MS-OFMSW	CSTR (T)	11.2	25	9.7	6.2	0.5		57.9	Pavan et al., 2000a
SS + OFMSW (50:50)	CSTR (M)	14.5	5.7	2.8		0.365	67.5	57	Mata-Alvarez et al., 1990
WAS + OFMSW (1:1, wb)	CSTR (M)	15	14.2	4.8	5.28	0.42 (CH_4)	55.6	70	Liu et al., 2012
WAS + OFMSW (NR)	CSTR (M)	23.5		1.6	0.53	0.21 (CH_4)	60.4	39.5	Cavinato et al., 2013
WAS + OFMSW (NR)	CSTR (T)	21.6		2.21	1.24	0.35 (CH_4)	62.2	47.2	Cavinato et al., 2013
SS + FVW (82.3:17.7, wb)	CSTR (M)	20	F: 6.4			0.57 (kg raw waste)			Edelmann et al., 2000
WAS + OFMSW (84.3:15.7)	CSTR (M)	20	1.8	1.21	0.32	0.17 (CH_4)	65		Bolzonella et al., 2006a
		22	3	0.78	0.34	0.28 (CH_4)	64		
WAS + OFMSW (59:41) SS + OFMSW (80:20)	CSTR (M) CSTR (M)	10 10		1 (VSS)	0.67	0.6–0.89 (g VSS)		81	Zupančič et al., 2008

Abbreviations: F = feed; FVW = fruit and vegetable waste; FW = food waste; GPR = gas production rate; IPW = industrial potato waste; M = mesophilic; NR = not reported; PW = press water; SS = sewage sludge; T = thermophilic.

in the literature, which are about 100 mg free acids L^{-1} and 3000 mg NH_4^+ L^{-1}. The optimum operating conditions were found to be an HRT of 13 days and a mixture of 50:50 in terms of the buffer capacity of the digester and the effluent VS concentration. The VS removal efficiency, biogas production rate (GPR), and specific methane production in this condition achieved 56.8%, 1.24 m^3 m^{-3} day^{-1}, and 0.321 m^3 CH_4 kg^{-1} VS fed with an OLR of 2.43 kg VS m^{-3} day^{-1}. It was speculated that the digestion of other types of organic waste with WAS could result in a synergistic enhancement. Recently, in Yun et al.'s work (2015), a slight amount of food waste was added to WAS such that the change of the substrate concentration and the C/N ratio could be neglected. The authors showed that as the amount of food waste input increased, the total amount of CH_4 produced and CH_4 yield increased. The calculation proved that at least 30% of the increased amount of CH_4 produced was derived from WAS, clearly signaling a synergistic enhancement. Measurements of the hydrolytic extracellular protease activity and ammonia concentration supported the finding of synergism in that the addition of easily biodegradable organics to WAS facilitated the degradation of protein, a major constituent of WAS.

Dai et al. (2013) reported possible inhibition by ammonia from sewage sludge and sodium from food waste, but the stability of the high-solids digester improved when both substrates were codigested due to the dilution effect. In CSTRs with a high solids content (>10%–15%), mixing at laboratory scale is difficult, and researchers prefer to mix intermittently. In some studies, mixing is provided right after feeding and just before effluent withdrawal (Zhang et al., 2013). In other studies, mixing is semicontinuous. For instance, Dai et al. (2013) used a helix-type mixer and the rotation speed was one rotation per minute for 10 minutes, followed by a 10-minute break.

Interestingly, there are few cases where OFMSW was added as a cosubstrate to an existing sewage sludge anaerobic digester. These digesters were designed typically for 3%–4% TS content for PS and WAS at low OLR, and therefore only a small increase in OLR was reported. Zupančič et al. (2008) reported an increase from 0.8 to 1 kg VSS m^{-3} day^{-1} when OFMSW was added to a conventional WAS digester, resulting in an 80% increase in biogas production. Edelmann et al. (2000) increased the OLR by 20% by adding size-reduced SC-OFMSW, which resulted in a 25% increase in biogas production. Similarly, Park et al. (2011) reported an increase in biogas production of 8%–17% and the stable full-scale operation of two digesters in series when fruit and vegetable waste were added. However, fruit and vegetable residues were present in the final dewatered sludge, suggesting the possibility of short circuiting in the process, probably due to the fact that the digester was not designed for this type of waste and that no pretreatment steps, such as size reduction and mixing with SS, were applied. Bolzonella et al. (2006a) reported two successful implementations of WAS + OFMSW codigestion with 50% and 240% increases in biogas production when 3–5 tons of SS-OFMSW and 8–9 tons of SC-OFMSW were added in a 100,000 population equivalent (PE) plant in Viareggio (Italy) and a 70,000 PE plant in Treviso (Italy), respectively. The operation of the Treviso digester at thermophilic temperatures improved the biogas production by 40% (Cavinato et al., 2013). These full-scale retrofitting experiences are to be distinguished from other full-scale plants designed explicitly for OFMSW. The plant normally operates at higher OLR and requires specific pretreatment units that are not found in existing WWTPs.

4.3.3 Codigestion of Sewage Sludge and Fats, Oils, and Grease

Table 4.9 lists the successful case studies of sewage sludge and FOG codigestion. FOG is an interesting substrate because of the high BMP in the range of 0.7–1.1 m^3 CH_4 kg^{-1} VS, but it should be dosed very carefully because of inhibition by LCFAs. Other issues includes clogging in the liquid or gas systems, foaming, floating biomass (Pereira et al., 2003), or increased amount of H_2S in the biogas (Pastor et al., 2013). The anaerobic codigestion of sewage sludge with FOG was almost all done in CSTRs, and methane production was always higher than that of sewage sludge monodigestion (between a 285% and 410% increase). However, this was achieved at a low OLR (about 2–3 kg VS m^{-3} day^{-1}), which indicates that FOG are difficult to treat, and that an acclimatized biomass is required. From the successful case studies, it seems that the FOG organic loading rate (OLR_{FOG}) of 2.5 kg VS_{FOG} m^{-3} day^{-1} is a critical threshold. In addition, it seems that the OLR_{FOG} limit was lower when raw sewage sludge, instead of WAS, was codigested. This is probably due to the higher lipid content of raw sewage than that of WAS (Mata-Alvarez et al., 2014). However, some failures were reported at OLR_{FOG} values as low as 0.48 kg VS_{FOG} m^{-3} day^{-1} (Razaviarani et al., 2013a). Successful operation was reported at 2–3 kg VS m^{-3} day^{-1} and 20–25 days' HRT and resulted in a wide range of methane yield (0.37–0.75) and a VS removal of usually lower than 60%. Methane percentages in biogas are usually very high, that is, near 70%. Some studies used FOG produced on the WWTP site, whereas others used industrial FOG, which can explain the different performance in the literature. Other important factors include the type of inoculum, characteristics of feedstock, operational conditions, and adaptation period.

4.3.4 Codigestion of Animal Manures and Agro-Industrial Waste

In the last 10–20 years, there has been a great interest in codigestion. This is because the C/N ratio of feedstocks is critically important to facilitate the conversion of lignocellulosic biomass to CH_4 (Wu et al., 2010). Dedicated energy crops are often rich in carbohydrates but low in nitrogen (Giuliano et al., 2013b; Ye et al., 2013). Thus, the monodigestion of energy crops alone may result in significantly low CH_4 yield (i.e., biogas high in CO_2 but low in CH_4 content) if an optimal C/N ratio of 20–30 is not properly maintained.

For instance, Wu et al. (2010) reported the best codigestion performance when a mixture between pig manure and cereal straws had a C/N of 20. Panichnumsin et al. (2010), who codigested cassava pulp and pig manure, reported a maximum methane yield when the feedstock contained a C/N ratio of 33, while Zhang et al. (2013) found an optimum C/N ratio of 16 when treating cow manure and OFMSW. Moreover, it has been shown that optimum values for the C/N ratio are within the range of 20–70 (Astals et al., 2012). The number of codigestion studies at the laboratory scale was echoed by an increase in codigestion plants at the industrial scale.

Another significant demerit of monodigestion of energy crops is the lack of essential trace elements, such as iron, cobalt, nickel, molybdenum, selenium, and tungsten.

TABLE 4.9 Laboratory- and Bench-Scale Anaerobic Codigestion Processes with Sewage Sludge and FOG

Feedstock (VS Basis Unless Stated Otherwise)	Process Configuration Reactors	Operational Parameters			Process Performance				References
		HRT (Days)	TS (%)	OLR (g VS or COD L⁻¹ Day⁻¹)	GPR (L$_{biogas}$ L⁻¹ Day⁻¹)	SGP (L$_{biogas\,[or\,methane]}$ g⁻¹ VS$_{feed}$)	CH$_4$ (%)	VS (or COD) Removal (%)	
SS + FOG (40:60)	CSTR (M)	15	F: 6.2	3.5 (unstable at 4.1)	3–4	0.49	70	59	Noutsopoulos et al., 2013
SS + FOG (40:60, COD)	CSTR (M)			11 (COD)	3.4 (CH$_4$)	0.31 (g COD)		65	Alanya et al., 2013
WAS/FOG (48:52)	CSTR (M)	25		1.2	0.972	0.55	69	44	Girault et al., 2012
SS + FOG (77:23)	CSTR (M)	20		1.6	0.8	0.37	70	52	Silvestre et al., 2011
WAS + FOG (36:64) failed at 74%	CSTR (M)	15		2.34	1.04	0.598 (CH$_4$)	66.8	57.1	Wan et al., 2011
SS + FOG (99.8:0.2, vb)	CSTR (M)	30	2	0.77	0.35	0.3 (CH$_4$)	65		Martinez et al., 2012
SS + used oil (80.6:19.4)	CSTR (M)	30	3.3	0.91	0.64	0.47 (CH$_4$)	65	41	Pastor et al., 2013
WAS + FOG (34.5:65.5)	CSTR (M)	20	5	2.16	2.28	0.75 (CH$_4$)	70		Wang et al., 2013
PS + WAS + FOG (21.1:31.3:47.6)	CSTR (M)	12		4.35	1.05	0.473 (CH$_4$)	65.8	45	Kabouris et al., 2009
PS + WAS + FOG (35:35:30)	CSTR (M)	13		2.4		0.344 (CH$_4$)	69		Davidsson et al., 2008
SS + FOG (54:46)	CSTR (M)	16		3.46	1.5	0.463 (CH$_4$)	62	70–72	Luostarinen et al., 2009
SS + SHW (95:5, vb)	CSTR (M)	22.5		2.68	1.67 (CH$_4$)	0.62 (CH$_4$)		63.7	Pitk et al., 2013
SS + SHW (7:1, wb)	CSTR (M)	20 (failed at 14)	6.3	2.8	1.575	0.43	65		Luste and Luostarinen, 2010
SS + GLY (99:1, vb)	CSTR (M)	23–25			0.78 (CH$_4$)				Fountoulakis et al., 2010
SS + FOG (77:23)	CSTR (M)	20		1.58	1.53	0.63	66	56	Razaviarani et al., 2013a

Abbreviations: F = feed; FW = food waste; GLY = glycerol; GPR = gas production rate; M = mesophilic; PS = primary sludge; SHW = slaughterhouse waste; SS = sewage sludge; T = thermophilic; vb = volume basis; wb = weight basis.

These metals, although present in miniscule quantities, are considered vital for sustaining methanogens. Thus, the supplementation of nutrients and trace elements enhances CH_4 yields, in addition to improving digester stability. For example, the CH_4 yield from Napier grass reportedly increased by 40% when nickel, cobalt, molybdenum, and selenium were added to the reactor (Demirel and Scherer, 2011). The monodigestion of conventional AD substrates, such as animal waste, is also not recommended, as it can result in digester instability caused by ammonia toxicity from the rapid degradation of organic nitrogen, such as urea and protein (Abouelenien et al., 2014). Thus, the codigestion of carbohydrate-rich lignocellulosic biomass with nitrogen-rich animal waste has significant implications in maintaining an optimal C/N ratio for commercial CH_4 production with renewable feedstocks (Giuliano et al., 2013). Several studies to date have demonstrated the successful anaerobic codigestion of livestock wastes and lignocelluloses, as illustrated in Tables 4.10 and 4.11.

Ye et al. (2013) reported the codigestion of rice straw and swine manure in a series of batch experiments. The CH_4 yield increased by 71% compared with the monodigestion of rice straw when the quantity of swine manure to rice straw was adjusted to 2:1 (on a VS basis). The C/N ratios of the cosubstrates (i.e., mixture of swine manure and rice straw) and monosubstrate (i.e., rice straw) were 21.7 and 47, respectively. In an earlier study, a C/N ratio of 20 resulted in the highest biogas yields during the codigestion of swine manure and three lignocellulosic substrates, namely, wheat straw, corn stalk, and oat straw (Wu et al., 2010). The volume of biogas produced reportedly increased by 11-, 8-, and 6-fold when compared with AD of swine manure as the control, respectively. Chicken manure is rich in organic nitrogen, compared with most other animal wastes. Consequently, during monodigestion, AD systems are often prone to suffer from ammonia toxicity. Li et al. (2014) investigated the codigestion of chicken manure and corn stover using batch and CSTRs, where the C/N ratio was adjusted to 20. The authors found that the CH_4 yield obtained from their batch experiment achieved 62% of the calculated theoretical yield. The CSTR was operated at an OLR of 4 kg VS m^{-3} day^{-1} with stable performance without VFA accumulation. This enhancement in biogas production strongly supports the importance of an optimal C/N ratio for better CH_4 yields, but other parameters, such as inhibitory or toxic compounds, heavy metals, macro- and micronutrients, pH, alkalinity, and biodegradable and solid contents, should not be overlooked.

A recent review by Mata-Alvarez et al. (2014) reported that between 2010 and 2013, the most frequent main substrates were animal manures (mostly cow and pig manures) (54%), sewage sludge (22%), and OFMSW (11%). At the same time, the most used cosubstrates were industrial waste (41%), agricultural waste (23%), and municipal waste (20%). Industrial waste included crude glycerol, cheese whey, and OMW. Manures are rich in nitrogen and have low C/N ratios and a high buffer capacity. Therefore, good cosubstrates for manures are wastes made of easily biodegradable compounds with high C/N ratios and poor buffering capacity. The codigestion will result in stable pH due to the buffer capacity of manure, and will provide enough nitrogen for bacterial metabolism. Furthermore, ammonia inhibition is avoided due to dilution by the cosubstrate (Astals et al., 2012; Wang et al., 2012).

TABLE 4.10 Laboratory- and Bench-Scale Anaerobic Codigestion Processes with Agricultural Residues and Manures or Municipal Solid Waste

Feedstock (VS Basis Unless Stated Otherwise)	Process Configuration Reactors	Operational Parameters			Process Performance				References
		HRT (Days)	TS (%)	OLR (g VS or COD L^{-1} Day^{-1})	GPR (L$_{biogas\ or\ (CH4)}$ L^{-1} Day^{-1})	SGP (L$_{biogas\ [or\ methane]}$ g^{-1} VS$_{fed}$)	CH$_4$ (%)	VS (or COD) Removal (%)	
cM + CS (1:1.4)	CSTR (M)	22.5	12	4	0.895 ± 0.023	0.223 (CH$_4$)	50	55 ± 4	Li et al., 2014
CM + CW (1:1, ww)	CSTR (M)	42		2.65	0.468	0.343 (CH$_4$)	55	82 (COD)	Comino et al., 2012
CM + sugar beet BP (1:1)	CSTR (T)	20		6.75	0.63–0.9	0.28 (CH$_4$)			Fang et al., 2011a
CM + GLY (95:5, wb)	CSTR (M)			2.6		0.82			Robra et al., 2010
PM + GLY (96:4, wb)	CSTR (M)	20	F: 4.7	1.9	1.4	0.78		74.1	Astals et al., 2012
CM + OMW (80:20, wb)	CSTR (M) CSTR (M) System	3 16 19	F: 6.8	3.63 (COD)	0.77 1.3–2 0.91 (CH$_4$)	0.25 (COD)	29 67–71	34.2	Dareioti et al., 2010a
CM + OMW (50:50, wb)	CSTR (T)	13		7.8 (COD)	1.25	0.16			Angelidaki and Ahring, 1997
OMW + PW (83:17, wb)	AF (M)	6–7		4.5–5.7 (COD)	1.7–2.1	0.31–0.34 (COD)	65–75	70–80 (COD)	Marques, 2001
CM + CW (50:50, wb)	CSTR (M) CSTR (M)	5 20	F: 3.6	1.7 (COD)	0.51	0.26	60	83 (SCOD)	Bertin et al., 2013
CM + GLY (94:6, db)	CSTR (T)	20	F: 7.2	2.91	1.9 (CH$_4$)	0.21 (CH$_4$)	63–67	80.7 (COD)	Castrillón et al., 2013
CM + GLY (94:6, db)	IBR (T)	18	F: 9	3.71	2 (CH$_4$)	0.59 (CH$_4$)	65	89.6	Castrillón et al., 2013
PM + winery ww (60:40, COD)	CSTR (M)	12	F: 0.65	0.85 (COD)	0.127	0.11 (COD)	64.4	53 (VSS)	Riaño et al., 2011

(Continued)

TABLE 4.10 (CONTINUED) Laboratory- and Bench-Scale Anaerobic Codigestion Processes with Agricultural Residues and Manures or Municipal Solid Waste

Feedstock (VS Basis Unless Stated Otherwise)	Process Configuration	Operational Parameters			Process Performance				References
	Reactors	HRT (Days)	TS (%)	OLR (g VS or COD L⁻¹ Day⁻¹)	GPR (L biogas or (CH4) L⁻¹ Day⁻¹)	SGP (L biogas [or methane] g⁻¹ VS fed)	CH₄ (%)	VS (or COD) Removal (%)	
PM + seaweed (52:48, wb)	CSTR (M)	30	F: 12.2	1.6	0.3 (CH₄)	0.126	53	32–34	Peu et al., 2011
PM + GLY (20:80, COD)	UASB (NR)	4 3 1	F: 0.6–0.8	1.3 (COD) 1.7 (COD) 5.4 (COD)		0.24 (COD) 0.15 (COD) 0.10 (COD)	62	80	Nuchdang and Phalakornkule, 2012
CM + CW (1:2, wb)	CSTR (M)	5 10 20	10		1.51 1.21 0.81		60 60 60	25 49 30	Kavacik and Topaloglu, 2010
CM + PM + ww (35:35:30, wb)	CSTR (M)	21	8	1.9		0.21		65–73	Zarkadas and Pilidis, 2011
CM + PM + ww (25:35:40, wb)	CSTR (T)	21	7	1.76		0.28		70–77	Zarkadas and Pilidis, 2011
CM + FW (1:2)	CSTR (M)			12 15 18		0.388 (CH₄) 0.317 (CH₄) 0.139 (CH₄)	62.3 60.2 39.7		Zhang et al., 2013
OFMSW + CM (1:1.4)	CSTR (T)	14–18	F: 35.6	3.34		0.63–0.71	59.7	69–74	Hartmann and Ahring, 2005

Abbreviations: AF = anaerobic filter; BP = by-products; cM = chicken manure; CM = cow manure; CS = corn stover; CW = cheese whey; db = dry weight basis; EC = energy crops; F = feed; FW = food waste; GLY = glycerol; GPR = gas production rate; IBR = induce bed reactor; M = mesophilic; PM = pig manure; PS = primary sludge; PW = piggery wastewater; S = silage; SCOD = soluble chemical oxygen demand; SHW = slaughterhouse waste; SS = sewage sludge; T = thermophilic; vb = volume basis; wb = weight basis; ww = wastewaters.

TABLE 4.11 Pilot- and Full-Scale Anaerobic Codigestion Processes with Agricultural Residues and Manures or Municipal Solid Waste

Feedstock (VS Basis Unless Stated Otherwise)	Process Configuration Reactors	Operational Parameters HRT (Days)	TS (%)	OLR (g VS or COD L⁻¹ Day⁻¹)	Process Performance GPR (L_{biogas} L⁻¹ Day⁻¹)	SGP ($L_{biogas \; [or \; methane]}$ g⁻¹ VS_{fed})	CH₄ (%)	VS (or COD) Removal (%)	References
CM + OMW (3:1 wb)	CSTR (M)	21.4	11	5.5 (COD)	0.687	0.18	62.8	53.4	Goberna et al., 2010
CM + OMW (3:1 wb)	CSTR (T)	21.4	11	5.5 (COD)	0.781	0.21	62	54	Goberna et al., 2010
CM + S (4:1, wb)	CSTR (100L, M)	43		5.15 (unstable at 7.8)	0.762	0.237–0.249 (CH₄)	52		Comino et al., 2010
PM + waste sardine oil (1.4:1.7–4.2)	CSTR (M)	16	F: 2.2	1.6–5.2 (COD)	2.3	0.43–0.5	70	90 (COD)	Ferreira et al., 2012
OFMSW + CM	LB AF	Batch (141)			0.1	0.1	73	78	Macias-Corral et al., 2008
OFMSW + FW + CM	BTA	8–14		3–5.4		0.27–0.34		47–64	Kübler et al., 2000
CM + FVW (27:73, wb)	CSTR (47°C)	34.5	10–12	5.7	2.9	0.32	59		Cavinato et al., 2010
CM + FVW (27:73, wb)	CSTR (T)	34.2	10–12	4.7	2.7	0.38	62		Cavinato et al., 2010
CM + EC + FVW (50:25:25)	CSTR (M)	60		2	0.98	0.25	54		Giuliano et al., 2013a
		30		4	2	0.25	52		
CM + EC + FVW (50:25:25)	CSTR (T)	60		2	1.13	0.29	53		Giuliano et al., 2013a
		30		4	2.17	0.28	54		
OFMSW + PS + WAS	CSTR (M)	20	F: 7	3.1	0.96	0.29	54 ± 4	53	Griffin et al., 1998

Abbreviations: AF = anaerobic filter; BTA = refers to process described in Section 4.4.5.4; CM = cow manure; EC = energy crops; F = feed; FVW = fruit and vegetable waste; FW = food waste; GPR = gas production rate; LB = leach bed; M = mesophilic; PM = pig manure; PS = primary sludge; S = silage; SHW = slaughterhouse waste; T = thermophilic; wb = weight basis.

Most studies used a conventional CSTR at an OLR below 6–7 kg VS m^{-3} day^{-1}, as shown in Table 4.11, but there is room for improvement, as demonstrated by Castrillón et al. (2013), who used an induced bed reactor, which allowed a higher OLR and organic matter removal than conventional CSTRs when codigesting cow manure and glycerol (Table 4.10).

4.4 Industrial-Scale Anaerobic Treatment of Solid Waste

4.4.1 Introduction

Anaerobic digesters for wastewater have considerably improved the performance in the last 20–30 years with a twofold increase in relative capacity: early-stage CSTRs could treat wastewater at <1 kg COD m^{-3} day^{-1}, while the most advanced internal recirculation reactor can now treat up to 40 kg COD m^{-3} day^{-1} (Luo et al., 2014). Anaerobic processes for solid waste use OLRs in terms of VS, instead of COD, and 1 g of VS is not necessarily equal to 1 g of COD. Nevertheless, the improvement in solid waste treatment has not followed the same trend in the last 20–30 years; the typical OLRs were around 4 kg VS m^{-3} day^{-1}, while the maximum threshold is about 15 kg VS m^{-3} day^{-1}. This is because of the long time required for the hydrolysis of solid waste, and also because COD concentrations can sometimes be higher than in wastewater.

The classification of anaerobic digesters treating solid wastes has been based on two parameters: the number of stages and the concentration (% TS), because these parameters have a great impact on the cost, performance, and reliability of the digestion process. One of the main challenges is the lack of a good digester for handling high-solids feedstock, such as lignocellulosic biomass. A significant part of knowledge of the current anaerobic digester design came from the WWTP and was mainly developed for handling low-solids feedstock (Sawatdeenarunat et al., 2015). For AD of solid waste, an important parameter is the solid residence time, which determines how long solids (VS or substrate) remain in the digester. This is determined in batch assay and is typically in the range of 20–40 days. The SRT should be long enough to allow for hydrolysis, but not too long, as solids would accumulate in the digester. This is largely dependent on the lignin content of solid waste. In other words, the SRT should be long enough to allow for a good digestate quality and to maintain a workable TS content in the digester.

Operating such a digester for digesting lignocellulosic biomass is energy-intensive, especially for mixing. Besides the digester, the process must include the pretreatment units and a post-treatment unit to handle the solid residue, which is usually rich in cellulose and lignin. The liquid effluent is usually rich in ammonia and is usually discharged to a local WWTP. According to Vandevivere et al. (2003), the large majority (about 90%) of industrial applications use a one-stage process, and these are evenly split between dry systems, where the wastes are digested as received, and wet systems, where the wastes are slurried with water to circa 12% TS. A review of current processes is presented below.

4.4.2 Batch Systems

4.4.2.1 Introduction
In a batch system, the dry waste (30%–40%) is allowed to go through all degradation steps sequentially in the digester. Very often, the leachate is recirculated in order to disperse the inoculant, nutrients, and acids. Batch reactors are simple and cheap and require a small quantity of water. On the other hand, the OLR is generally low. In addition, they require a surface comparable to that of aerobic composting, and there is a risk of explosion during emptying of the reactors. The three basic designs of batch digesters are depicted in Figure 4.1, and are commented on the following sections.

4.4.2.2 Biocel Process
The Biocel process is based on the single-stage batch design (Figure 4.1a). From a technical viewpoint, the perforated floor often becomes clogged, which prevents the leaching process. To alleviate that, the thickness of the fermenting wastes is limited. Dewatered digested wastes are added (1 ton · ton^{-1} fresh wastes) in order to inoculate and dilute the fresh wastes. With regards to biological performances, the biogas yield is poor due to the channeling of the percolate, that is, the lack of uniform spreading of the leachate, which tends to flow along preferential paths. The OLR of the Biocel process at 37°C falls between 3.6 and 5.1 kg VS m^{-3} day^{-1}.

4.4.2.3 Sequential Batch Anaerobic Composting Reactor
The SEBAC process (Figure 4.1b) was extensively studied during the nineties, mainly for MSW treatment. It consists of the recycling of the leachate between new and mature waste beds in order to provide moisture, nutrients, and inocula for rapid start-up of the new bed (Chynoweth et al., 1992). The process requires a freshly filled reactor and a "mature" reactor containing anaerobically stabilized wastes. The fresh waste is moistened, and water in excess of bed saturation requirements is added. The leachate from the new reactor is recirculated to the top of the bed after flushing it through the mature bed, where methanogenesis takes place. This process is continued until the waste has degraded. The bed of anaerobically stabilized waste is used to start up another fresh waste bed, and the process is repeated.

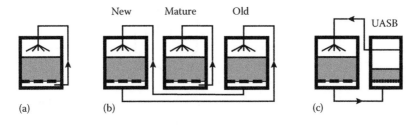

Figure 4.1 Basic design for batch digesters. (a) Single-stage, (b) sequential batch, and (c) hybrid batch-UASB. (Reproduced from Vandevivere, P. et al., in *Biomethanization of the Organic Fraction of Municipal Solid Wastes*, ed. J. Mata-Alvarez, IWA Publishing, London, 2003, pp. 111–140, Figure 5.6, p. 136. With permission.)

This configuration allows us to reach a biogas yield equivalent to 80%–90% of the maximal yield at an OLR of 3.2 kg VS m^{-3} day^{-1} (O'Keefe et al., 1992), which is considerably more than that of the Biocel process. This can be explained by not only the thermophilic temperatures applied in the SEBAC, but also because the yields obtained in the SEBAC were obtained with unsorted MSW or mechanically sorted MSW at 60% TS, compared with the poorly structured source-sorted biowaste at 40% TS in the Biocel process. In the former case, the coarser structure and lesser degree of compaction render these wastes less conducive to the channeling phenomena responsible for poor biogas yields. The main drawback of this process involves the huge amount of water required to saturate each batch of fresh wastes to provide an appropriate quantity of leachate for recirculation. Moreover, the performance is limited because the acclimatized methanogens are not retained as the reactor is emptied to load a new batch.

4.4.2.4 Hybrid Batch–UASB

Finally, the hybrid batch–UASB reactor (Figure 4.1c) is characterized by an UASB reactor as a mature reactor where the biomass forms flocks. This is very similar to the two-stage systems with biomass retention, except that the first stage is a batch configuration.

4.4.3 One-Stage Continuous Systems

4.4.3.1 Wet Complete Mix Systems

This kind of process requires that the feed is diluted to less than 15% TS (often about 1 m^3 tap water ton^{-1} solid waste), and a pulper is used to shred, homogenize, and dilute the wastes in sequential batches. A typical design of a one-stage wet process is depicted in Figure 4.2.

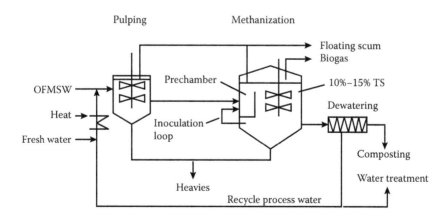

Figure 4.2 Typical design of a one-stage wet system. (Reproduced from Vandevivere, P. et al., in *Biomethanization of the Organic Fraction of Municipal Solid Wastes*, ed. J. Mata-Alvarez, IWA Publishing, London, 2003, pp. 111–140, Figure 5.2, p. 116. With permission.)

This process often results in the formation of three distinct layers; the heavies accumulate at the bottom and may damage the propellers, while a floating scum layer will hamper effective mixing. Some still fermentable matter can also be removed, together with the floating scum layer and the heavies, leading to an incomplete digestion. Additional equipment, such as screens, pulpers, drums, presses, breakers, and flotation units, are therefore necessary. These pretreatment steps inevitably incur a 15%–25% loss of VS, with a proportional drop in biogas yield. Another technical drawback is the possible short circuiting of a fraction of the feed through the reactor with a shorter retention time than the average retention time of the bulk stream. This short circuiting is, however, somewhat alleviated by injecting the feed in a prechamber within the main reactor in order to ensure a few days' retention time. Nevertheless, steam still remains necessary to pasteurize the waste (70°C for 1 hour) to get rid of pathogenic viruses and bacteria prior to land application.

When treating biowastes, inhibitory substances (ammonia) can accumulate within the completely mixed reactor. This in turn can eliminate spatial niches where bacteria may be protected from transitory high concentrations of inhibitors. However, this disadvantage can be compensated for by freshwater added to the incoming waste. Weiland (1992) stated that in substrates with a C/N ratio below 20% and 60% biodegradable VS, the ammonium concentration will reach inhibitory values (even with dilution) and a one-stage process will therefore fail in this case.

4.4.3.2 Krüger Process

Krüger (Denmark, 37,000–100,000 tpa) initially specialized in wastewater and sewage sludge treatment, and then diversified its portfolio by integrating the digestion of manures and industrial waste. The Krüger approach is to retrofit a conventional anaerobic digester used for sewage sludge into a codigestion system. The average size of Krüger's system is 62,000 tpa, which is large. This is due to the low energy content of animal manures, and large-scale systems are therefore required to be economically viable. One example of the process is the Grindsted wastewater and biowaste treatment plant (shown in Figure 4.3; Denmark), where food waste (35% on dry basis), industrial waste (20%), and sewage sludge (45%) are codigested. The plant is integrated in to the municipal WWTP and receives 39,000 tpa sewage sludge (2.5% TS), 12,000 tpa industrial organic waste (50% TS), 250 tpa food waste from supermarkets (20% TS), and 1,200 tpa food waste from household (10% TS). The process therefore requires high-quality waste with minimum contamination. Plastic contaminants may float in the digester, and manual removal may be required. Plastics may also stick to the walls of the digester and pipes, which may eventually lead to blockage (Al Saedi et al., 2001).

Household wastes are first mixed with sewage sludge in a pulper; then they go through a fine cutter and a buffer tank with constant mixing to avoid settling. Industrial wastes are fed to a heat exchanger for hygienization (70°C for 1 hour), together with the household waste and sewage sludge mixture, before entering the single-stage wet mesophilic digester operating at 20 days' HRT, pH in the range of 6.5–7, and 4.5% TS content. The effluent goes through a screw press and a buffer tank and is finally dewatered, and the wastewater goes back to the WWTP. The process

Figure 4.3 Anaerobic digester and wastes processing buildings at Grindsted plant (Denmark). (With permission from Anne Abraham, Krüger A/S, Søborg, Denmark.)

offers some flexibility regarding the ratio of the various wastes, as long as the TS content allows for pumping. There are no chemical additions for pH control, and the ratios of industrial and food waste can be lowered if the pH drops until the process regulates itself. The digester produced about 150 m^3 biogas tonne^{-1} of waste, resulting in 248 kW of electricity and 344 kW of calorific energy, which covers all the on-site heating requirements. As in most WWTPs, iron chloride is added to sewage sludge to improve settling and sedimentation. This removes phosphorus from wastewater, but also fixes sulfates in the digester so that there is no H$_2$S in the biogas. A desulfurization unit is therefore not required, which results in more positive plant economics.

4.4.3.3 Citec Process

The Citec process (Finland, capacity 3,000–230,000 tpa) can treat OFMSW, as well as other organic waste, such as slaughterhouse waste, fish waste, industrial liquid waste, sewage, and household waste. It is similar to the Krüger process, except that the digester operates at thermophilic temperatures. One of the first plant in Finland was operational in 1990, and several other successful cases were implemented in Italy, the Netherlands, and Japan. Large contaminants are removed manually or with a crane in the waste pit. The waste is transported to a trommel sieve with a 50 mm screen size. Heavy and light fractions of waste more than 50 mm are separated in an air separator. The light fraction is compacted and sent to an incinerator, while the heavy fraction passes through a ferrous metal separator before going to a landfill. Waste under 50 mm is passed through a ferrous metal separator to the next process step, which is the mix separator. The undersized particles (<50 mm) are injected with steam at a ratio of about 1:1 in order to homogenize the waste, reduce particle size, dissolve organics, lower the TS content, and raise the temperature to thermophilic range. Inerts such as stones and glass are removed from the bottom of the mix separator. Recycled process water and some fresh makeup water is used in the dilution, as well as steam. The waste stream now at 15% TS, passes through a screw press, which separates paper, plastic, and wood from the other organics. This is necessary otherwise a thick floating scum layer will form in the digester, which would impede the gas and liquid flow. It would also reduce the working volume, and ultimately, the digester would need to be shut down to remove it.

The remaining organics are now at 6% TS, a maximum size of 12 mm, and 60°C, and are pumped to the digester. The Citec digester is a single-stage wet thermophilic CSTR operating at 30–35 days' HRT. The operating pH is 7.5–7.8 and is self-regulating. Mixing is achieved by biogas pumping and impellers. To minimize risk of a short circuit, a prechamber is located within the digester to retain the waste for 1–2 days before it reaches the digester. Inoculation with some digestate is also done in the prechamber. The biogas production is usually in the range of 100–150 m^3 ton^{-1} waste anaerobically digested. The gas is 60%–70% methane. The digestate is dewatered to 30% TS using centrifuge. The thermophilic process does not kill all pathogens, and there is still some short circuiting even with the prechamber. However, the digestate can be used for landfill cover. Due to the wet nature of the process, it requires 0.97 m^3 freshwater ton^{-1} of waste, which is high. Another disadvantage is the production of wastewater, which may put an additional load on the WWTP.

4.4.3.4 Ros Roca Process

In the Ros Roca (Germany and Spain) process, the digester is a wet mesophilic CSTR operating at 20 days' HRT, but the mixing is provided by compressed biogas injection. Numerous industrial-scale plants with a capacity of more than 100,000 tpa are in operation in different European countries. The plant in Barcelona (Spain) treats 400,000 tpa of MSW. It can be applied for codigestion; for instance, the plant in Våsterås (Sweden) treats food waste from households (14,000 tpa, 30% TS), grease from restaurants and industries (4,000 tpa, 4% TS), and energy crop (ley) from local farmers (5,000 tpa, 35% TS).

Compared with other processes, there are differences in the pretreatment. The first step is a wet pretreatment in a turbomixer shown in Figure 4.4, which produces a suspension at 15% TS and separates out stones, glass, and bones by a grit system.

The suspension then flows to a screen that removes floating material like plastics and wood. It then flows through an aeration sand trap, which removes small inert particles like stones, sand, and glass, resulting in a high-quality organic suspension almost free of impurities. The suspension goes through a crushing unit that reduces the size to below 12 mm, and it is then pasteurized in batch mode in a mixed tank, ensuring no short circuiting (Figure 4.5).

Figure 4.4 Turbomixers for the pretreatment in the Ros Roca process in Våsterås (Sweden). (With permission from Per-Erik Persson, VafabMiljö Kommunalförbund, Västerås, Sweden.)

Figure 4.5 Pasteurization tanks in Våsterås (Sweden). (With permission from Per-Erik Persson, VafabMiljö Kommunalförbund, Västerås, Sweden.)

Figure 4.6 Digestate storage tank, buffer tank, and anaerobic digester in Västerås (Sweden). (With permission from Per-Erik Persson, VafabMiljö Kommunalförbund, Västerås, Sweden.)

Excess heat from the co-generation unit is used for the sanitation step. The suspension is then fed to a predigestion mixing or buffer tank, where it is held prior to AD, as shown in Figure 4.6.

As it is a wet process, one would expect that this process would generate wastewater requiring further treatment, as previously explained. However, in some circumstances, the process water can be sprayed on land, as is the case in the Västerås experience. This was made possible by the farmers wishing to improve their soil due to a lack of manure (Monson et al., 2007). So in this particular context, the choice to go for a wet process was justified and economically viable.

4.4.4 One-Stage Dry Systems

In dry systems, the TS content of the reactor is kept in the range of 20%–40%. This requires robust pumps to handle impurities such as stones, glass, or wood. The only pretreatment is the removal of particles larger than circa 40 mm in a drum screen or a shredder. In contrast with their wet counterparts, the pretreatment in dry systems is somewhat simple because the heavy inert material, such as stones and glass, need not be removed from the influent.

4.4.4.1 Dranco Process
In the Dranco (Belgium, 11,000–100,000 tpa) process, a recirculation (one part fresh waste for six to eight parts digested wastes) ensures good inoculation (Figure 4.7).

Figure 4.7 Dranco anaerobic digester (Bergheim-Siggerwiesen, Austria). (With permission from Winfried Six, OWS, Gent, Belgium.)

Mechanical pretreatment is relatively easier than with wet processes, and particles with a maximum 60 mm are allowed in the digester. The mix is also heated to 48°C–55°C by steam injection. The steam is produced from excess heat, and there is no mixer or heating element in the digester. The mix is then inserted into the top of the fermenter using a heavy-duty concrete pump.

The design can treat SS-OFMSW and MS-OFMSW at 30%–45% TS content, 20 days' HRT, and an OLR of 5–8 kg VS m^{-3} day^{-1}. The digestion takes place at thermophilic temperatures due to the steam injection in the feed. The pH is around 8, which is high in dry solids AD because of the lack of water to dissolve the carbon dioxide in headspace, but no pH adjustments are made.

Due to the high viscosity, the fermenting wastes move by gravity via plug flow inside the reactors at an HRT of 20 days. As a result, no stirrer is needed. This, in turn, may cause problems of mixing, which can lead to local overloading and acidification. The lack of mixing can result in a loss of biogas of 20%–30%. There is no wastewater produced. The design has been reported to operate successfully on several large-scale plants. For each ton of waste, there is approximately 120–170 m^3 of biogas produced and 400 kg of digestate. The electricity production is 200 kWh ton^{-1} waste and the plant uses 30%–40% of it and exports the rest. The digestate is sent to the dewatering equipment by a screw pump and is then left to stabilize for 2–3 weeks in an in-vessel composting building.

4.4.4.2 Kompogas Process
The Kompogas design (Switzerland, 5,000–100,000 tpa) is similar, except that the plug flow is horizontal in a steel vessel (Figure 4.8) and enhanced by slowly rotating impellers, which also serve to degas and resuspend heavier particles (Figure 4.9b).

Figure 4.8 Kompogas anaerobic digester at Oetwil Am See (Switzerland). (With permission from Martina Gschwandl, Hitachi Zosen Inova AG, Zurich, Switzerland.)

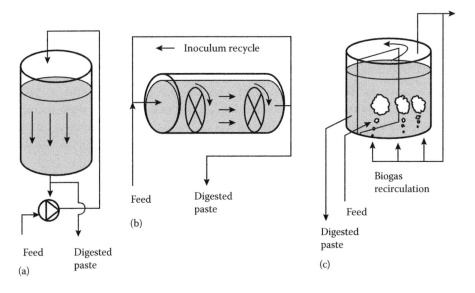

Figure 4.9 Different digester designs used in dry systems: (a) Dranco design, (b) Kompogas design, and (c) Valorga design. (Reproduced from Vandevivere, P. et al., in *Biomethanization of the Organic Fraction of Municipal Solid Wastes*, ed. J. Mata-Alvarez, IWA Publishing, London, 2003, pp. 111–140, Figure 5.3, p. 122. With permission.)

It therefore has a greater footprint than a vertical digester with a similar volume. Nevertheless, at TS percentages below 23%, the heavies sink and tend to accumulate in the reactor, while at values higher than 30%, the excessive resistance becomes a hindrance to the flow. Therefore, the TS content must carefully be controlled at 30% TS. As the plug flow is guaranteed, there is no short circuiting of the waste possible (where some of the waste remains less than the designed retention time and thus avoids being adequately treated). Because of this, there is no need for a pasteurization step because 14 days at 55°C is sufficient to produce a pathogen-free digestate. This provides an economic advantage, although the thermophilic regime requires more energy than mesophilic ones.

There are more than 25 plants in Europe using this design to treat OFMSW and biowaste (food waste and garden waste). Codigestion of 80% municipal biowaste with 20% industrial organic waste from industries, restaurants, and supermarkets is also commonly applied by Kompogas (Monson et al., 2007). All Kompogas systems operate in the thermophilic range of 55°C–60°C at 14 days' HRT, 30% TS, and maximum particle size of 50 mm. The digestate is then dewatered using a screw press, and part of the liquid is recirculated and mixed with the incoming waste. The solid fraction is composted in covered composting bays with aerated floors for 2–3 weeks.

4.4.4.3 Valorga Process

In the Valorga system (France, 28,000–268,000 tpa), after mechanical pretreatment and sorting, the waste stream (<30 mm) is mixed with recirculated digestate, process water, and steam in the mixing tank until a temperature of 42°C and a TS content of 35% are reached prior to AD. The digester operates at 30 days' HRT and 40°C, and pH is normally in the range of 7.8–8, and no chemicals are required. The VFA concentration is usually in the range of 3–10 g L^{-1}. If pH drops, the feeding regime stops until the process regulates itself. Mixing occurs by biogas injection at high pressure (5 bar) at the bottom of the cylindrical reactor (Figure 4.9c). Unfortunately, the pores often get clogged and need maintenance.

It has been reported that the Valorga and Dranco processes remain stable at ammonia concentrations equal to 3 and 2.5 g L^{-1}, respectively. It has been speculated (Vandevivere et al., 2003) that the sturdiness of the dry systems toward ammonia inhibition is due to the fact that microorganisms are better shielded against toxicants; only restricted zones are affected by temporary shock loads, leaving other pockets little exposed to transient high levels of inhibitors.

4.4.5 Two-Stage Systems

4.4.5.1 Introduction

Two-stage systems are the most complex and most expensive of all systems (Vandevivere et al., 2003). However, another benefit of the two-stage process, highlighted by Held et al. (2002), is the successful elimination of the odor in the overall output of the process, which cannot be achieved in a CSTR alone. The following processes have no biomass retention mechanism.

There are various two-stage configurations without biomass retention, including two CSTRs in series or two plug-flow reactors in series, such as the Strabag process and the Schwarting–Uhde process, which operate, respectively, in the dry–dry and wet–wet mode.

The Saschenhagen plant in Germany is an example of a two-stage wet process treating 55,000 tpa of residual MSW. After mechanical pretreatment, water is added to the waste (to 10% TS) before it enters the hydrolysis stage (4–5 days' HRT, 20°C–25°C). The slurry (<16 mm particle size) then goes to a mixing or buffer for 2 days' residence time before the mesophilic anaerobic digester for 15 days. Rainwater is harvested to adjust the TS content in the process. The Saschenhagen has two process lines with two hydrolytic reactors and two anaerobic digesters. This allows for flexibility and continuous operation even when one unit is under maintenance. Sand is a common problem in wet digestion and has been reported to be problematic at this plant because of pump and pipe abrasion. A sand separation unit (centrifuge) was added after pulping and prior to the hydrolytic tanks.

Two-phase systems designed with biomass retention devices display a larger resistance toward toxicants and inhibiting substances, such as ammonia, due to the higher biomass concentrations. Another substantial advantage is the higher OLR that can be applied in the methanogenic reactor in two-stage processes with biomass retention. Kubler and Wild (1992) reported values up to 10 and 15 kg VS m^{-3} day^{-1} for the BTA and ISKA processes, respectively.

A biomass retention device will allow the uncoupling of the HRT and the SRT. There are two basic ways to achieve this. The first way is to filter the effluent through a membrane. There are several ways to alleviate plugging of the membranes: high cross-flow velocity of the effluent in an external membrane module, high cross-flow achieved by biogas reinjection, and intermittent liquid or gas back-flushing. Another method is to provide an inert medium on which the microorganisms can grow. However, a prerequisite is that the influent must be little charged in suspended solids. The BTA and ISKA processes are two full-scale plants based on this principle.

4.4.5.2 Strabag Process

The Strabag (Germany, 15,000–300,000 tpa) wet digestion system for industrial waste products, sewage sludge, food waste, and OFMSW is comparable to the BTA design (Section 4.4.5.4). Depending on the waste input and the location conditions, plants may be designed in either single- or two-stage configurations, and can be operated in both mesophilic and thermophilic modes. The following systems proved to be reliable for the preparation of pumpable slurry out of waste, with a higher content of solid matter, which can be fed to the digester:

A robust hammer mill allows continuous crushing down of organic waste (e.g., food waste) to a defined grain size, which is necessary for hygienization by means of a screen. Strabag's hammer mill additionally allows the separation of hard particles. The pulper system or the proven pulper–trommel screen combination prepares purified slurry out of feedstock (e.g., source-separated biowaste), containing a considerable amount of impurities. This technology allows the removal of heavy solids (e.g., stones, sand, and glass), as well as the separation of larger particles, like plastic film. Finally, an impact crusher selectively crushes down individual fractions and components of the waste to a defined size. This selective treatment ensures optimum conditions for the removal of impurities after the crushing step.

After the pulper, the mashed waste stream is stored in a buffer tank, where it undergoes a first hydrolysis step (in the two-stage configuration). From the hydrolysis tank, the waste stream is fed into the AD reactor that is operated at thermophilic temperatures, an OLR of 6 kg VS m^{-3} day^{-1}, and 16–20 days' HRT. The plant in the Lisbon region (Portugal) has implemented this process to treat 40,000 tpa of food waste (30%–50%), fresh market waste (30%–45%), canteen waste (10%–20%), and garden waste (0%–10%). The same process was implemented in Wels (Austria) to treat 185,000 tpa of OFMSW, and biogas yields in the range of 88–137 m^3 ton^{-1} incoming waste were reported (Beck, 2004).

Depending on the type of input material, Strabag wet digestion processes can be run at either thermophilic or mesophilic temperatures. The characteristic feature of the process is how the digestion reactor (Figure 4.10) is fitted with a gas recirculation system using a centrally located recirculation tube. Small amounts of oxygen are added in gas recirculation to suppress hydrogen sulfide-producing bacteria, while at the same time not affecting the other microorganisms in the anaerobic consortium. The vortex created by the recirculation pumps helps to settle down the solids at the bottom, where it can be removed.

In the Lemgo plant (Germany) treating 30,000 tpa of kitchen waste, 6,000 tpa of garden waste, and 2,000 tpa of industrial organic waste, the waste is reduced in size by a screw mill (<40 mm), and undergoes a 2- to 4-day period of anaerobic hydrolysis. The

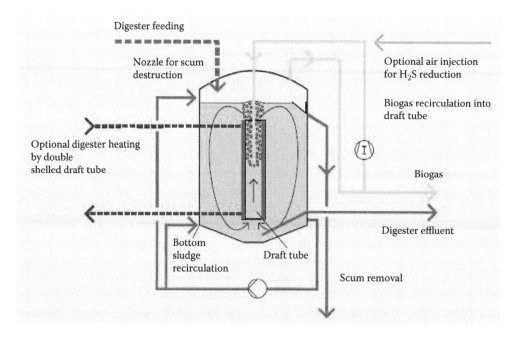

Figure 4.10 Digester in the Strabag wet digestion process. (With permission from Axel Huber, Strabag-Umwelttechnik GmbH, Düsseldorf, Germany.)

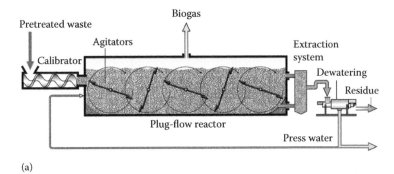

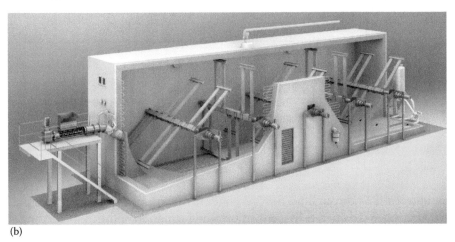

(b)

Figure 4.11 Strabag dry digestion system. (a) Side view of the process. (b) 3-D view of the reactor. (From Garaffa, C., and Gröll, K., *Biocycle*, 54(10), 40, 2013. With permission from Axel Huber, Strabag-Umwelttechnik GmbH, Düsseldorf, Germany.)

second reactor (Figure 4.11) is a horizontal vessel operating at 22% TS with transverse paddle stirrers and moving floor. Before the treated material is fed to the digester, it is chopped by a calibrator into 1.5-inch pieces. After thermophilic digestion with an HRT of about 21 days, the digestate is separated into a liquid fraction with a 20% TS content and a solid fraction having a >45% TS content. The liquid fraction is recycled to dilute the incoming fresh waste, and to moisten the compost windrows.

The Strabag dry digestion system is similar to the Kompogas system, with a few minimal design differences. For example, some of the reactor heating is done outside the digester with a short heat exchanger, but primarily, heating occurs within the digester walls using a heat exchanger. After solid separation, only the liquid fraction is recycled, which leads to a lower inoculation rate, and hence a little longer HRT (Beck, 2004).

4.4.5.3 Schwarting–Uhde Process

The Schwarting–Uhde process (Figure 4.12) treats source-sorted biowaste diluted to 12% TS in two successive vertical plug-flow reactors. The process can sustain an OLR up to 6 kg VS m^{-3} day^{-1}.

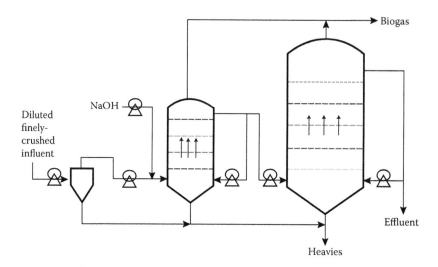

Figure 4.12 The Schwarting–Uhde process is a two-phase wet–wet plug-flow system applicable to the source-sorted fraction of the municipal solid wastes. (Reproduced from Vandevivere, P. et al., in *Biomethanization of the Organic Fraction of Municipal Solid Wastes*, ed. J. Mata-Alvarez, IWA Publishing, London, 2003, pp. 111–140, Figure 5.4, p. 128. With permission.)

The upward flow through perforated plates is imparted by pulsating pumps. This ensures mixing and complete hygienization, and formation of the thick floating scum layer is avoided. Clogging of the perforated plates, however, limits the applicability of this process.

4.4.5.4 BTA Wet–Wet Process

The BTA (Germany) process follows a two-stage wet–wet design with a biomass retention scheme in the second stage that allows a higher OLR than in one-stage processes. The core element of the BTA process is the hydropulper shown in Figure 4.13, where the preshredded feedstock is diluted to 8%–10% TS (maximum 12% TS) and chopped.

It can treat OFMSW alone or codigested with industrial waste, sewage, or agricultural waste. In Figure 4.14, the 10% TS sludge from the pasteurization unit is dewatered and the liquid is fed to the methanogenic reactor, while the solid fraction is resuspended in water to feed a hydrolyzer operating at 2–3 days' retention time. Its effluent is dewatered and the liquid fraction sent to the methanizer, which is a fixed film reactor. The solid fraction is composted, causing a 20%–30% loss of VS (Kubler and Wild, 1992). The drawbacks are the same as those of one-stage wet systems, that is, short circuiting, foaming, and sinking of heavies. The main drawback is its complexity because it needs four reactors to achieve what other systems do in a single reactor.

A variant of the BTA process was implemented in Ypres (Belgium) to treat 45,000–55,000 tpa SS-OFMSW. Blischke (2004) reported the successful operation after some initial difficulties during start-up and highlighted the flexibility of the system, which

(a)

Figure 4.13 (a) Hydropulpers used at Ypres (Belgium). (*Continued*)

is an important feature of the process due to the considerable seasonal variation; in winter months, the throughput may fall as low as 40,000 tpa. In the Ypres plant (Figure 4.15), the waste is first loaded onto a trommel screen equipped with knives to cut open the bags. Materials more than 15 cm are discarded and sent for composting. The undersize fraction is sent to the hydropulper (32 m^3) operating in batch mode (60–75 minutes) (Figure 4.13).

(b)

Figure 4.13 (Continued)   (b) BTA waste pulper used in Valorlis (Portugal). (With permission of Stephan Schulte, BTA International GmbH, Pfaffenhofen, Germany.)

Using process water, the waste is vigorously mixed at 8%–10% TS. At the end of the batch, floating and sinking materials are automatically removed, leaving a suspension of organic-rich materials. The AD takes place in two mesophilic digesters (12–15 days' HRT, 2500 m^3 each, 15 m tall) and can be operated in series or in parallel, depending on the type and amount of waste. This design provides flexibility, which is very important when waste availability, volumes, and contracts are subject to change. The digesters are mixed continuously using compressed biogas to avoid the formation of a floating layer. Biogas yields in the range of 73–109 m^3 ton^{-1} of incoming waste have been reported, depending on the type of waste (Blischke, 2004). The plant covers all its electrical and calorific requirements and exports 50% of its electrical energy (Figure 4.15).

4.4.5.5 ISKA GmbH Process
The ISKA process (Germany, 30,000–175,000 tpa) follows the same principles except that the first stage takes place in a dry microaerophilic leach bed with recirculation of the leachate. The flush water is fed to an anaerobic plug-flow reactor with a biofilm. The pulsating flow prevents plugging and channeling of the support medium, which improves mass transfer and degasification. The overall retention time is only 7 days due to the innovative mixing systems in the first reactor; percolation occurs in a large slow-rotating (1 rpm) sieve drum with 1 mm mesh openings.

The patented percolation system is designed to extract the soluble fraction from the OFMSW. The leachate is then treated in an anaerobic reactor, while the solid

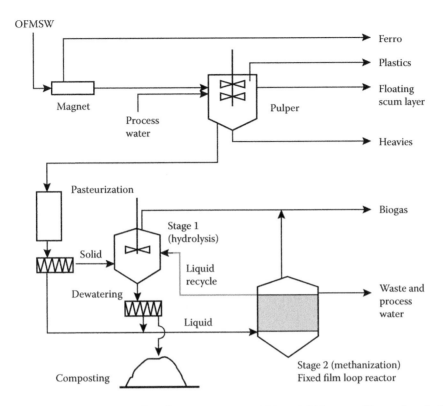

Figure 4.14 BTA process flowsheet to treat 8,000–150,000 tpa. (Reproduced from Vandevivere, P. et al., in *Biomethanization of the Organic Fraction of Municipal Solid Wastes*, ed. J. Mata-Alvarez, IWA Publishing, London, 2003, pp. 111–140, Figure 5.5, p. 132. With permission.)

Figure 4.15 Anaerobic digester at Ypres (Belgium). (With permission of Stephan Schulte, BTA International GmbH, Pfaffenhofen, Germany.)

fraction is in-vessel composted and landfilled; There are several plants in Germany and one in Australia using this technology.

The ISKA process can be flexibly engineered to meet client requirements, and the modular nature of the system facilitates this. In the ISKA process, bags containing MSW are fed to a coarse (300 mm) slow-speed shredder and then the bags are cut open. The waste is mechanically sorted: ferrous metals and aluminum are removed for recycling, and paper, cardboard, and plastics are sent to the incinerator for energy recovery. The remaining fraction enters the percolation drums where water (38°C) is added. The process water, after AD, is recirculated to the percolation tank, resulting in a water-neutral process. A small amount of fresh makeup water is required to avoid the buildup of recalcitrant and toxic substances in the process. Rainwater is used as makeup water (Figure 4.16).

There are several percolation tanks. For instance, at the Buchen plant (Germany) there are five of them operating in parallel. There are typically horizontal cylindrical vessels made of stainless steel operating at 3 days' HRT.

It leads to a COD extraction of 250 kg COD ton^{-1} waste, and the resulting slurry is easily pumped to a high-rate AD process (Monson et al., 2007). The pretreatment in these drums results in reduced time in the anaerobic digester and in the composting unit. There is a central mixer, and water is introduced from the top and removed through screens at the bottom, which also serve as aeration channels. After

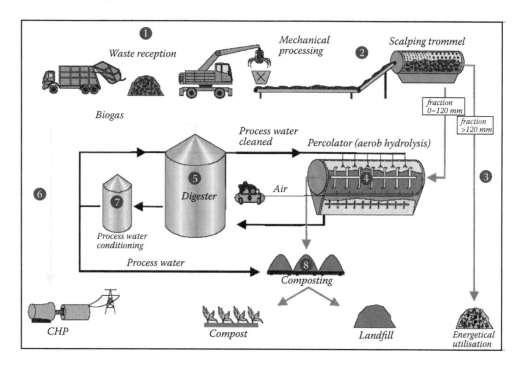

Figure 4.16 Process flow diagram of Buchen plant (Germany). (From Archer, E. et al., Mechanical-biological treatment: a guide for decision makers. Processes, policies and markets. March 2005. Published by Juniper Consultancy Services Ltd. Available at http://www.cti2000.it /Bionett/BioG-2005-007%20MBT_AnnexD0.0009%20G-J%20(Final).pdf, accessed in May 2017.)

percolation, dewatering, and degritting in a sand cyclone, the liquid phase goes to AD and the solid waste is sent to a trommel (60 mm). Large particles are sent to incineration, and small particles are sent to composting for stabilization and are ultimately landfilled.

The mesophilic digester (35°C–40°C) operates in a fixed bed- or fluidized bed-type mode, also known as a hybrid filter. In the Buchen plant in Germany, the digester is 14 m tall and the bottom zone (4 m) consists of a plastic packing material designed to retain bacteria. Above that, the reactor operates in the fluidized bed mode. Biogas is flushed for 2–3 minutes every 3 days to shake the packing material and reduce risks of blocking and channeling. Retention time in the digester is only 4 days, and pH is maintained between 6.5 and 8. pH is adjusted by manually dosing citric acid in the percolator. The plant in Buchen has five percolators and three digesters, which gives flexibility to the operators and also allows for continuous operation even during maintenance. The pumping and piping system are all integrated, which provides a big advantage. The biogas production is about 100 m³ ton⁻¹ OFMSW, which is low, probably due to the percolation system that cannot recover all the organics, therefore resulting in a loss of about 20%–30% biogas (Vandevivere et al., 2003). However, the ISKA process operates at an overall HRT of 7 days, that is, 3 days in the percolator and 4 days in the digester compared with 15–30 days in other AD systems.

4.5 Conclusions

Between 2000 and 2007, at least 35 plants treating SS-OFMSW were built while 40 were designed with an integrated pretreatment or mechanical biological treatment (MBT) to remove the nonbiodegradable fraction prior to AD. From the 35 plants treating SS-OFMSW, 12 treated only SS-OFMSW and 23 plants applied codigestion. Most codigestion substrates were organic industrial waste (10 plants), agricultural waste (4 plants), and sewage sludge (2 plants), showing clearly the trend towards codigestion in the last 15 years. From the 35 plants, half used dry AD (17 systems) and half used wet AD (18 systems). This even distribution in Europe mirrors the fact that both strategies have proven successful on a large scale. Regardless of the substrate, SS- or SC-OFMSW, Monson et al. (2007) reported that 77% applied mesophilic AD, while 23% applied thermophilic AD for treating solid waste. Thermophilic digestion of SS-OFMSW is advantageous to reduce pathogens, depending on whether the digestate is to be used as a soil conditioner.

The waste treated by mechanical pretreatment is also called centrally separated OFMSW (CS-OFMSW). From the 168 plants installed in Europe between 1980 and 2007, 48 treat CS-OFMSW and 120 treat SS-OFMSW. This underlines the flexibility of AD systems available on the market. From the 40 plants treating CS-OFMSW, 75% do not codigest with other waste. This was due to the fact that large MBT was preferred, in particular in Germany and Spain. In the remaining 25% (10 plants), the most common cosubstrate was sewage sludge. Of the 40 AD systems treating CS-OFMSW between 2000 and 2007, 29 were wet AD systems and 11 were dry AD systems. Furthermore, 90% applied mesophilic AD and 10% applied thermophilic AD. This is probably due to the fact that most plants operate in the wet mode and heating to thermophilic temperatures would be more economical for dry systems due to the lower moisture content. Monson et al. (2007) identified 50 companies that have built AD systems for solid waste, but the top 10 suppliers in terms of capacity were Valorga (15 plants), Linde (17), Kompogas (25), Ros Roca (11), BTA (13), Haase (8), OWS Dranco (14), CiTec (8), Krüger (4), and Alkane Biogas (2). Most of the AD plants were built after 2000.

References

Abouelenien, F., Namba, Y., Kosseva, M.R., Nishio, N., Nakashimada, Y. 2014. Enhancement of methane production from co-digestion of chicken manure with agricultural wastes. *Bioresour. Technol.*, 159, 80–87.

Akassou, M., Kaanane, A., Crolla, A., Kinsley, C. 2010. Statistical modelling of the impact of some polyphenols on the efficiency of anaerobic digestion and the co-digestion of the wine distillery wastewater with dairy cattle manure and cheese whey. *Water Sci. Technol.*, 62(3), 475–483.

Alanya, S., Yilmazel, Y.D., Park, C., Willis, J.L., Keaney, J., Kohl, P.M., Hunt, J.A., Duran, M. 2013. Anaerobic co-digestion of sewage sludge and primary clarifier skimmings for increased biogas production. *Water Sci. Technol.*, 67(1), 174–179.

Al Saedi, T., Holm-Nielsen, J.B., Lindberg, A., Wheeler, P. 2001. Good practice in quality management of AD residues from biogas production. http://213.229.136.11/bases/ainia _probiogas.nsf/0/70996A6A88900B70C125753F005B70AD/$FILE/IEA%20BUENAS %20PR%C3%81CTICAS%20DA.pdf

Anderson, G.K., Donnelly, T., McKeown, K.J. 1982. Identification and control of inhibition in the anaerobic treatment of industrial wastewaters. *Process Biochem.*, 17(4), 28.

Anderson, G.K., Saw, C.B. 1992. Leach-bed two-phase anaerobic digestion of municipal solid waste. Presented at International Symposium on Anaerobic Digestion of Solid Waste, Venice, April 14–17.

Anderson, G.K., Yang, G. 1992. pH control in anaerobic treatment of industrial wastewater. *J. Environ. Eng.*, 118(4), 551–567.

Angelidaki, I., Ahring, B.K. 1993. Thermophilic anaerobic digestion of livestock waste: The effect of ammonia. *Appl. Microbiol. Biotechnol.*, 38, 560–564.

Angelidaki, I., Ahring, B.K. 1994. Anaerobic thermophilic digestion of manure at different ammonia loads: Effect of temperature. *Water Res.*, 28(1), 727–731.

Angelidaki, I., Ahring, B.K. 1997. Co-digestion of olive oil mill wastewaters with manure, household waste or sewage sludge. *Biodegradation*, 8(4), 221–226.

Angelidaki, I., Ahring, B.K. 1999. Methods for increasing the biogas potential from the recalcitrant organic matter contained in manure. In *Second International Symposium on Anaerobic Digestion of Solid Wastes*, Barcelona, June 15–18, pp. 375–380.

Angelidaki, I., Chen, X.X., Cui, J.B., Kaparaju, P., Ellegaard, L. 2006. Thermophilic anaerobic digestion of source-sorted organic fraction of household municipal solid waste: Start-up procedure for continuously stirred tank reactor. *Water Res.*, 40(14), 2621–2628.

Anthonisen, A.C., Loehr, R.C., Prakasam, T.B.S., Srinath, E.G. 1976. Inhibition of nitrification by ammonia and nitrous-acid. *J. Water Pollut. Control Fed.*, 48(5), 835–852.

Archer, E., Baddeley, A., Klein, A., Schwager, J., Whiting, K. 2005. Mechanical-biological treatment: A guide for decision makers. Processes, policies and markets. Juniper Consultancy Services Ltd. Available at http://www.cti2000.it/Bionett/BioG-2005 -007%20MBT_AnnexD0.0009%20G-J%20(Final).pdf, accessed May 2017.

Astals, S., Ariso, M., Galí, A., Mata-Alvarez, J. 2011. Co-digestion of pig manure and glycerine: Experimental and modelling study. *J. Environ. Manag.*, 92(4), 1091–1096.

Astals, S., Esteban-Gutiérrez, M., Fernández-Arévalo, T., Aymerich, E., García-Heras, J.L., Mata-Alvarez, J. 2013. Anaerobic digestion of seven different sewage sludges: A biodegradability and modelling study. *Water Res.*, 47(16), 6033–6043.

Astals, S., Nolla-Ardèvol, V., Mata-Alvarez, J. 2012. Anaerobic co-digestion of pig manure and crude glycerol at mesophilic conditions: Biogas and digestate. *Bioresour. Technol.*, 110, 63–70.

Babel, S., Fukushi, K., Sitanrassamee, B. 2004. Effect of acid speciation on solid waste liquefaction in an anaerobic acid digester. *Water Res.*, 38(9), 2417–2423.

Baere, L.D. 2006. Will anaerobic digestion of solid waste survive in the future? *Water Sci. Technol.*, 53, 187–194.

Bailey, J.E., Ollis, D.F. 1986. *Biochemical Engineering Fundamentals*. 2nd ed. Singapore: McGraw-Hill Book Co.

Banks, C.J., Humphreys, P.N. 1998. The anaerobic treatment of a ligno-cellulosic substrate offering little natural pH buffering capacity. *Water Sci. Technol.*, 38(4–5), 29–35.

Baronofsky, J.J., Schreurs, W.J.A., Kashket, E.R. 1984. Uncoupling by acetic-acid limits growth of and acetogenesis by *Clostridium thermoaceticum*. *Appl. Environ. Microbiol.*, 48(6), 1134–1139.

Beccari, M., Carucci, G., Majone, M., Torrisi, L. 1999. Role of lipids and phenolic compounds in the anaerobic treatment of olive oil mill effluents. *Environ. Technol.*, 20(1), 105–110.

Beck, R.W. 2004. Anaerobic digestion feasibility study for the Bluestem Solid Waste Agency and Iowa Department of Natural Resources. Seattle, WA. https://www.scribd.com /document/51290973/bluestem (accessed July 2017).

Bertin, L., Grilli, S., Spagni, A., Fava, F. 2013. Innovative two-stage anaerobic process for effective codigestion of cheese whey and cattle manure. *Bioresour. Technol.*, 128, 779–783.

Blischke, J. 2004. Sustainability for the 21st century. Combining anaerobic digestion with enclosed tunnel composting. *Biocycle*, 45(4), 49.

Boari, G., Brunetti, A., Passino, R., Rozzi, A. 1984. Anaerobic digestion of olive oil mill wastewaters. *Agric. Wastes*, 10(3), 161–175.

Bolzonella, D., Battistoni, P., Susini, C., Cecchi, F. 2006a. Anaerobic codigestion of waste activated sludge and OFMSW: The experiences of Viareggio and Treviso plants (Italy). *Water Sci. Technol.*, 53, 203–11.

Bolzonella, D., Pavan, P., Mace, S., Cecchi, F. 2006b. Dry anaerobic digestion of differently sorted organic municipal solid waste: A full scale experience. *Water Sci. Technol.*, 53(8), 23–32.

Borja, R., Banks, C.J., Sanchez, E. 1996. Anaerobic treatment of palm oil mill effluent in a two-stage up-flow anaerobic sludge blanket (UASB) system. *J. Biotechnol.*, 45(2), 125–135.

Borja, R., Martín, A., Alonso, V., García, I., Banks, C.J. 1995a. Influence of different aerobic pretreatments on the kinetics of anaerobic digestion of olive mill wastewater. *Water Res.*, 29(2), 489–495.

Borja, R., Martín, A., Banks, C.J., Alonso, V., Chica, A. 1995b. A kinetic study of anaerobic digestion of olive mill wastewater at mesophilic and thermophilic temperatures. *Environ. Pollut.*, 88(1), 13–18.

Borowski, S. 2015. Co-digestion of the hydromechanically separated organic fraction of municipal solid waste with sewage sludge. *J. Environ. Manag.*, 147, 87–94.

Braber, K. 1995. Anaerobic digestion of municipal solid waste: A modern waste disposal option on the verge of breakthrough. *Biomass Bioenergy*, 9(1–5), 365–376.

Braun, R., Hubert, P., Meyrath, J. 1981. Ammonia toxicity in liquid piggery manure digestion. *Biotechnol. Lett.*, 3(4), 159–164.

Callaghan, F.J., Wase, D.A.J., Thayanithy, K., Forster, C.F. 1999. Codigestion of waste organic solids: Batch studies. *Bioresour. Technol.*, 67, 117–122.

Castillo, M.E.F., Cristancho, D.E., Victor Arellano, A. 2005. Study of the operational conditions for anaerobic digestion of urban solid wastes. *Waste Manag.*, 26(5), 546–556.

Castrillón, L., Fernández-Nava, Y., Ormaechea, P., Marañón, E. 2011. Optimization of biogas production from cattle manure by pre-treatment with ultrasound and co-digestion with crude glycerin. *Bioresour. Technol.*, 102(17), 7845–7849.

Castrillón, L., Fernández-Nava, Y., Ormaechea, P., Marañón, E. 2013. Methane production from cattle manure supplemented with crude glycerin from the biodiesel industry in CSTR and IBR. *Bioresour. Technol.*, 127, 312–317.

Cavinato, C., Bolzonella, D., Pavan, P., Fatone, F., Cecchi, F. 2013. Mesophilic and thermophilic anaerobic co-digestion of waste activated sludge and source sorted biowaste in pilot- and full-scale reactors. *Renewable Energy*, 55, 260–265.

Cavinato, C., Fatone, F., Bolzonella, D., Pavan, P. 2010. Thermophilic anaerobic co-digestion of cattle manure with agro-wastes and energy crops: Comparison of pilot and full scale experiences. *Bioresour. Technol.*, 101(2), 545–550.

Cecchi, F., Bolzonella, D. 2005. Full scale experiences of anaerobic digestion of the organic fraction of municipal solid waste in Italy—From collection to energy and valuable products. In *Proceedings of ADSW 2005*, Copenhagen, August 31–September 2, pp. 60–71.

Cecchi, F., Bolzonella, D., Pavan, P., Macé, S., Mata-Alvarez, J. 2011. Anaerobic digestion of the organic fraction of municipal solid waste for methane production: Research and industrial application. In *Comprehensive Biotechnology*, ed. M. Moo-Young. 2nd ed. Burlington, MA: Academic Press, pp. 463–472.

Cecchi, F., Pavan, P., Mata Alvarez, J., Bassetti, A., Cozzolino, C. 1991. Anaerobic digestion of municipal solid waste: Thermophilic vs. mesophilic performance at high solids. *Waste Manag. Res.*, 9(4), 305–315.

Cecchi, F., Vallini, G., Pavan, P., Bassetti, A. 1992. Management of macroalgae from Venice lagoon through anaerobic co-digestion and co-composting with MSW. Presented at International Symposium on Anaerobic Digestion of Solid Waste, Venice, April 14–17.

Chanakya, H.N., Borgaonkar, S., Rajan, M.G.C., Wahi, M. 1993. Two phase fermentation of whole leaf biomass to biogas. *Biomass Bioenergy*, 5(5), 359–367.

Cherubini, F. 2010. The biorefinery concept: Using biomass instead of oil for producing energy and chemicals. *Energy Convers. Manag.*, 51(7), 1412–1421.

Cho, J.K., Park, S.C., Chang, H.N. 1995. Biochemical methane potential and solid-state anaerobic-digestion of Korean food wastes. *Bioresour. Technol.*, 52(3), 245–253.

Chynoweth, D.P., Haley, P., Owens, J., Teiweira, A., Welt, B., Rich, E., Townsend, T., Choi, H.-L. 2002. Anaerobic digestion for reduction and stabilization of organic solid wastes during space missions: Laboratory studies. SAE Technical Paper 2002-01-2351. Warrendale, PA: Society of Automotive Engineers.

Chynoweth, D.P., Jerger, D.E. 1985. Anaerobic digestion of woody biomass. *Dev. Ind. Microbiol.*, 265, 235–246.

Chynoweth, D.P., Owens, J., Okeefe, D., Earle, J.F.K., Bosch, G., Legrand, R. 1992. Sequential batch anaerobic composting of the organic fraction of municipal solid-waste. *Water Science and Technology*, 25(7), 327–339.

Chynoweth, D.P., Turick, C.E., Owens, J.M., Jerger, D.E., Peck, M.W. 1993. Biochemical methane potential of biomass and waste feedstocks. *Biomass Bioenergy*, 5(1), 95–111.

Cirne, D.G., Paloumet, X., Björnsson, L., Alves, M.M., Mattiasson, B. 2007. Anaerobic digestion of lipid-rich waste—Effects of lipid concentration. *Renewable Energy*, 32(6), 965–975.

Clarkson, W.W., Xiao, W. 2000. Bench-scale anaerobic bioconversion of newsprint and office paper. *Water Sci. Technol.*, 41(3), 93–100.

Comino, E., Riggio, V.A., Rosso, M. 2012. Biogas production by anaerobic co-digestion of cattle slurry and cheese whey. *Bioresour. Technol.*, 114, 46–53.

Comino, E., Rosso, M., Riggio, V. 2010. Investigation of increasing organic loading rate in the co-digestion of energy crops and cow manure mix. *Bioresour. Technol.*, 101(9), 3013–3019.

D'Addario, E., Pappa, R., Pietrangeli, B., Valdiserri, M. 1992. The acidogenic digestion of the organic fraction of municipal solid waste for the production of liquid fuels. Presented at International Symposium on Anaerobic Digestion of Solid Waste, Venice, April 14–17.

Dai, X., Duan, N., Dong, B., Dai, L. 2013. High-solids anaerobic co-digestion of sewage sludge and food waste in comparison with mono digestions: Stability and performance. *Waste Manag.*, 33(2), 308–316.

Dareioti, M.A., Dokianakis, S.N., Stamatelatou, K., Zafiri, C., Kornaros, M. 2010a. Exploitation of olive mill wastewater and liquid cow manure for biogas production. *Waste Manag.*, 30, 1841–1848.

Dareioti, M.A., Dokianakis, S.N., Stamatelatou, K., Zafiri, C., Kornaros, M. 2010b. Exploitation of olive mill wastewater and liquid cow manure for biogas production. *Waste Manag.*, 30(10), 1841–1848.

Davidsson, A., Gruvberger, C., Christensen, T.H., Hansen, T.L., Jansen, J.L. 2007. Methane yield in source-sorted organic fraction of municipal solid waste. *Waste Manag.*, 27(3), 406–414.

Davidsson, Å., Lövstedt, C., la Cour Jansen, J., Gruvberger, C., Aspegren, H. 2008. Co-digestion of grease trap sludge and sewage sludge. *Waste Manag.*, 28(6), 986–992.

De Baere, L. 1999. Anaerobic digestion of solid waste: State-of-the-art. In *2nd International Symposium on Anaerobic Digestion of Solid Waste*, Barcelona, June 15–17, pp. 290–299.

de Bok, F.A.M., Plugge, C.M., Stams, A.J.M. 2004. Interspecies electron transfer in methanogenic propionate degrading consortia. *Water Res.*, 38(6), 1368–1375.

Del Borghi, A., Converti, A., Palazzi, E., Del Borghi, M. 1999. Hydrolysis and thermophilic anaerobic digestion of sewage sludge and organic fraction of municipal solid waste. *Bioprocess Eng.*, 20(6), 553–560.

Demirel, B., Scherer, P. 2011. Trace element requirements of agricultural biogas digesters during biological conversion of renewable biomass to methane. *Biomass Bioenergy*, 35, 992–998.

Dolfing, J., Bloemen, W.G.B.M. 1985. Activity measurements as a tool to characterize the microbial composition of methanogenic environments. *J. Microbiol. Methods*, 4(1), 1–12.

Droste, R.L. 1997. *Theory and Practice of Water and Wastewater Treatment*. New York: John Wiley & Sons.

Eastman, J.A., Ferguson, J.F. 1981. Solubilization of particulate organic carbon during the acid phase of anaerobic digestion. *J. Water Pollut. Control Fed.*, 53(3), 352–366.

Edelman, W. 2008. Digestione anaerobica di RSU separati alla fonte e altri co-substrati: Stato ed esperienze in Svizzera. In *Biogas da rifiuti solidi urbani*, ed. R. Vismara, M, Centemero. Milano: Flaccovio Publishing.

Edelman, W., Engeli, H. 2005. More than 12 years of experience with commercial anaerobic digestion of the organic fraction of municipal solid waste in Switzerland. In *Proceedings of ADSW 2005*, Copenhagen, August 31–September 2, pp. 19–26.

Edelmann, W., Engeli, H., Gradenecker, M. 2000. Co-digestion of organic solid waste and sludge from sewage treatment. *Water Sci. Technol.*, 41, 213–221.

Elango, D., Pulikesi, M., Baskaralingam, P., Ramamurthi, V., Sivanesan, S. 2007. Production of biogas from municipal solid waste with domestic sewage. *J. Hazard. Mater.*, 141(1), 301–304.

Fang, C., Boe, K., Angelidaki, I. 2011a. Anaerobic co-digestion of by-products from sugar production with cow manure. *Water Res.*, 45(11), 3473–3480.

Fang, C., Boe, K., Angelidaki, I. 2011b. Anaerobic co-digestion of desugared molasses with cow manure; focusing on sodium and potassium inhibition. *Bioresour. Technol.*, 102, 1005–1011.

Fernández-Rodríguez, J., Pérez, M., Romero, L.I. 2014. Dry thermophilic anaerobic digestion of the organic fraction of municipal solid wastes: Solid retention time optimization. *Chem. Eng. J.*, 251, 435–440.

Ferreira, L., Duarte, E., Figueiredo, D. 2012. Utilization of wasted sardine oil as co-substrate with pig slurry for biogas production—A pilot experience of decentralized industrial organic waste management in a Portuguese pig farm. *Bioresour. Technol.*, 116, 285–289.

Field, J.A., Lettinga, G. 1987. The methanogenic toxicity and anaerobic degradability of a hydrolyzable tannin. *Water Res.*, 21(3), 367–374.

Forster-Carneiro, T., Pérez, M., Romero, L.I. 2008. Thermophilic anaerobic digestion of source-sorted organic fraction of municipal solid waste. *Bioresour. Technol.*, 99(15), 6763–6770.

Fountoulakis, M.S., Dokianakis, S.N., Kornaros, M.E., Aggelis, G.G., Lyberatos, G. 2002. Removal of phenolics in olive mill wastewaters using the white-rot fungus *Pleurotus ostreatus*. *Water Res.*, 36(19), 4735–4744.

Fountoulakis, M.S., Petousi, I., Manios, T. 2010. Co-digestion of sewage sludge with glycerol to boost biogas production. *Waste Manag.*, 30(10), 1849–1853.

Gallert, C., Winter, J. 1997. Mesophilic and thermophilic anaerobic digestion of source-sorted organic wastes: Effect of ammonia on glucose degradation and methane production. *Appl. Microbiol. Biotechnol.*, 48(3), 405–410.

Garaffa, C., and Gröll, K. 2013. Anaerobic digestion of compostable bags. *Biocycle*, 54(10), 40.

Gellens, V., Boelens, J., Verstraete, W. 1995. Source separation, selective collection and in-reactor digestion of biowaste. *Antonie Van Leeuwenhoek*, 67(1), 79–89.

Girault, R., Bridoux, G., Nauleau, F., Poullain, C., Buffet, J., Peu, P., Sadowski, A.G., Béline, F. 2012. Anaerobic co-digestion of waste activated sludge and greasy sludge from flotation process: Batch versus CSTR experiments to investigate optimal design. *Bioresour. Technol.*, 105, 1–8.

Giuliano, A., Bolzonella, D., Pavan, P., Cavinato, C., Cecchi, F. 2013. Co-digestion of livestock effluents, energy crops and agro-waste: Feeding and process optimization in mesophilic and thermophilic conditions. *Bioresour. Technol.*, 128, 612–618.

Goberna, M., Schoen, M.A., Sperl, D., Wett, B., Insam, H. 2010. Mesophilic and thermophilic co-fermentation of cattle excreta and olive mill wastes in pilot anaerobic digesters. *Biomass Bioenergy*, 34(3), 340–346.

Golueke, C.G., Oswald, W.J., Gotaas, H.B. 1957. Anaerobic digestion of algae. *Appl. Microbiol.*, 5, 47–55.

Gomez, X., Cuetos, M.J., Cara, J., Moran, A., Garcia, A.I. 2005. Anaerobic co-digestion of primary sludge and the fruit and vegetable fraction of the municipal solid wastes: Conditions for mixing and evaluation of the organic loading rate. *Renewable Energy*, 31(12), 2017–2024.

Gonçalves, M.R., Costa, J.C., Marques, I.P., Alves, M.M. 2012. Strategies for lipids and phenolics degradation in the anaerobic treatment of olive mill wastewater. *Water Res.*, 46(6), 1684–1692.

González-Fernández, C., Molinuevo-Salces, B., García-González, M.C. 2011. Evaluation of anaerobic codigestion of microalgal biomass and swine manure via response surface methodology. *Appl. Energy*, 88(10), 3448–3453.

Griffin, M.E., McMahon, K.D., Mackie, R.I., Raskin, L. 1998. Methanogenic population dynamics during start-up of anaerobic digesters treating municipal solid waste and biosolids. *Biotechnol. Bioeng.*, 57(3), 342–355.

Gujer, W., Zehnder, A.J.B. 1983. Conversion processes in anaerobic digestion. *Water Sci. Technol.*, 15(8–9), 127–167.

Hajarnis, S.R., Ranade, D.R. 1994. Effect of propionate toxicity on some methanogens at different pH values and in combination with butyrate. In *7th International Symposium on Anaerobic Digestion*, Capetown, South Africa, p. 46.

Hamdi, M. 1992. Toxicity and biodegradability of olive mill wastewaters in batch anaerobic digestion. *Appl. Biochem. Biotechnol.*, 37(2), 155–163.

Han, S.K., Shin, H.S. 2004. Performance of an innovative two-stage process converting food waste to hydrogen and methane. *J. Air Waste Manag. Assoc.*, 54(2), 242–249.

Hanaki, K., Matsuo, T., Nagase, M. 1981a. Mechanism of inhibition caused by long chain fatty acids in anaerobic digestion process. *Biotechnol. Bioeng.*, 23, 1591–1610.

Hanaki, K., Matsuo, T., Nagase, M. 1981b. Mechanism of inhibition caused by long-chain fatty acids in anaerobic digestion process. *Biotechnol. Bioeng.*, 23(7), 1591–1610.

Hansen, K.H., Angelidaki, I., Ahring, B.K. 1998. Anaerobic digestion of swine manure: Inhibition by ammonia. *Water Res.*, 32(1), 5–12.

Hansen, K.H., Angelidaki, I., Ahring, B.K. 1999. Improving thermophilic anaerobic digestion of swine manure. *Water Res.*, 33(8), 1905–1810.

Hartmann, H., Ahring, B.K. 2005. Anaerobic digestion of the organic fraction of municipal solid waste: Influence of co-digestion with manure. *Water Res.*, 39(8), 1543–1552.

Hartmann, H., Angelidaki, I., Ahring, B.K. 1999. Increase of anaerobic degradation of particulate organic matter in full-scale biogas plants by mechanical maceration. In *Second International Symposium on Anaerobic Digestion of Solid Wastes*, Barcelona, June 15–18, pp. 129–136.

He, R., Shen, D.S., Wang, J.Q., He, Y.H., Zhu, Y.M. 2005. Biological degradation of MSW in a methanogenic reactor using treated leachate recirculation. *Process Biochem.*, 40(12), 3660–3666.

Held, C., Wellacher, M., Robra, K.-H., Gubitz, G.M. 2002. Two-stage anaerobic fermentation of organic waste in CSTR and UFAF-reactors. *Bioresour. Technol.*, 81(1), 19–24.

Hendriks, A.T.W.M., Zeeman, G. 2009. Pretreatments to enhance the digestibility of lignocellulosic biomass. *Bioresour. Technol.*, 100(1), 10–18.

Heo, N.H., Park, S.C., Kang, H. 2004. Effects of mixture ratio and hydraulic retention time on single-stage anaerobic co-digestion of food waste and waste activated sludge. *J. Environ. Sci. Health A Tox. Hazard. Subst. Environ. Eng.*, 39(7), 1739–1756.

Hill, D.T., Cobb, S.A., Bolte, J.P. 1987. Using volatile fatty-acid relationships to predict anaerobic digester failure. *Trans. ASAE*, 30(2), 496–501.

Hills, D.J., Nakano, K. 1984. Effects of particle size on anaerobic digestion of tomato solid wastes. *Agric. Wastes*, 10(4), 285–295.

Jarvis, A., Nordberg, A., Mathisen, B., Svensson, B.H. 1995. Stimulation of conversion rates and bacterial activity in a silage-fed two-phase biogas process by initiating liquid recirculation. *Antonie Van Leeuwenhoek*, 68(4), 317–327.

Kabouris, J.C., Tezel, U., Pavlostathis, S.G., Engelmann, M., Dulaney, J.A., Todd, A.C., Gillette, R.A. 2009. Mesophilic and thermophilic anaerobic digestion of municipal sludge and fat, oil, and grease. *Water Environ. Res.*, 81(5), 476–485.

Kavacik, B., Topaloglu, B. 2010. Biogas production from co-digestion of a mixture of cheese whey and dairy manure. *Biomass Bioenergy*, 34(9), 1321–1329.

Kayhanian, M. 1994. Performance of a high-solids anaerobic-digestion process under various ammonia concentrations. *J. Chem. Technol. Biotechnol.*, 59(4), 349–352.

Kayhanian, M. 1995. Biodegradability of the organic fraction of municipal solid-waste in a high-solids anaerobic digester. *Waste Manag. Res.*, 13(2), 123–136.

Kayhanian, M., Rich, D. 1995. Pilot-scale high solids thermophilic anaerobic digestion of municipal solid waste with an emphasis on nutrient requirements. *Biomass Bioenergy*, 8(6), 433–444.

Kayhanian, M., Tchobanoglous, G. 1992. Pilot investigations of an innovative two-stage anaerobic digestion and aerobic composting process for the recovery of energy and compost from the organic fraction of MSW. In *International Symposium on Anaerobic Digestion of Solid Waste*, Venice, April 14–17, p. 181.

Kettunen, R.H., Hoilijoki, T.H., Rintala, J.A. 1996. Anaerobic and sequential anaerobic-aerobic treatments of municipal landfill leachate at low temperatures. *Bioresource Technology*, 58(1), 31–40.

Khoufi, S., Aloui, F., Sayadi, S. 2006. Treatment of olive oil mill wastewater by combined process electro-Fenton reaction and anaerobic digestion. *Water Res.*, 40(10), 2007–2016.

Kim, H.-W., Nam, J.-Y., Shin, H.-S. 2011. A comparison study on the high-rate co-digestion of sewage sludge and food waste using a temperature-phased anaerobic sequencing batch reactor system. *Bioresour. Technol.*, 102(15), 7272–7279.

Kim, M., Ahn, Y.H., Speece, R.E. 2002. Comparative process stability and efficiency of anaerobic digestion; mesophilic vs. thermophilic. *Water Res.*, 36(17), 4369–4385.

Kristensen, J.B., Borjesson, J., Bruun, M.H., Tjerneld, F., Jorgensen, H. 2007. Use of surface active additives in enzymatic hydrolysis of wheat straw lignocellulose. *Enzyme Microb. Technol.*, 40(4), 888–895.

Krupp, M., Schubert, J., Widmann, R. 2005. Feasibility study for co-digestion of sewage sludge with OFMSW on two wastewater treatment plants in Germany. *Waste Manag.*, 25(4), 393–399.

Kübler, H., Hoppenheidt, K., Hirsch, P., Kottmair, A., Nimmrichter, R., Nordsieck, H., Mücke, W., Swerev, M. 2000. Full scale co-digestion of organic waste. *Water Sci. Technol.*, 41(3), 195–202.

Kubler, H., Wild, M. 1992. The BTA-process high rate biomethanisation of biogenous solid wastes. In *International Symposium on Anaerobic Digestion of Solid Waste*, Venice, April 14–17, p. 535.

Kumar, R., Singh, S., Singh, O. 2008. Bioconversion of lignocellulosic biomass: Biochemical and molecular perspectives. *J. Ind. Microbiol. Biotechnol.*, 1125, 308–321.

Lagerkvist, A., Chen, H. 1992. Control of anaerobic degradation of MSW by enzyme addition. In *International Symposium on Anaerobic Digestion of Solid Waste*, Venice, April 14–17, p. 75.

Lay, J.J., Lee, Y.J., Noike, T. 1999. Feasibility of biological hydrogen production from organic fraction of municipal solid waste. *Water Res.*, 33(11), 2579–2586.

Lay, J.J., Noike, T., Endo, G., Ishimoto, S. 1997. Analysis of environmental factors affecting methane production from high-solids organic waste. *Water Sci. Technol.*, 36(6–7), 493–500.

Lettinga, G., Hulshoff Pol, L.W. 1991. UASB-process design for various types of wastewaters. *Water Sci. Technol.*, 24(8), 87–107.

Li, Y., Zhang, R., He, Y., Zhang, C., Liu, X., Chen, C., Liu, G. 2014. Anaerobic codigestion of chicken manure and corn stover in batch and continuously stirred tank reactor (CSTR). *Bioresour. Technol.*, 156, 342–347.

Limayem, A., Ricke, S.C. 2012. Lignocellulosic biomass for bioethanol production: Current perspectives, potential issues and future prospects. *Progress Energy Combust. Sci.*, 38(4), 449–467.

Liu, X., Li, R., Ji, M., Han, L. 2013. Hydrogen and methane production by co-digestion of waste activated sludge and food waste in the two-stage fermentation process: Substrate conversion and energy yield. *Bioresour. Technol.*, 146, 317–323.

Liu, X., Wang, W., Shi, Y., Zheng, L., Gao, X., Qiao, W., Zhou, Y. 2012. Pilot-scale anaerobic co-digestion of municipal biomass waste and waste activated sludge in China: Effect of organic loading rate. *Waste Manag.*, 32(11), 2056–2060.

Lopes, W.S., Leite, V.D., Prasad, S. 2004. Influence of inoculum on performance of anaerobic reactors for treating municipal solid waste. *Bioresour. Technol.*, 94(3), 261–266.

Luo, J., Zhou, J., Qian, G., Liu, J. 2014. Effective anaerobic biodegradation of municipal solid waste fresh leachate using a novel pilot-scale reactor: Comparison under different seeding granular sludge. *Bioresour. Technol.*, 165, 152–157.

Luostarinen, S., Luste, S., Sillanpää, M. 2009. Increased biogas production at wastewater treatment plants through co-digestion of sewage sludge with grease trap sludge from a meat processing plant. *Bioresour. Technol.*, 100(1), 79–85.

Luste, S., Luostarinen, S. 2010. Anaerobic co-digestion of meat-processing by-products and sewage sludge—Effect of hygienization and organic loading rate. *Bioresour. Technol.*, 101(8), 2657–2664.

Mace, S., Dosta, J., Bolzonella, D., Mata-Alvarez, J. 2005. Full scale implementation of AD technology to treat the organic fraction of municipal solid waste in Spain. In *Proceedings of ADSW 2005*, Copenhagen, August 31–September 2, pp. 409–416.

Macias-Corral, M., Samani, Z., Hanson, A., Smith, G., Funk, P., Yu, H., Longworth, J. 2008. Anaerobic digestion of municipal solid waste and agricultural waste and the effect of co-digestion with dairy cow manure. *Bioresour. Technol.*, 99(17), 8288–8293.

Marques, I.P. 2001. Anaerobic digestion treatment of olive mill wastewater for effluent re-use in irrigation. *Desalination*, 137(1–3), 233–239.

Martínez, E.J., Fierro, J., Sánchez, M.E., Gómez, X. 2012. Anaerobic co-digestion of FOG and sewage sludge: Study of the process by Fourier transform infrared spectroscopy. *Int. Biodeterior. Biodegradation*, 75, 1–6.

Massey, M.L., Pohland, F.G. 1978. Phase separation of anaerobic stabilization by kinetic controls. *J. Water Pollut. Control Fed.*, 50(9), 2204–2222.

Mata-Alvarez, J., Cecchi, F., Pavan, P., Bassetti, A. 1992. Semi-dry thermophilic anaerobic digestion of fresh and pre-composted organic fraction of MSW. Digester performance. Presented at International Symposium on Anaerobic Digestion of Solid Waste, Venice, April 14–17.

Mata-Alvarez, J., Cecchi, F., Pavan, P., Llabres, P. 1990. The performances of digesters treating the organic fraction of municipal solid-wastes differently sorted. *Biol. Wastes*, 33(3), 181–199.

Mata-Alvarez, J., Dosta, J., Romero-Güiza, M.S., Fonoll, X., Peces, M., Astals, S. 2014. A critical review on anaerobic co-digestion achievements between 2010 and 2013. *Renewable Sustainable Energy Rev.*, 36, 412–427.

McCarty, P.L., Brousseau, M.H. 1963. Effect of high concentrations of individual volatile acids in anaerobic treatment. In *18th Annual Purdue Industrial Waste Conference*, Purdue, IN, p. 283.

McFarlane, P.N., Pfeffer, J.T. 1981. Biological conversion of biomass to methane. SERI/TR-98357-1. Department of Civil Engineering, University of Illinois, Urbana.

Mizuki, E., Akao, T., Saruwatari, T. 1990. Inhibitory effect of *Citrus unshu* peel on anaerobic digestion. *Biol. Wastes*, 33(3), 161–168.

Møller, H.B., Sommer, S.G., Ahring, B.K. 2004. Methane productivity of manure, straw and solid fractions of manure. *Biomass Bioenergy*, 26(5), 485–495.

Monson, K.D., Esteves, S.R., Guwy, A.J., Dinsdale, R.M. 2007. *Anaerobic Digestion of Biodegradable Municipal Solid Waste: A Review*. Trefforest, Wales: University of Glamorgan.

Morillo, J.A., Antizar-Ladislao, B., Monteoliva-Sánchez, M., Ramos-Cormenzana, A., Russell, N.J. 2009. Bioremediation and biovalorisation of olive-mill wastes. *Appl. Microbiol. Biotechnol.*, 82(1), 25–39.

Mosier, N., Hendrickson, R., Ho, N., Sedlak, M., Ladisch, M.R. 2005. Optimization of pH controlled liquid hot water pretreatment of corn stover. *Bioresour. Technol.*, 96(18), 1986–1993.

Nagao, N., Tajima, N., Kawai, M., Niwa, C., Kurosawa, N., Matsuyama, T., Yusoff, F.M., Toda, T. 2012. Maximum organic loading rate for the single-stage wet anaerobic digestion of food waste. *Bioresour. Technol.*, 118, 210–218.

Nair, K., Kannan, V., Sebastian, S. 1983. Biogas generation using microalgae and macrophytes. *Indian J. Environ. Health*, 24(4), 277–284.

Nallathambi, G.V. 1997. Anaerobic digestion of biomass for methane production: A review. *Biomass Bioenergy*, 13(1–2), 83–114.

Nayono, S.E., Gallert, C., Winter, J. 2010. Co-digestion of press water and food waste in a biowaste digester for improvement of biogas production. *Bioresour. Technol.*, 101(18), 6998–7004.

Neves, L., Oliveira, R., Alves, M.M. 2004. Influence of inoculum activity on the biomethanization of a kitchen waste under different waste/inoculum ratios. *Process Biochem.*, 39(12), 2019–2024.

Neves, L., Pereira, M.A., Mota, M., Alves, M.M. 2009. Detection and quantification of long chain fatty acids in liquid and solid samples and its relevance to understand anaerobic digestion of lipids. *Bioresour. Technol.*, 100(1), 91–96.

Nopharatana, A., Pullammanappallil, P.C., Clarke, W.P. 2006. Kinetics and dynamic modelling of batch anaerobic digestion of municipal solid waste in a stirred reactor. *Waste Manag.*, 27(5), 595–603.

Nordberg, A., Jarvis, A., Stenberg, B., Mathisen, B., Svensson, B.H. 2006. Anaerobic digestion of alfalfa silage with recirculation of process liquid. *Bioresour. Technol.*, 98(1), 104–111.

Noutsopoulos, C., Mamais, D., Antoniou, K., Avramides, C., Oikonomopoulos, P., Fountoulakis, I. 2013. Anaerobic co-digestion of grease sludge and sewage sludge: The effect of organic loading and grease sludge content. *Bioresour. Technol.*, 131, 452–459.

Novak, J.T., Carlson, D.A. 1970. Kinetics of anaerobic long chain fatty acid degradation. *J. Water Pollut. Control Fed.*, 42(11), 1932–1970.

Nuchdang, S., Phalakornkule, C. 2012. Anaerobic digestion of glycerol and co-digestion of glycerol and pig manure. *J. Environ. Manag.*, 101, 164–172.

Nurliyana, M.Y., H'ng, P.S., Rasmina, H., Kalsom, M.S.U., Chin, K.L., Lee, S.H., Lum, W.C., Khoo, G.D. 2015. Effect of C/N ratio in methane productivity and biodegradability during facultative co-digestion of palm oil mill effluent and empty fruit bunch. *Ind. Crops Products*, 76, 409–415.

O'Keefe, D.M., Chynoweth, D.P., Barkdoll, A.W., Nordstedt, R.A., Owens, J.M., Sifontes, J. 1992. Sequential batch anaerobic composting. In *International Symposium on Anaerobic Digestion of Solid Waste*, Venice, April 14–17, p. 117.

O'Keefe, D.M., Owens, J.M., Chynoweth, D.P. 1996. Anaerobic composting of crab-picking wastes for byproduct recovery. *Bioresour. Technol.*, 58(3), 265–272.

Oleszkiewicz, J.A., PoggiVaraldo, H.M. 1997. High-solids anaerobic digestion of mixed municipal and industrial waste. *J. Environ. Eng.*, 123(11), 1087–1092.

Owen, W.F., Stuckey, D.C., Healy, J.B., Young, L.Y., McCarty, P.L. 1979. Bioassay for monitoring biochemical methane potential and anaerobic toxicity. *Water Res.*, 13(6), 485–492.

Owens, J.M., Chynoweth, D.P. 1993. Biochemical methane potential of municipal solid waste (MSW) components. *Water Sci. Technol.*, 27(2), 1–14.

Pahl, O., Firth, A., MacLeod, I., Baird, J. 2008. Anaerobic co-digestion of mechanically biologically treated municipal waste with primary sewage sludge—A feasibility study. *Bioresour. Technol.*, 99(9), 3354–3364.

Panichnumsin, P., Nopharatana, A., Ahring, B., Chaiprasert, P. 2010. Production of methane by co-digestion of cassava pulp with various concentrations of pig manure. *Biomass Bioenergy*, 34(8), 1117–1124.

Park, N.D., Thring, R.W., Garton, R.P., Rutherford, M.P., Helle, S.S. 2011. Increased biogas production in a wastewater treatment plant by anaerobic co-digestion of fruit and vegetable waste and sewer sludge—A full scale study. *Water Sci. Technol.*, 64(9), 1851–1856.

Pastor, L., Ruiz, L., Pascual, A., Ruiz, B. 2013. Co-digestion of used oils and urban landfill leachates with sewage sludge and the effect on the biogas production. *Appl. Energy*, 107, 438–445.

Pavan, P., Battistoni, P., Cecchi, F., Mata-Alvarez, J. 2000a. Performance of thermophilic semi-dry anaerobic digestion process changing the feed biodegradability. *Water Sci. Technol.*, 41(3), 75–81.

Pavan, P., Battistoni, P., Cecchi, F., Mata-Alvarez, J. 2000b. Two-phase anaerobic digestion of source sorted OFMSW (organic fraction of municipal solid waste): Performance and kinetic study. *Water Sci. Technol.*, 41(3), 111–118.

Pereira, M.A., Cavaleiro, A.J., Mota, M., Alves, M.M. 2003. Accumulation of long chain fatty acids onto anaerobic sludge under steady state and shock loading conditions: Effect on acetogenic and methanogenic activity. *Water Sci. Technol.*, 48, 33–40.

Pereira, M.A., Sousa, D.Z., Mota, M., Alves, M.M. 2004. Mineralization of LCFA associated with anaerobic sludge: Kinetics, enhancement of methanogenic activity, and effect of VFA. *Biotechnol. Bioeng.*, 88(4), 502–511.

Peu, P., Sassi, J.F., Girault, R., Picard, S., Saint-Cast, P., Béline, F., Dabert, P. 2011. Sulphur fate and anaerobic biodegradation potential during co-digestion of seaweed biomass (*Ulva* sp.) with pig slurry. *Bioresour. Technol.*, 102(23), 10794–10802.

Pfeffer, J.T., Liebman, J.C. 1974. Biological conversion of organic refuse to methane. UILU-ENG-74-2019 NSF/RANN/SE/GI/39191/PR/75/2. Department of Civil Engineering, University of Illinois, Urbana.

Pitk, P., Kaparaju, P., Palatsi, J., Affes, R., Vilu, R. 2013. Co-digestion of sewage sludge and sterilized solid slaughterhouse waste: Methane production efficiency and process limitations. *Bioresour. Technol.*, 134, 227–232.

Popescu, M.-C., Popescu, C.-M., Lisa, G., Sakata, Y. 2011. Evaluation of morphological and chemical aspects of different wood species by spectroscopy and thermal methods. *J. Mol. Struct.*, 988(1–3), 65–72.

Raposo, F., Banks, C.J., Siegert, I., Heaven, S., Borja, R. 2006. Influence of inoculum to substrate ratio on the biochemical methane potential of maize in batch tests. *Process Biochem.*, 41(6), 1444–1450.

Raynal, J., Delgenes, J.P., Moletta, R. 1998. Two-phase anaerobic digestion of solid wastes by a multiple liquefaction reactors process. *Bioresour. Technol.*, 65(1–2), 97–103.

Razaviarani, V., Buchanan, I.D., Malik, S., Katalambula, H. 2013a. Pilot-scale anaerobic co-digestion of municipal wastewater sludge with restaurant grease trap waste. *J. Environ. Manag.*, 123, 26–33.

Razaviarani, V., Buchanan, I.D., Malik, S., Katalambula, H. 2013b. Pilot scale anaerobic co-digestion of municipal wastewater sludge with biodiesel waste glycerin. *Bioresour. Technol.*, 133, 206–212.

Riaño, B., Molinuevo, B., García-González, M.C. 2011. Potential for methane production from anaerobic co-digestion of swine manure with winery wastewater. *Bioresour. Technol.*, (5), 4131–4136.

Richards, B.K., Cummings, R.J., Jewell, W.J., Herndon, F.G. 1991. High solids anaerobic methane fermentation of sorghum and cellulose. *Biomass Bioenergy*, 1(1), 47–53.

Rintala, J.A., Ahring, B.K. 1994. A 2-stage thermophilic anaerobic process for the treatment of source sorted household solid-waste. *Biotechnol. Lett.*, 16(10), 1097–1102.

Rinzema, A., Alphenaar, A., Lettinga, G. 1989. The effect of lauric acid shock loads on the biological and physical performance of granular sludge in UASB reactors digesting acetate. *J. Chem. Technol. Biotechnol.*, 46, 257–266.

Robra, S., Serpa da Cruz, R., de Oliveira, A.M., Neto, J.A.A., Santos, J.V. 2010. Generation of biogas using crude glycerin from biodiesel production as a supplement to cattle slurry. *Biomass Bioenergy*, 34(9), 1330–1335.

Sawatdeenarunat, C., Surendra, K.C., Takara, D., Oechsner, H., Khanal, S.K. 2015. Anaerobic digestion of lignocellulosic biomass: Challenges and opportunities. *Bioresour. Technol.*, 178, 178–186.

Schmit, K.H., Ellis, T.G. 2001. Comparison of temperature-phased and two-phase anaerobic co-digestion of primary sludge and municipal solid waste. *Water Environ. Res.*, 73(3), 314–321.

Schober, G., Schafer, J., Schmid-Staiger, U., Trosch, W. 1999. One and two-stage digestion of solid organic waste. *Water Res.*, 33(3), 854–860.

Schwartz, R.D., Keller, F.A. 1982. Acetic-acid production by *Clostridium thermoaceticum* in pH-controlled batch fermentations at acidic pH. *Appl. Environ. Microbiol.*, 43(6), 1385–1392.

Sharma, S.K., Saini, J.S., Mishra, I.M., Sharma, M.P. 1989. Biogasification of woody biomass: *Ipomoea fistulosa* plant stem. *Biol. Wastes*, 28(1), 25–32.

Shea, T.G., Pretorius, W.A., Cole, R.D., Pearson, E.A. 1968. Kinetics of hydrogen assimilation in the methane fermentation. *Water Res.*, 2(12), 833.

Siles, J.A., Martín, M.A., Chica, A.F., Martín, A. 2010. Anaerobic co-digestion of glycerol and wastewater derived from biodiesel manufacturing. *Bioresour. Technol.*, 101(16), 6315–6321.

Siles, J.A., Martín, M.D.L.A., Chica, A.F., Martin, A. 2009. Anaerobic digestion of glycerol derived from biodiesel manufacturing. *Bioresour. Technol.*, 100, 5609–5615.

Silvestre, G., Rodríguez-Abalde, A., Fernández, B., Flotats, X., Bonmatí, A. 2011. Biomass adaptation over anaerobic co-digestion of sewage sludge and trapped grease waste. *Bioresour. Technol.*, 102(13), 6830–6836.

Smith, D.P. 1986. H2 in anaerobic processes. Chemical Engineering, Stanford University, Stanford.

Sosnowski, P., Wieczorek, A., Ledakowicz, S. 2003. Anaerobic co-digestion of sewage sludge and organic fraction of municipal solid wastes. *Adv. Environ. Res.*, 7(3), 609–616.

Sousa, D.Z., Pereira, M.A., Smidt, H., Stams, A.J.M., Alves, M.M. 2007. Molecular assessment of complex microbial communities degrading long chain fatty acids in methanogenic bioreactors. *FEMS Microbiol. Ecol.*, 60(2), 252–265.

Speece, R.E. 1996. *Anaerobic Biotechnology for Industrial Wastewaters.* Nashville, TN: Archae Press.

Tong, X.G., Smith, L.H., McCarty, P.L. 1990. Methane fermentation of selected lignocellulosic materials. *Biomass*, 21(4), 239–255.

Triolo, J.M., Pedersen, L., Qu, H., Sommer, S.G. 2012. Biochemical methane potential and anaerobic biodegradability of non-herbaceous and herbaceous phytomass in biogas production. *Bioresour. Technol.*, 125, 226–232.

Trzcinski, A.P., Ray, M.J., Stuckey, D.C. 2010. Performance of a three-stage membrane bioprocess treating the organic fraction of municipal solid waste and evolution of its archaeal and bacterial ecology. *Bioresour. Technol.*, 101(6), 1652–1661.

Trzcinski, A.P., Stuckey, D.C. 2009. Continuous treatment of the organic fraction of municipal solid waste in an anaerobic two-stage membrane process with liquid recycle. *Water Res.*, 43(9), 2449–2462.

Trzcinski, A.P., Stuckey, D.C. 2010. Treatment of municipal solid waste leachate using a submerged anaerobic membrane bioreactor at mesophilic and psychrophilic temperatures: Analysis of recalcitrants in the permeate using GC-MS. *Water Res.*, 44(3), 671–680.

Trzcinski, A.P., Stuckey, D.C. 2012a. Denaturing gradient gel electrophoresis analysis of archaeal and bacterial populations in a submerged anaerobic membrane bioreactor treating landfill leachate at low temperatures. *Environ. Eng. Sci.*, 29(4), 219–226.

Trzcinski, A.P., Stuckey, D.C. 2012b. Determination of the hydrolysis constant in the biochemical methane potential test of municipal solid waste. *Environ. Eng. Sci.*, 29(9), 848–854.

Vandevivere, P., De Baere, L., Verstraete, W. 2003. Types of anaerobic digesters for solid wastes. In *Biomethanization of the Organic Fraction of Municipal Solid Wastes.* London: IWA Publishing, pp. 111–140.

Van Lier, J.B., Martin, J.L.S., Lettinga, G. 1996. Effect of temperature on the anaerobic thermophilic conversion of volatile fatty acids by dispersed and granular sludge. *Water Res.*, 30(1), 199–207.

Veeken, A., Hamelers, B. 1999. Effect of temperature on hydrolysis rates of selected biowaste components. *Bioresour. Technol.*, 69(3), 249–254.

Vieitez, E.R., Ghosh, S. 1999. Biogasification of solid wastes by two-phase anaerobic fermentation. *Biomass Bioenergy*, 16(5), 299–309.

Vieitez, E.R., Mosquera, J., Ghosh, S. 2000. Kinetics of accelerated solid-state fermentation of organic-rich municipal solid waste. *Water Sci. Technol.*, 41(3), 231–238.

Wan, C., Zhou, Q., Fu, G., Li, Y. 2011. Semi-continuous anaerobic co-digestion of thickened waste activated sludge and fat, oil and grease. *Waste Manag.*, 31, 1752–1758.

Wang, G., Wang, D.I.C. 1984. Elucidation of growth-inhibition and acetic acid production by *Clostridium thermoaceticum. Appl. Environ. Microbiol.*, 47(2), 294–298.

Wang, L., Aziz, T.N., de los Reyes, F.L. 2013. Determining the limits of anaerobic co-digestion of thickened waste activated sludge with grease interceptor waste. *Water Res.*, 47(11), 3835–3844.

Wang, X., Yang, G., Feng, Y., Ren, G., Han, X. 2012. Optimizing feeding composition and carbon–nitrogen ratios for improved methane yield during anaerobic co-digestion of dairy, chicken manure and wheat straw. *Bioresour. Technol.*, 120, 78–83.

Wang, Y.S., Byrd, C.S., Barlaz, M.A. 1994. Anaerobic biodegradability of cellulose and hemicellulose in excavated refuse samples using a biochemical methane potential assay. *J. Ind. Microbiol.*, 13(3), 147–153.

Weiland, P. 1992. One- and two-step anaerobic digestion of solid agroindustrial residues. Presented at International Symposium on Anaerobic Digestion of Solid Waste, Venice, April 14–17.

Weiland, P. 2010. Biogas production: Current state and perspectives. *Appl. Microbiol. Biotechnol.*, 85, 849–860.

Wellinger, A., Wyder, K., Metzler, A.E. 1992. Kompogas—A new system for the anaerobic treatment of source separated waste. Presented at International Symposium on Anaerobic Digestion of Solid Waste, Venice, April 14–17.

Wittmann, C., Zeng, A.P., Deckwer, W.D. 1995. Growth inhibition by ammonia and use of pH controlled feeding strategy for the effective cultivation of *Mycobacterium chorophenolicum*. *Appl. Microbiol. Biotechnol.*, 44, 519–525.

Wu, X., Yao, W., Zhu, J., Miller, C. 2010. Biogas and CH_4 productivity by co-digesting swine manure with three crop residues as an external carbon source. *Bioresour. Technol.*, 101(11), 4042–4047.

Wujcik, W.J., Jewell, W.J. 1979. Dry anaerobic fermentation. Presented at 2nd Symposium on Biotechnology in Energy Production and Conservation, Gatlinburg, TN.

Xiao, W., Clarkson, W.W. 1997. Acid solubilization of lignin and bioconversion of treated newsprint to methane. *Biodegradation*, 8, 61–66.

Xie, R., Xing, Y., Yahya, A.G., Ooi, K.-E., Ng, S.-W. 2007. Full-scale demonstration of an ultrasonic disintegration technology in enhancing anaerobic digestion of mixed primary and thickened secondary sewage sludge. *J. Environ. Eng. Sci.*, 6(5), 533–541.

Ye, J., Li, D., Sun, Y., Wang, G., Yuan, Z., Zhen, F., Wang, Y. 2013. Improved biogas production from rice straw by co-digestion with kitchen waste and pig manure. *Waste Manag.*, 33, 2653–2658.

Yu, H.Q., Fang, H.H.P. 2003. Acidogenesis of gelatin-rich wastewater in an upflow anaerobic reactor: Influence of pH and temperature. *Water Res.*, 37(1), 55–66.

Yun, Y.-M., Cho, S.-K., Kim, H.-W., Jung, K.-W., Shin, H.-S., Kim, D.-H. 2015. Elucidating a synergistic effect of food waste addition on the enhanced anaerobic digestion of waste activated sludge. *Kor. J. Chem. Eng.*, 32(8), 1542–1546.

Zarkadas, I.S., Pilidis, G.A. 2011. Anaerobic co-digestion of table olive debittering & washing effluent, cattle manure and pig manure in batch and high volume laboratory anaerobic digesters: Effect of temperature. *Bioresour. Technol.*, 102(8), 4995–5003.

Zeeman, G., Wiegant, W.M., Koster-Treffers, M.E., Lettinga, G. 1985. The influence of the total ammonia concentration on the thermophilic digestion of cow manure. *Agric. Wastes*, 14, 19–35.

Zhang, C., Xiao, G., Peng, L., Su, H., Tan, T. 2013. The anaerobic co-digestion of food waste and cattle manure. *Bioresour. Technol.*, 129, 170–176.

Zhang, R., El-Mashad, H.M., Hartman, K., Wang, F., Liu, G., Choate, C., Gamble, P. 2007. Characterization of food waste as feedstock for anaerobic digestion. *Bioresour. Technol.*, 98, 929–935.

Zhu, L., O'Dwyer, J.P., Chang, V.S., Granda, C.B., Holtzapple, M.T. 2007. Structural features affecting biomass enzymatic digestibility. *Bioresour. Technol.*, 99(9), 3817–3828.

Zoetemeyer, R.J., Arnoldy, P., Cohen, A., Boelhouwer, C. 1982a. Influence of temperature on the anaerobic acidification of glucose in a mixed culture forming part of a 2-stage digestion process. *Water Res.*, 16(3), 313–321.

Zoetemeyer, R.J., van den Heuvel, J.C., Cohen, A. 1982b. pH influence on acidogenic dissimilation of glucose in an anaerobic digestor. *Water Res.*, 16(3), 303–311.

Zouari, N., Ellouz, R. 1996. Toxic effect of coloured olive compounds on the anaerobic digestion of olive oil mill effluent in UASB-like reactors. *J. Chem. Technol. Biotechnol.*, 66(4), 414–420.

Zupančič, G.D., Uranjek-Ževart, N., Roš, M. 2008. Full-scale anaerobic co-digestion of organic waste and municipal sludge. *Biomass Bioenergy*, 32(2), 162–167.

5

Microbial Biomethane Production from Municipal Solid Waste Using High Solids Anaerobic Digestion

Gregory R. Hinds, Piet N.L. Lens,
Qiong Zhang, and Sarina J. Ergas

Contents

5.1 Introduction

Municipal solid waste (MSW) includes all substances and items that are discarded into trash cans and dumpsters from residential, commercial, institutional, and industrial sources. Worldwide, the most common MSW management strategies are disposal in landfills and open dumps, incineration, and aerobic composting; however, these alternatives incur significant environmental, public health, and economic costs (Hoornweg and Bhada-Tata, 2012). Ongoing efforts to minimize these costs include the development of technologies that recover resources from the organic fraction of MSW (OFMSW), which is primarily composed of food and yard waste and makes up approximately 50% by mass of the solid waste generated globally.

Anaerobic digestion (AD) is a microbiologically facilitated technology that enables recovery of biogas (a mixture of methane and CO_2) from waste materials. High-solids AD (HS-AD) is a specific application of AD that is particularly applicable to OFMSW and organic crop residues (e.g., rice straw) due to the characteristics of these waste streams. HS-AD is easily paired with composting to enable the recovery of nutrients, thus making HS-AD a more environmentally sustainable OFMSW management technology relative to other alternatives, such as landfilling with landfill gas to energy, incineration waste to energy (WtE), and advanced thermal treatment (ATT), using gasification and pyrolysis systems or composting alone. This chapter provides an overview of the HS-AD process, including process microbiology and design and operating parameters. Trends in HS-AD implementation in Europe and the United States are presented, along with critical economic, policy, and other factors needed for successful implementation of HS-AD systems. Lastly, the sustainability of HS-AD is assessed through a review of the life cycle assessment (LCA) literature related to the topic.

Figure 5.1 shows photographs from the Zero Waste Energy HS-AD facility in Marina, California. Operational since 2013, the 5000-ton-per-year (TPY) facility receives source-separated OFMSW (SS-OFMSW) from nearby commercial (e.g., restaurants and grocery stores) and institutional (e.g., schools and hospitals) sources, generates combined heat and power (100 kW), sells recovered electricity to a neighboring wastewater treatment facility, and sells recovered nutrients as compost to nearby farms (ZWE, 2013, 2015).

Figure 5.1 Photographs of Marina (California) HS-AD facility. Top row from the left: Fresh yard waste, fresh food waste, digester units, and combined heat and power (CHP) unit (with emergency flare, biofilter, and control building). Bottom row from the left: Fresh digestate, trommel screen, posttrommel digestate composting, and final compost product.

5.2 Background and Terminology

In municipal and industrial waste management applications, AD technologies have been widely used for treating and recovering energy from high-strength wastewaters and organic sludges. HS-AD technologies (those designed to operate with a total solids [TS] content of >15%) were developed in Europe in the late 1980s and early 1990s following increased landfill taxation, banning of organics disposal in landfills, and mandated source separation of organic waste. Since then, HS-AD for processing OFMSW has developed more rapidly in Europe than any other alternative OFMSW management technology. It is expected that as the costs of environmental degradation are further incorporated into the decision-making processes of governments around the world, legislative incentives to source-separate and recycle OFMSW will steadily increase and HS-AD will continue to emerge as a leading OFMSW technology.

Several different terms are used in place of *high solids*, including *solid-state*, *dry*, and *solid-substrate* AD. In addition, the terms *fermentation* and *anaerobic composting* have often been used in place of *anaerobic digestion*, resulting in many alternative ways to refer to HS-AD (Hinds, 2015). There is also a lack of agreement on the cutoff TS content between liquid AD (L-AD) and HS-AD conditions (Li et al., 2011; De Baere and Mattheeuws, 2014). For example, Kothari et al. (2014) stated that the cutoff point between L-AD and HS-AD is 22% TS and further separated AD into low (<15%), medium (15%–20%), and high (>20%) solids categories. Tchobanoglous et al. (2003) separated AD into three categories, but with L-AD constituting AD with <10% TS, while Karagiannidis and Perkoulidis (2009) defined L-AD as 10%–25% TS and HS-AD as 30%–40% TS.

OFMSW, especially the food waste fraction, is often integrated into L-AD systems at municipal, agricultural, or industrial wastewater treatment facilities. However, HS-AD technologies have largely been preferred over L-AD technologies in standalone systems designed specifically for processing OFMSW because of the many advantages they offer (Table 5.1) and the ease of pairing them with aerobic composting operations. In Europe, approximately 70% of the installed capacity for AD since 2009 has been HS-AD, and in the Netherlands and Belgium, approximately 80% of all composting operations incorporate AD as a primary treatment technology (De Baere and Mattheeuws, 2014).

A typical process flow diagram of HS-AD for the recovery of resources from OFMSW is shown in Figure 5.2. Feedstocks for these systems are stackable porous waste mixtures (Table 5.2), which can be moved using screw augers, high-power pumps, conveyer belts, or front-end loaders. Some systems are designed to operate in a specific TS range, and water or leachate is added to feedstock to adjust the TS content when necessary. However, the operational TS content can change within multistage systems or single-stage systems as feedstock changes or with changes in season. The physical and chemical characteristics of each potential substrate make some substrates more valuable and appropriate than others for processing via HS-AD. Protocols have been developed for assessing the value of various substrates with respect to their energy recovery potential in AD (Owen et al., 1979; Owens and Chynoweth, 1993;

TABLE 5.1   Benefits of Anaerobic Digestion and Advantages and Disadvantages of High-Solids Anaerobic Digestion vs. Liquid Anaerobic Digestion

Benefits of AD	Summary	References
Enables energy recovery	Energy positive. Production of biogas enables direct combustion for heating, lighting, cooking, conversion to electricity, production of compressed natural gas for vehicles, or injection into the natural gas grid.	Owens and Chynoweth, 1993; Tchobanoglous et al., 2003; Khanal, 2008; Li et al., 2011; Kothari et al., 2014
Enables nutrient recovery	Nutrients, N and P, present in digestate and can be recovered through postprocessing (e.g., trommel and composting/curing).	Owens and Chynoweth, 1993; Khanal, 2008; Li et al., 2011; Kothari et al., 2014
Mass/volume reduction	Up to 50% substrate mass and volume reduction can be achieved. Because anaerobic microorganisms are slow growing, low excess biomass is produced.	Tchobanoglous et al., 2003; Li et al., 2011; Kothari et al., 2014
Destruction of pathogens	Long-term exposure to high temperatures in a microbiologically competitive anaerobic environment enhances pathogen destruction/inactivation.	Wilkie, 2005; Khanal, 2008
Reduced GHG emissions	CH_4, which otherwise would be emitted through degradation of organic waste in uncontrolled environments or as fugitive emissions in landfills, is captured. Additional reduction achieved by offsetting fossil fuel-derived electricity consumption.	Owens and Chynoweth, 1993; Tchobanoglous et al., 2003; Edelmann et al., 2005; Li et al., 2011; Kothari et al., 2014
Reduced odors	Enclosed reactors with biogas capture yield little odor.	Wilkie, 2005; Khanal, 2008
Advantages of HS-AD vs. L-AD		
Reduced energy consumption	Less energy used for heating and internal mixing yields lower parasitic energy losses and higher overall energy efficiency.	Li et al., 2011; Kothari et al., 2014
Reduced water use	Zero or minimal water addition required in HS-AD; percolate often recirculated; minimal excess percolate production results in reduced side-stream treatment.	Li et al., 2011; Kothari et al., 2014
Reduced reactor size	Reduced liquid volume allows HS-AD systems to handle greater OLRs, resulting in lower required reactor volumes for given waste volumes.	Guendouz et al., 2010; Li et al., 2011; Kothari et al., 2014
Reduced postprocessing	The compost-like digestate by-product of HS-AD requires only minor postprocessing (trommel/sieve and composting/curing), whereas L-AD by-product requires dewatering.	Li et al., 2011; Kothari et al., 2014
No waste stratification	In L-AD, stratification of FOG and fibrous materials can create operational challenges. This does not occur in HS-AD systems.	Guendouz et al., 2010

(Continued)

TABLE 5.1 (CONTINUED)   Benefits of Anaerobic Digestion and Advantages and Disadvantages of High-Solids Anaerobic Digestion vs. Liquid Anaerobic Digestion

Benefits of AD	Summary	References
Disadvantages of HS-AD vs. L-AD		
More inoculum required	Lower moisture content reduces microbe–substrate contact, resulting in greater inoculation requirements.	Li et al., 2011
Reduced homogeneity	Lower moisture content reduces mixing capabilities and homogeneity of digester contents, yielding spatial variations.	Kothari et al., 2014
Longer retention times	Up to three times longer retention times are needed in HS-AD in some cases due to slower mass transport.	Li et al., 2011; Kothari et al., 2014

Angelidaki et al., 2009), and a range of potential substrates have been tested. Table 5.2 lists the specific methane yields achieved in select studies focusing on common OFMSW-derived substrates used in HS-AD, as well as common HS-AD codigestion strategies.

5.3 Process Microbiology

The metabolic progression of AD involves four primary steps (Figure 5.3): hydrolysis, acidogenesis, acetogenisis, and methanogenesis (Adekunle and Okolie, 2015). Two additional steps, or pathways, are sometimes cited as independent and critical components of the process: solubilization and anaerobic oxidation (Massé and Droste, 2000; Khanal, 2008). Fermentation is sometimes cited as a separate metabolic step in the place of acidogenesis (Li et al., 2011).

5.3.1 Metabolic Phases

The four major microbial processes, or metabolic phases, of AD can be summarized as follows (Khanal, 2008; Li et al., 2011; Adekunle and Okolie, 2015):

- *Hydrolysis*: Conversion of insoluble organics and complex molecules, such as lipids, proteins, carbohydrates, polysaccharides, and nucleic acids, into amino acids, monosaccharides, fatty acids, alcohols, and other simple organics suitable to serve as energy and/or carbon sources for subsequent groups of microorganisms. Hydrolysis is carried out by strict anaerobes and facultative bacteria, certain species of which secrete extracellular enzymes that aid in solubilizing complex molecules. Hydrolysis is generally considered the rate-limiting step in HS-AD, especially in cases where lignocellulosic wastes are a primary feedstock (Veeken and Hamelers, 1999).
- *Acidogenesis*: Conversion of monomers, simple sugars, amino acids, and fatty acids into short-chain volatile organic acids, such as butyric, propionic, and acetic acids,

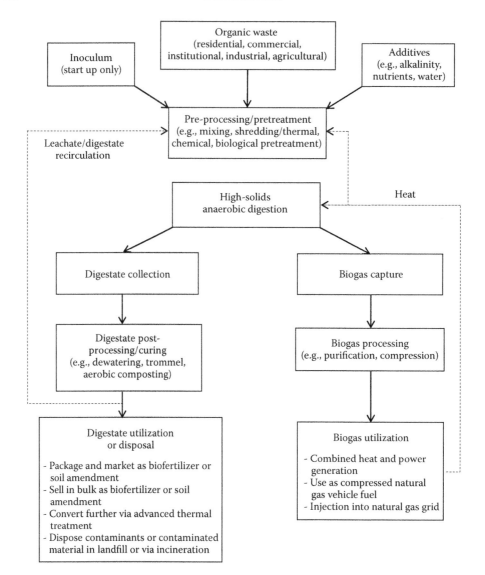

Figure 5.2 General process flow diagram for HS-AD for the recovery of resources from organic waste. (Adapted from Zupančič, G.D., and Grilc, V., in *Management of Organic Waste*, ed. S. Kumar, InTech, Rijeka, Croatia, Figure 3. Copyright 2012. Under CC BY 3.0 license. Available from http://dx.doi.org/10.5772/32756.)

and hydrogen and carbon dioxide gases. Acidogenesis is carried out by obligate anaerobic and facultative bacteria.

- *Acetogenisis*: Conversion of volatile fatty acids (VFAs) to acetate and hydrogen and carbon dioxide gases, which are direct substrates for methanogenesis. Acetogenesis is carried out by strict anaerobes.
- *Methanogenesis*: Conversion of acetate, hydrogen, and carbon dioxide into methane and carbon dioxide. Methanogenesis is performed by strict obligate anaerobic archaea and is considered the rate-limiting step in most L-AD applications.

TABLE 5.2   Specific Methane Yields from Common Substrates in High-Solids Anaerobic Digestion and in Codigestion

Substrate	Notes	Methane Yield (L CH$_4$/kg VS$_{added}$)	References
Mechanically separated OFMSW	Samples from Sumter County, Florida; ground to 1.53 mm	186–222	Owens and Chynoweth, 1993
Yard waste	Turf grass and laurel oak leaves and branches at a 1:1:1 ratio (by VS); ground to 1.53 mm	143	
Leaves	Laurel oak; ground to 1.53 mm	123	
Branches	Laurel oak; ground to 1.53 mm	134	
Grass	Turf grass; ground to 1.53 mm	209	
Office paper	Mixed; ground to 1.53 mm	369	
Newspaper	Spruce with print; ground to 1.53 mm	203	
Magazine paper	Ground to 1.53 mm	100	
Black alder	Samples described as "woody biomass"; all samples ground to 0.8 mm; CH$_4$ yields became negligible after 120 days	240	Jerger et al., 1982
Cottonwood		220	
Eucalyptus		14	
Hybrid poplar		320	
Loblolly pine		63	
Sycamore		320	
Grass	Bermuda grass	137–226	Sharma et al., 1998
Food waste	Source-separated, multiple commercial sources; ground to unspecified size	440	Zhang et al., 2007
Food waste	Source-separated from the University of Cadiz restaurant; ground to 2 mm	530	Forster-Carneiro et al., 2007
Fruit and vegetable waste	Source-separated from food markets, shredded to "small" size	420	Bouallagui et al., 2005
Food waste and waste activated sludge	90% food waste and 10% waste activated sludge (by VS)	489	Heo et al., 2004
Food waste and green (yard) waste	Food waste from a student was blended; green waste (mostly grass and leaves) from Zhejiang University; sieved to 5 mm	160–320 (varying FW/YW ratios) 176–272 (varying TS content)	Chen et al., 2014
Food waste and yard waste	Food waste from hopper feeding L-AD system (Quasar, Ohio) and yard waste from OARDC Wooster campus (leaves and branches); milled to 5 mm	0–120 (varying S/I ratio and FW/YW ratio)	Brown and Li, 2013

Note: FW, food waste; OARDC, Ohio Agricultural Research and Development Center; YW, yard waste.

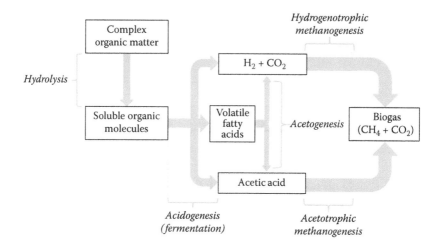

Figure 5.3   Metabolic pathways in the AD process.

On an operational basis, AD processes are often separated into two primary phases: the acid phase (hydrolysis, acidogenesis, and acetogenesis) and the gas phase (methanogenesis) (Deublein and Steihauser, 2008; Adekunle and Okolie, 2015). This convention parallels the method commonly used in two-stage commercial AD systems that carry out hydrolysis, acidogenesis, and acetogenesis in one reactor and methanogenesis (or, in some cases, acetogenesis and methanogenesis) in a separate reactor. Additional microbial processes that take place in AD systems include sulfate (SO_4^{2-}) and ferric iron (Fe^{3+}) reduction (Madigan et al., 2014).

The relationships between the microbial groups, or trophic groups, active in AD are complex, and each interaction is essential to the stability of the community. In particular, the syntrophic relationship between methanogens and acetogens is critical (Khanal, 2008; Madigan et al., 2014). As acetogens metabolize VFAs and generate hydrogen gas, the partial pressure of hydrogen gas in the system increases and can become inhibitory to the acetogens themselves (Madigan et al., 2014). It is the conversion of the hydrogen gas to methane by hydrogenotrophic methanogens that ensures that the partial pressure of hydrogen remains below inhibitory levels. In this way, the efficiency of acetogens in converting volatile acids to methanogenic substrates is dependent on the efficiency of the methanogens in converting hydrogen gas to methane.

The rate of methanogenesis is dependent on the rate of acetogenesis, which supplies the methanogens with a significant fraction of their carbon and energy source. The relationship between fermentative bacteria and methanogens is similar: methanogens depend on fermentative bacteria to provide them with their energy and carbon sources. However, if the fermentative bacteria in an AD system produces VFAs more rapidly than the methanogens can consume them, the system pH can plummet and the pH-sensitive methanogens can be inhibited, resulting in greater accumulation of VFAs and an eventual total acidification and failure of the AD system (Amani et al., 2010). When this occurs, the fermentative bacteria can also no longer thrive.

5.3.2 Stoichiometric Relationships

The general stoichiometric relationships in AD are helpful for conducting carbon and nutrient balances and for understanding the dynamics of alkalinity and other operational parameters (i.e., VFAs and ammonia). A simplified stoichiometry of fermentation, in which organic matter (with an assumed empirical formula of $C_5H_7O_2N$) is converted into acetate (CH_3COO^-), is shown in Equation 5.1 (van Haandel and Lubbe, 2007).

$$C_5H_7O_2N + 3H_2O \rightarrow 2.5CH_3COO^- + 1.5H^+ + NH_4^+ \qquad (5.1)$$

Note that in the fermentation phase, ammonium is produced and protons are released. Ammonium is not an energy source for microorganisms active in the AD process and is consumed only in the synthesis of new cells (as a nitrogen source) by the microbial populations present in the AD process (Kayhanian, 1994; Mussoline et al., 2013). Thus, the majority of the ammonium that is produced accumulates at a rate that is proportional to the fermentation rate (hydrolysis or acidogenesis). Note also that as protons are released, alkalinity is depleted. Equation 5.2 shows the representative stoichiometric reaction for the conversion of acetic acid to methane and carbon dioxide (van Haandel and Lubbe, 2007).

$$CH_3COO^- + H^+ \rightarrow CO_2 + CH_4 \qquad (5.2)$$

Other methanogenic reactions include the conversion of ethanol to acetic acid and methane (Equation 5.3) and the conversion of carbon dioxide and hydrogen gas to methane and water (Equation 5.4).

$$2C_2H_5OH + CO_2 \rightarrow CH_4 + 2CH_3COOH \qquad (5.3)$$

$$CO_2 + 4H_2 \rightarrow CH_4 + 2H_2O \qquad (5.4)$$

Theoretically, greater methane production can be achieved from the reduction of carbon dioxide than from the reaction shown in Equation 5.2; however, in AD the majority of methane comes from the acetic acid reaction carried out by acetoclastic methanogens because hydrogen is typically limiting (Amani et al., 2010). For every mole of acetic acid consumed in the methanogenic phase of AD, a mole of protons is consumed (Equation 5.2). As shown in Equation 5.1, 1.5 moles of protons are produced per 2.5 moles of acetic acid. Hence, when all processes are in balance, the overall AD process is an alkalinity-producing process (more protons consumed than produced). This can be seen in Equation 5.5, which shows a stoichiometric representation of the overall AD process of microbial biomass (assumed empirical formula of $C_5H_7O_2N$). In this reaction, alkalinity as bicarbonate (HCO_3^-) is produced at a one-to-one molar ratio with the biomass digested (van Haandel and Lubbe, 2007).

$$C_5H_7O_2N + 4H_2O \rightarrow HCO_3^- + 1.5CO_2 + 2.5CH_4 + NH_4^+ \qquad (5.5)$$

In accordance with the above stoichiometry, a well-functioning AD process will produce methane, CO_2, alkalinity, and ammonium. Based on this observation, these parameters can be used as a qualitative measure for the overall health or stability of an AD system. For example, in batch AD processes, the greater the process efficiency, the more organic matter that is degraded and the more alkalinity, ammonium, and methane that is generated. However, due to the complex biochemical nature of AD, changes in alkalinity and ammonium concentrations can be dynamic, leaving methane generation as the best measure of AD process efficiency (Khanal, 2008).

5.4 Operation and Design Parameters

The major operating parameters monitored in HS-AD can be separated into physical and chemical categories. Physical parameters include TS content, substrate-to-inoculum ratio (S/I ratio), substrate-to-substrate ratios in codigestion, temperature, retention time, organic loading rate (OLR), volatile solids (VS) reduction, and methane generation rate. Chemical parameters include the carbon-to-nitrogen ratio (C/N ratio), pH and alkalinity, VFA concentration, and concentrations of micronutrients and inhibitory compounds, such as free NH_3 and H_2S. Many of these parameters are interrelated. A summary of the major operational parameters is provided in Table 5.3, along with optimal and inhibitory ranges where applicable.

5.4.1 Particle Size

Reduction of substrate particle size has been shown to increase substrate biodegradability and methane yields from solid wastes in AD systems (Sharma et al., 1988; Delgenés et al., 2002; Kaparaju et al., 2002; Bruni et al., 2010; Kreuger et al., 2011). In general, reducing the particle size increases the specific surface area of a substrate, which leads to increased hydrolysis and enhanced biogas production rates, especially for substrates for which hydrolysis is the rate-limiting step, such as lignocellulosic wastes (Sharma et al., 1988; Izumi et al., 2010; Veeken, 2014). This also leads to reduced reactor volume requirements (Gollakota and Meher, 1988; Moorhead and Nordstedt, 1993).

As a result of these findings, many HS-AD systems incorporate some form of size reduction (e.g., grinders) before loading wastes into reactors. Typically, substrate particle size is reduced to ≤40 mm (Veeken, 2014). However, in HS-AD systems that incorporate percolate recirculation, the presence of larger particles (>40 mm) is essential for maintaining adequate porosity of the waste mixture to improve permeability

TABLE 5.3   Summary of Anaerobic Digestion Operating Parameters, Optimal Ranges, and Inhibitory Concentrations

Parameter	Optimal Ranges/Inhibitory Concentrations	References
TS content	TS content can vary widely in different applications (3%–40% TS). Two general categories: HS-AD (>15% TS) and L-AD (<15% TS). CH_4 yields generally increase with TS content; however, TS contents of >25%–40% are inhibitory.	Rapport et al., 2008; Li et al., 2011; De Baere and Mattheuws, 2014; Kothari et al., 2014
Substrates, inocula, S/I ratio, and codigestion ratios	S/I significantly influences process efficiency and CH_4 yields. Optimal S/I ranges widely (3/1–1/7). Ratios of substrates in codigestion are critical to process efficiency and range widely.	Li et al., 2011; De Baere and Mattheuws, 2014; Deublein and Steinhauser, 2008
Waste particle size and feedstock porosity	Reduction of waste particle size can increase CH_4 yields; wastes often ground to ≤40 mm. In systems with percolate recirculation, presence of particles of >40 mm is important for feedstock structure and porosity and overall digestion efficiency. Thus, particles retained in the trommel process are often recirculated.	Sharma et al., 1988; Izumi et al., 2010; Veeken, 2014
Temperature	Common operating ranges of temperature are 35°C–40°C (mesophilic) and 50°C–55°C (thermophilic). AD microorganisms are capable of surviving temperatures from 0°C to 82°C. Thermophilic digestion shown to generate greater biogas yields, but mesophilic digestion considered more stable.	Amani et al., 2010; De Baere and Mattheuws, 2014
Retention time	Defined in practice as time that the waste material undergoes digestion. Function of feedstock biodegradability and system operating conditions. Selected to optimize system economics. Retention times in HS-AD range from 10 to 30 days.	Tchobanoglous et al., 2003; Amani et al., 2010; Kothari et al., 2014; PIS, 2008
Organic loading rate	Mass of organic matter loaded to AD system per unit volume and time. Biogas yields directly affected by OLR. If OLR is too high, inhibition will occur. Maximum recommended OLRs for mesophilic and thermophilic AD systems are 5 and 8 kg VS/m^3-day.	Vandenburgh and Ellis, 2002; Zupančič and Grilc, 2012; Chen et al., 2014
VS reduction and methane yield	Most commonly expressed in percent by mass and L CH_4/kg VS, respectively. Direct measures of digester performance and degree of digestion. Both decrease logarithmically with time and depend on feedstock biodegradability and system operating parameters. The maximum VS reduction achievable in AD is often cited as 60%.	Burton and Turner, 2003; Kaparaju and Rintala, 2005; Li et al., 2011

(Continued)

TABLE 5.3 (CONTINUED)   Summary of Anaerobic Digestion Operating Parameters, Optimal Ranges, and Inhibitory Concentrations

Parameter	Optimal Ranges/Inhibitory Concentrations	References
C/N ratio	Critical parameter in AD. Optimal C/N range from 20/1 to 30/1, depending on the bioavailability of N and C in the feedstocks. If C/N ratio is too low, ammonia inhibition can result, and if too high, nitrogen can become limiting and slow biogas production or VFAs can accumulate, leading to inhibition.	Li et al., 2011; Kothari et al., 2014; Yang et al., 2015
pH and alkalinity	Sufficient alkalinity essential for efficient digestion. Alkalinity of >1,000 mg/L as calcium carbonate ($CaCO_3$) and pH of >6.5 recommended to ensure methanogenic populations not inhibited.	Tchobanoglous et al., 2003; Amani et al., 2010
Volatile fatty acids	Crucial AD intermediate, which can accumulate at high OLRs and cause inhibition. Total VFA concentrations as acetic acid of >10,000 mg/L generally considered inhibitory to methanogenesis.	Khanal, 2008; Amani et al., 2010
Ammonia (NH_3), hydrogen sulfide (H_2S), and micronutrients	NH_3 important for cell synthesis but inhibitory at concentrations of ≥1,500–1,700 mg/L. H_2S produced when sulfates present in feedstocks, odorous, corrosive, degrades biogas quality, and inhibitory at concentrations of ≥200 mg/L. Certain micronutrients required in trace amounts for AD, some of which can become inhibitory beyond certain concentrations (Tables 5.4 and 5.5).	Gerardi, 2003; Chen et al., 2008; Amani et al., 2010; Yenigün and Demirel, 2013

and mass transfer efficiency (Veeken, 2014). Thus, these systems often add large particles from incoming waste and/or recycle large particles from digestate or compost retained in the trommel process with incoming wastes to improve the "structure" of the feedstock (Veeken, 2014). However, in the case of AD of food waste, overreduction of particle size has been shown to lead to VFA accumulation and reduced methane yields (Izumi et al., 2010). Optimization of particle size, especially in percolate-recirculating HS-AD systems, is a parameter that deserves further investigation at various scales (Veeken, 2014).

5.4.2 Temperature

Temperature is a significant parameter in all biochemical reactions. The two common operational temperature ranges of commercial AD systems are mesophilic (35°C–40°C) and thermophilic (50°C–55°C), although AD microorganisms are capable of surviving temperatures ranging from 0°C to 82°C (Amani et al., 2010). Mesophilic digesters are currently more common in HS-AD applications because they are considered less sensitive to toxicants, process fluctuations, and temperature variations, and they require less energy input for heating than thermophilic digesters (De Baere and Mattheeuws, 2014). However, it has been shown that thermophilic digestion accelerates digestion efficiency (del Real Olvera and Lopez-Lopez, 2012), and in recent years, thermophilic HS-AD has been proven reliable in numerous full-scale systems with high degrees of process control, yielding 30%–50% increases in biogas production rates compared with rates observed in mesophilic digestion (De Baere and Mattheeuws, 2014). Thus, thermophilic digestion is becoming the preferred mode of operation, especially in HS-AD, where heating plays a smaller role in the net energy balance (compared with L-AD), and in warm climates, where less energy input is required for heating to thermophilic ranges (Amani et al., 2010; De Baere and Mattheeuws, 2014). In addition to increased digestion rates, thermophilic digestion provides the advantage of improved pathogen inactivation compared with mesophilic digestion (Li et al., 2011).

5.4.3 Retention Time

Retention times in L-AD are separated into two components: hydraulic retention time (HRT) (or liquid residence time) and solids retention time (SRT) (or mean cell residence time) (Amani et al., 2010). It is important in L-AD systems to decouple HRT and SRT because of the slow growth rates of anaerobic microorganisms and the demand for treating large volumes of water quickly and economically (Kato et al., 1994). However, in single-stage HS-AD systems, retention time is generally expressed as a single value (Cecchi et al., 1988) and is defined as the average amount of time that the substrate is digested (Kothari et al., 2014).

For multistage systems, different retention time conventions may be used. For example, residence time in stage 1 can be differentiated from the residence time in stage 2 (Veeken, 2014). For single-stage batch HS-AD systems with percolate recirculation, SRT and HRT can be decoupled, with the SRT describing the time that the solid waste material is digested and the HRT describing the average amount of time that the percolate remains in the system. However, the volume of percolate in these systems is regulated such that it does not change significantly over time, and therefore, the HRT is equal to the SRT, which is approximately equal to the system volume (V) divided by the volumetric feedstock throughput (Q):

$$SRT = HRT = V/Q \qquad (5.6)$$

In cases of percolate depletion, water can be added to incoming feedstock to replenish percolate in the system (Veeken, 2014). In cases when percolate accumulates, excess percolate can be removed from the percolate storage tank and added to the compost (evaporation occurs) (ZWE, 2015).

Because long retention times are required to hydrolyze the more slowly degrading fraction of the feedstock, AD systems are normally designed to optimize economics (Cecchi et al., 1988). Higher retention times increase reactor volume requirements for a given volumetric loading rate. Due to the exponential nature of biogas generation in AD, systems are often designed such that the operational retention time maximizes the energy recovery efficiency. For example, a batch system can be designed so that digestion ends when the biogas production rate decreases below a certain percentage of the maximum (Rapport et al., 2008). Retention times in full-scale HS-AD systems range from 10 to 30 days, depending on the system and the feedstock being processed (PIS, 2008; Kothari et al., 2014). Thermophilic systems generally require shorter retention times than mesophilic systems (Tchobanoglous et al., 2003; Rapport et al., 2008).

5.4.4 Organic Loading Rate

OLR is an important operating parameter, especially in continuous systems, as it is related to the raw substrate available to microorganisms, the production rate of intermediate compounds (VFAs and H_2), biogas generation, and overall system performance, cost, and stability (Chen et al., 2014). OLR is expressed in units of mass of VS or chemical oxygen demand (COD) loaded to the digester per unit volume per unit time (e.g., kg VS/m^3-day), and is calculated as follows:

$$OLR = \%VS \times Q \times \delta/V \qquad (5.7)$$

where %VS is the percent VS fraction of the feedstock by weight, and δ is the density of the feedstock. It should be emphasized that %VS, density, moisture content, and all other physical and chemical parameters vary with time during the digestion process.

When the OLR is very low and increases, the rate of biogas generation increases. However, beyond a certain threshold, biogas generation plummets with increased OLR as a result of inhibition due to accumulation of intermediates (Vandenburgh and Ellis, 2002). Some studies suggest that HS-AD systems have greater capacity to handle higher OLR than L-AD systems (Schievano et al., 2010) due to reduced rate of mass transfer of toxicants (Abbassi-Guendouz et al., 2012). The maximum OLR varies significantly depending on feedstock biodegradability, but is typically cited as 5 kg VS/m^3-day for mesophilic AD and 8 kg VS/m^3-day for thermophilic AD (Zupančič and Grilc, 2012). Schievano et al. (2010) suggest that in HS-AD, the putrescibility (short-term digestibility) is a particularly critical indicator of potentially inhibitory OLRs. Thus, with highly putrescible waste (e.g., food waste) inhibitory OLRs can be lower than the typically cited values.

5.4.5 Volatile Solids Reduction and Methane Yield

VS reduction and methane yield are the most direct measures of AD performance. VS reduction is calculated on a percent by mass basis as follows:

$$VS \text{ reduction } (\%) = (VS_i - VS_f)/VS_i \qquad (5.8)$$

where VS_f is the final mass of VS present in the digestate and VS_i is the initial mass of VS present in the feedstock.

Methane yield can be expressed in terms of methane flow rate (m³ CH₄/day) or methane production rate per unit reactor volume (m³ CH₄/m³reactor-day), but is most commonly expressed as specific methane yield, which is the volume of methane produced normalized by the mass of VS or COD loaded to the digester (L CH₄/kg VS or m³ CH₄/kg COD). Reporting methane yields as opposed to biogas yields is preferred because methane yield is a more direct measure of the energy recovery potential. Specific methane yield is calculated as follows:

$$\text{Specific methane yield} = V_{CH4} / VS_i \qquad (5.9)$$

where V_{CH4} is the cumulative volume of CH₄ generated from a known mass of feedstock.

Expressing methane yields on a per mass VS basis is preferred over expressing methane yields on a per mass COD basis, especially in HS-AD, because COD concentrations measured from a solid material depend on the solubility of the COD in the material, which can be significantly altered during digestion (Khanal, 2008). Both VS reduction and methane yield are a function of feedstock biodegradability and operating parameters. The maximum VS destruction in full-scale AD systems is generally cited as 60%, but decreases depending on feedstock composition, for example, the presence of recalcitrant compounds (Kaparaju and Rintala, 2005). Some laboratory experiments have achieved up to 90% VS reduction in HS-AD of food waste (Cho et al., 1995).

5.4.6 Carbon-to-Nitrogen Ratio

The C/N ratio is one of the most critical parameters pertaining to AD performance and is defined as

$$C/N = C_{TC} / C_{TN} \qquad (5.10)$$

where C_{TC} is the concentration of total carbon present in a feedstock (mg/L) and C_{TN} is the concentration of total nitrogen present in a feedstock (mg/L).

C/N ratios in the range of 20/1 and 30/1 are considered good, with ~25/1 being optimal (Li et al., 2011; Kothari et al., 2014). C/N ratios below this range contain too much nitrogen relative to the amount of carbon present, resulting in high ammonia release and increased likelihood of ammonia inhibition (Yang et al., 2015). C/N ratios above this range can lead to biogas decreases due to insufficient N availability and/or excessive VFA production due to inhibition of methanogens from a lack of N for cell synthesis (Yang et al., 2015).

The bioavailability of carbon and nitrogen compounds present in a given substrate affects the optimal C/N ratio. The optimal C/N ratio may also be a function of other operating parameters, such as pH and temperature (Yang et al., 2015). Codigestion strategies are often employed to balance C/N ratios in HS-AD of OFMSW. For example, mixed food wastes have C/N ratios ranging from 10/1 to 20/1 (Brown and Li, 2013; Chen et al., 2014), and yard wastes can have C/N ratios exceeding 50/1 (Yang et al., 2015). Therefore, codigestion of a mixture of food and yard waste can result in a favorable C/N ratio (Rapport et al., 2008; Li et al., 2011; Chen et al., 2014).

5.4.7 Alkalinity

Alkalinity is a measure of the buffering capacity of a system to changes in pH. The alkalinity of an AD system is critical to ensure that the pH remains above inhibitory levels as acids are being produced (Amani et al., 2010). Alkalinity is generally expressed as an equivalent concentration of calcium carbonate (mg/L as $CaCO_3$) and is defined as

$$\text{Alkalinity (eq/L)} = [HCO_3^-] + 2[CO_3^{2-}] + [OH^-] - [H^+] \tag{5.11}$$

As discussed previously, the symbiotic relationship between fermenters and methanogens aids in maintaining a healthy pH and alkalinity balance, with methanogens consuming acids (generated by the fermenters) and producing alkalinity. A pH exceeding 6.5 and an alkalinity of more than 1000 mg/L as $CaCO_3$ is recommended to ensure that methanogenic populations are not inhibited (Fabián and Gourdon, 1999; Tchobanoglous et al., 2003; del Real Olvera and Lopez-Lopez, 2012). For two-stage AD, the optimal pH range for the acid phase is 5.2–6.3 and the optimal pH range for the gas phase is 6.7–7.5 (Deublein and Steinhauser, 2008). However, high pH is not normally a concern in HS-AD, except in cases where high pH contributes to free ammonia inhibition (Chen et al., 2008).

It is worth noting that the alkalinity and pH are affected by concentrations of CO_2 in the headspace of the digester. As headspace CO_2 concentrations increase, the partial pressure of CO_2 increases and concentrations of CO_2 as carbonic acid (H_2CO_3) in solution increase, resulting in reduced pH. As alkalinity is depleted, the pH will begin to plummet and digester failure will follow. Ideally, enough alkalinity will be provided by the substrates, and inocula and alkalinity levels will remain sufficient throughout the digestion process. However, in the case that alkalinity becomes depleted (e.g., when fermentation is significantly outpacing methanogenesis and VFAs are accumulating), additional alkalinity sources can be added to prevent acidification of a digester. Alkalinity sources that are commonly used include lime (calcium hydroxide), soda ash (sodium carbonate or sodium bicarbonate), and limestone (calcium carbonate) (Tchobanoglous et al., 2003).

5.4.8 Volatile Fatty Acids

VFAs are the most important and descriptive intermediate chemicals produced in the AD process (Li et al., 2011). As discussed in Section 5.3.1, VFA accumulation is an indicator that methanogenesis is the rate-limiting step rather than hydrolysis. Reduction in VFA concentrations, on the other hand, indicates that methanogenesis is occurring at a more rapid pace than acidogenesis and acetogenesis, indicating that hydrolysis is the rate-limiting step. In the case of batch AD processes, decreases in VFA concentrations can also be an indicator that bioavailable raw substrate (e.g., proteins, carbohydrates, and lipids) has been metabolized and the AD process is coming to an end (Mussoline et al., 2013). Total VFA concentrations are typically expressed in terms of acetic acid equivalents because acetic acid is the most common VFA produced in natural systems (Tchobanoglous et al., 2003). VFA concentrations exceeding 10,000 mg/L as acetic acid are generally considered inhibitory to methanogenesis (Amani et al., 2010). However, methanogens have been acclimated to tolerate high-VFA environments, although certain populations are prone to inhibition at lower VFA concentrations (Khanal, 2008).

5.4.9 Inhibitory Compounds

Ammonia inhibition of methanogens can lead to an accumulation of VFAs, which compounds the inhibition of the AD process (Chen et al., 2008). The principle compound of concern is free ammonia, which increases with increasing total ammonia nitrogen (TAN) concentration, temperature, and pH (Yenigün and Demirel, 2013). However, TAN concentration values are more commonly measured and reported in the literature than free ammonia. Reported threshold ammonia inhibition concentrations vary widely, and depend on feedstock and system characteristics (Yenigün and Demirel, 2013). Chen et al. (2008) reported that TAN concentrations exceeding 1500–1700 mg/L are considered inhibitory. Kayhanian (1994) found inhibition began at a TAN concentration of 1000 mg/L for HS-AD of OFMSW and found optimal performance between 600 and 800 mg/L. Ammonia inhibition is a particular concern in the digestion of manures and is less common in AD of OFMSW (Yenigün and Demirel, 2013).

H$_2$S inhibition can be just as problematic as VFA and ammonia inhibition. The presence of sulfate (SO$_4^{2-}$) in feedstocks and/or inocula results in the production of sulfides (H$_2$S and HS$^-$) in AD, which can accumulate and become toxic to the methanogens (Chen et al., 2008). At neutral pH, H$_2$S concentrations greater than 200 mg/L have been reported as inhibitory (Gerardi, 2003). Measuring input COD-to-sulfate ratios (COD/SO$_4^{2-}$) is an effective strategy for protecting AD systems against sulfide inhibition (Chen et al., 2008). Input COD/SO$_4^{2-}$ ratios greater than 2.7 are unlikely to result in problems with sulfide inhibition (Chen et al., 2008). Another concern is that H$_2$S production by sulfate-reducing bacteria (SRB) is more thermodynamically favorable than methane production by CO$_2$-reducing methanogens (Tchobanoglous et al., 2003; Chen et al., 2008). Thus, SRB will outcompete CO$_2$-reducing methanogens for available H$_2$.

A number of other substances have been shown to inhibit AD, including various salts, organic compounds, and metals (Chen et al., 2008; Deublein and Steinhauser, 2008; Zupančič and Grilc, 2012). Optimal, moderately inhibitory, and inhibitory concentrations of common inorganic salts are shown in Table 5.4 (Zupančič and Grilc, 2012). Inhibition due to high salinity is more commonly reported in HS-AD than L-AD because the use of percolate recirculation in HS-AD can result in steadily increasing salt concentrations. Several trace metals are essential for AD, especially for methanogenesis, but these elements can also be inhibitory at high concentrations (Deublein and Steinhauser, 2008). A list of trace metals, their minimum required concentrations, and their inhibitory concentrations is shown in Table 5.5 (Deublein and Steinhauser, 2008).

TABLE 5.4   Optimal, Moderately Inhibitory, and Inhibitory Concentrations of Inorganic Salts in Anaerobic Digestion

Inorganic Salt	Optimal Concentration (mg/L)	Moderate Inhibition (mg/L)	Severe Inhibition (mg/L)
Sodium (Na)	100–200	3,500–5,500	16,000
Potassium (K)	200–400	3,500–4,500	12,000
Calcium (Ca)	100–200	2,500–4,500	8,000
Magnesium (Mg)	75–150	1,000–1,500	3,000

Source: Adapted from Zupančič, G.D., and Grilc, V., in *Management of Organic Waste*, ed. S. Kumar, InTech, Rijeka, Croatia.

TABLE 5.5   Micronutrient Requirements in Anaerobic Digestion and Potentially Inhibitory Concentrations

Micronutrient	Minimum Required Concentration (mg/L)	Inhibition Concentration as Ion (mg/L)	Inhibitory Concentration as Carbonate (mg/L)
Chromium (Cr)	0.005–50	28–300	530
Iron (Fe)	1–10	N/A	1750
Nickel (Ni)	0.005–0.5	10–300	N/A
Copper (Cu)	>0	5–300	170
Zinc (Zn)	>0	3–400	160
Lead (Pb)	0.02–200	8–340	N/A
Cobalt (Co)	0.06	N/A	N/A
Molybdenum (Mo)	0.05	N/A	N/A
Selenium (Se)	0.008	N/A	N/A
Manganese (Mn)	0.005–50	1500	N/A

Source: Adapted from Deublein, D., and Steinhauser, A., *Biogas from Waste and Renewable Resources*, Willey-VCH Verlag, Weinheim, Germany, 2008.

5.5 Process Engineering and Full-Scale System Descriptions

HS-AD technologies are most often classified according to three key characteristics: loading conditions (continuous or batch), number of stages (single stage or multistage), and operating temperature (mesophilic or thermophilic) (Rapport et al., 2008). In addition to these classifications, HS-AD systems are often classified by feedstock (whether they are processing SS-OFMSW, mechanically separated OFMSW [MS-OFMSW], or mixed MSW) and whether they are processing a single substrate (e.g., yard waste) or are codigesting (e.g., OFMSW with biosolids) (De Baere and Mattheeuws, 2014). Figure 5.4 illustrates the many possible AD system process configurations based on these classifications, and Table 5.6 summarizes their advantages and disadvantages. Tracking trends in the development of HS-AD with respect to the relative prevalence of these AD system types can be helpful for understanding industry preferences and identifying appropriate technologies.

Continuous HS-AD systems are normally loaded on a daily basis, with fresh material going in one end and digested material coming out the other. These systems are generally configured as large plug-flow-type reactors. Batch systems normally consist of multiple "garage"- or "shipping container"-type reactors that are loaded, sealed,

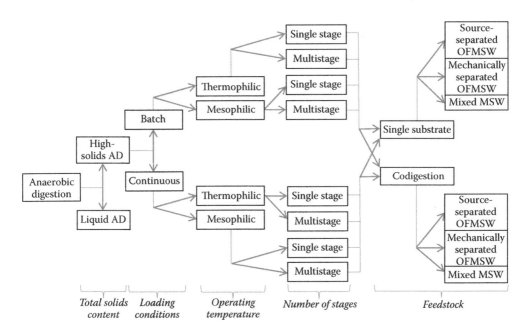

Figure 5.4   AD system process configurations based on predominant system classification.

TABLE 5.6   Technical, Biological, and Environmental/Economic Advantages and Disadvantages of Anaerobic Digestion Technologies for Organic Fraction of Municipal Solid Waste by Classification

System	Criteria	Advantages	Disadvantages
Batch vs. continuous	Technical	Simplifies material handling; reduced preprocessing requirements	Compaction within digester can reduce percolation and percolate recirculation
	Biological	Separation of hydrolysis and methanogenesis	Variable biogas production with time; reduced process control
	Economic and environmental	Low capital cost; low O&M costs; lower overall impact	Less complete degradation
Multistage vs. single stage	Technical	More operationally flexible	Complex design and materials handling
	Biological	Can tolerate high OLRs and fluctuations in OLRs	Can be difficult to achieve true separation of phases in digesters
	Economic and environmental	Can yield higher digestion efficiencies	Increases capital and O&M costs
Thermophilic vs. mesophilic	Technical	Requires minimal change in design (heat transfer systems)	Requires more heat transfer equipment
	Biological	Improves digestion efficiency; improves pathogen destruction	Greater risk of process inhibition with thermophilic systems
	Economic and environmental	Improves bioenergy production rate and marketability of compost	Thermophilic systems require greater heat input
Codigestion vs. single substrate	Technical	Requires no change in design	Requires increased preprocessing
	Biological	Enables optimization of environmental conditions; can improve bioconversion rates	Greater potential for variation in feedstock characteristics and shock inhibition
	Economic and environmental	Can yield significant enhancements in bioenergy generation	Can increase the economic and environmental costs of waste collection
Source-separated OFMSW vs. mixed MSW	Technical	Collection simple with mixed MSW; feedstock contamination of little importance	Collection schemes can be challenging; minor contamination (e.g., glass) can pose a problem for digestate reuse
	Biological	Source separation reduces variation in feedstock characteristics and yields more consistent conditions and performance	Processing mixed MSW poses threats of contamination with strong inhibitory compounds

(Continued)

TABLE 5.6 (CONTINUED)   Technical, Biological, and Environmental/Economic Advantages and Disadvantages of Anaerobic Digestion Technologies for Organic Fraction of Municipal Solid Waste by Classification

System	Criteria	Advantages	Disadvantages
	Economic and environmental	Less energy is needed for mixed MSW collection; less energy is needed for processing source-separated waste and more energy and nutrients are recovered	Processing mixed MSW increases energy input requirements and reduces bioenergy yields and nutrient recovery potential

Source: Adapted from Rapport, J. et al., Current anaerobic digestion technologies used for treatment of municipal organic solid waste, contractor report to the California Integrated Waste Management Board, University of California, Davis, 2008.

and left to digest for a specified amount of time until being unloaded (Rapport et al., 2008). Single-stage systems use a single reactor for the entire AD process, whereas multistage systems use two or more reactors with varying environmental conditions and retention times to separately optimize different phases of the AD process, for example, hydrolysis and acidogenesis in one reactor and acetogenesis and methanogenesis in a subsequent reactor (Deublein and Steinhauser, 2008). Multistage systems also sometimes feature both HS-AD and L-AD (e.g., hydrolysis, acidogenesis, and acetogenesis via HS-AD and methanogenesis via L-AD) (Deublein and Steinhauser, 2008). Mesophilic AD systems have operating temperatures ranging from 35°C to 40°C, whereas thermophilic systems have operating temperatures ranging from 50°C to 55°C. Some multistage systems have stages with varying temperatures, for example, mesophilic first stage and thermophilic second stage (Yunqin et al., 2010).

The number of HS-AD systems available on the market today is rapidly increasing. A summary of the characteristics of systems marketed by different HS-AD vendors based on their marketing materials, discussions with vendors, and field visits to full-scale facilities is provided in Table 5.7. The schematics of Kompogas, Dranco, and Valorga systems are shown in Figure 5.5.

5.6 Trends in HS-AD Implementation for OFMSW in Europe and the United States

Trends in the development of AD of OFMSW in Europe (Table 5.8) were reviewed by De Baere and Mattheeuws (2014). As of 2014, there were 244 full-scale AD plants for processing OFMSW, with a total capacity of approximately 8 million TPY; 62% of installed AD in Europe was HS-AD, and the remaining 38% was L-AD. HS-AD is generally preferred over L-AD for processing OFMSW due to its economic and environmental advantages, and this trend is expected to continue in the future.

The majority of AD systems in Europe as of 2014 were continuous. However, batch systems have been increasing since 2009 due to their simplicity and low cost. Single-stage systems made up approximately 93% of the AD capacity in 2014, with only 7% being multistage (two stage). Implementation of multistage systems has been

TABLE 5.7 Characteristics of Commercially Available High-Solids Anaerobic Digestion Systems

Vendor	Technology	TS Content	Loading Conditions	Number of Stages	Retention Time	Other
Kompogas AG[1]	Kompogas	23%–28%	Continuous	1	15–20 days	Horizontal configuration; mixing via internal rotors; digestate recirculation
Waste Recovery Systems[2]	Valorga	25%–35%	Continuous	1	21 days	Vertical configuration; mixing and inoculation via gas injection
Eggersmann Group; Zero Waste Energy, LLC[3]	SmartFerm; KompoFerm	<50%	Batch	1	21 days	Underground leachate storage with recirculation; PED = 20%
CleanWorld Corporation[4]	CleanWorld HS-AD	~10%	Continuous	3	20–30 days	Separate hydrolysis, biogasification, and biostabilization reactors
Orbit Energy, Inc.[5]	Orbit Energy HS-AD	<45%	Continuous	1	NR	PED = 8%
Viessmann Group; BIOFerm Energy Systems[6]	BIOFerm dry fermentation	25%–35%	Batch	1	28 days	Leachate or digestate recirculation; PED = 5%–10%
Organic Waste Systems, Inc.[7]	Dranco	<50%	Continuous	1	20 days	Vertical configuration; digestate recirculation
GICON Bioenergie GmbH; Harvest Power, Inc.[8]	GICON	NR	Batch	2	≥14 days	Gas-phase digestion occurs under L-AD conditions
Eisenmann Corporation[9]	BIOGAS-GW	NR	Continuous	1	NR	Horizontal configuration
Solum Group; Aikan A/S; Turning Earth, LLC[10]	Aikan dry AD	NR	Batch	2	21 days	Acid-phase digestion units temporarily converted to in-vessel composting units after digestion
EcoCorp, Inc.[11]	EcoCorp dry AD	35%–40%	Continuous	1	20 days	Leachate recirculation; PED = 20%

Note: Additional available commercial HS-AD systems may exist. NR = not reported; PED = parasitic energy demand. Each technology operates under thermophilic conditions except BIOFerm dry fermentation (mesophilic), Valorga (mesophilic or thermophilic), and Dranco (mesophilic or thermophilic). References are as follows: [1]Lissens et al., 2001; Nichols, 2004; [2]Bolzonella et al., 2006; [3]Zhang, 2013, 2015; [4]ZWE, 2013, 2015; CleanWorld, 2015a, 2015b; [5]Greer, 2011; Orbit Energy, 2015; [6]BIOFerm, 2014; [7]De Baere, 2012; [8]Harvest Power, 2014; [9]Eisenmann, 2014; [10]Aikan, 2015; [11]EcoCorp, 2015.

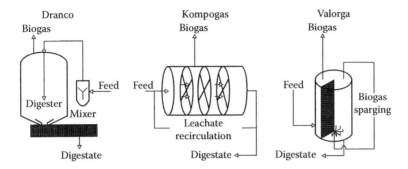

Figure 5.5 Schematics of Dranco, Kompogas, and Valorga HS-AD processes. (Drawn by Phillip Dixon, University of South Florida, Tampa.)

TABLE 5.8   Characterization of Anaerobic Digestion of Organic Fraction of Municipal Solid Waste in Europe

Classification	% of Installed Capacity	Trends	Expected Future Trends
TS content	62% HS-AD, 38% L-AD	HS-AD preferred over L-AD for OFMSW for >20 years; more than 70% installed since 2009 are HS-AD.	HS-AD will increase in prevalence due to the economic and environmental advantages over L-AD.
Loading conditions	>50% continuous	Continuous systems have dominated the industry, but batch systems have been catching on since 2009.	Batch systems expected to continue to increase in popularity due to their simplicity and low cost.
Number of stages	93% single stage, 7% two stage	Multistage systems have been in decline since the 1990s.	No immediate changes in this trend are expected due to the higher investment and operating costs for multistage systems.
Operating temperature	67% mesophilic, 33% thermophilic	Thermophilic digestion has been becoming increasingly common in the last decade.	Thermophilic expected to surpass mesophilic capacity because they are well proven and yield net economic benefits in most cases.
Codigestion	89% single substrate, 11% codigestion	Most current systems are dedicated to OFMSW; however, recently there has been a slight rise in codigestion.	Potential economic advantages of codigestion may make it become increasingly common.
Feedstock	55% source separated, 45% mixed MSW	Increases in capacity for processing SS-OFMSW have been in direct proportion to increases in legislation regulating the source separation.	Source-separated regulations are expected to continue to increase, and therefore digestion of SS-OFMSW will continue to increase.

Source: Adapted from De Baere, L., and Mattheeuws, B., Anaerobic digestion of the organic fraction of municipal solid waste in Europe—Status, experience and prospects, *Waste Manag.*, 3(TK), 517–526, 2014.

continuously declining because their benefits may not justify their higher capital and operating costs. Mesophilic digestion accounted for 67% of AD in Europe in 2014, but thermophilic digestion is becoming increasingly common and is expected to surpass mesophilic digestion, as it is now considered mature and has been shown to yield net economic benefits.

With respect to feedstock, single-substrate digestion (OFMSW) accounted for 89% of AD in Europe in 2014, with codigestion (e.g., OFMSW with wastewater biosolids or livestock wastes) representing only 11% of the installed capacity (De Baere and Mattheeuws, 2014). The long-standing trend has been from codigestion to single-substrate digestion, as stand-alone systems tailored to process OFMSW have become increasingly common. More recently, there has been a slight increase in codigestion, as facilities in the agro-industrial sector have demonstrated the potential economic advantages of codigestion (De Baere and Mattheeuws, 2014). With respect to source separation, 55% of European AD systems in 2014 were processing source-separated OFMSW, while 45% were processing mixed MSW. Increases in capacity for processing source-separated OFMSW have followed promulgation of regulations on source separation of OFMSW in commercial, institutional, and residential settings (De Baere and Mattheeuws, 2014).

In the United States, several pilot-scale and/or demonstration-scale HS-AD projects were constructed prior to 2002, as described by Rapport et al. (2008). The first full-scale demonstration HS-AD system in the United States was constructed in Clinton, North Carolina, in 2002 (Greer, 2011). The 3380 TPY facility employs an HS-AD technology (now marketed by Orbit Energy, Inc.) developed by the U.S. Department of Energy National Renewable Energy Laboratory (Greer, 2011). The first commercial HS-AD system in the United States was constructed in 2011 at the University of Wisconsin (Oshkosh) and began operation in 2012 (UW Oshkosh, 2015). Currently, eight full-scale HS-AD facilities are operating in the United States, with a total capacity of 189,600 TPY. As of this writing, another 19 or more HS-AD projects are in the planning, permitting, or construction phases. The majority of these facilities are located in California and utilize the SmartFerm technology marketed by Zero Waste Energy, LLC (ZWE) (U.S. affiliate of the German company Eggersmann Group), including the largest HS-AD facility in the United States (90,000 TPY in San Jose, California). However, several other vendors have established themselves in the North American HS-AD market, and several other states have implemented or are planning to implement HS-AD.

The current status and trends in the development of AD of OFMSW in the United States are shown in Table 5.9. According to a recent report by the Environmental Research and Education Foundation (EREF, 2015), there are currently 181 AD facilities in the United States processing OFMSW, with a total OFMSW throughput of 780,000 TPY. Of these facilities, 81 are wastewater treatment plant digesters accepting some food waste or fats, oils, and grease (FOG), with a total throughput of 226,000 TPY (29%); 75 are on-farm digesters accepting food and/or yard waste, with a total throughput of 140,000 TPY (18%); and 25 are stand-alone facilities (designed specifically for processing OFMSW) with a total capacity of 406,000 TPY (52%). It follows that approximately 47% of the existing stand-alone capacity for AD of OFMSW is

TABLE 5.9   Characterization of Anaerobic Digestion of Organic Fraction of Municipal Solid Waste in the United States

Classification	Current Status	Expected Future Trends
TS content	Since 2011, the fraction of the stand-alone capacity for AD of OFMSW has increased from nearly 0% to ~48% (189,600 of 406,000 TPY).	HS-AD will become the dominant form of AD of OFMSW by 2017 due to the economic and environmental advantages it offers over L-AD.
Loading conditions	Approximately 61% of HS-AD capacity is currently of the batch variety (116,200 of 189,600 TPY).	14 of the 27 HS-AD systems expected to be in operation by 2017 will be batch systems; no clear trend exists.
Number of stages	Around 63% of HS-AD capacity is currently of the single-stage variety (119,600 of 189,600 TPY).	Only 6 of the 27 HS-AD systems expected to be in operation by 2017 will be multistage, suggesting that single-stage systems are preferred due to their simplicity and cost.
Operating temperature	Thermophilic digestion represents the vast majority (>95%) of the existing capacity for HS-AD of OFMSW.	Thermophilic digestion is expected to remain dominant due to the increased efficiency and demonstrated stability.
Codigestion	Currently, 47% of the capacity for AD of OFMSW is codigestion, with 29% being at wastewater treatment plants and 18% at farms.	The stand-alone capacity for AD of OFMSW is expected to quadruple to 2.5 million tons by 2017 (EREF, 2015) surpassing codigestion as the dominant form.
Feedstock	Limited information on whether existing facilities are processing mixed, mechanically separated, or source-separated OFMSW.	Increased mandates and studies showing economic advantages associated with SS-OFMSW over MS-OFMSW suggest that processing source-separated feedstock will be the dominant form of AD of OFMSW.

Source: Adapted from EREF (Environmental Research & Education Foundation), Anaerobic digestion of municipal solid waste: Report on the state of practice, EREP, Raleigh, NC, 2015, www.erefdn.org; Hinds, G.H., High-solids anaerobic digestion of the organic fraction of municipal solid waste state of the art, outlook in Florida, and enhancing methane yields from lignocellulosic wastes, thesis, University of South Florida, Tampa, 2015, http://scholarcommons.usf.edu/etd/5883.

HS-AD (189,600 of 406,000 TPY). However, if all planned AD facilities for OFMSW come online, by 2017 HS-AD will be the dominant AD technology type for processing OFMSW in the United States, which parallels trends in Europe. With respect to the prevalence of HS-AD systems by other classification categories, 61% of the capacity (on a TPY basis) is batch, 63% is single stage, and 95% is thermophilic.

5.7 Socioeconomic Factors Affecting Project Success

Decades of successful OFMSW HS-AD projects in Europe have provided investors and industry leaders with confidence in the process stability and economic benefits of this technology (RWI, 2013; De Baere and Mattheeuws, 2014). In the United States, however, the economics or cost–benefits of advanced OFMSW management technologies, such as AD, have not yet been fully evaluated (EREF, 2013). Critical factors associated with the economics of HS-AD in MSW management include local tipping fees, local energy market, and quality and consistency of feedstock (RWI, 2013). A primary economic factor associated with capital costs is related to greater economies of scale (Rapport et al., 2008). Other variable economic factors include permitting, engineering planning and design, construction, labor, and insurance. Accordingly, cost estimates for various MSW management technologies and projects vary substantially. According to the World Bank (Hoornweg and Bhada-Tata, 2012), the costs of landfilling, incineration, composting, and AD in high-income countries per ton of waste processed range from $40 to $100, $70 to $200, $35 to $90, and $65 to $150, respectively. These values include the sale of electricity, but exclude the sale of compost and digestate. According to a business analysis of AD in the United States conducted by Renewable Waste Intelligence (RWI, 2013), AD projects can cost up to $600 per ton of annual processing capacity and operating costs can range from $40 to $150 per ton of waste processed. Table 5.10, adapted from PIS (2008), shows estimated capital and operations and maintenance (O&M) costs of various management technologies and processing capacities, and the estimated time requirements from planning to commission.

A detailed cost analysis specifically for HS-AD in the United States was conducted by Rogoff and Clark (2014). In the analysis, the authors estimated the capital investment required for a 5000 TPY capacity HS-AD system based on ZWE's SmartFerm design. This estimate was used to develop a *pro forma* model for estimating required tipping fees for various HS-AD projects and assessing economic feasibility. Required tipping fees were estimated for four different scenarios: with and without electricity production and at two different capacities (5,000 and 10,000 TPY). The results suggest that when electricity sales are included, the required break-even tipping fees range from $4 to $32 per ton. Without electricity sales, required break-even tipping fees range from $41 to $53 per ton.

TABLE 5.10   Estimated Capital and Operations and Maintenance Costs for a Select Processing Capacity or Capacity Range and Estimated Time Requirements from Planning to Commission of MSW Management Technologies

Technology	Plant Capacity (tons/day)	Capital Cost ($/ton)	O&M Cost ($/ton)	Time to Commission (Months)
Landfill	500	5–15	10–30	9–18
Incineration	1300	30–180	80–120	54–96
Pyrolysis	70–270	16–90	80–150	12–30
Gasification	900	15–170	80–150	12–30
Composting (in-vessel)	500	50–80	30–60	9–15
Anaerobic digestion	300	20–80	60–100	12–24

Source: Adapted from PIS (Pytheas Investors Service), Treatment of municipal solid waste—Anaerobic digestion technologies, PIS, Nicosia, Cyprus, 2008, www.pytheas.net.

Based on conclusions from Rogoff and Clark (2014), observations from the development of HS-AD in Europe (De Baere and Mattheeuws, 2014), business analyses of HS-AD in the United States (PIS, 2008, Rapport et al., 2008; RWI, 2013), and feasibility studies for HS-AD implementation in the United States (RIS, 2005), a number of factors, either singularly or in combination, are critical for the economic sustainability and competitiveness of HS-AD, including

- High local electricity costs, high on-site or nearby electricity demand, and/or economic incentives for utility companies to purchase the renewable bioenergy
- Significant centralized sources of source-separated organic wastes, such as from food processing and packaging plants, hospitals, schools, prisons, or other institutional facilities with large cafeterias, or from large agricultural operations with crop residues
- Limited land suitable for composting and/or landfilling and/or lack of conventional WtE facilities
- Markets for the produced compost
- Public–private partnerships, for example, between municipalities, waste management companies and haulers, utility companies, and local organizations
- Grants for funding renewable energy and/or recycling projects
- Regulatory drivers, such as bans on organics disposal in landfills, regulated source separation of OFMSW, renewable energy incentives, air quality regulations increasing the costs of composting and/or WtE operations, and incentives for nutrient recovery and compost use

5.8 Life Cycle Assessment

OFMSW management significantly contributes to a number of environmental impact categories of critical concern, including global climate change and eutrophication. An estimated 1%–5% of total greenhouse gas (GHG) emissions, an anthropogenic contributor to global climate change, in the United States result from waste degradation and waste management practices (EPA, 2009). On a global scale, degradation of waste in open dumps and landfills accounts for 10%–12% of methane (one of the major GHGs) emissions (Hoornweg and Bhada-Tata, 2012). Additionally, nutrient loading from landfill leachate is considered a significant point source of nutrients contributing to eutrophication and negatively impacting wastewater treatment plants, to which landfill leachates are commonly discharged (Townsend et al., 2015). Several LCAs have been conducted comparing the environmental impacts of various OFMSW management methods. Although there are many inconsistencies in the studies, the results have shown a strong preference toward AD with respect to overall environmental impacts (Edelmann et al., 2005; Haight, 2005; Sundqvist, 2005; Zaman, 2009; Kim and Kim, 2010; Levis and Barlaz, 2011; Bernstad and la Cour Jansen, 2012; Morris et al., 2013).

A review by Morris et al. (2013) of the LCA literature compared the impacts of AD, landfilling with flaring, landfilling with landfill gas to energy, incineration WtE, aerobic composting, and home composting. The average climate change impact, expressed in metric tons of CO_2 equivalents, of processing 1 metric ton of OFMSW is far less for AD than for other waste management technologies considered (Table 5.11). On average, 0.25 metric tons of CO_2 equivalents are offset for each metric ton of OFMSW processed by AD due to the avoidance of direct emissions from natural degradation and offsets from energy recovery via biogas utilization. Landfilling and incineration both result in added climate change impacts compared with natural degradation. This is primarily due to fugitive methane emissions in the case of landfilling and carbon dioxide and nitrous oxide emissions (approximately 310 times greater in global warming potential than carbon dioxide) in the case of incineration (Morris et al., 2013).

In addition, AD was shown to have the lowest or nearly lowest average life cycle impacts in the categories of human carcinogenicity, ecotoxicity, acidification, and ground-level smog (Table 5.12). Although AD was reported as among the worst

TABLE 5.11   Climate Change Data from Life Cycle Assessment Literature Review

Management Method	Studies Reviewed	Minimum	Maximum	Median	Mean
		(Metric Tons CO_2 Equivalents/Metric Tons Organic Waste)			
Anaerobic digestion	5	−0.74	−0.06	−0.14	−0.25
Aerobic composting	30	−0.76	0.22	0.04	−0.07
Incineration WtE	9	−0.24	0.63	−0.02	0.02
Home aerobic composting	8	−0.69	0.29	0.14	0.05
Landfill with gas to energy	9	−0.31	1.00	0.11	0.16
Landfill with gas flaring	2	−0.06	−0.05	−0.06	−0.06

Source: Adapted from Morris, J. et al., *Waste Manag.*, 33(3), 545–551, 2013.

TABLE 5.12   Nonclimate Impacts of Management Technologies

Management Method	Human Carcinogenicity	Ecotoxicity	Acidification	Eutrophication[a]	Ground-Level Smog
	Average Ranking[b]				
Anaerobic digestion	1.5 (2)	1.3 (3)	1.0 (2)	3.0 (1)	1.0 (2)
Aerobic composting	1.2 (5)	1.3 (6)	1.8 (6)	1.6 (5)	2.0 (2)
Incineration WtE	2.4 (5)	2.3 (6)	2.2 (6)	2.3 (4)	1.7 (3)
Home aerobic composting	NR	1.0 (1)	1.5 (2)	1.5 (2)	1.0 (1)
Landfill with gas to energy	1.8 (5)	2.0 (5)	1.8 (5)	2.0 (3)	1.5 (2)
Landfill with gas flaring	2.5 (2)	3.5 (2)	2.5 (2)	3.5 (2)	NR

Source: Adapted from Morris, J. et al., *Waste Manag.*, 33(3), 545–551, 2013.

Note: Numbers in parentheses indicate the number of studies that included that impact category. NR = not ranked because that category was not taken into account in any study.

[a]Eutrophication ranking assumes L-AD without nutrient removal from leachate.

[b]The lowest possible ranking is 1.0; higher rankings are relative to the lowest ranking.

options with respect to eutrophication, the one study included in the Morris et al. (2013) review that evaluated the eutrophication impact of AD assumed L-AD without nutrient recovery from the side stream generated in the digestion process (Morris et al., 2013). In this respect, HS-AD is environmentally superior to L-AD for processing OFMSW due to minimal excess leachate generation and the low impact associated with on-site management (recirculation, addition to compost, and evaporation) at most HS-AD facilities (Rapport et al., 2008; Li et al., 2011).

HS-AD paired with composting enables the efficient recovery of nutrients from SS-OFMSW and results in reduced nutrient loading to wastewater treatment facilities and aquatic ecosystems (De Baere and Mattheeuws, 2014). Nutrient recovery also offsets the environmental impacts associated with phosphate mining and inorganic nitrogen production via the Haber–Bosch process. In an LCA study conducted by Edelmann et al. (2005), energy generation and nutrient recovery through AD of OFMSW and combinations of AD and composting were shown to yield significant environmental benefits when compared with energy generation through incineration or nutrient recovery through composting alone. In a study comparing emissions to soil from composting fresh OFMSW versus composting digested OFMSW, emissions of volatile organic compounds (VOCs) and ammonia were found to decrease by a factor of 195 (from 588 g/ton to 3 g/ton) and 1.6 (from 159 g/ton to 98 g/ton), respectively (De Baere, 1999).

A final consideration that is relevant to the environmental aspects of HS-AD of OFMSW is the comparison of the life cycle environmental impacts of AD relative to ATT technologies, such as gasification and pyrolysis. The one LCA study that directly compared AD with these technologies (Zaman, 2009) suggests that ATT offers the potential to reduce life cycle impacts of MSW management relative to incineration WtE and landfilling, but still incurs significantly greater impacts than AD. Another study reviewing 250 LCAs comparing thermal technologies to incineration WtE and landfills concluded that more comprehensive analyses are necessary (Astrup et al., 2014). The study further concluded that the impacts of alternative thermal technologies are generally comparable to traditional incineration WtE, depending on the effectiveness of the air pollution control systems (Astrup et al., 2014), supporting the findings of Zaman (2009) that AD likely provides additional environmental benefits compared with ATTs.

LCA studies have established that AD offers greater potential to enhance the environmental sustainability of OFMSW management than other leading MSW management technologies. Furthermore, because of the benefits that HS-AD offers over L-AD with regard to water consumption, wastewater generation, and energy requirements (parasitic energy demand), it can be deduced that HS-AD is the most environmentally friendly OFMSW management method, especially when feedstocks are of high quality (uncontaminated organics) and HS-AD is used in combination with composting for final curing to improve the quality of the digestate as a soil amendment for nutrient recovery. However, there are a number of inconsistencies in system boundaries, allocation methods, data quality and assumptions, impact assessment methods, and results in LCA studies. Additional research is needed that specifically examines the sensitivities and uncertainties of environmental impacts of HS-AD and the break-even points relative to costs of HS-AD with consideration of key parameters, such as feedstock composition, energy conversion efficiency, recovery of materials for recycling, beneficial offsets for end product alternatives, distance to liquid fuel markets, and market prices for energy products (EPA, 2012).

5.9 Case Study: Codigestion of OFMSW and Biosolids

As discussed in Section 5.4.6, codigestion of food and yard waste has been shown to provide an appropriate C/N balance for efficient digestion. However, little is known about the effect of incorporation of dewatered biosolids (i.e., sewage sludge) as a cosubstrate with OFMSW in HS-AD systems. Biosolids management is increasingly expensive for U.S. wastewater facilities due to the relatively limited L-AD capacity for biosolids in the United States, the increasingly stringent regulations on land application of biosolids, and the high cost of biosolids disposal in landfills (Forbes, 2011). In this study, we carried out biochemical methane potential (BMP) assays (Angelidaki et al., 2009) to investigate the effects of biosolids addition on codigestion of OFMSW with different inoculum sources. Materials and methods are described in detail in Hinds (2015). Briefly, yard waste and food waste (60% fruit and vegetable; 10% eggs and dairy; 15% bread, rice, and grain; and 15% meat) were set up in triplicate in 250 mL glass microcosms, with and without biosolids addition. Two different inoculum sources were used: (1) sludge from an L-AD that processed pulp and paper mill wastewater (P&P) and (2) sludge from an L-AD at a wastewater treatment plant (WW-AD). The P&P sludge was chosen as an alternative inoculum based on prior studies showing that the use of P&P as an inoculum for HS-AD could enhance methane production from yard waste (Mussoline et al., 2013; Hinds et al., 2016), possibly because P&P contains microbial communities adapted to AD of lignocellulosic wastes. WW-AD was used as a conventional inoculum source.

Methane production over a 60-day digestion cycle for digesters with and without biosolids addition and with different inoculum sources is shown in Figure 5.6. The pH decreased from 6.4 to 5.4, and little methane was produced over the first 4 days in digesters that were inoculated with WW-AD. The observed pH decline and inhibition of methanogenesis was likely due to the low alkalinity present in WW-AD

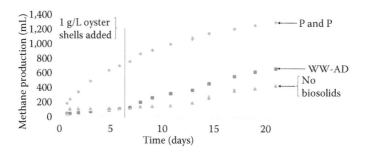

Figure 5.6 Effects of codigestion of biosolids and OFMSW and inoculum source on methane production in HS-AD BMP assays.

(~600 mg/L) compared with P&P (~2000 mg/L). Crushed oyster shells were added to WW-AD inoculated digesters on day 6, resulting in a recovery in methane production (Figure 5.6). The average final methane yields were 65 and 60 L CH_4/kg VS for digesters inoculated with P&P and WW-AD, respectively. The results indicate that inoculation with P&P sludge can lead to improved system stability and methane yield enhancement. Although differences in final average methane yield between digesters inoculated with WW-AD with and without biosolids addition were not significant, biosolids addition resulted in accelerated system recovery after oyster shell addition.

Results from this study show that inoculation with P&P can enhance methane production in HS-AD, although the enhancement observed (15%) was not as pronounced as was observed in studies with yard waste alone (73%) (Hinds et al., 2016). In addition, the results highlight the importance of providing sufficient alkalinity in HS-AD systems. As discussed in Section 5.4, hydrolysis of readily biodegradable organic matter, such as food waste, can result in the release of large quantities of VFAs, which can result in inhibition of methanogenesis if sufficient alkalinity is not provided. Solid-phase alkalinity sources commonly used in AD include lime (calcium hydroxide), soda ash (sodium carbonate or sodium bicarbonate), and limestone (calcium carbonate) (Tchobanoglous et al., 2003). Crushed oyster shells are a low-cost waste material that are composed of calcium carbonate in a protein matrix. Sengupta et al. (2007) showed that the rate of dissolution of oyster shells is relatively slow, and that the waste material could provide long-term buffering against system acidification in autotrophic denitrification systems. Additional experiments are currently underway in our laboratories to investigate this material further.

5.10 Conclusions

Source separation and recycling of OFMSW is likely to become the norm in cities around the world, and HS-AD is expected to become the primary centralized recycling technology for OFMSW because of the many environmental and economic advantages it offers. HS-AD efficiently recovers energy from OFMSW and is easily paired with composting to enable the recovery of nutrients. In the process, GHG emissions that would result from uncontrolled or partially controlled degradation of OFMSW are avoided. GHG emissions are also offset by the substitution of fossil fuel-derived energy with biomethane, which can be used for heating, electricity generation, and/or vehicle fuel. Diversion of OFMSW from landfills to HS-AD facilities reduces eutrophication impacts on the environment or additional energy and chemical inputs needed for removing N and P from leachate streams at wastewater treatment facilities. The recovery and use of nutrients as fertilizer also reduces the impacts of inorganic fertilizer production on the nitrogen cycle (Haber–Bosch process) and depletion of mineral nutrient reservoirs. Trends in the development of HS-AD in Europe and more recently in the United States reveal that the optimization of HS-AD technologies, expansion of regulatory drivers, and development of public–private partnerships are necessary for accelerating the transition from current waste management practices to HS-AD of OFMSW.

Acknowledgments

The authors would like to thank Matthew Dawley for his assistance with this research. Funding for this research was provided by the William W. "Bill" Hinkley Center for Solid and Hazardous Waste Management. Additional funding was provided by the National Science Foundation (NSF) S-STEM Scholarship Program (grant 0965743) and the NSF Research Experience for Teachers (RET) Program (grant 1200682). Any opinions, findings, and conclusions or recommendations expressed in this material are those of the authors and do not necessarily reflect the views of the funding agencies.

References

Abbassi-Guendouz, A., Brockmann, D., Trably, E., Dumas, C., Delgenes, J.P., Steyer, J.P., Escudie, R. 2012. Total solids content drives high solid anaerobic digestion via mass transfer limitation. *Biores Technol* 111:55–61.

Adekunle, K.F., Okolie, J.A. 2015. A review of biochemical process of anaerobic digestion. *Adv Biosci Biotechnol* 6:205–212.

Aikan (Aikan North America). 2015. Batch processing. How it works. Atlanta: Aikan. http://www.aikantechnology.com/56/.

Amani, T., Nosrati, M., Sreekrishnan, T.R. 2010. Anaerobic digestion from the viewpoint of microbiological, chemical, and operational aspects—A review. *Environ Rev* 18:255–278.

Angelidaki, I., Alves, M., Bolzonella, D., Borzacconi, L., Campos, J.L., Guwy, A.J., Kalyuzhnyi, S., Jenicek, P., van Lier, J.B. 2009. Defining the biomethane potential (BMP) of solid organic wastes and energy crops: A proposed protocol for batch assays. *Water Sci Technol* 59(5):927–934.

Astrup, T.F., Tonini, D., Turconi, R., Boldrin, A. 2014. Life cycle assessment of thermal waste-to-energy technologies: Review and recommendations. *Waste Manag* 37:104–115.

Bernstad, A., la Cour Jansen, J. 2012. Review of comparative LCAs of food waste management systems—Current status and potential improvements. *Waste Manag* 32:2439–2455.

BIOFerm (BIOFerm Energy Systems). 2014. BIOFerm dry fermentation. Anaerobic digestion systems for solid input materials. Madison, WI: BIOFerm. http://www.biofermenergy.com/wp-content/uploads/2014/03/2014_BIOFerm-Dry-Fermentation_Batch-System-Digester_BIOFerm-Energy-Systems.pdf.

Bolzonella, D., Pavan, P., Mace, S., Cecchi, F. 2006. Dry anaerobic digestion of differently sorted organic municipal solid waste: A full-scale experience. *Water Sci Technol* 53(8):23–32.

Brown, D., Li, Y. 2013. Solid state anaerobic co-digestion of yard waste and food waste for biogas production. *Bioresour Technol* 127:275–280.

Bruni, E., Jensen, A.P., Angelidaki, I. 2010. Comparative study of mechanical, hydrothermal, chemical and enzymatic treatments of digested biofibers to improve biogas production. *Bioresour Technol* 101:8713–8717.

Cecchi, F., Traverso, P.G., Mata-Alvarez, J., Clancy, J., Zaror, C. 1988. State of the art of R&D in the anaerobic digestion process of municipal solid waste in Europe. *Biomass* 16:257–284.

Chen, X., Yan, W., Sheng, K., Sanati, M. 2014. Comparison of high-solids to liquid anaerobic co-digestion of food waste and green waste. *Bioresour Technol* 154:215–221.

Chen, Y., Cheng, J.J., Creamer, K.S. 2008. Inhibition of anaerobic digestion process: A review. *Bioresour Technol* 99(10):4044–4064.

Cho, J.K., Park, S.C., Chang, H.N. 1995. Biochemical methane potential and solid-state anaerobic-digestion of Korean food wastes. *Bioresour Technol* 52(3):245–253.

De Baere, L. 1999. Anaerobic digestion of solid waste: State of the art. *Water Sci Technol* 41(3):283–290.

De Baere, L., Mattheeuws, B. 2014. Anaerobic digestion of the organic fraction of municipal solid waste in Europe—Status, experience and prospects. *Waste Manag* 3(TK):517–526.

Delgenés, J.P., Penaud, V., Moletta, R. 2002. Pretreatments for the enhancement of anaerobic digestion of solid wastes. In *Biomethanization of the Organic Fraction of Municipal Solid Wastes*, ed. J. Mata-Alvarez, 201–228. London: IWA Publishing.

del Real Olvera, J., Lopez-Lopez, A. 2012. Biogas production from anaerobic treatment of agroindustrial wastewater. In *Biogas*, ed. S. Kumar, 91–112. Rijeka, Croatia: InTech.

Deublein, D., Steinhauser, A. 2008. *Biogas from Waste and Renewable Resources*. Weinheim, Germany: Willey-VCH Verlag.

EcoCorp. 2015. Technology. Topanga, CA: EcoCorp. http://www.ecocorp.com/Technology .html.

Edelmann, W., Baier, U., Engeli, H. 2005. Environmental aspects of the anaerobic digestion of the organic fraction of municipal solid wastes and of agricultural wastes. *Water Sci Technol* 52(1–2):203–208.

EPA (U.S. Environmental Protection Agency). 2009. Opportunities to reduce greenhouse gas emissions through materials and land management practices. Washington, DC: Office of Solid Waste and Emergency Response. https://www.epa.gov/region-9-documents /opportunities-reduce-greenhouse-gas-emissions-through-materials-and-land.

EPA (U.S. Environmental Protection Agency). 2012. State of practice of emerging waste conversion technologies. EPA 600/R-12/705. Research Triangle Park, NC: RTI International. http://nepis.epa.gov/Adobe/PDF/P10 0FBUS.pdf.

EREF (Environmental Research & Education Foundation). 2013. Summary of waste conversion technologies. Prepared for NEWMOA by Bryan Staley. Raleigh, NC: EREP. http:// www.newmoa.org/events/docs/112_109/EREF_MSW_Conversion_Techs_Aug2013.pdf.

EREF (Environmental Research & Education Foundation). 2015. Anaerobic digestion of municipal solid waste: Report on the state of practice. Raleigh, NC: EREP. www.erefdn.org.

Fabián, R.M., Gourdon, R. 1999. Effect of baling on the behavior of domestic wastes: Laboratory study on the role of pH in biodegradation. *Bioresour Technol* 69:15–22.

Forbes, R.H. 2011. The changing landscape of biosolids management in Florida: The 21st century's first decade & predictions for the next one. *Florida Water Res J* 54–60.

Gerardi, M.H. 2003. *The Microbiology of Anaerobic Digesters*. Hoboken, NJ: Wiley-Interscience.

Gollakota, K.G., Meher, K.K. 1988. Effect of particle size, temperature, loading rate and stirring on biogas production from castor cake. *Biol Wastes* 24:243–249.

Greer, D. 2011. Digester developers target mixed organic waste streams. *Biocycle* 52(8):53.

Haight, M. 2005. Assessing the environmental burdens of anaerobic digestion in comparison to alternative options for managing the biodegradable fraction of municipal solid wastes. *Water Sci Technol* 52(1–2):553–559.

Harvest Power. 2014. Clean energy fund: Urban renewal bioenergy production. Public project report. Richmond, BC: Fraser Richmond Soil & Fibre. http://www.harvestpower.com /wp-content/uploads/2014/10/Harvest-Fraser-Richmond-Organics-CEF-Outreach -Report-EN.pdf.

Hinds, G.H. 2015. High-solids anaerobic digestion of the organic fraction of municipal solid waste state of the art, outlook in Florida, and enhancing methane yields from lignocellulosic wastes. Thesis, University of South Florida, Tampa. http://scholarcommons.usf .edu/etd/5883.

Hinds, G.R., Mussoline, W., Casimir, L., Dick, G., Yeh, D.H., Ergas, S.J. 2016. Enhanced methane production from yard waste in high-solids anaerobic digestion through inoculation with pulp and paper mill anaerobic sludge. *Environmental Engineering Science* 33(11):907–917.

Hoornweg, D., Bhada-Tata P. 2012. What a waste: A global review of solid waste management. Urban Development Series Knowledge Papers No. 15. Urban Development & Local Government Unit. Washington, DC: World Bank.

Izumi, K., Okishio, Y.K., Nagao, N., Niwa, C., Yamamoto, S., Toda, T. 2010. Effects of particle size on anaerobic digestion of food waste. *Int Biodeterior Biodegrad* 64(7):601–608.

Kaparaju, P., Luostarinen, S., Kalmari, E., Kalmari, J., Rintala, J. 2002. Co-digestion of energy crops and industrial confectionery by-products with cow manure: Batch scale and farm-scale evaluation. *Water Sci Technol* 45:275–280.

Kaparaju, P., Rintala, J. 2005. Anaerobic co-digestion of potato tuber and its industrial by-products with pig manure. *Resour Conserv Recycling* 43(2):175–188.

Karagiannidis, A., Perkoulidis, G. 2009. A multi-criteria ranking of different technologies for the anaerobic digestion for energy recovery of the organic fraction of municipal solid wastes. *Bioresour Technol* 100:2355–2360.

Kato, M.T., Field, J.A., Versteeg, P., Lettinga, G. 1994. Feasibility of expanded granular sludge bed reactors for the anaerobic treatment of low-strength soluble wastewaters. *Biotechnol Bioeng* 44(4):469–479.

Kayhanian, M. 1994. Performance of a high-solids anaerobic digestion process under various ammonia concentrations. *Chem Tech Biotechnol* 59:349–352.

Khanal, S. 2008. *Anaerobic Biotechnology for Bioenergy Production Principles and Applications*. Hoboken, NJ: John Wiley & Sons.

Kim, M.H., Kim, J.W. 2010. Comparison through a LCA evaluation analysis of food waste disposal options from the perspective of global warming and resource recovery. *Sci Total Environ* 408(19):3998–4006.

Kothari, R., Pandey, A.K., Kumar, S., Tyagi, V.V., Tyagi, S.K. 2014. Different aspects of dry anaerobic digestion for bio-energy: An overview. *Renewable Sustainable Energy Rev* 39:174–195.

Kreuger, E., Sipos, B., Zacchi, G., Svensson, S.E., Björnsson, L. 2011. Bioconversion of industrial hemp to ethanol and methane: The benefits of steam pretreatment and co-production. *Bioresour Technol* 102:3457–3465.

Levis, J.W., Barlaz, M.A. 2011. What is the most environmentally friendly way to treat commercial food waste? *Environ Sci Technol* 45(17):7438–7444.

Li, Y., Park, S., Zhu, J. 2011. Solid-state anaerobic digestion for methane production from organic wastes. *Renewable Sustainable Energy Rev* 15:821–826.

Lissens, G., Vandevivere, P., De Baere, L., Biey, E.M., Verstraete, W. 2001. Solid waste digestors: Process performance and practice for municipal solid waste digestion. *Water Sci Technol* 44(8):91–102.

Madigan, M.T., Martinko, J.M., Bender, K.S., Buckley, D.H., Stahl, D.A. 2014. *Brock Biology of Microorganisms*. 14th ed. Boston: Pearson.

Massé, D., Droste, R. 2000. Comprehensive model of anaerobic digestion of swine manure slurry in a sequencing batch reactor. *Water Res* 34(12):3087–3106.

Moorhead, K.K., Nordstedt, R.A. 1993. Batch anaerobic digestion of water hyacinth: Effects of particle size, plant nitrogen content and inoculum volume. *Bioresour Technol* 44(1):71–76.

Morris, J., Matthews, H.S., Morawski, C. 2013. Review and meta-analysis of 82 studies on end-of-life management methods for source separated organics. *Waste Manag* 33(3):545–551.

Mussoline, W., Esposito, G., Lens, P., Spagni, A., Giordano, A. 2013. Enhanced methane production from rice straw co-digested with anaerobic sludge from pulp and paper mill treatment process. *Bioresour Technol* 148:135–143.

Nichols, C.E. 2004. Overview of anaerobic digestion technologies in Europe. *Biocycle* 45(1):47–53.

Owen, W.F., Stuckey, D.C., Healy, J.B., Young, L.Y., McCarty, P.L. 1979. Bioassay for monitoring biochemical methane potential and anaerobic toxicity. *Water Res* 13:485–492.

Owens, J.M., Chynoweth, D.P. 1993. Biochemical methane potential of municipal solid waste (MSW) components. *Water Sci Technol* 27(2):1–14.

PIS (Pytheas Investors Service). 2008. Treatment of municipal solid waste—Anaerobic digestion technologies. Nicosia, Cyprus: PIS. www.pytheas.net.

Rapport, J., Zhang, R., Jenkins, B.M., Williams, R.B. 2008. Current anaerobic digestion technologies used for treatment of municipal organic solid waste. Contractor report to the California Integrated Waste Management Board. Davis: University of California.

RIS (RIS International Ltd., in association with MacViro Consultants Inc.). 2005. Feasibility of generating green power through anaerobic digestion of garden refuse from the Sacramento area. SMUD Advanced Renewable and Distributed Generation Program. Sacramento: RIS.

Rogoff, M.J., Clark, B.J. 2014. Anaerobic digestion—What are the economics? *Waste Advantage Magazine*, July, pp. 22–25.

RWI (Renewable Waste Intelligence). 2013. Business analysis of anaerobic digestion in the USA. London: RWI.

Schievano, A., D'Imporzano, G., Malagutti, L., Fragail, E., Ruboni, G., Adani, F., 2010. Evaluating inhibition conditions in high-solids anaerobic digestion of organic fraction of municipal solid waste. *Bioresour Technol* 101:5728–5732.

Sengupta, S., Ergas, S.J., Lopez-Luna, E. 2007. Investigation of solid-phase buffers for sulfur-oxidizing autotrophic denitrification. *Water Environment Research* 79(13):2519–2526.

Sharma, S.K., Mishra, I.M., Sharma, M.P., Saini, J.S. 1988. Effect of particle size on biogas generation from biomass residues. *Biomass* 17(4):251–263.

Sundqvist, J.O. 2005. How should municipal solid waste be treated—A system study of incineration material recycling, anaerobic digestion and composting. Stockholm: Swedish Environmental Research Institute.

Tchobanoglous, G., Burton, F.L., Stensel, H.D., revs. 2003. *Wastewater Engineering: Treatment and Reuse*. 4th ed. Boston: McGraw-Hill.

Townsend, T.G., Powell, J., Jain, P., Xu, Q., Tolaymat, T., Reinhart, D. 2015. Waste and landfill fundamentals. In *Sustainable Practices for Landfill Design and Operation*, 13–34. New York: Springer.

UW Oshkosh (University of Wisconsin Oshkosh). 2015. UW Oshkosh urban anaerobic dry biogas systems. Innovation in sustainability and renewable energy. Oshkosh: UW Oshkosh. http://www.uwosh.edu/biodigester/About/uw-oshkosh-biodigester.

van Haandel, A.V., Lubbe, J.V.D. 2007. *Handbook of Biological Waste Water Treatment: Design and Optimisation of Activated Sludge Systems*. Leidschendam, the Netherlands: Quist Publishing.

Vandenburgh, S.R., Ellis, T.G. 2002. Effect of varying solids concentration and organic loading on the performance of temperature phased anaerobic digestion process. *Water Environ Res* 74(2):142–148.

Veeken, A. 2014. Personal communication with Adrie Veeken, European Federation of Waste Management and Environmental Services (FEAD), Attero, Moerdijk, the Netherlands. October 13.

Veeken, A., Hamelers, B. 1999. Effects of temperature on hydrolysis rates of selected biowaste components. *Bioresour Technol* 69:249–254.

Yang, L., Xu, F., Ge, X., Li, Y. 2015. Challenges and strategies for solid-state anaerobic digestion of lignocellulosic biomass. *Renewable Sustainable Energy Rev* 44:824–834.

Yenigün, O., Demirel, B. 2013. Ammonia inhibition in anaerobic digestion: A review. *Process Biochem* 48:901–911.

Yunqin, L., Dehan, W., Lishang, W. 2010. Biological pretreatment enhances biogas production in the anaerobic digestion of pulp and paper sludge. *Waste Manag Res* 28:800–810.

Zaman, A.U. 2009. Life cycle environmental assessment of municipal solid waste to energy technologies. *Global J Environ Res* 3(3):155–163.

Zhang, R. 2013. U.S. Patent application publication. US 2013/0260433. patentimages.storage .googleapis.com/pdfs/US20130260433.pdf.

Zupančič, G.D., Grilc, V. 2012. Anaerobic treatment and biogas production from organic waste. In *Management of Organic Waste*, ed. S. Kumar, 3–28. Rijeka, Croatia: InTech.

ZWE (Zero Waste Energy). 2013. Our solutions. Dry anaerobic digestion. Lafayette, CA: ZWE. http://zerowasteenergy.com/our-solutions/dry-anaerobic-digestion/.

ZWE (Zero Waste Energy). 2015. Personal communication with Chris Axton, ZWE Monterey facility manager. May 5. Monterey, CA: ZWE.

6

Microbial Biomethane Production from Agricultural Solid Wastes

Rémy Bayard and Pierre Buffière

Contents

6.1 Introduction

In the context of increasing long-term fossil fuel prices, the European Union (EU) is developing policies that aim to ensure its energy security. According to the European regulations, the proportion of renewable energies is supposed to reach 20% of the total energy consumption in 2020, and biofuel will increase to 10% of the total fuel consumption. In this context, EU countries are developing strategies to produce biofuels (Bauer et al., 2010). Among others, methane production by anaerobic digestion of agricultural and assimilated resources is considered one of the sectors of the future to help reduce our fossil fuel dependence. Indeed, methane is an energy carrier that has the advantage of being produced from renewable resources and used as a substituent of natural gas.

The primary energy production in Europe from biogas was 13.4 million tons of oil equivalent (MTOE) of primary energy in 2013, with an annual growth of more than 10% over the past 5 years (EurObserv'ER, 2014). Anaerobic digestion was historically developed for the treatment of industrial effluents and sewage sludge. Today, agricultural resources represent a strong potential for biogas production, with an annual output of agricultural residues estimated at nearly 1000 Mt (Bidlingmaier et al., 2004). Agricultural biogas has experienced strong growth in some European countries, in particular in Germany under the leadership of an incentive policy, often through a high purchase price of the electricity produced from biogas. Germany stands out for its major development of individual units using agricultural energy crops to produce first-generation biofuels, with several drawbacks. The massive use of energy crops leads to some questions about the competition between food crops and the environmental point of view when it comes to intensive cultivation of biomass for energy purposes.

In view of the energy transition driven by European regulations, the forecasts agree on the continued deployment of agricultural biogas. This will give way to anaerobic digestion in smaller units at the farm scale dedicated to the treatment and recovery of agricultural residues and to centralized biogas plants dedicated to the treatment of organic residues recovered locally (agricultural residues, agro-food industry waste, waste from hypermarkets and supermarkets, etc.).

This chapter presents an overview of the situation today. We first explore the available resources for biogas production and give global figures about their ability to be fermented into methane. We then focus on the current practices for biogas processes used for anaerobic digestion of agricultural resources. The next section is dedicated to future trends in terms of technologies and research requirements. The chapter ends with a focus on actual operating systems in the form of a case study.

6.2 Exploring Resources for Biogas Production

6.2.1 Eligible Resources for Agricultural Biogas

The use of biomass residues is of great interest for producing biogas with economic and environmental interests. Among the organic residues "eligible" for the production of biogas, there are many resources that can be mobilized, including agricultural residues; household, supermarket, and catering food waste; waste from forestry; and waste from food industries. The evaluation of these resources at the European level in terms of quantity and availability is still difficult to estimate (Bidlingmaier et al., 2004). According to the average composition of elements given by Davidsson et al. (2007), the annual potential recoverable in the form of biogas would be close to 18 MTOE.

As eligible resources for agricultural biogas, we consider any type of organic matter that can be converted into methane. These residual biofuels are currently the subject of a quite remarkable interest in the goal of energy recovery by anaerobic digestion. This explains the development of "farm biogas plants" in Germany (with around 10,000 installations) and the development of centralized systems in the Nordic countries, like Denmark. These facilities are primarily dedicated to livestock waste (animal manure) and crop residues, but can also treat waste from the agrofood industry, waste from super- and hypermarkets, and the direct digestion of crops (also called energy crops). Table 6.1 shows the main resources that can be mobilized for agricultural digestion. The ability of a given product to produce methane during its anaerobic digestion is called the biochemical methane potential (BMP). It is defined as the maximal amount of methane produced during its degradation under controlled conditions (Angelidaki and Sanders, 2004).

6.2.1.1 Forestry Residues
Wood wastes correspond to the portion of forest products that have no value to energy, furniture, and construction. The logging operations provide the raw material for the wood industry. These are branches, tops, and stumps, but also entire trees from thinning and stems that do not meet the quality requirements for the wood industry. This residual biomass is potentially recoverable for further use. These by-products are rarely collected because of their weak economic interest: they are moist and have a low density. In addition, their BMP without pretreatment is too low to be considered a valuable resource for anaerobic digestion. Indeed, these residues have the characteristics of being predominantly composed of lignocellulose, whose complex structure makes them difficult to be degraded biologically (Hendriks and Zeeman, 2009). For

TABLE 6.1   Main Biomass Resources Potentially Available for Agricultural Anaerobic Digestion

Source	Origin	Residues
Forestry Residues		
Waste from forest cultivation	Wood logging activities	Crowns, branches, stumps
Waste from the wood industry	Industry primary wood	Bark, sawdust, wood chips
	Industry secondary wood	Sawdust, wood scraps, machining chips, sanding dust
Agricultural Residues		
Residues from livestock	Breeding activities	Slurries and manures
Residues from crops	Agricultural activities	Plant parts left behind: straws, canes, etc.
Industrial Residues		
Agricultural processing waste	Food processing industries (primary and secondary processings)	Molasses, slaughterhouse wastes, fruit and vegetable pulp and peelings, whey, shells, etc.
Market Residues		
Waste from retail	Outdoor markets, supermarkets, hypermarkets	Food waste
Household and Catering Waste		
Postconsumer residues	Domestic waste and catering waste	Organic food waste

forestry residues and the wood industry, destructuring the lignocellulosic complex is a major challenge for biogas recovery (Taherzadeh and Karimi, 2008; Hendriks and Zeeman, 2009; Chandra et al., 2012).

6.2.1.2 Agricultural Residues: Waste from Livestock
The main livestock wastes are slurries and manure. Slurries are mixtures of feces, urine (sometimes with litter residues), washing water, and food residue. Manures are mixtures of feces and urine with a large amount of litter (straw). From a practical point of view, the difference between the two is their consistency, that is, their ability to be pumped (slurries) or to be handled as solids (manure). Today, farming practices are largely based on animal housing, that is, the breeding of animals in a building or in an enclosed park: 80% of cattle are kept on litter (manure production), and 100% of calves and 90% of pigs are raised aboveground without litter (production of manure). Consequently, breeding conditions lead to the generation of potentially mobilized residues for biogas, particularly in areas where there is excess production, which makes it difficult to spread on agricultural land through systematic recourse (Védrenne, 2007). Moreover, these deposits are composed of very different products ranging from 2%–3% total solids (TS) for slurries to 30%–40% TS for certain poultry manures. This variation comes from the type of farming and feeding animals (e.g., the difference between finishing pigs and lactating sows). In addition, for the same

TABLE 6.2   Methanogenic Potential of Different Types of Animal Residues (m³ CH₄/kg$_{VS}$)

	Manures	Slurries
Cattle	0.15–0.45	0.15–0.3
Pigs	0.15–0.2	0.25–0.4
Poultry	ND	0.3–0.5
Rabbits	0.15–0.3	0.2–0.3
Sheep and goats	0.2–0.3	ND

Note: ND, no data.

breeding category, the contents vary from one operation to another. It is therefore difficult to obtain consolidated data that would estimate the overall biomethane potential. However, we found a few ranges of values in the literature (Table 6.2).

6.2.1.3 Agricultural Residues: Waste from Crops
Crop residues correspond to the portion of the crop that has no nutritional value. It consists primarily of straw and grain growing rods (wheat, barley, oats, rye, rice, triticale, etc.), oilseeds (sunflower and rapeseed), and vine shoots. The waste is partly left on the harvesting site as structuring and organic material, making it difficult to estimate the gross waste of crops. The cereal sector produces straw, which is the woody stem of the plant harvested at maturity. For instance, wheat straw production in France is about 22 Mt/year of wet material, mainly produced from crops of wheat and barley, respectively, 70% and 20% of the cereal production. A total of 16.5 Mt/year is collected and used as bedding for animals kept indoors. Only 5.5 Mt/year would thus be left on agricultural soils, corresponding to a rate of return on the ground below the recommended average. The corn and sunflower sectors produce canes and crop residues that are left on the ground after harvesting. They consist of stems, leaves, and husks of the plant harvested when the grains are mature: 50% stems, 22% leaves, 15% corn, and 13% envelopes. The canes are usually crushed after the grain threshing.

Residual crops are relatively dry (TS rate of about 30% by weight or higher), rich in lignocellulose, and low in nitrogen (0.5%–0.8% TS). These properties make them difficult to use as such in a biogas plant, but they can be incorporated as cosubstrates (typical case of manure). Many of these compounds have been investigated for biogas production, and some results for methane potential are given in Table 6.3.

6.2.1.4 Distribution and Retail Losses, and Domestic Food Waste
Several European countries have put in place regulations for sorting and recycling biowaste. This is, for example, the case of France, where the waste from food industries, catering, and food trade are also involved. This obligation for source sorting is an important innovation that brings significant changes in the organization of the producers of biowaste as actors making their collection or processing. These bioresources are likely to be treated in agricultural biogas and recovered as energy in the form of biogas (Table 6.4).

TABLE 6.3   Biomethane Potentials of Residues from Crops (m^3 CH$_4$/kg$_{VS}$)

	VS (% TS)	BMP (m^3 CH$_4$/kg$_{VS}$)	Reference
Barley straw	94.9	0.264	Liu et al., 2015
Corn stover	95.5	0.292	Liu et al., 2015
Corn stover	ND	0.275	Chandler et al., 1980
Corn stover	93.2	0.360	Tong et al., 1990
Corn leaves	ND	0.255	Chandler et al., 1980
Fruits (rotten)	98.0	0.270	Buffière et al., 2008
Grapes	88.9	0.180	Gunaseelan, 2004
Grass (cutting lawn)	88.1	0.209	Owens and Chynoweth, 1993
Grass (cutting lawn)	85.1	0.258	Veeken and Hamelers, 1999
Grass (fodder)	95.8	0.291	Tong et al., 1990
Rice straw	86.3	0.200	Zhang and Zhang, 1999
Rice straw	89.0	0.250	Ghosh and Bhattacharyya, 1999
Sugarcane bagasse	96.8	0.244	Liu et al., 2015
Sugarcane bagasse	93.8	0.211	Liu et al., 2015
Wheat straw	91.5	0.324	Liu et al., 2015
Wheat straw	ND	0.276	Bauer et al., 2010
Wheat straw	94.8	0.203	Veeken and Hamelers, 1999
Wheat straw	91.3	0.302	Tong et al., 1990
Wheat straw	86.0	0.304	Hashimoto, 1986
Sorghum leaves	94.7	0.367	Gunaseelan, 2004
Sorghum bagasse	91.7	0.407	Gunaseelan, 2004
Sunflower stalks	88.0	0.192	Monlau et al., 2013
Sugar beet leaves	81.4	0.231	Gunaseelan, 2004

Note: ND, no data.

TABLE 6.4   Biomethane Potentials of Food Waste Production and Consumption (m^3 CH$_4$/kg$_{VS}$)

	BMP (m^3 CH$_4$/kg$_{VS}$)	Reference
Food waste (from cooking preparation)	0.540	Chynoweth et al., 1992
Food waste (from separation collection)	0.290	Liu et al., 2015
Mixed food waste	0.470	Cho and Park, 1995
Biowaste (from separate collection)	0.360–0.400	Mace et al., 2003
Biowaste (mechanical separation from residual municipal solid waste)	0.230–0.260	Mace et al., 2003

6.2.1.5 Energy Crops

From a purely technical point of view, agricultural land can be specifically dedicated to this type of culture. Indeed, most crops can be used directly in agricultural biogas plants as a renewable energy source (Amon et al., 2007). The biomethane potentials of several energy crops are presented in Table 6.5. These crops can be classified into three categories. The intermediate crops are grown in between two

TABLE 6.5   Biomethane Potentials of Energy Crops

	BMP (m^3 CH$_4$/kg$_{VS}$)	Reference
Alfalfa	0.320	Lehtomäki, 2006
Clover	0.290–0.390	Amon et al., 2005; Lehtomäki, 2006
Faba bean	0.370	Pakarinen et al., 2012
Forage beet	0.360–0.460	Weiland, 2003
Hemp (whole crop)	0.310–0.380	Pakarinen et al., 2012
Maize (whole crop)	0.400	Pouech et al., 1998; Pakarinen et al., 2012
Oat (grain)	0.295	Pouech et al., 1998
Potatoes	0.570–0.400	Weiland, 2003
Rye grass	0.400	Pouech et al., 1998
Sorghum (sweet cultivar)	0.350–0.400	Jerger et al., 1987
Sorghum (grain cultivar)	0.280–0.310	Jerger et al., 1987
Sugar beet	0.230	Gunaseelan, 2004
Sunflower	0.150–0.330	Amon et al., 2005
Triticale	0.410–0.440	Plöchl and Heiermann, 2006
Wheat (grain)	0.380	Pouech et al., 1998

main crops, during summer or winter, and use plants with a short growing cycle. This is very often forages being bred for animal feed during the winter by farmers and generally characterized by a good nutritional quality and high biomethane yield. Many of them are leguminous crops such as alfalfa, ryegrass, mustard, and oilseed radish. They require little fertilization and have the ability to bind nitrogen from the atmosphere with their root nodules. The use for energy of "conventional" dual-purpose crops (food and energy) poses direct competition issues in relation to food. Among the most cited in the literature are corn; sorghum (grain and fodder); cereals such as wheat, triticale, and rye, and generally full and ensiled plants; sugar beet; potato; the oilseeds (sunflower and rapeseed); and protein crops (peas, faba beans, etc.). The dedicated energy crops are strictly cultivated for energy purposes, which are not usable for food production, but are in competition for the use of arable land. Among them, biomass sorghum, cane of Provence, switchgrass, and miscanthus are widely studied because of their high biomethane potential, biomass production per hectare yield, and low requirement in terms of moisture and supply of fertilizers.

The European energy potential of energy crops is around 750 million tons of biomass; in some countries, like Finland, 25% of agricultural land can be used in this way (Lehtomäki, 2006). This option is seriously considered in some European states, including Austria, Finland, Germany, Benelux, and the United Kingdom. Recently, global analyses of production runs indicated that the mobilization of agricultural land for energy purposes could have a significant effect on the balance of world food markets. The recent rise in food prices, which has resulted in numerous supply problems in developing countries, is largely due to the diversion of agricultural production to biofuels in developed countries. Should this effect be confirmed, energy crops would probably be less encouraged at the European level.

6.2.2 Characteristics

The eligibility of these resources also depends on their biophysicochemical properties. The structure of materials, presence of undesirable materials, biochemical composition, and anaerobic biodegradability (BMP) are sources of technical limitations to the conversion into biogas. Indeed, as noted in Section 6.2.1, all organic resources do not have the same methane production potential. This depends on their structural (particle size heterogeneity and water holding capacity) and biochemical (soluble carbohydrates or as insoluble polymers, lipids, and proteins) properties. Among these organic resources, we find easily degradable products, such as simple sugars, simple alcohols, and short-chain organic acids. But other compounds are less readily biodegradable, such as fats and certain polymers, such as cellulose and lignin.

Agricultural and forest biomass have the common characteristic of being mainly made of lignocellulose, whose complex structure makes them difficult to degrade biologically, especially at the hydrolysis step (Hendriks and Zeeman, 2009). The structure of the lignocellulosic complex is a major issue for the bioconversion of lignocellulosic biomass to bioethanol or methane (Mosier et al., 2005; Taherzadeh and Karimi, 2008; Chandra et al., 2012; Monlau et al., 2013). We will analyze later the main characteristics of lignocellulosic biomass that affect the anaerobic biodegradability, including lignin, cellulose, and hemicellulose, since their accessibility is the most crucial feature.

6.2.2.1 Lignin Content

It is widely accepted by the scientific community that the resistance of lignocellulosic biomass to anaerobic biodegradation increases with the lignin content. The presence of lignin leads to a reduction in the conversion to methane of carbohydrate polymers. However, the correlation between the lignin content and anaerobic biodegradability is not clearly established, as shown by the disparity of available data (Liew et al., 2012). Cellulose, hemicellulose, and lignin are present in varying proportions depending on the origin of the treated material (Jeffries, 1990). Several studies have tried to establish the link between BMP and lignin content on agricultural residues (Chandler et al., 1980; Gunaseelan, 2004, 2007; Monlau et al., 2013), and the biomass of various origins (Tong et al., 1990; Eleazer et al., 1997; Buffière et al., 2006). It is not always easy to compare the results with another (variables include analytical techniques tested from one publication to another), and sometimes they are contradictory when it comes to explaining the influence of the lignocellulosic structure on the BMP. In addition, the quantitative relationship between the BMP and the lignin content varies, depending on the origin and characteristics of the investigated biomass (Barlaz, 1996). Indeed, Tong et al. (1990) were able to demonstrate very different food waste anaerobic biodegradation rates, while their lignin contents were identical. The work of Eleazer et al. (1997) on various organic waste (seed, grass, leaves, and branches) has confirmed that the lignin content is not a sufficient parameter to predict the anaerobic biodegradation.

6.2.2.2 Carbohydrate Accessibility

A hurdle to recovery by anaerobic digestion of the majority of plant biomass is the predominant lignocellulosic structure. This complex structure combines three closely linked polymers: cellulose, hemicellulose, and lignin. Cellulose and hemicellulose are polysaccharides, while the lignin is an aromatic polymer. Thus, lignin is a three-dimensional skeleton that protects the cellulose and hemicellulose. These three components are present in variable proportions, depending on the origin of the material considered. The long polysaccharide chains of cellulose are organized in the form of combined microfibrils, together with the hemicellulose polymers, the assembly being incorporated into the cross-linked network of the lignin.

The hydrolysis of polysaccharides requires the concerted action of several exoenzymes, carried out in aqueous phase. Therefore, direct contact between the substrate and the exoenzyme is required for hydrolysis (Taherzadeh and Karimi, 2008). Several studies have highlighted a good correlation between the surface, the pore volume, and the enzymatic digestibility of lignocellulosic biomass (Chang and Holtzapple, 2002; Sun and Cheng, 2002). This accessibility is also associated with the cellulose crystallinity, and the organization of the lignin and hemicellulose with the micro- and macrofibrils of cellulose. Thus, pretreatment operations are often needed to modify the lignocellulosic complex in order to enhance the accessibility of cellulose, increasing the methane potential.

6.3 Current Digestion Practices

The implementation of anaerobic digestion in a reactor aims to convert the organic matter content into biogas. Organic matter is only a part of the reactor inlet, the rest being mostly composed of inorganic matter and water. A schematic mass balance is proposed in Figure 6.1. Anaerobic digestion is a biological process with a low biomass

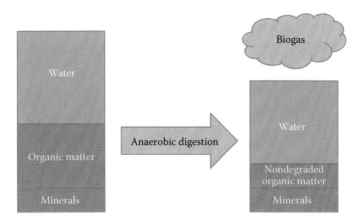

Figure 6.1 Schematic mass balance for the biological conversion of organic resources by anaerobic digestion.

growth yield; this means that 5%–10% of the consumed organic matter will be converted into biological cells. The rest of the degraded organic matter is thus converted into biogas. In Figure 6.1, the block named "nondegraded organic matter" thus refers to (1) the recalcitrant organic matter, (2) the untouched biodegradable matter, and (3) the cellular mass due to biomass growth.

During the process, only the biodegradable fraction of the organic matter can be converted. The conversion efficiency can thus be defined as the amount of biodegradable organic matter converted into biogas. This efficiency may vary according to the design parameters of the process.

6.3.1 Standard Setup for an Agricultural Biogas Plant

The standard setup of an agricultural biogas system can be reduced to a simplified block diagram on which the main mass fluxes of material can be represented (Figure 6.2). A typical installation consists of three major unit operations: a pretreatment (or conditioning unit), the reactor itself, and a posttreatment. The conditioning unit may be a simple mixing system with the aim of homogenizing the incoming fluxes before transferring them to the reactor. In other cases, it can provide additional advantages, such as a mechanical pretreatment (pulping, comminution, and shredding) or even more sophisticated biochemical, thermal, or chemical pretreatments (Taherzadeh and Karimi, 2008). The reactor itself may have different features in terms of operating mode (perfectly mixed, plug flow, wet or dry digestion, single- or multistage system, fed batch or continuous, and mesophilic or thermophilic), which are discussed in the next sections. The type of posttreatment depends on the mode of disposal of the digestate. The reactor output is commonly stored in a postdigester

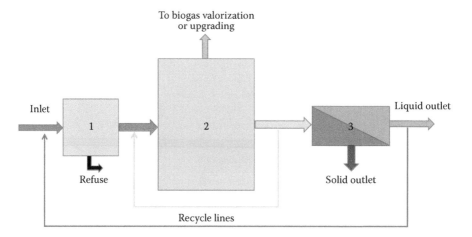

Figure 6.2 Standard setup of a biogas plant: (1) conditioning or pretreatment, (2) digester, and (3) liquid–solid separator and/or posttreatment.

(at ambient temperature) in order to recover residual biogas production. A liquid–solid separator may also be used, depending on the moisture content and land-spreading requirements.

It is important to identify each input and output flow. Inputs are mainly the organic matter, as well as water, which can optionally be added to the inlet to adjust the solid content in the reactor. The outputs are more numerous. We first note the refusal of pretreatment, the output of the solid after separation, the liquid outlet, and of course, the biogas produced. Knowing all these flows enables us to get a good overview of the process. It is also essential to characterize these different streams (total and volatile solids [VS] content in each stream, biogas flow, and composition).

6.3.1.1 Main Design Parameters

The reaction time is a common design parameter. In continuous systems (most cases), the reaction time is defined as the hydraulic retention time (HRT), that is, the ratio between the reactor volume and the average input flow. Common values of HRT are between 30 and 40 days in mesophilic conditions (35°C–40°C), and can be reduced at higher temperatures (Parkin and Owen, 1986).

The volumetric organic loading rate (OLR) is a second important design parameter. It is defined as the amount of organic matter entering the unit per cubic meter of reactor per day. At the industrial scale, the applied OLRs vary between 5 and 15 $kg_{VS}/m^3/day$ (Vandevivere et al., 2003). This important design parameter depends on the nature of the compounds to be treated and the type of technology used.

Both HRT and loading rate can be used to estimate the necessary reactor volume once the nature (amount and characteristics) of the incoming organic matter is known.

6.3.1.2 Important Criteria for Quantifying the Performance of a Digester

The objective of an anaerobic digester is to convert as much organic matter as possible into methane. This is the reason why the performances of a system can be estimated from two main parameters: the conversion of organic matter and the methane production. In addition, the productivity of the reactor itself can be characterized by its specific biogas production.

6.3.1.2.1 Conversion Efficiency. The removal efficiency (conversion efficiency) of an anaerobic digester is defined as the fraction of the incoming matter that has been removed. It can be calculated from dry or, more likely, organic matter abatement. The amount of organic matter is often estimated as the amount of VS.

$$\eta_{VS} = 1 - \frac{VS_{out}}{VS_{in}} \qquad (6.1)$$

In a continuous system, the parameters VS_{in} and VS_{out} can be considered the average VS fluxes in the inlet and outlet of the reactor over a given period of time.

6.3.1.2.2 Methane Production. The first criterion for quantifying the methane production is the methane yield (Y_{CH4}). It is defined as the volume of methane produced per unit mass of VS entering the reactor (e.g., as m^3 STP·tVS_{in}):

$$Y_{CH4} = \frac{Q_{CH4}}{VS_{in}} \qquad (6.2)$$

The expression (and unit) of the methane yield is similar to that of the BMP. The difference lies in the fact that the methane yield is the achieved methane production under process conditions, while the BMP is the supposed maximal methane production that can be expected from the full conversion of the biodegradable matter into methane. It is thus convenient to define the efficiency of the process as a BMP yield, the ratio of the achieved methane yield to the BMP:

$$\eta_{BMP} = \frac{Y_{CH4}}{BMP} \qquad (6.3)$$

In many cases, the productivity of a given process is characterized as the specific gas production (SGP), which is the amount of biogas produced per unit volume of reactor. This indicator does not directly refer to the conversion of biomass into methane, but rather to the intensity of the biological treatment. Hence, compact and highly loaded systems will have a higher SGP than extensive or low-load systems.

$$SGP = \frac{Q_{biogas}}{V_{reactor}} \qquad (6.4)$$

6.3.2 Technologies for Biogas Production from Organic Products

The objective of this section is to propose an overview of the main types of digesters available for the agricultural sector. There are many kinds of systems that can adapt to the wide diversity of incoming organic resources or even to various agricultural practices.

6.3.2.1 Mode of Operation: Batch versus Continuous

In continuous systems, the reactor is fed on a regular basis (e.g., once or twice a day), and an equivalent amount of digested matter is extracted. It is a common mode of operation for easily flowing products, since they can be directly pumped into the reactor. For higher consistencies, the incoming products can be shredded and

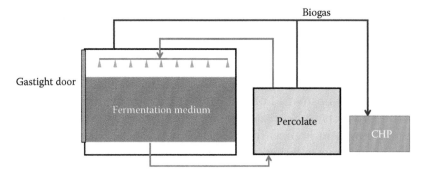

Figure 6.3 Schematic representation of a batch digestion system for high-solids agricultural products.

homogenized and can be introduced by a screw pump. The advantages of a continuous system are

- Regular and smooth biogas production
- Limited daily handling operation (filling the hopper with a wheel loader)
- Less fluctuations of operating parameters

A large majority of the existing plants at the farm scale operate in continuous mode. Nevertheless, batch systems have also been developed. This type of technology generally applies to semisolid products, such as manure or other agricultural products having a high solids content. The biogas fermentation takes place in a closed tank (garage-shaped fermenter), where the organic products are initially mixed with an inoculum (resulting from the previous fermentation). The process water (leachate) produced percolates through the fermenting medium. It is collected and recovered in a tank, before being recycled to the fermentation zone to adjust the moisture content. A schematic view of this type of system is presented in Figure 6.3. The biogas is produced both in the fermenter itself and in the leachate storage tank. The heat produced by the combined heat and power (CHP) engine can be used for heating the percolate or the reactor itself. More sophisticated designs of this type of system can be imagined, with several fermentation tanks and a specific strategy for leachate recycling in each tank (Chynoweth et al., 1992). The main advantages of this kind of technology are

- Low (or no) water requirements
- The nonfermented solid part can be easily postcomposted
- Simple operations

6.3.2.2 Wet (Low-Solids) or Dry (High-Solids) Digestion
The fermentation of solid or semisolid products can be performed in wet digestion systems. Wet digestion processes require the addition of water in order to reach a flowing consistency. The TS content (dry matter) is commonly below 10% (Vandevivere

et al., 2003). The fermentation medium is continuously mixed by a mechanical device. Mixing is very important for the homogeneous distribution of temperature (most reactors are heated) and to prevent the formation of a scum layer that can hamper or stop biogas outflow.

In Europe, wet digestion technologies are massively used at the farm scale (Germany, Italy, and northern Europe) because they are simple and adapted to the nature of the products to be treated (cattle or pig slurries). Nevertheless, in some cases (very common in France) the breeding mode for animals can differ significantly, with more straw addition in stalling, and lead to much more solid products (manure). Hence, processes able to handle solid products have been developed. These kinds of processes are usually named "dry" or "high-solids" digestion processes. They are able to operate at a dry matter content between 15% and 35% (according to the kind of incoming organic products). This is the case of the batch systems described in Section 6.3.2.1. Actually, the commonly used term *dry digestion* is not appropriate since the consistency of the medium is that of a paste. It is saturated with water, even if the amount of free-flowing water in it is very low. Up to now, continuous high-solids digestion systems were specifically used for the digestion of the organic fraction of municipal solid waste (Mata-Alvarez, 2003). Only recently have they been developed for continuous biogas fermentation in the agricultural sector. The major advantage of this type of technology is the limited water requirements. The characterization and understanding of high-solids digestion has given way to very challenging scientific questions since 2010, since both biological and physical effects are important. An example of this technology is presented as a case study at the end of this chapter.

6.3.2.3 Digestion Temperature

Temperature is an important parameter for anaerobic digestion. As for all chemical or biochemical reactions, the rate of the processes leading to methane production increases with temperature according to Arrhenius's law (El-Fadel et al., 1996; Veeken and Hamelers, 1999). For instance, the upper temperature limit for acetate conversion into methane has been found to be close to 65°C. Above this value, the biological activity drops dramatically due to a pasteurization effect (Van Lier et al., 1996). It is thus interesting to provide heat to the digestion systems in order to achieve an adequate methane production rate. The thermal energy used for heating the digester usually comes from the CHP unit.

Installed systems usually operate under two temperatures. The mesophilic mode (35°C–40°C) is mostly used. It is known to be very stable for the overall digestion process. The thermophilic mode (50°C–60°C) is also current. It enables higher biological rates and, consequently, reduced reactor size and retention times. Higher temperatures (>60°C) may also promote the rate and extension of hydrolysis of particulate organic matter, and increase the removal of microbial contaminants, such as pathogenic microbes. Reactors operating at intermediate temperatures between the mesophilic and thermophilic modes are less common, probably because this range has been less extensively investigated (Pfeffer, 1974).

6.3.3 Biogas and Digestate Management

6.3.3.1 Energy Production from Biogas

Methane is commonly used as an energy source since it is the major component of natural gas. Its energy content (lower heating value) is 10.5 kWh/m³ in standard conditions. The energy produced from biogas thus depends on two factors:

1. The methane content of the biogas itself. Biogas is composed of variable amounts of methane (50%–70%), with the rest being carbon dioxide and other minor compounds (such as hydrogen sulfide).
2. The energy yield of the system that produces the valuable form of energy.

Heat is the simplest way to produce energy from biogas. Boilers are able to burn biogas even at low methane content (20%). This technique is interesting in plants with important heat requirements (for heating buildings and greenhouses, drying forages or digestate, or use as hot water and steam in food transformation facilities). In most, cases however, it is more interesting to produce electricity and heat through CHP engines. These systems present an electricity yield of circa 35% and a thermal energy yield of 50%. Engines are designed for the production of electricity ranging from 50 to 3000 kW. These systems require a limited level of pretreatment (desulfuration) in order to protect the engines. At the farm scale, a 50 kW engine corresponds (roughly) to a rather important amount of organic resources (e.g., around 2000 tons/year of cattle manure). Engines with lower capacity could be interesting for the development of anaerobic digestion in small farms. Heat and electricity production is now the major method of energy production at the farm scale in Europe. The reason for such success is the obligation of the energy network holder to buy the electricity produced at a very advantageous rate.

Biogas purification and upgrading is another method of energy utilization. In this case, there are two possibilities for getting value out of the methane produced:

1. Biofuels for vehicles (tractors or cars). Indeed, many vehicles are now able to use compressed natural gas as fuel.
2. Injection in the natural gas network.

In both cases, the biogas has to be cleaned and purified to a higher grade (>95% methane), similar to that of commercial natural gas (the term *biomethane* is used to qualify the purified biogas). The cost for biogas upgrading is still elevated, but it is constantly decreasing due to the recent developments of simple technologies for CO_2 absorption (Benizri, 2015).

6.3.3.2 Digestate Management and Disposal

Digestate is the common name for the digester output after anaerobic digestion. During anaerobic digestion, complex organic matter is transformed into smaller molecules. Some biomolecules, such as lignin and cellulose, are difficult to degrade. The fate of inorganic molecules is different, since very little amounts are metabolized.

Nevertheless, the anaerobic conditions in the digester are responsible for the presence of inorganic products in their reduced form. Nitrogen will thus tend to mineralize and be present under the form of ammonium ions (NH) in the digestate. Compared with undigested manure, nitrogen in a digestate will be more available for plants (Möller and Müller, 2015). Other interesting fertilizing elements, such as phosphorus, potassium, or magnesium, are known to be almost unaffected by anaerobic digestion in the sense that their availability for plants is the same (with or without digestion). For all these reasons, the digestate in its liquid form can be directly spread on agricultural land. In some cases, it can replace chemical fertilization. The remaining organic matter may also be of interest for agriculture, since undegraded products (mostly lignin and cellulose) are precursors of humic substances. In some cases, the separation of the solid fraction may be useful in order to generate an organic soil amendment in a solid form.

Anaerobic digestion at the farm scale is thus an appropriate option not only for getting value out of the production of energy, but also for reducing the farming inputs through reuse of the fertilizing properties of the digestate.

6.4 Future Trends

6.4.1 Pretreatments

The bioconversion of cellulose and hemicellulose requires breaking the structure of the lignocellulosic complex in order to increase the bioavailability, and thus the biodegradability, of these substrates. Much research has been carried out over 30 years to assess the optimal thermal, physical, chemical, and/or biological pretreatment conditions of lignocellulosic biomass. There are many literature reviews on these techniques for optimizing methane production from agricultural resources (Taherzadeh and Karimi, 2008; Hendriks and Zeeman, 2009; Frigon and Guiot, 2010; Chandra et al., 2012; Monlau et al., 2013). The choice of pretreatment depends on economical and technical criteria. We will only discuss technical criteria, preferably considering energy-efficient pretreatment that can be competitive in the future. These techniques are generally classified into several categories.

6.4.1.1 Mechanical Pretreatment
Reducing the particle size is applied as a mechanical pretreatment to facilitate material handling and increase the ratio of the surface area to volume (Harmsen et al., 2010). This reduction can be carried out by grinding, crushing, and shredding (Palmowski and Müller, 2000). According to these authors, the mechanical operations also lead to changes in the structure and chemical composition of lignocellulosic biomass. Increasing the available surface area, a lower degree of polymerization, and the disintegration of fibers are very favorable for increasing the accessibility of polysaccharide substrates for exoenzymes. Several authors have highlighted the increased rate of hydrolysis of cellulose with a reduction in particle size (Chang and

Holtzapple, 2000; Hartmann et al., 2000). Consequently, the reduction in particle size results in a significant increase in the biomethane potential of forestry and agricultural residues.

6.4.1.2 Thermal Pretreatment

Thermal pretreatments are used in a wide range of technologies based on the principle of hydrolysis and partial solubilization of the organic compounds at high temperature. Treatments are usually performed in the presence of liquid water at high temperature (150°C–260°C) and high pressure (10–50 bar), which, in a second step, can be quickly relaxed as vapor (steam explosion). Most of the work cited in the literature focuses on the production of paper or ethanol, but this technique has been poorly studied for biogas production from agricultural and other resources. However, the energy cost and the risk of producing inhibitory by-products suggest that such pretreatments are not possible in the biogas sector.

6.4.1.3 Acid Pretreatment

Acid pretreatment of lignocellulosic biomass has been the subject of much research for the optimization of operating conditions (Mosier et al., 2005; Monlau et al., 2013). The principle is to hydrolyze hemicellulose (and potentially cellulose) without degrading lignin, which can be separated, leading to a very significant increase in the enzymatic digestibility of cellulose. In the vast majority of studies, acid hydrolysis is carried out in a few minutes with sulfuric acid (H_2SO_4), nitric acid (HNO_3), or hydrochloric acid (HCl), diluted (at high temperature) or, more generally, at high concentrations (72% H_2SO_4 and 40% HCl). However, even if they are very effective, concentrated acids are toxic, corrosive, and dangerous. Their use requires reactors to be resistant to corrosion. Their cost requires their recovery after hydrolysis so that they can be recycled and reused in the process (Sun and Cheng, 2002).

6.4.1.4 Alkaline Pretreatment

The alkaline pretreatment is one of the most commonly used techniques. It is generally carried out at low temperature and pressure, with different bases, such as sodium hydroxide (NaOH), potassium hydroxide (KOH), calcium hydroxide–lime ($CaOH_2$), or ammonia (NH_3). Its duration is of the order of hours to days. The alkaline hydrolysis mechanism is based on the rupture of ester bonds in hemicellulose and other components, such as lignin (Sun and Cheng, 2002). Unlike acid pretreatment, the basics are easily recoverable, as incorporated in the form of salts in the treated biomass (Mosier et al., 2005). The alkali treatment generally leads to hydrolysis of lignin and hemicelluloses (Monlau et al., 2013).

6.4.1.5 Biological and Enzymatic Pretreatments

Fungal and enzymatic pretreatments are attractive on these criteria for energy-efficient. Biological and enzymatic pretreatments require reasonable operating conditions in terms of reagent consumption. These techniques are based on the hydrolytic properties of many fungal species, which are characterized by a broad spectrum of enzymes with the ability to hydrolyze the constitutive lignocellulosic polymers.

6.4.2 Storage

The diversification of anaerobic digestion inputs is quite wide, as energy can be recovered from almost all types of organic wastes, forages, or catch or energy crops. Otherwise, although there is a need for continuous feeding of biogas plants throughout the year, some of these agricultural and industrial wastes or crops are seasonally produced, leading to storage requirements, in some cases even of extended durations. Consequently, the storage mode is a key factor for the development of agricultural anaerobic digestion: storage, pretreatment or preconditioning, and so forth. Indeed, agricultural residues, such as manure or crop residues, require temporary storage before injection into the digester. Beyond agricultural residues, other available resources for agricultural biogas require storage, such as wastes from the food industry, organic waste, or waste from unsold products in hypermarkets. Depending on their origin and characteristics, the products must be stored more or less long term, in aerated or confined conditions, before injection into the digester. In addition, strategies for storage (mechanical pretreatment, such as milling and silage starter additions) can be implemented and affect the stability of the waste. Nowadays, regarding the storage types, several options should be mentioned.

The first one is open-air storage, mostly used for agricultural residues such as animal manure, since it is nonexpensive and this kind of substrate is regularly produced, normally with no need for prolonged storage. Nevertheless, even during short periods, open-air storage can lead to substantial losses in terms of organic matter and methane potential, due to air–material contact and aerobic degradation.

Storage with controlled preaeration before anaerobic digestion combines storage with the pretreatment of the biomass prior to injection into the digester. The positive effect of short-term ventilation on the biomethane potential was highlighted in various categories of organic waste, such as the biodegradable fraction of household waste (ten Brummeler and Koster, 1990; Hasegawa et al., 2000). The phenomenon is still not well explained and is probably related to several mechanisms, such as increased bioavailability of the cellulose and the elimination of the easily biodegradable organic fraction, reducing the risk of inhibition of methanogenesis by accumulation of volatile fatty acids. But other assumptions, such as the expression of enzyme activities stimulating the breakdown of lignocellulose compounds, are being put forward to describe these mechanisms (Hasegawa et al., 2000; Johansen and Bakke, 2006). Finally, microaeration could play an important role in the important compounds' cycle, such as sulfur.

For energy crops and agriculture that is seasonally produced, farmers have developed two main storage and preservation technologies that can be adapted for anaerobic digestion purposes: hay and silage systems. Hay storage consists of field drying and inhibiting detrimental microbial activity, followed by the use of large round bales stored outdoors (Egg et al., 1993). Even though this system minimizes both labor and storage costs, it leads to high losses in terms of dry matter, and it is restricted to crops that can dry quickly and uniformly, also being limited by rainfall during harvest (Weinberg and Ashbell, 2003). In contrast to the physical transformations in hay systems, ensiling provides a biochemical process based on preservation under an anaerobic environment, using bacterial fermentation to prevent further degradation. This process, used to preserve forages for animal feed during centuries, minimizes weight and energy losses, and therefore appears to be the logical choice for the storage of biomass in view of methane production.

Ensiling can be divided into four phases, according to the main biochemical and microbiological events present in the system (Herrmann et al., 2011). In the initial aerobic period, after filling and sealing the silo, biomass respiration occurs due to the presence of oxygen trapped in the system. Respiration continues for several hours, consuming sugars and producing carbon dioxide and water, until all oxygen is removed. Once oxygen has been depleted, the anaerobic period starts and microorganisms capable of anaerobic growth (e.g., lactic acid bacteria, enterobacteria, clostridia, and yeasts) begin to proliferate and compete for available matter. Changes in the first days are critical to the success or failure of subsequent fermentation (McDonald et al., 1991). If the conditions are suitable, lactic acid bacteria (LAB) will produce lactic acid for several weeks, decreasing the pH to around 4.0. The third phase is stable, where the pH remains relatively stable and wherein minimal enzymatic and microbial activity occur under anaerobic conditions. Finally, biomass once again enters into the aerobic environment during the feed-out operation (unloading the silo for transportation or biodigester feeding), which may spoil the silage and lead to mass and biomethane potential losses.

6.4.3 Improving Anaerobic Digestion Processes

At first sight, anaerobic digestion can be considered a mature technology in the agricultural sector. With more than 10,000 plants in operation in Europe (mostly in Germany), many equipment providers are present on the market. Nevertheless, the exportation of the German technical standards in other countries is sometimes difficult. In France, for instance, 80% of the farm-scale digestion units have experienced technical problems. This is mostly due to a high diversity of the products incoming the digester since farm-scale plants in France look like small codigestors (E-Cube, 2015). The actual processes (mostly based on wet digestion units) are not always fitted to this diversity. As a consequence, anaerobic digestion processes require new adjustments and designs in order to achieve a sustainable development.

6.4.3.1 Emerging Technologies

As mentioned in Section 6.4.1, the future trends of research on anaerobic digestion will focus on the optimization of energy recovery from biological resources. Pretreatments and their coupling with the digester are crucial steps to achieving this objective. Recent research has also proposed interesting options based on two- or multistage reactors with the aim of optimizing each reaction step (hydrolysis and methanogenesis). Temperature-phased anaerobic digestion (TPAD) reactors have been investigated and exhibit very interesting results. The principle is to propose a first stage at high temperature (thermophilic or higher), followed by a mesophilic stage. This configuration obviously increases the conversion of organic matter into biogas by accelerating hydrolysis even in the mesophilic compartment (Ge et al., 2010; Lv et al., 2010).

6.4.3.2 Understanding Physical Effects

The development of high-solids systems (in both batch and continuous modes) has raised several questions associated with the influence of the physical parameters on the reactor design and on operational performances. All these questions will probably support the emergence of new reactor designs and operation modes in the next 10 years, and more specifically, the development of anaerobic technologies based on high-solids contents.

6.4.3.2.1 Mixing. A common problem in wet (low-solids) digestion systems is the importance of the sedimentation of heavy particles (and the flotation of light particles). In high-solids digesters, the higher consistency and viscosity considerably limit this problem (Benbelkacem et al., 2013). Efforts are underway for better characterization of the rheological properties in high-solids digesters in order to be able to design appropriate mixing devices (Garcia-Bernet et al., 2011b).

In counterparts, the limited mixing may affect the efficiency of the systems through mass transfer limitations, either in the liquid phase (diffusion limitation) or at the liquid–gas interface (Bollon et al., 2013; Xu et al., 2014).

6.4.3.2.2 Water Distribution. In high-solids systems, the water content is close to the saturation limit (Garcia-Bernet et al., 2011a). The biological activity can thus be reduced at low water content. A fourfold decrease has been observed for the methanogenic activity when the TS content increased from 15% to 35% (Le Hyaric et al., 2012). In the case of a batch of high-solids fermentation, the distribution of water in the fermenting medium is also a crucial aspect that determines the biological activity (André et al., 2016).

6.4.3.3 Sensors and Monitoring

The lack of reliable systems for estimating the state of a process in operation is a major limitation in anaerobic digestion. In agricultural biogas plants, the online sensors used are limited to the quantification of biogas output and composition. Since biogas is the end product of the anaerobic digestion process, the detection of an operational problem with biogas data generally occurs too late. Other state parameters, such as the pH, the volatile fatty acid content, or the ammonia content, are not easily measured online. Anaerobic digestion technologies are still poorly instrumented, even if a wide number of solutions exist at the research level (Jimenez et al., 2015). The development of cheaper and easier sensors and monitoring systems is thus a challenging issue. Recently, electronic noses appeared to be rather intriguing for the detection of a large variety of compounds in the biogas and could be an interesting option (Adam et al., 2015).

6.5 Case Studies

6.5.1 Case Study 1: The ARKOMETHA® Technology in Napagèse, Tarn (France)

6.5.1.1 Main Project Characteristics

The ARKOMETHA technology has been developed by the company ARKOLIA ENERGIES. This technology is based on a tanks in series (plug-flow) high-solids

TABLE 6.6   Characteristics of the Incoming Products

Incoming Products	Tons/Year
Cattle manure	350
Grass silage	500
Cereal offal	300
Cattle slurry	600
Poultry manure	400
Hay and straw	50
Total	2200

(dry) anaerobic digestion system. The mixing is achieved by injection of pressurized biogas. The reactor is composed of a series of compartments separated by walls. Openings in the walls allow the digestion medium to flow from one compartment to another during feeding. Mixing is provided by the injection of pressurized biogas in each compartment individually.

The system was designed in order to treat 2200 tons/year of organic products, mostly cattle manure and slurry, grass silage, cereal offal, and poultry manure (Table 6.6). The installed CHP engine is 100 kWh (electrical power). The size of the system is small since this was the first prototype for the technology.

6.5.1.2 Layout
The digestion unit is described in Figure 6.4. It is composed of a heated shredder (65°C), followed by the reactor itself (a hydrolysis zone and a methanogenic zone, both operating at 55°C). The TS content at the inlet of the reactor is adjusted to around 18%–20% TS.

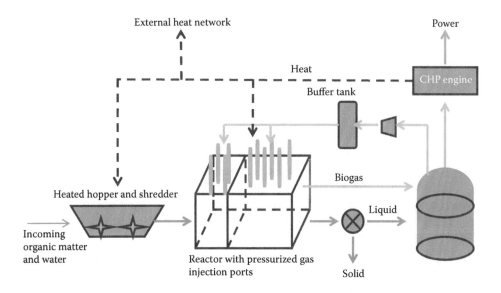

Figure 6.4   Layout of the ARKOMETHA system in NAPAGESE (Tarn, France).

TABLE 6.7   Summary of 12-Month Average Results of the Plant

OLR ($kg_{VS}/m^3/day$)	7.9
HRT hydrolysis (days)	4.9
HRT total (days)	20.8
VS removal efficiency ($\eta_{VS}\%$)	46
Methane yield $\left(m^3_{STP,CH4}/t_{VS} \right)$	245
BMP yield (%)	99
SGP $\left(m^3_{STP}\, m^3_{Reactor}/day \right)$	3.25

The total reactor volume is 205 m³ (170 m³ of useful volume), divided into a hydrolysis sector representing 25% of the reactor volume and a methanogenic sector for the remaining part. Due to high-pressure injection, each sector is homogeneously mixed. Since they are placed in series without recycling, a plug-flow effect is expected. The total HRT in the reactor is around 20 days.

6.5.1.3 Operational Results

The start-up of the plant was in October 2014, and the performances were characterized between December 2014 and December 2015. The reactor performances averaged over this period are reported in Table 6.7. The plant exhibits excellent performances. Over the investigated period, the actual OLR was 20% below the design value. This is mostly due to an overestimation of the incoming amount of products (1800 tons/year achieved vs. 2200 tons/year in the design). An interesting output is that the fraction of the BMP recovered (BMP yield) was closed to 100%. This result can be explained by the design of the treatment line itself, which could promote an extended amount of conversion in comparison due to the mechanical pretreatment and plug-flow effect in thermophilic conditions. The average monthly biogas production was more than 10,000 m³, with a methane content of 52%–55%. The electric power produced was 52.1 MWh/month.

6.5.2 Case Study 2: BERT® Technology by Bio4gas Express, GAEC P2MN, Ain, France

6.5.2.1 Main Project Characteristics

The development of farm biogas needs to have robust technical solutions, be simple and economical, and require minimal maintenance. On these basic principles, two academics have developed, one of which is the BERT technology, whose first model was built in 2006 at the Rotholz Agricultural High School in Austria. Since 2010, the Austrian patent has been operated by the company Bio4gas Express, which is in charge of technology commercialization in German-speaking countries and sales of the BERT distribution license around the world.

This liquid-state digestion technology consists of a circular digester with two separated circular chambers, the inner cylinder and the outer ring, connected with a pneumatic flip-flop valve. The outlet located under the top sealing of the digester allows the digestion medium to flow from one compartment to another during feeding. When biogas production starts, gas pressure in the headspace of the external chamber (C1) displaces liquid substrate below the baffle to the inner chamber (C2). To optimize mixing and avoid settlement, annual flow is induced intermittently by small stirrers. The pressure of the headspaces of the two chambers is regularly balanced to guarantee recycle flow between the two compartments (flip-flop valve opened when $\Delta P > 50$ mb) (Figure 6.5). A second feature for promoting continuous mixing, but also heating of the substrate, is the thermo-gas-lift (TGL) pipe, which is installed in each chamber. Based on the principle of thermoconvection, heating the substrate causes gas bubbles to form and these rise to the top of the digester.

Located in the Department of Ain in France, the GAEC P2MN is a specialized breeding farm with nearly 120 Holstein breed dairy cows and a production of 1.2 million L of milk per year. The farm has 254 hectares divided into meadow, cereals, alfalfa, and canola. In 2012, the BERT system was selected because the technique is able to treat a large range of agricultural residues, including residues from livestock, crops, and energy crops. The system has been designed in order to treat 5000 tons of organic feedstocks per year (12%–13% TS), mostly cattle manure and cattle slurry (95%). The project also includes 180 tons of corn wastes (5%) per year from an agricultural

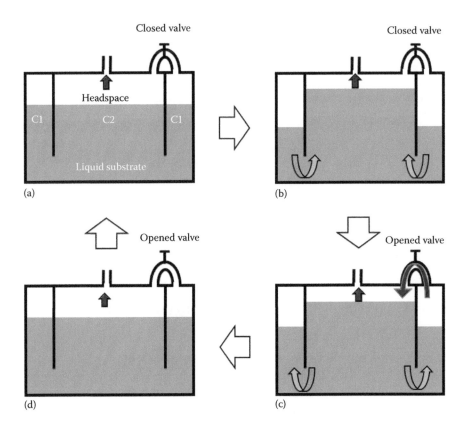

Figure 6.5　BERT mixing system by gas–liquid displacement. (a) Initial state with the flip-flop valve closed; (b) biogas generation; (c) flip-flop valve opened when $\Delta P > 50$ mb; (d) pressure balance between the two chambers C1 and C2.

cooperative, a methanogenic substrate that increases the profitability of the installation while maintaining a secure deposit. The installed CHP engine is 50 kW (electrical power).

6.5.2.2 Layout
The BERT digestion unit is shown in Figure 6.6.

The installation is composed of a concrete premixing pit (50 m³) to receive manure and slurry from the farm storage, and to shred and macerate the mixture prior to injection. The tank is entirely sunk into the ground and surrounded by sloped concrete to collect any spillages. The digester, which is a circular design, is made of in situ concrete. The digester is operated under mesophilic conditions (37°C–39°C). The TS content at the inlet of the reactor is adjusted to be around 12%–13% TS. The total reactor volume is 700 m³ (600 m³ of useful volume and 100 m³ of gas space), divided into two chambers, as previously explained and shown in Figure 6.5. The total HRT in the reactor is around 30 days. Two small stirrers are intermittently operated (6 min/h, 15 kW) to prevent surface crusting. A dome-shaped gas storage bag is placed on a section of the digester roof or in a container for protection. The CHP unit is preassembled in a steel container, housing the engine and controls.

6.5.2.3 Operational Results
The start-up of the plant was in August 2014. The reactor performances are shown in Table 6.8. The BMP recovered (BMP yield) was closed to 91%. The average biogas

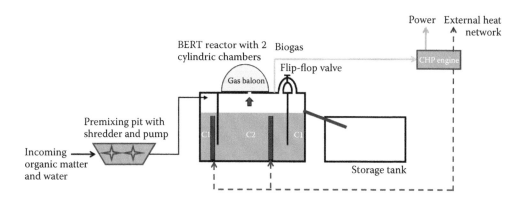

Figure 6.6   Layout of the BERT system in GAEC P2MN (Ain, France).

TABLE 6.8   Average Results of the Plant

OLR (kg$_{VS}$/m³/day)	4.2
HRT total (days)	33
VS removal efficiency (η_{VS}%)	60
Methane yield $\left(m^3_{STP/CH4} \cdot t_{VS} \right)$	23
BMP yield (%)	91
SGP $\left(m^3_{STP} \cdot m^3_{Reactor}/day \right)$	0.71

production was 175,000 m³ during the year, with a methane content of 52%–55%. The average electric power produced in 2015 was 32,000 kWh/month, and the average thermal power 23,000 kWh/month, 100% used on site.

Thanks to anaerobic digestion, the farm was able to improve its environmental footprint and reduce its consumption of fertilizers. Furthermore, a multiproduct dryer was set up to develop heat. Drying of forage and corn will improve the nutritional quality of food products by the GAEC. A partnership with a dynamic local player in the timber industry will create a wafer drying activity for optimal recovery of heat, as well as additional income for farmers.

6.6 Conclusion

Agricultural biomass can be easily converted into biogas. The resource is very important and could give way to a considerable amount of renewable energy production. It is thus an excellent contributor to energy transition. In addition, the digestion residue can be used as fertilizer, which could minimize the cost of farming inputs.

Even if the anaerobic digestion sector can be considered mature, the technologies still require research and development in order to increase its profitability, for example,

- Optimization of the methane recovery from organic matter
- Optimization of the design and operation of plants
- Optimization of the energy production from biogas

Biogas production at the farm scale could thus not only be an alternative way toward energy self-sufficiency, but also have a positive impact on the environment through the limitations of chemical inputs for fertilization. The risk is indeed that the agricultural sector could look for more economical benefits through energy production, and that agriculture would turn into "energy-culture." This is why farm-scale digestion should be kept within an eco-friendly framework. Growing crops only for energy is thus an option that has to be discussed in terms of the economy, energy, and environmental impacts. Since agriculture is a sector responsible for many environmental issues, such as soil and water pollution, a rationale use of biogas plants in this sector could contribute to reducing these impacts and improving the image of the sector.

References

Adam, G., Lemaigre, S., Romain, A.-C., Nicolas, J., Delfosse, J. (2012). Evaluation of an electronic nose for the early detection of organic overload of anaerobic digesters. *Bioprocess and Biosystems Engineering* 36: 23–33.

Amon, T., Amon, B. et al. (2007). Methane production through anaerobic digestion of various energy crops grown in sustainable crop rotation. *Bioresource Technology* 98: 3204–3212.

Amon, T., Kryvoruchko, V., Bodiroza, V., Amon, B. (2005). Methane production from cereals, grass and sun flowers: Effect of harvesting time, and pre-treatment on the methane yield. In *KTBL*, 7. Braunschweig, Germany: Internationale Tagung Bau, Technik und Umwelt in der Nutztierhaltung.

André, L., Lamy, E., Lutz, P., Pernier, M., Lespinard, O., Pauss, A., Ribeiro, T. (2016). Electrical resistivity tomography to quantify in situ liquid content in a full-scale dry anaerobic digestion reactor. *Bioresource Technology* 201: 89–96.

Angelidaki, I., Sanders, W. (2004). Assessment of the anaerobic biodegradability of macropollutants. *Reviews in Environmental Science and Biotechnology* 3: 117–129.

Barlaz, M.A. (1996). Microbiology of solid waste landfills. In *Microbiology of Solid Waste*, ed. A.C. Palmisano, M.A. Barlaz, Chap. 2. Boca Raton, FL: CRC Press.

Bauer, A., Leonhartsberger, C., Bosch, P., Amon, B., Friedl, A., Amon, T. (2010). Analysis of methane yields from energy crops and agricultural by-products and estimation of energy potential from sustainable crop rotation systems in EU-27. *Clean Technologies and Environmental Policy* 12: 153–161.

Benbelkacem, H., Garcia-Bernet, D., Bollon, J., Loisel, D., Bayard, R., Steyer, J.P., Gourdon, R., Buffière, P., Escudié, R. (2013). Liquid mixing and solid segregation in high-solid anaerobic digestion. *Bioresource Technology* 147: 387–394.

Benizri, D. (2015). Epuration du biogaz à la ferme: EPUROGAS, une solution économique et énergétique d'avenir [Biogas upgrading at farm-scale: EPUROGAS, a future energetic and economic solution]. PhD thesis, Université Fédérale de Toulouse, INSA de Toulouse.

Bidlingmaier, W., Sidaine, J.-M. et al. (2004). Separate collection and biological waste treatment in the European Community. *Reviews in Environmental Science and Biotechnology* 3: 307–320.

Bollon, J., Benbelkacem, H., Gourdon, R., Buffière, P. (2013). Measurement of diffusion coefficients in dry anaerobic digestion media. *Chemical Engineering Science* 89: 115–119.

Buffiere. P., Frederic. S., Marty, B., Delgenes. J.P. (2008). A comprehensive method for organic matter characterization in solid wastes in view of assessing their anaerobic biodegradability. *Water Science and Technology* 58(9): 1783–1788.

Buffière, P., Loisel, D., Bernet, N., Delgenes, J.P. (2006). Towards new indicators for the prediction of solid waste anaerobic digestion properties. *Water Science and Technology* 53(8): 233–241.

Chandler, J.A., Jewell, W.J., Gossett, J.M., Van Soest, P.J., Robertson, J.B. (1980). Predicting methane fermentation biodegradability. *Biotechnology and Bioengineering Symposium Series* 10: 93–107.

Chandra, R., Takeuchi, H., Hasegawa, T. (2012). Methane production from lignocellulosic agricultural crop wastes: A review in context to second generation of biofuel production. *Renewable and Sustainable Energy Review* 16(3): 1462–1476.

Chang, V.S., Holtzapple, M.T. (2000). Fundamental factors affecting biomass enzymatic reactivity. *Applied Biochemistry and Biotechnology* 84: 5–37.

Cho, J.K., Park, S.C. (1995). Biochemical methane potential and solid state anaerobic digestion of Korean food wastes. *Bioresource Technology* 52(3): 245–253.

Chynoweth, D.P., Owens, J., O'Keefe, D., Earle, J.F.K., Bosch, G., Legrand, R. (1992). Sequential batch anaerobic composting of the organic fraction of municipal solid waste. *Water Science and Technology* 25(7): 327–339.

Davidsson, A., Gruvberger, C., Christensen, T.H., Hansen, T.L., Jansen, J. (2007). Methane yield in source-sorted organic fraction of municipal solid waste. *Waste Management* 27: 406–414.

E-Cube. (2015). Etat des lieux des lieux de la filière biogaz en France [State of the art of biogas technologies in France]. Association Technique Energie Environnement. http://atee.fr /biogaz/etude-etat-des-lieux-de-la-filiere-biogaz-en-france-2015.

Egg, R., Coble, C., Engler, C., Lewis, D. (1993). Feedstock storage, handling and processing. *Biomass and Bioenergy* 5(1): 71–94.

Eleazer, W.E., Odle, W.S., Wang, Y.S., Barlaz, A. (1997). Biodegradability of municipal solid waste components in laboratory scale landfills. *Environmental Science and Technology* 31(3): 911–917.

El-Fadel, M., Findikakis, A.N., Leckie, J.O. (1996). Temperature effects in modelling solid waste biodegradation. *Environmental Technology* 17: 915–935.

EurObserv'ER. (2014). Biogas barometer. *EurObserv'ER.* http://www.energies-renouvelables .org/observ-er/stat_baro/observ/baro224_Biogas_en.pdf.

Frigon, J.C., Guiot, S.R. (2010). Biomethane production from starch and lignocellulosic crops: A comparative review. *Biofuels, Bioproducts and Biorefining* 4(4): 447–458.

Garcia-Bernet, D., Buffière, P., Latrille, E., Steyer, J.P., Escudié, R. (2011a). Water distribution in biowastes and digestates of dry anaerobic digestion technology. *Chemical Engineering Journal* 72: 924–928.

Garcia-Bernet, D., Loisel, D., Guizard, G., Buffière, P., Steyer, J.P., Escudié, R. (2011b). Rapid measurement of the yield stress of anaerobically-digested solid waste using slump tests. *Waste Management* 31: 631–635.

Ge, H., Jensen, P.D., Batstone, D.J. (2010). Pre-treatment mechanisms during thermophilic-mesophilic temperature phased anaerobic digestion of primary sludge. *Water Research* 44: 123–130.

Ghosh, S., Bhattacharyya, B.C. (1999). Biomethanation of white rotted and brown rotted rice straw. *Bioprocess Engineering* 20: 297–302.

Gunaseelan, V.N. (2004). Biochemical methane potential of fruits and vegetable solid waste feedstocks. *Biomass and Bioenergy* 26(4): 389–399.

Gunaseelan, V.N. (2007). Regression models of ultimate methane yields of fruit and vegetable solid wastes, sorghum and napier grass on chemical composition. *Bioresource Technology* 98: 1270–1277.

Harmsen, P., Huijgen, W., Bermudez, L., Bakker, R. (2010). Literature review of physical and chemical pretreatment processes for lignocellulosic biomass. Wageningen, the Netherlands: UR Food & Biobased Research.

Hartmann, H., Angelidaki, I., Ahring, B.K. (2000). Increase of anaerobic degradation of particulate organic matter in full-scale biogas plants by mechanical maceration. *Water Science and Technology* 41(3): 145–153.

Hasegawa, S., Shiota, N., Katsura, K., Akashi, A. (2000). Solubilization of organic sludge by thermophilic aerobic bacteria as a pretreatment for anaerobic digestion. *Water Science and Technology* 41(3): 163–169.

Hashimoto, A.G. (1986). Pretreatment of wheat straw for fermentation to methane. *Biotechnology and Bioengineering* 28: 1857–1866.

Hendriks, A., Zeeman, G. (2009). Pretreatments to enhance the digestibility of lignocellulosic biomass. *Bioresource Technology* 100(1): 10–18.

Herrmann, C., Heiermann, M., Idler, C. (2011). Effects of ensiling, silage additives and storage period on methane formation of biogas crops. *Bioresource Technolology* 102: 5153–5161.

Jeffries, T.W. (1990). Biodegradation of lignin-carbohydrate complexes. *Biodegradation* 1: 163–176.

Jerger, D.E., Chynoweth, D.R., Isaacson, H.R. (1987). Anaerobic digestion of sorghum biomass. *Biomass* 14: 99–113.

Jimenez, J., Latrille, E. et al. (2015). Instrumentation and control of anaerobic digestion processes: A review and some research challenges. *Reviews in Environmental Science and Bio/Technology* 14: 615–648.

Johansen, J.E., Bakke, R. (2006). Enhancing hydrolysis with microaeration. *Water Science and Technology* 53(8): 43–50.

Lehtomäki, A. (2006). Biogas Production from Energy Crops and Crop Residues. Doctoral dissertation, Pedro Aphalo Edt, Department of Biological and Environmental Science, University of Jyväskylät, Finland.

Le Hyaric, R., Benbelkacem, H., Bollon, J., Bayard, R., Escudié, R., Buffière, P. (2012). Influence of moisture content on the specific methanogenic activity of dry mesophilic municipal solid waste digestate. *Journal of Chemical Technology and Biotechnology* 87: 1032–1035.

Liew, L.N., Shi, J., Li, Y. (2012). Methane production from solid-state anaerobic digestion of lignocellulosic biomass. *Biomass and Bioenergy* 46: 125–132.

Liu, X., Bayard, R., Benbelkacem, H., Buffiere, P., Gourdon, R. (2015). Evaluation of the correlations between biodegradability of lignocellulosic feedstocks in anaerobic digestion process and their biochemical characteristics. *Biomass and Bioenergy* 81: 534–543.

Lv, W., Schanbacher, F.L., Yu, Z. (2010). Putting microbes to work in sequence: Recent advances in temperature phased anaerobic digestion processes. *Bioresource Technology* 101: 9409–9414.

Mace, S., Bolzonella, D., Cecchi, F., Mata-Alvarez, J. (2003). Comparison of the biodegradability of the grey waste fraction of MSW of Barcelona in mesophilic and thermophilic conditions. *Water Science and Technology* 48(4): 21–28.

Mata-Alvarez, J. (2003). *Biomethanization of the Organic Fraction of Municipal Solid Waste.* London: IWA Publishing.

McDonald, P., Henderson, A., Heron, S.J. (1991). *The Biochemistry of Silage.* 2nd ed. Marlow, Bucks, UK: Chalcombe Publications.

Möller, K., Müller, T. (2012). Effects of anaerobic digestion on digestate nutrient availability and crop growth: A review. *Engineering and Life Science* 12: 242–257.

Monlau, F., Latrille, E., Da Costa, A.C., Steyer, J.-P., Carrère, H. (2013). Enhancement of methane production from sunflower oil cakes by dilute acid pretreatment. *Applied Energy* 102: 1105–1113.

Mosier, N., Wyman, C., Dale, B., Elander, R., Lee, Y., Holtzapple, M., Ladisch, M. (2005). Features of promising technologies for pretreatment of lignocellulosic biomass. *Bioresource Technology* 96(6): 673–686.

Owens, J.M., Chynoweth, D. (1993). Biochemical methane potential of MSW components. *Water Science and Technology* 27(2): 1–14.

Pakarinen, A. (2012). Evaluation of fresh and preserved herbaceous field crops for biogas and ethanol production. Doctoral dissertation, Department of Agricultural Sciences, University of Helsinki, Finland.

Palmowski, L.M., Müller, J.A. (2000). Influence of the size reduction of organic waste on their anaerobic digestion. *Water Science and Technology* 41(3): 155–162.

Parkin, G.F., Owen, W.F. (1986). Fundamentals of anaerobic digestion of wastewater sludges. *Journal of Environmental Engineering* 112: 867–920.

Pfeffer, J.T. (1974). Temperature effects on anaerobic fermentation of domestic refuse. *Biotechnology and Bioengineering* 16: 771–787.

Plöchl, M., Heiermann, M. (2006). Biogas Farming in Central and Northern Europe: A Strategy for Developing Countries? *Agricultural Engineering International: The CIGR E-Journal* VIII(8).

Pouech, P., Fruteau, H., Bewa, H. (1998). Agricultural crops for biogas production on anaerobic digestion plants. In *Proceedings of the 10th European Conference and Technology Exhibition*, Würzburg, Germany, 8–11 June 1998. pp. 163–165.

Sun, Y., Cheng, J. (2002). Hydrolysis of lignocellulosic materials for ethanol production: A review. *Bioresource Technology* 83(1), 1–11.

Taherzadeh, M.J., Karimi, K. (2008). Pretreatment of lignocellulosic wastes to improve ethanol and biogas production: A review. *International Journal of Molecular Sciences* 9(9): 1621–51.

ten Brummeler, E., Koster, I.W. (1990). Enhancement of dry anaerobic batch digestion of the organic fraction of municipal solid waste by an aerobic pretreatment step. *Biological Wastes* 31: 199–210.

Tong, X., Smith, L.H. et al. (1990). Methane fermentation of selected lignocellulosic materials. *Biomass* 21: 239–255.

Vandevivere, P., De Baere, L., Verstraete, W. (2003). Types of anaerobic digester for solid wastes. In *Biomethanization of the Organic Fraction of Municipal Solid Wastes*, 111–137. London: IWA Publishing.

Van Lier, J.B., Sanz Martin, J.L., Lettinga, G. (1996). Effect of temperature on the anaerobic thermophilic conversion of volatile fatty acids by dispersed and granular sludge. *Water Research* 30: 199–207.

Védrenne, F. (2007). Etude des processus de dégradation anaérobie et de production de méthane au cours du stockage des lisiers. Science de l'Environnement. Thèse de doctorat [doctoral thesis], ENSA Rennes, Ecole Nationale Supérieure d'Agronomie.

Veeken, A., Hamelers, B. (1999). Effect of temperature on hydrolysis rate of selected biowaste components. *Bioresource Technology* 69: 249–254.

Weiland, P. (2003). Production and energetic use of biogas from energy crops and wastes in Germany. *Applied Biochemistry and Biotechnology* 109(1–3): 263–274.

Weinberg, Z.G., Ashbell, G. (2003). Engineering aspects of ensiling. *Biochemical Engineering Journal* 13: 181–188.

Xu, F., Wang, Z., Tang, L., Li, Y. (2014). A mass diffusion-based interpretation of the effect of total solids content on solid-state anaerobic digestion of cellulosic biomass. *Bioresource Technology* 167: 178–185.

Zhang, R., Zhang, Z. (1999). Biogasification of rice straw with an anaerobic-phased solids digester system. *Bioresource Technology* 68(3): 235–245.

7

Dark Fermentative Hydrogen Production:
From Concepts to a Sustainable Production

*Patrícia Madeira da Silva Moura, Joana Resende Ortigueira,
Idania Valdez-Vazquez, Ganesh Dattatray Saratale,
Rijuta Ganesh Saratale, and Carla Alexandra Monteiro da Silva*

Contents

7.1 Hydrogen as Energy Source

The use of renewable sources and environmentally friendly processes is considered a priority for the construction of a sustainable energy future. The harmful impact of fossil fuels and the fact that we are reaching a disrupting point regarding environmental damage require the rapid implementation of new energy systems and a substantial increase in the use of alternative, unconventional energy sources. Hydrogen (H_2) is considered one of the most promising sources as a clean energy vector, because of its high energy density (120 MJ/kg) and carbon-free combustion (Argun and Kargi, 2011). Hydrogen is the simplest and most abundant element on earth; however, it barely exists in nature in its molecular state. Instead, it is almost always found as part of other compounds from which it should be separated, either by thermochemical processes or through biological conversion.

Currently, the dominant technology for direct H_2 production consists in the steam reforming of hydrocarbons, as well as additional thermochemical methods, such as electrolysis and thermolysis (Saratale et al., 2013). The major obstacles of the current H_2-producing methods are the high production costs and energy demand involved in the production process. Therefore, there is an imperative need to come up with strategies that can make the process more sustainable and economically feasible.

7.1.1 Biohydrogen

The production of H_2 from renewable resources, such as water, organic wastes or biomass, either biologically or photobiologically, is termed "biohydrogen" (bioH_2). Compared with the thermochemical route, biological H_2 production is more favorable, owing to its environmentally friendly and energy-saving production process. The biological conversion occurs at near-ambient temperature and pressure, at far less extreme conditions, which lower the energy requirements substantially. Additionally, it is well suited for decentralized energy production in small-scale installations located where biomass or wastes are available, thus avoiding energy expenditure and diminishing transport costs (Saratale et al., 2008, 2013; Wang and Wan, 2009). Methods adopted to produce H_2 biologically are based on biophotolysis of water by algae and cyanobacteria, photodecomposition of organic compounds by photosynthetic bacteria, dark fermentation by facultative and anaerobic bacteria, hybrid systems using sequential two-stage dark and photofermentation processes, and bioelectrohydrogenesis or microbial fuel cells (Das and Veziroğlu, 2001; Chen et al., 2008; Saratale et al., 2008, 2015; Lee et al., 2014). From all these biological methods, the dark fermentation has achieved the highest production rates (>1 m^3/h $\cdot$ m^3), positioning itself as the leading production method for bioH$_2$ (Ren et al., 2011). Consequently, dark fermentation may overcome barriers to the commercialization of bioH$_2$ in the short term.

7.1.2 Hydrogen Gas Production by Dark Fermentation

Dark fermentation is performed by facultative and strictly anaerobic bacteria that evolve H_2 as a means of eliminating the excess electrons generated during the degradation of carbon compounds (Rupprecht et al., 2006). Dark fermentative H_2 production offers several advantages in comparison with the other biological counterparts, including process simplicity, low energy requirements, and ability to generate H_2 from a wide variety of economic organic substrates, such as food wastes, wastewater, agricultural residues, forestry wastes, wastepaper, algal biomass, and energy crops. The fermentative process for H_2 production is carried out by both pure isolated microorganisms and microbial consortia acclimated from different sources, such as sewage sludge (Wang and Wan, 2009). Also, the H_2 fermentation is governed by several environmental conditions, mainly pH, temperature, and type and concentration of substrate. It is therefore essential to control the type of microorganisms and environmental conditions prevailing in the fermentation for obtaining a competitive process.

Addressing the above-mentioned issues, this chapter presents an overview of dark fermentative H_2 production by pure cultures and microbial consortia, along with a critical review on the processes involved in biological conversion of lignocellulosic-based and waste biomass to H_2, and the application of life cycle analysis to bioH$_2$ production.

7.2 Biochemistry of Hydrogen Production in Dark Fermentation

Obligate anaerobic bacteria carry out fermentative metabolism by degrading organic substrates, through oxidation, in order to gain energy for growth (in the form of ATP). This oxidation generates electrons, which need to be disposed of in order to maintain electron neutrality (Nath and Das, 2004). Therefore, they are transferred to internal electron carriers (NADH and ferredoxin [Fd]), and further to a variety of compounds that act as final electron sinks. Hydrogen generation occurs when protons act as terminal electron acceptors in anoxic environments. When NADH regeneration is not coupled to proton reduction, reduced end products are produced (Nath and Das, 2004; Kalia and Purohit, 2008).

In this process, H_2 is produced by hydrogenases that catalyze the reversible oxidation of H_2 into protons and electrons (according to the reaction depicted in Equation 5.1) (Vignais et al., 2001).

$$2H^+ + 2e^- \leftrightarrow H_2 \qquad (7.1)$$

Hydrogenases can be found in several prokaryotic and lower eukaryotic biological systems (Mertens and Liese, 2004) and are usually involved in either H_2 uptake or evolution, depending on the organism's metabolic demands (Vignais et al., 2001). Hydrogenases can be categorized into three major groups, according to the metal content of their active sites: nickel–iron hydrogenases ([NiFe]-H_2ases), iron-only hydrogenases ([FeFe]-H_2ases), and metal-free H_2ases (Vignais et al., 2001; Mertens and Liese, 2004; Calusinska et al., 2010). In particular, the expression levels of genes encoding for [FeFe]-H_2ases (*hyd*) have been considered a good index for optimal H_2 production in *Clostridium butyricum* fermentations (Wang et al., 2009).

Hydrogen can also be produced by nitrogenases. This enzymatic multiprotein complex is responsible for nitrogen fixation, catalyzing nitrogen reduction, and generating H_2. Nitrogenases act as ATP-dependent hydrogenases; however, the H_2 produced is, on its hand, an inhibitor of nitrogen reduction (Burgess et al., 1981). Genes encoding nitrogen fixation are widespread in cyanobacteria, but were also found in dark fermentative bacteria, where the expression of these genes was related to H_2 production under specific culture conditions (Calusinska et al., 2015). Key nitrogen fixation genes (*nif*) were found in *C. pasteurianum*, *C. butyricum*, and solventogenic clostridia in coexistence with the butyrate production pathway (Wang et al., 1990; Chen et al., 2001; Calusinska et al., 2015), as well as in marine bacteria with a mixed-acid fermentation type (Ma et al., 2015).

The biochemical reactions currently known for H_2 production in dark fermentation are depicted in Figure 7.1. The reactions and enzymes that are more directly involved in the H_2 production steps are highlighted. These include the following:

1. Hydrogen production via pyruvate–ferredoxin oxireductase (Pfor). Strict anaerobes break down glucose into pyruvate and NADH, through the Embden–Meyerhof–Parnas (EMP) pathway, leading to ATP formation (Kim and Gadd, 2008). Pyruvate is then further dehydrogenated to acetyl coenzyme A (acetyl-CoA) and CO_2 by Pfor, producing reduced ferredoxin (Fd_{H2}) (Equation 7.2). Ferredoxin is an electron carrier, which needs to be regenerated. Fd_{H2} is oxidized by cytosolic [FeFe]-H_2ase and molecular H_2 is produced (Equation 7.3). This reaction is responsible for generating 2 mol H_2/mol glucose consumed (Nath and Das, 2004; Hallenbeck and Ghosh, 2009).

$$\text{Pyruvate} + \text{CoA} + \text{Fd} \rightarrow \text{Acetyl-CoA} + Fd_{H2} + CO_2 \qquad (7.2)$$

$$Fd_{H2} \rightarrow \text{Fd} + H_2 \qquad (7.3)$$

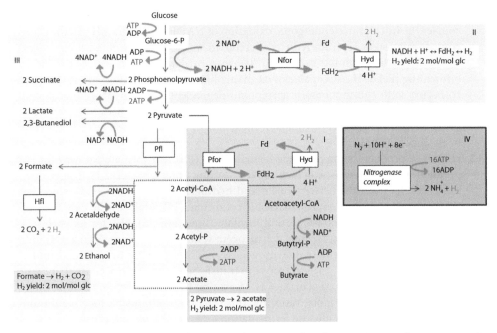

Figure 7.1 Metabolic pathways for H_2 production. Subpathway I: H_2 production via Pfor. Glucose is broken into pyruvate and NADH through the EMP pathway, pyruvate is dehydrogenated to acetyl-CoA and CO_2 by Pfor, producing Fd_{H2}, which is regenerated by [FeFe]-H_2ase, and molecular H_2 is produced. Subpathway II: H_2 production via Nfor. At low H_2 partial pressures, the NADH produced during glycolysis is reoxidized and Fd_{H2} is produced, which in turn is regenerated by [FeFe]-H_2ase to yield H_2. Subpathway III: H_2 production via Pfl. Pfl catalyzes the oxidation of pyruvate to produce acetyl-CoA and formate. Hfl is a complex of two enzymes, formate dehydrogenase and hydrogenase; the former oxidizes formate to CO_2, and under fermentative conditions, the electrons are transferred to the hydrogenase that catalyzes proton reduction and generates H_2. Subpathway IV: Nitrogenase-mediated H_2 production. Nitrogenase is a complex protein consisting of azoferredoxin and molybdoferredoxin. Separately, azoferredoxin is reduced by accepting the electrons from ferredoxins or flavodoxins, followed by ATP binding, and molybdoferredoxin binds to nitrogen. The ATP-reduced azoferredoxin and nitrogen-bound molybdoferredoxin form a complex, and the electrons are transferred from azoferredoxin to molybdoferredoxin, with nitrogen reduction and H_2 production.

2. Hydrogen production via NADH–ferredoxin oxireductase (Nfor). Strict anaerobes are also capable of producing additional H_2 molecules through reoxidation of the NADH produced during glycolysis, at low H_2 partial pressures ($P_{H2} < 60$ Pa) (Hallenbeck and Ghosh, 2009). NADH reoxidation is mediated by Nfor, to produce Fd_{H2} (Equation 7.4), which in turn is reoxidized by [FeFe]-H_2ase to yield H_2 (Equation 7.5) (Saint-Amans et al., 2001; Hallenbeck and Ghosh, 2009). This reaction is responsible for generating 2 mol H_2/mol glucose consumed (Zheng et al., 2014).

$$NADH + Fd \rightarrow NAD^+ + Fd_{H2} \quad (7.4)$$

$$Fd_{H2} \rightarrow Fd + H_2 \quad (7.5)$$

A ferredoxin- and NAD-dependent electron-bifurcating [FeFe]-hydrogenase that couples the unfavorable endergonic oxidation of NADH, to produce H_2, to the exergonic formation of H_2 from reduced ferredoxin was found in *Ruminococcus albus*. In the hyperthermophilic bacterium *Thermotoga maritima*, in which a trimeric bifurcating [FeFe]-H_2ase capable of oxidizing both NADH and Fd_{H2} under low p_{H2} co-occurs with the monomeric ferredoxin-dependent [FeFe] hydrogenase (Schut and Adams, 2009). The presence of similar trimeric [FeFe]-H_2ases were reported in the genome of various members of the genus *Clostridium*, such as *C. butyricum* (Calusinska et al., 2010).

3. Hydrogen production via pyruvate formate lyase (Pfl). The formate pathway for H_2 production is characteristic of facultative anaerobic enteric bacteria. Hydrogen evolution results from the activity of Pfl and hydrogen formate lyase (Hfl), enzymes that are only expressed under anaerobic conditions. Pfl catalyzes the oxidation of pyruvate to produce acetyl-CoA and formate (Equation 7.6). Hfl, for its part, is a complex of two enzymes, formate dehydrogenase and hydrogenase (Kim and Gadd, 2008). Subsequently, formate dehydrogenase oxidizes formate to CO_2, and under fermentative conditions, the electrons are transferred to the hydrogenase that catalyzes proton reduction and generates H_2 (Equation 7.7). Hydrogen and carbon dioxide are produced at a 1:1 molar ratio, and this reaction is responsible for a maximum H_2 yield of 2 mol/mol in glucose fermentation by *Enterobacter aerogenes*.

$$\text{Pyruvate} + \text{CoA} \rightarrow \text{Acetyl-CoA} + \text{Formate} \tag{7.6}$$

$$\text{Formate} \rightarrow CO_2 + 2H^+ + 2e^- \rightarrow CO_2 + H_2 \tag{7.7}$$

4. Nitrogenase-mediated H_2 production. Nitrogenase is a complex protein consisting of azoferredoxin and molybdoferredoxin, where the former acts as dinitrogenase reductase and the latter as dinitrogenase (Kim and Gadd, 2008). Separately, azoferredoxin is reduced by accepting the electrons from ferredoxins or flavodoxins, followed by ATP binding, and molybdoferredoxin binds to nitrogen. The ATP-reduced azoferredoxin and nitrogen-bound molybdoferredoxin form a complex, and the electrons are transferred from azoferredoxin to molybdoferredoxin with ATP hydrolysis, nitrogen reduction, and H_2 production as a by-product of nitrogen fixation (Equation 7.8). This process is highly energy-intensive due to the molecular stability of molecular nitrogen and impacts negatively on the microbial growth rate (Chen et al., 2001). This reaction is responsible for generating 1 mol H_2/mol N_2 reduced.

$$N_2 + 8H^+ + 8e^- \rightarrow 2NH_3 + H_2 \tag{7.8}$$

7.2.1 Hydrogen Yield in Acetate and Butyrate Fermentation

In practice, the amount of H_2 produced by facultative or strict anaerobes never achieves the theoretical maximum yield of 4 mol H_2/mol hexose (Thauer et al., 1977). Besides determined by the fermentation pathway, H_2 production yields are also dependent on the oxidation level of organic end products. Higher H_2 yields are associated with the formation of acetate, whereas the formation of more reduced metabolites, such as butyrate and lactate, implicates lower H_2 production (Levin et al., 2004, 2009; Hallenbeck and Ghosh, 2009).

In the acetate and butyrate fermentation, acetyl-CoA is an important metabolic intermediate, constituting a branching point for acetate and butyrate formation pathways (Liu et al., 2006). In the acetate pathway, acetyl-CoA and 2 mol H_2/mol glucose are produced from the phosphoroclastic reaction catalyzed by Pfor from pyruvate oxidation. Acetate production from acetyl-CoA is then catalyzed by phosphotransacetylase (Pta) and acetate kinase (Ack), with the intermediate formation of acetylphosphate and two additional moles of ATP (Liu et al., 2006; Zhang et al., 2009). In this pathway, however, more H_2 is evolved than is pyruvate oxidized (Thauer et al., 1977). When the H_2 partial pressure is low, hydrogenase oxidizes reduced ferredoxin, producing H_2. This increases the Fd/Fd_{H2} ratio, and Nfor couples NADH oxidation to ferredoxin reduction (Kim and Gadd, 2008). Acetyl-CoA is no longer needed as the electron acceptor for NADH reoxidation, and it is converted to acetyl-phosphate and further to acetate and ATP. Since 2 mol NADH are generated during glycolysis, a maximum of two additional moles of H_2 can be produced per mole of glucose (Hallenbeck and Ghosh, 2009). Hence, obligate anaerobes are theoretically able to produce a maximum of 4 mol H_2/mol glucose consumed, with reductant provided by 2 mol NADH and 4 mol reduced Fd, when glucose is catabolized solely to acetate (Schut and Adams, 2009) (Equation 7.9).

$$\text{Glucose} \rightarrow 2\,\text{acetate} + 2CO_2 + 4H_2 + 4ATP \tag{7.9}$$

At high H_2 partial pressure, NADH is reoxidized to produce more reduced organic compounds, such as butyric and lactic acid (Hallenbeck and Ghosh, 2009). Pyruvate can yield lactate, through the reaction catalyzed by lactate dehydrogenase (Ldh), accompanied by the reoxidation of NADH to NAD^+ (Zhang et al., 2009). Lactate formation does not favor H_2 production because NADH is coupled to lactate production instead of proton reduction.

In the reaction sequence to butyrate, the NADH generated is quantitatively consumed in butyrate formation. In the butyrate formation branch, two acetyl-CoA molecules are condensed and subsequently reduced, to form butyryl coenzyme A (butyryl-CoA) in a four-step reaction with concomitant reoxidation of NADH (Louis et al., 2004). In the last stage of butyrate biosynthesis, the enzymes phosphotransbutyrylase (Ptb) and butyrate kinase (Buk) play a key role in the conversion of butyryl-CoA into the intermediate butyryl-phosphate (butyryl-P), and the subsequent formation of butyrate plus an additional ATP molecule (Louis et al., 2004; Liu et al., 2006; Zhang et al., 2009). During the conversion of glucose to butyrate, 3 mol ATP are generated and a maximum of 2 mol H_2 are produced (Equation 7.10).

$$\text{Glucose} \rightarrow \text{Butyrate} + 2CO_2 + 2H_2 + 3ATP \tag{7.10}$$

In most cases, the catabolic metabolism in anaerobic bacteria is branched, and the relative rates of each branch are adjusted to optimize ATP gains and thermodynamic efficiencies of ATP synthesis (Thauer et al., 1977). This optimization governs the disposal of reductant power to the reduction of protons or to the production of more reduced end products, and consequently regulates H_2 formation. When acetate and butyrate are coproduced, the H_2 production averages 2.6 mol H_2/mol glucose consumed (Equation 7.11) (Thauer et al., 1977).

$$\text{Glucose} \rightarrow 0.6\,\text{Acetate} + 0.7\,\text{Butyrate} + 2CO_2 + 2.6H_2 + 3.3\text{ATP} \qquad (7.11)$$

7.2.2 Overview of Mesophilic, Thermophilic, and Psychrophilic Hydrogen-Producing Bacteria

The mesophilic bacteria most extensively studied for H_2 production are affiliated to the families of Clostridiaceae and Enterobacteriaceae. Among the former, the genus *Clostridium* covers anaerobic, Gram-positive, rod-shaped, endospore-forming bacteria that are able to degrade organic material to acids, alcohols, CO_2, H_2, and minerals (Gottshalk et al., 1981). Due to their ability to form spores, clostridia are highly resistant to dryness and acidity, to oxygen exposure and high temperatures, which makes them very robust microorganisms. The members of this genus are widespread in nature, and possess the ability to grow on a wide range of naturally occurring and complex substrates, such as, for example, polymeric structures like cellulose and starch. Moreover, clostridia exhibit a diversity of fermentation metabolic pathways that lead to the production of varied co-metabolites with industrial applications that may add potential economic revenue within the framework of a H_2 biorefinery, for example, butyric acid (Tracy et al., 2012; Sarma et al., 2015).

A number of clostridial strains have been isolated from H_2-producing sludge sources, including *C. butyricum*, *C. pasteurianum*, and *C. bifermentans* (Wang et al., 2003; Lo et al., 2008; Wang et al., 2008a). Other clostridia have been recognized as efficient H_2-producing bacteria, such as *C. beijerinckii* and the thermophile *C. thermocellum*, including from complex substrates like cellulose, such as *C. cellulolyticum* and *C. populeti* (Levin et al., 2006; Ren et al., 2007; Islam et al., 2009; Skonieczny and Yargeau, 2009). *C. acetobutylicum* and *C. beijerinckii* have also been studied for H_2 production, but mostly due to their solventogenic properties, in particular for butanol production through dark fermentation (Lee et al., 2008, 2009; Z. Liu et al., 2010). *C. tyrobutyricum*, *C. paraputrificum*, and *C. saccharoperbutylacetonicum* have been successfully engineered to increase H_2 production by either the inactivation of genes involved in competitive metabolic pathways, the overexpression of hydrogenase evolving genes, or the downregulation of hydrogenase uptake genes (Morimoto et al., 2005; Liu et al., 2006; Nakayama et al., 2008).

The most common enteric bacteria used for H_2 production include *Enterobacter aerogenes*, *E.cloacae*, and *Escherichia coli* that are capable of fermenting sugars to lactate, acetate, formate, succinate, butanediol, and ethanol by the mixed-acid fermentation pathway (Kim and Gadd, 2008). Although these microorganisms are facultative anaerobes, a fact that is regarded as advantageous over clostridia in the implementation and scale-up of fermentative processes, H_2 formation is dependent on the activity of anaerobic enzymes, such as Pfl and Hfl of the formate pathway and the oxygen-sensible NADH-dependent hydrogenase in *E. aerogenes* (Zhang et al., 2011; Zhao et al., 2015). The possibility of directing the NADH formed aerobically in this microorganism to the anaerobic H_2 production, the screening of hydrogenases that maintain their activity in the presence of oxygen, and the overcoming of the thermodynamic limitation by the H_2 partial pressure have been objects of investigation, to pursue the maximization of the theoretical yield up to 10 mol/mol glucose in the NADH-dependent pathway (Zhang et al., 2011). Other mesophilic microorganisms extensively studied for H_2 production include the genera *Bacillus*, *Klebsiella*, and *Citrobacter* (Minnan et al., 2005; Maru et al., 2012; P. Kumar et al., 2013), and also *Halanaerobium saccharolyticum* strains that combine the ability to support hypersaline conditions with the possibility of using crude glycerol as fermentation substrate (Kivisto et al., 2010).

Hydrogen production by thermophilic and extreme thermophilic bacterial and archaeal strains isolated from hot and extreme environments has been widely studied due to the high H_2 production yields that can be achieved. In terms of fermentation process, the higher temperatures tend to decrease the H_2 partial pressure and gas solubility, favoring H_2 evolution and providing a more efficient separation of the gas product. *Caldicellulosiruptor saccharolyticus*, *Pyrococcus furiosus*, *Thermococcus kodakaraensis*, and *Thermotoga maritima* are examples of such microorganisms presenting high H_2 production yields and also very high specific H_2 production rates (Chou et al., 2008). In particular, the archaea *Thermococcus kodakaraensis* and *P. furiosus* exhibit a modified Embden–Meyerhof pathway to which these high specific rates may be correlated. This modified pathway includes ADP- instead of ATP-dependent glucokinases and is based on the direct conversion of glyceraldehyde 3-phosphate to 3-phosphoglycerate by glyceraldehyde 3-phosphate–ferredoxin oxidoreductase (Gapor), utilizing only ferredoxin as electron carrier (Siebers and Schönheit, 2005). Additional advantages of hydrogenesis at elevated temperature include better protection against contamination, the formation of fewer metabolic by-products, and the ability to perform catalysis at high temperature and hydrolyze and ferment complex polymeric carbohydrates (Levin et al., 2004, 2006; Pawar and van Niel, 2013). This last property has been intensively explored in the conversion of lignocellulosic substrates by *Caldicellulosiruptor saccharolyticus* and *T. neapolitana*, and for the study of consolidated bioprocessing (CBP) possibilities (de Vrije et al., 2009; Talluri et al., 2013). *Caldicellulosiruptor saccharolyticus*, an extreme thermophile that was isolated from a New Zealand hot spring, gained relevance in this context. This microorganism is able to utilize cellulose and hemicellulose, fermenting C5 and C6 sugars simultaneously and producing H_2 at yields close to 4 mol/mol glucose (Van De Werken et al., 2008). The ability to degrade a wide range of simple and complex carbohydrates and produce fermentative H_2 at high yields was also described for the genus *Thermotoga*. The relevant xylanolytic activity of *T. maritima* is relevant and is based on hyperthermostable enzymes, one xylanase and one arabinofuranosidase, that have great interest for the pulp and paper and functional food industries, besides its possible use in CBP for H_2 production from lignocellulosic biomass (Jiang et al., 2004; Frock et al., 2010).

A prevailing problem in H_2 production by thermophiles and extreme thermophiles is the low cell biomass level generally attained that tends to limit the volumetric H_2 production rate (Chou et al., 2008). The immobilized or aggregated cell systems that promote biomass retention and low hydraulic retention times (HRTs) can be used to prevent washout of the H_2-producing cells and increase the bioconversion efficiency (Pawar and van Niel, 2013). Porous glass beads and thermostable hydrogel-based systems that provide additional pH buffering proved to be suitable immobilization options for *T. neapolitana* cells (Basile et al., 2012; Ngo and Bui, 2013). Moreover, the optimization of the culture media composition to circumvent the redirection of H_2 for the reduction of elemental sulfur to form hydrogen sulfide, and the removal of the H_2 accumulated to avoid microbial inhibition are also indicated as possible strategies to increase the H_2 production yields and rates (Schut and Adams, 2009).

The search for new enzymes that are stable and effective under extreme environmental conditions has stimulated the exploration of the natural biodiversity of the polar regions. Very recently, several psychrophilic microorganisms were isolated from samples collected in a glacier of Antarctica and screened for their potential as H_2 producers (Alvarado-Cuevas et al., 2015). Two isolates closely related to *Janthinobacterium agaricidamnosum* and *Polaromonas jejuensis* produced H_2, respectively, at 16.64 mL/L h and 1.57 mol H_2/mol glucose. These microorganisms may hold a promise for future research options of low-temperature H_2 production.

7.2.3 Clostridium butyricum *as Model Microorganism for Hydrogen Production*

C. butyricum can be considered a model microorganism for H_2 production. Accordingly, *C. butyricum* strains have been extensively characterized in their hydrogenogenic ability and subject to strain improvement. These strategies encompass different approaches, targeting (1) the characterization and optimization of the technology process conditions, by adjusting the operation mode to best fit each combination of microorganism–substrate, and creating conditions that promote shifts in the metabolic pathways toward H_2 evolution; (2) the implementation of two-stage dark and photofermentation systems where the effluent of one process is used to feed a second bioconversion; (3) the use of improved strains, by exploring more efficient microorganisms and/or cocultures with complementary catabolic routes mostly to increase the extension of substrate utilization and production yields; (4) metabolic engineering by gene upregulation of H_2-evolving native pathways or gene downregulation of competitive pathways in dark fermentation; and (5) the use of easily accessible and low-cost substrates.

7.2.3.1 *Characterization and Optimization of the Technology Process Conditions*
The studies with *C. butyricum* CGS5 have extensively addressed the kinetics of H_2 production. This strain was isolated from an effluent sludge of H_2-producing bioreactors (Chen et al., 2005). The comparison of the nearly full-length 16S rDNA sequence

produced the highest similarity with the sequence of *C. butyricum* ATCC 19398 (type strain), which is an efficient H_2-producing strain. In batch experiments, it was observed that H_2 evolution followed a sigmoidal profile, starting when the cells were at the midstage of exponential growth, after a former period of biosynthesis where the value of the cell biomass almost doubled. Hydrogen production increased sharply afterwards, and the maximum production rate was achieved at the beginning of the stationary phase. Correspondingly, the copy number of the hydrogenase gene *hyd*A in the strain CGS5 increased during the late exponential growth phase (Wang et al., 2008b). *C. butyricum* CGS5 was adapted to sucrose supplemented media, and the H_2 content reached 64% (vol) in the produced biogas. It was observed that the pH of the medium strongly influenced H_2 production, since at pH 5.5 the maximum H_2 yield was 2.78 mol/mol sucrose, but at pH 5.0, the cell growth was completely inhibited (Chen et al., 2005). The accumulation of acidic fermentation products that induce a decrease in the pH of the medium to values near or below the respective pKa values (pKa [acetate] = 4.82, pKa [butyrate] = 4.75) leads to the accumulation of the undissociated forms of the acids. Due to their hydrophobicity and permeability to the cytoplasmic membrane, these are toxic to cells, dissipating the proton motive force (Kim and Gadd, 2008). An uncommon alkaline-tolerant *C. butyricum* strain (TM-9A) was isolated from estuarine sediment samples by Junghare et al. (2012). This strain achieved the maximum and minimum H_2 production at pH values 8 and 5, respectively, and tolerated NaCl concentrations up to 16 g/L without a major decrease in H_2 production, from 61 to approximately 40 mmol/L.

The strain *C. butyricum* CWBI 1009 was isolated by Masset et al. (2010), and is closely related to *C. butyricum* DSM 523. CWBI 1009 produced lactate, acetate, butyrate, formate, and ethanol, from glucose, sucrose, and starch, at 30°C. The optimum pH was 5.2, and the control within this pH range determined a significant increase in the H_2 production yield from 0.97 to 2.10 mol/mol glucose (Masset et al., 2012). In eight sequences of sequence batch reactor (SBR) operation, the strain was able to achieve a H_2 yield of 2.3 mol H2/mol glucose (Masset et al., 2010). However, when using raw starch as the fermentation substrate, the H_2 production rate seemed to be limited by the activity of amylases and starch hydrolysis at pH levels below 5.6 (Masset et al., 2010). In the same study, a sharp decrease of the H_2 production rate was observed, which appeared to be related to product inhibition due to the accumulation mainly of acetate and butyrate in concentrations approaching 90 mM (Masset et al., 2010). Studies with the strain *C. butyricum* CWBI 1009 at different fermentation operation modes that promoted a higher liquid-to-gas transfer, such as strong stirring, N_2 degassing, and different reactor configurations, were conducted by Hiligsmann et al. (2014), Puhulwella et al. (2014), and Beckers et al. (2015). Lower concentrations of dissolved H_2 in the culture medium determined significant increases in the H_2 production yield (up to 3.1 mol/mol glucose) and rate (up to 278 mL/h) by CWBI 1009.

The strain *C. butyricum* EB6 was isolated from a palm oil mill effluent (POME) and presents high 16s rDNA sequence identity with the strain CGS5 (Chong et al., 2009). The interest in using raw POME as a cheap fermentation substrate by this strain led to process optimization studies encompassing cell immobilization. *C. butyricum* EB6 was entrapped in polyethylene glycol (PEG) and repeatedly cultivated in POME, and this enabled the H_2 production rate to increase from 12.3 to 22.7 mmol/L h (Singh et al., 2013). Smaller PEG beads (3 mm) proved to be more advantageous than 4 or 5 mm beads, possibly because of the flow pattern that was generated inside the reactor and the promotion of a better contact between the particles and the liquid medium. In line with this, the retention of *C. butyricum* CWBI 1009 as a biofilm in a polypropylene immobilization support of an anaerobic biodisc-like reactor, as a means to facilitate the adherence of the cells and simultaneously increase the liquid–gas exchange surface, produced 2.4 mol H_2/mol glucose and attained a H_2 production rate of 703 mL H_2/L h (Hiligsmann et al., 2014).

The effect of combining nanomaterials in cultures of *C. butyricum* for H_2 production has also been evaluated. A system where iron oxide nanoparticles of 2–3 nm in diameter encapsulated inside a porous silica matrix were suspended with *C. butyricum* CWBI 1009 enabled a very significant increase of the H_2 production rate (Beckers et al., 2013). Besides the advantage of the large surface area of the nanoparticles, the authors suggested a catalytic effect of the immobilized iron oxide active sites that would facilitate the electrons' transference to outside of the cells.

7.2.3.2 Two-Stage Dark and Photofermentative Sequential Hydrogen Production

The use of syntrophic cultures involving *C. butyricum* strains was also tested, mainly with the objective of extending the substrate utilization and improving the bioconversion yields. The so-called dual-culture systems with *C. butyricum* can take profit from the fact that the organic acids produced in dark fermentation constitute intermediate substrates for H_2-producing photofermentative bacteria. The sequential or simultaneous fermentation by these microorganisms has been explored to increase the H_2 production targeting the maximum stoichiometric yield of 12 mol/mol glucose (Hallenbeck and Ghosh, 2009). Fang et al. (2006a) prepared cocultures of *C. butyricum* DSM 10702 and *Rhodobacter sphaeroides* DSM 158 and examined in detail the syntrophic interaction between both microorganisms at specific dark–photo cell number ratios. The highest H_2 production yield was 0.60 mL/mL medium, and although *R. sphaeroides* in pure culture consumed glucose, the much lower rate at which this occurred guaranteed that the photofermentative bacteria did not compete with *C. butyricum* for glucose. This was corroborated by Laurinavichene and Tsygankov (2015) using a coculture of *R. sphaeroides* N7 with a *C. butyricum* strain isolated from a silo pit, with which it was possible to attain a H_2 production yield of 4.9 mol/mol hexose from starch. Similarly, under optimized conditions of dark–photo bacteria ratio, pH, and light intensity, and a concentration of substrate and phosphate that ensured glucose depletion, the coculture of a strain of *C. butyricum* from the China General Microbiological Culture Collection and immobilized by *Rhodopseudomonas faecalis* RLD-53 produced 4.134 mol H_2/mol glucose in a 240 h batch culture (Ding et al., 2009). However, the assimilation of acetate and butyrate by the photofermentative bacteria was slower than its production rate by *C. butyricum*, resulting in acidification of the culture medium, which limited the microbial growth and H_2 production (B. F. Liu et al., 2010). A combined coculture and sequential system using sweet potato starch as fermentation substrate, and where the supernatant of *C. butyricum* IFO13949 in association with *E. aerogenes* HO-39 culture broth was used to grow *Rhodobacter* sp. M-19, was developed by Yokoi et al. (2001). With the integration of the photofermentation stage, it was possible to increase the H_2 production yield from 2.4 to 7 mol/mol glucose, mostly at the expense of the starch residue that remained from the dark fermentation.

With the objective to increase the H_2 production yield and implement a CO_2-free production system in which both the liquid and gas effluents of the fermentation are reutilized, the strain CGS5 was used in a three-stage fermentation process that included dark and photofermentation, and autotrophic culture of *Chlorella vulgaris* (Lo et al., 2010). The biogas produced in both fermentative processes was fed to the microalga culture for the consumption of the CO_2, whereas the liquid effluent from the dark fermentation served as a substrate for the growth of *Rhodopseudomonas palutris*, a H_2-producing purple bacteria. A maximum H_2 yield of 11.61 mol H_2/mol sucrose and a H_2 content of 91% (vol) in the final biogas were achieved. Kim et al. (2006) explored the use of *Chlamydomonas reinhardtii* biomass as fermentation substrate for *C. butyricum* NCIB 9576 and the subsequent photofermentative H_2 production by *R. sphaeroides* KD131. This combination produced 8.0 mL H_2/mL broth, but the *C. butyricum* fermentate had to be sevenfold diluted so that *R. sphaeroides* would not be inhibited by the concentration of organic acids.

7.2.3.3 Cocultures with C. butyricum

The main objective of using cocultures of *C. butyricum* with other microorganisms for H_2 production is associated with attempts to increase the extension of substrate utilization and production yields and rates. However, the coculture with facultative anaerobes has also been used to help create the necessary anoxic conditions for *Clostridium* growth, exempting the addition of reducing agents to the culture media. Accordingly, Yokoi and coworkers (1998, 2001) studied the association of *C. butyricum* IFO13949 with *E. aerogenes* in continuous culture. The amount of H_2 evolved by the former in the presence of reducing agent was about 57% of the one produced by the coculture without cysteine hydrochloride. Moreover, after stable H_2 production (more than 100 h) it was possible to restart the continuous production within 0.5 h without the need for replacing the reactor headspace with argon. The coculture of *C. butyricum* CWBI 1009 with *C. pasteurianum* was tested in large lab-scale bioreactors for a time period of 13 days (Masset et al., 2012). This combination produced a stable coculture and increased the biogas production rate from 0.35 to 1.05 L/h, indicating a more efficient breaking down and consumption of the starch molecules. However, the H_2 production yield did not exceed the one obtained with *C. butyricum* CWBI 1009 as pure culture (2.32 vs. 2.91 mol H_2/mol hexose).

7.2.3.4 Metabolic Engineering of C. butyricum

The common metabolic engineering approaches applied to *C. butyricum* include gene upregulation of H_2-evolving native pathways or gene downregulation of competitive pathways in dark fermentation (Oh et al., 2011; Goyal et al., 2013). *C. butyricum* W5 is a H_2-producing strain that was isolated from a heat-treated activated sludge (Wang et al., 2007). This strain produced H_2 from starch and molasses, along with acetic, butyric, and lactic acid and ethanol (Wang and Jin, 2009, Cai et al., 2011).

The suppression of the butyrate and the ethanol formation pathways in *C. butyricum* W5, which are competing metabolic pathways to H_2 production, was tested (Cai et al., 2011). The strain was first engineered by the inactivation of the gene *hbd*, which encodes β-hydroxybutyryl-CoA dehydrogenase. As a result, the production of ethanol was increased in the *hbd*-deficient strain and the H_2 production decreased significantly. In turn, the disruption of the *aad* gene, which encodes a bifunctional aldehyde–alcohol dehydrogenase involved in the ethanol production pathway, led to a redirection of the metabolic flux towards lactate production. In this case, an increase of the H_2 yield (from 0.94 to 1.65 mol/mol glucose) was only possible by the addition of sodium acetate to the medium at pH 5.5. The authors hypothesized that the added acetate could act as an indirect proton acceptor that would promote lactate reutilization and stimulate butyrate and H_2 production by the mutant strain (Cai et al., 2013). Alternative strategies of metabolic engineering to improve H_2 production include genetic manipulation to increase the expression of hydrogenases (Hallenbeck and Ghosh, 2009). Klein et al. (2010) tested the heterologous expression of the gene *hyd*ACb of *C. butyricum* DSM 10702 in *C. acetobutylicum*, under controlled pH conditions. This gene encodes the [FeFe]-hydrogenase (*hyd*A) responsible for the transfer of electrons to oxidized ferredoxin. An increase in the hydrogenase activity of the recombinant strain (pSOShydACb) was detected, but it was not reflected in additional H_2 production, suggesting that *hyd*A activity is not limiting for H_2 production.

7.2.3.5 Low-Cost Substrates

Chen and co-workers (2008) observed that *C. butyricum* CGS5 was able to ferment raw starch. This feature is of interest for the production of low-cost H_2 at a commercial scale, for example, from potato peels (Yokoi et al., 2001; Foglia et al., 2010). Hydrogen production could be based on the use of this abundant and easily available food industry waste, and simultaneously, the application of expensive and environmental harmful pretreatment of the raw biomass would be avoided. The production of H_2 by *C. butyricum* CGS5 from cellulose, xylan and xylose, and lignocellulosic materials, such as rice straw and sugarcane bagasse, was reported (Lo et al., 2008, 2010; Saratale et al., 2010). *C. butyricum* CGS5 was able to utilize the sugars from the microalga *Chlorella vulgaris* to produce H_2 without any additional organic carbon source, after acid hydrolysis of the biomass or pretreatment with NaOH combined with enzymatic hydrolysis (Liu et al., 2012). However, the NaCl that resulted from the neutralization after the alkaline pretreatment was shown to partially inhibit the H_2 production. The use of low-cost substrates for H_2 production is discussed in detail in Section 7.4.

7.2.4 Case Study I: Microalgal Biomass as Fermentation Substrate for Hydrogen Production by Pure Cultures

Microalgae are largely recognized as the prime substrate for the so-called third-generation bioenergy production. This term applies to inedible types of biomass, capable of sequestering carbon dioxide in its gaseous form and wherein the main component of interest for conversion is not cellulose (Lee and Lavoie, 2013). Microalgae are highly productive photosynthetic organisms with the capability to store both carbohydrate and lipidic compounds inside their cellular structure, being easily grown in both saltwater and freshwater with minimum nutrient requirements (Gouveia and Oliveira, 2009). Microalgae have the following average composition: 6.7%–68.4% carbohydrates, 14.9%–84.0% proteins, and 0.8%–63.2% lipids per percentage of dry biomass (Chen et al., 2015).

The process of dark fermentation requires carbohydrates as the carbon and energy source. This necessity implies that the chosen type of biomass for a fermentative process should be, ideally, rich in this compound. *Scenedesmus*, *Chlorella*, *Spirogyra*, and *Dunaliella* are examples of microalgae that, besides possessing polymeric sugar cellular walls, are also able to store energy intracellularly in the form of starch (Becker, 2007). The major difficulty with the use of this type of biomass is the highly resistant cellular walls. Thermophilic bacteria are preferably chosen for the fermentation of this biomass, as the vast majority possess cellulose-degrading enzymes and are capable of achieving high productivity yields even with more recalcitrant substrates. The higher temperature is also advantageous at an operational level, as it increases the hydrolysis and fermentation rate while diminishing the probability of contamination (Abdel-Banat et al., 2010). Basile et al. (2012) and Nguyen et al. (2010b) cultured the thermophilic *T. neapolitana* strain with the microalgae *Thalassiosira weissflogii* and *Chlamydomonas reinhardtii* as the carbon and energy source. Both studies achieved similar H_2 yields of 3.3 and 2.5 mol/mol reducing sugars, respectively. However, the high amount of energy required at thermophilic conditions increases the overall process costs considerably. Consequently, *Clostridium* and *Enterobacter* are usually selected for mesophilic H_2 production. Experimental assays using *Chlorella vulgaris*, *Chlamydomonas reinhardtii*, *Scenedesmus obliquus*, and *Spyrogira* sp. as substrate for H_2 production registered particularly interesting results. *C. butyricum* was able to convert these different types of microalgal biomass, achieving H_2 yields that averaged 0.9–2.7 mol/mol reducing sugars, and a productivity as high as 325 mL H_2/L in some cases (Liu et al., 2012, 2013; Chen et al., 2015; Ortigueira et al., 2015). The best results were achieved when the microalgae had the capability to store starch intracellularly, as it is a preferred substrate by *C. butyricum* (Ortigueira et al., 2015). *C. butyricum* DSM 10702 was not capable of further degradation than 60% of the sugar fraction of *Spirogyra* sp., which suggested difficulties of the bacteria to convert pectose under the tested conditions (Ortigueira et al., 2015). In a similar manner, assays conducted with *E. aerogenes* resulted in sucessful fermentations, even when the biomass was solely treated thermically prior to conversion. Batista et al. (2015) registered yields of up to 56.8 mL H_2/g biomass in the fermentation of *S. obliquus* and, more interestingly, of 46.8 mL H_2/g biomass when using the biomass of a small consortium of microalgae grown in municipal wastewater (Batista et al., 2015). However, it is important to note that the fermentation of microalgal biomass may require the previous weakening of the cellular walls, as mesophilic bacteria are generally unable to produce cellulases. Pretreatment processes are intended to break down the algal biomass into sugar-rich fractions, much more easily fermentable than the raw material. These processes can be mechanical or chemical, varying from simple operations, such as milling or ultrasonication, to more advanced hydrolysis processes, such as acid, alkaline, or enzymatic hydrolysis. Kim et al. (2006) proposed a simple thermal treatment as adequate for the breakdown of the polymeric structure of *C. reinhardtii* in order to permit its conversion by *C. butyricum*. However, Ferreira et al. (2012) analyzed the conversion of *Anabaena* sp. biomass by *E. aerogenes* after a 15 min thermal treatment at 121°C and registered solely a maximum production of 0.14 mL H_2/g biomass. This result suggests that the thermal pretreatment might not be enough for the complete, or even partial, degradation of the polymeric structure of the cellular walls. Acid hydrolysis is largely considered to be the most successful alternative for the degradation of the polymeric structure of the microalgal cellular walls, as it is easier and less costly than enzymatic hydrolysis. Chen et al. (2015) improved the fermentation performance of *C. butyricum* from *C. vulgaris* by the introduction of a dilute acid hydrolysis, incrementing the production from 43 to 64 mL H_2/g biomass. K. Kumar et al. (2013) introduced a two-step hydrolysis stage that furthered the efficiency of the polymeric breakdown of *Chlorella sorokiniana* biomass, obtaining a solution easily fermentable by *E. cloacae* IIT-BT 08 and reaching a production yield of 9 mol H_2/kg chemical oxygen demand (COD) reduced.

Even with the limitations inherent to the use of real substrates in the fermentive H_2 production by pure cultures, small-scale anaerobic fermentations have been undertaken successfully, attaining yields comparable to those achieved in fermentations with simple monomers (2.2–2.7 mol H_2/mol sugar) (Liu et al., 2013; Batista et al., 2015; Chen et al., 2015; Ortigueira et al., 2015). While the need for pretreatment might be necessary in order to efficiently convert microalgal biomass, studies are already being performed in order to simplify pretreatments or make their necessity moot by incrementing the capacity of mesophilic cultures to degrade polymeric sugars through genetic engineering processes (Lynd et al., 2005).

7.3 Hydrogen Production by Microbial Consortia

A microbial consortium is defined as that in which microorganisms coexist with different taxonomic affiliation, each one performing a specific task that benefits the entire community (Valdez-Vazquez and Poggi-Varaldo, 2009). Microbial consortia from various origins have been used for producing H_2 via fermentation. In contrast to pure cultures, the most significant advantage of a microbial consortium is the conversion of complex nonsterile substrates to H_2. Microbial consortia extend the range of feedstocks to be used for biological H_2 production. For example, with microbial consortia it is feasible to convert wastewater (Prasertsan et al., 2009; Lin et al., 2012a; Yossan et al., 2012; Moreno-Andrade et al., 2015) and food wastes (Valdez-Vazquez et al., 2005; Kim et al., 2009; Ohnishi et al., 2010; Sreela-or et al., 2011; Moreno-Andrade and Buitrón, 2015) into H_2, a task economically restrictive with pure cultures. Also, microbial consortia act cooperatively to hydrolyze and ferment polysaccharides with different grades of complexity, from starch to lignocellulosic biomass (Akutsu et al., 2008; Lin and Hung, 2008; Nissilä et al., 2011a, 2011b; Chen et al., 2012; Lay et al., 2012; Gadow et al., 2013; Valdez-Vazquez et al., 2015a). On the other side of the coin, H_2-producing microbial consortia have significant drawbacks that may limit their performance and application in full-scale bioreactors. For example, the substrate consumption by biochemical pathways without the H_2 formation reduces the H_2 productivity and yields (Koskinen et al., 2007; Kim and Shin, 2008; Pérez-Rangel et al., 2015).

What determines the H_2 performance and robustness of the H_2-producing consortia? This question has been addressed from different points of view. Some authors propose that the microbial structure may define the performance (Akutsu et al., 2008; Baghchehsaraee et al., 2010; Sivagurunathan et al., 2013; Pérez-Rangel et al., 2015), while others postulate that the environmental conditions, such as the substrate concentration, temperature, and pH, determine the performance (Sreela-or et al., 2011; Yossan et al., 2012; Etchebehere et al., 2016). The most recent advances in our understanding of how to select H_2 producers, the structure of H_2-producing consortia with different origins, and their performance are presented below, as well as a case study of the design of a biorefinery for H_2 production.

7.3.1 Selection of Hydrogen Producers in Consortia

Authors have sampled different environments, such as anaerobic sludge (Koskinen et al., 2007; Davila-Vazquez et al., 2009; Baghchehsaraee et al., 2010; Yossan et al., 2012; Sivagurunathan et al., 2013), compost (Akutsu et al., 2008; Huang et al., 2010; Nissilä et al., 2011a, 2011b), and wastewater treatment plants (Akutsu et al., 2009; Baghchehsaraee et al., 2010; Chaganti et al., 2012; Lay et al., 2012; Sivagurunathan et al., 2013) with the aim of selecting an active microbial community able to produce H_2. Since in these consortia non-H_2-producing bacteria, H_2-producing bacteria, and H_2-consuming bacteria coexist, the literature is full of methods and strategies for enriching the inocula with H_2-producing bacteria (Baghchehsaraee et al., 2008; Kim and Shin, 2008; Ren et al., 2008a; Adav et al., 2009; Hafez et al., 2009; Kim et al., 2009; O-Thong et al., 2009; Ohnishi et al., 2010; Nasr et al., 2011; Nissilä et al., 2011a, 2011b; Rossi et al., 2011; Li et al., 2012b; Ning et al., 2012; Jeong et al., 2013; Wan et al., 2013).

One of these methods, the heat-shock treatment (HST), has become a routine procedure applied to the inocula for selecting H_2 producers (Jo et al., 2007; Akutsu et al., 2009; Davila-Vazquez et al., 2009; Kim et al., 2009; O-Thong et al., 2009; Baghchehsaraee et al., 2010; Huang et al., 2010; Nissilä et al., 2011a, 2011b; Sreela-or et al., 2011; Chen et al., 2012; Lay et al., 2012; Yossan et al., 2012; Sivagurunathan et al., 2013; Wang et al., 2013; Moreno-Andrade and Buitrón, 2015; Moreno-Andrade et al., 2015). The HST relies on the ability of *Clostridium* to form endospores (Valdez-Vazquez and Poggi-Varaldo, 2009); thus, all the vegetative cells are eliminated by this procedure while *Clostridium* survives. The efficacy of the HST has been widely documented, with H_2-producing consortia achieving H_2 yields between 5 and 20 times higher than the untreated inocula (Baghchehsaraee et al., 2008; Kim et al., 2009).

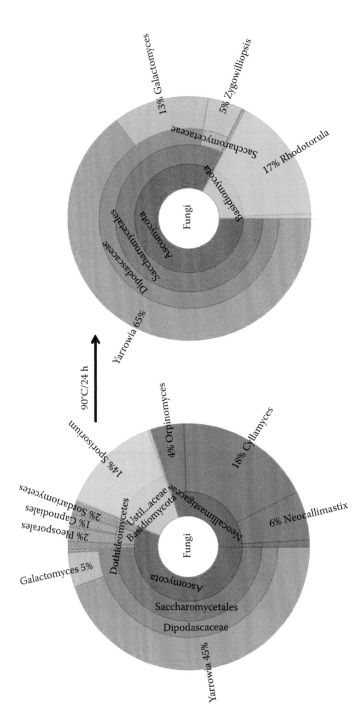

Figure 7.2   Impact of the HST on the diversity of anaerobic fungi in cow manure revealed by 454 pyrosequencing. Each krona chart represents the percentages of fungal species before and after the HST was applied to the inoculum (cow manure). Before the HST, anaerobic hydrolytic fungi, such as *Orpinomyces*, *Cyllamyces aberensis*, and *Neocallimastix frontalis*, represented 28% of the total of fungal community. After the HST was applied, these anaerobic hydrolytic fungi were eliminated.

Also important to consider is the temperature applied during the HST, since temperatures higher than 80°C may be detrimental to the H_2 process.

In spite of the apparent efficacy of the HST, it reduces the microbial diversity of the treated inocula dramatically, and this fact has detrimental results in bioreactors fed with complex substrates, such as cellulose and stillage (Lin et al., 2008; Nasr et al., 2011; Nissilä et al., 2011a, 2011b; Alamilla-Ortiz et al., 2015). Direct conversion of complex substrates requires a great diversity of microorganisms, fungi, and bacteria, acting cooperatively to hydrolyze and ferment the substrate ending up in H_2 and soluble metabolites. A recent study revealed that the HST eliminates all fungi involved in solubilization of complex polysaccharides, such as *Neocallimastix* and *Orpinomyces* (Figure 7.2). Thus, the H_2 production potential from cellulose decreased until 90% for the heat-shocked inoculum in comparison with that obtained with the untreated inoculum (Alamilla-Ortiz et al., 2015).

Other conventional methods for selecting inocula with H_2 producers without the HST application include biokinetic control through pH values between 5.5 and 7.0, short HRTs (<1.6 days), and high organic load rates (OLRs) (20 g/L.d) (Valdez-Vazquez and Poggi-Varaldo, 2009; Mariakakis et al., 2011; O-Thong et al., 2011). Finally, some other methods to enrich H_2 producers include acid treatments (Kim et al., 2009; O-Thong et al., 2011; Rossi et al., 2011), alkali treatments (Kim and Shin, 2008; Kim et al., 2009; O-Thong et al., 2011; Rossi et al., 2011), 2-bromoethanesulfonic acid (O-Thong et al., 2011), and chloroform (Ning et al., 2012), as well as physical treatments, such as freezing and thawing (Rossi et al., 2011), aeration, and application of an electric field (Jeong et al., 2013).

7.3.2 Bacterial Composition as a Function of the Inoculum Origin

Assessing the community structure, diversity, and abundance of the different species responsible for the H_2 production, and understanding how they are affected by the environmental conditions have been the major challenges in the last decades. Molecular methods, such as restriction fragment length polymorphism (RFLP) analysis and denaturing gradient gel electrophoresis (DGGE), have been the most commonly used to address these questions. These methods have provided useful information about how the H_2-producing communities are composed and how these species interact with each other.

Figure 7.3 displays most of the bacterial species revealed by DGGE in H_2-producing consortia with different origins: anaerobic sludge, compost, and wastewater treatment plants. The H_2-producing consortia derived from anaerobic digesters show a higher richness of species than those derived from compost and wastewater treatment plants. *Clostridium* ssp. have been ubiquitous in these H_2-producing consortia. The predominance of clostridia in these consortia is due to the effect of the HST (Baghchehsaraee et al., 2008; Kim et al., 2009; Huang et al., 2010; Nissilä et al., 2011a, 2011b). In spite of most studies having been focused on clostridia selection for carrying out the fermentative H_2 production, there are some drawbacks related to these species. They are strict anaerobes that require reducing compounds in the culture

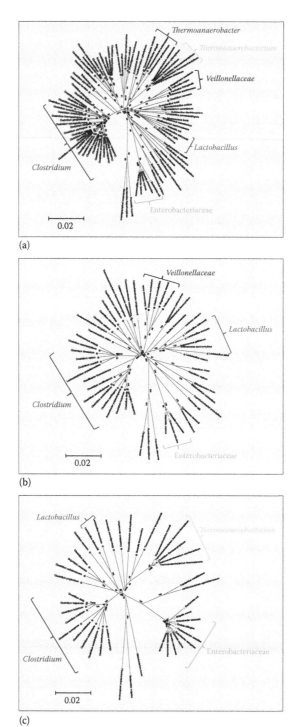

Figure 7.3 Comparison of the microbial diversity of H_2-producing consortia from different origins. Anaerobic digesters (a), compost (b), and wastewater treatment plants (c). Unrooted phylogenetic trees were built in Mega6 using 16S rRNA gene sequences.

medium or using an O_2-consuming bacterium to maintain the anaerobic environment (e.g., see Chang et al., 2008). Also, the growth of clostridia is inhibited by the bacteriocin-producing *Lactobacillus* (Jo et al., 2007). The most important factor is that there are several cases of bacteriophages infecting *Clostridium* spp. in full-scale and laboratory-scale bioreactors (Jones et al., 2000). For all these reasons, reducing the diversity to a single bacterial family may not be the best strategy for large-scale bioreactors. Contrary to this, when alternative methods to the HST are applied to select for H_2 producers, other nonclostridial H_2-producing species emerge, for example, *Caldoanaerobacter subterraneus* and *Caloramator fervidus* (Yokoyama et al., 2007); *Thermoanaerobacterium* (Akutsu et al., 2008); *Megasphaera elsdenii* (Ohnishi et al., 2010); *Ethanoligenes harbinense* (Mariakakis et al., 2011); *Klebsiella pneumonia* (Nasr et al., 2011); *Klebsiella, Enterobacter asburiae*, and *Escherichia coli* (Lay et al., 2012; Varrone et al., 2013); *Ruminobacillus xylanolyticum* and *Acetanaerobacterium elongatum* (Li et al., 2012a); *Janthinobacterium* and *Megasphaera paucivorans* (Ning et al., 2012); and *Enterococcus* (Valdez-Vazquez et al., 2015a, 2015b). These species, in conjunction with clostridia, may be a more robust system for H_2 production at a large scale.

Molecular approaches have also given insights into the microbial interactions that led to the instability of H_2 production, detecting changes in the microbial structure toward *Lactobacillus* (Jo et al., 2007), non-H_2-producing acidogens (Koskinen et al., 2007; Kim and Shin, 2008; D.-H. Kim et al., 2008), and H_2-consuming bacteria, such as *Desulfovibrio desulfuricans* (Koskinen et al., 2006).

7.3.3 Ecological Functions of Bacterial Populations

The number of reported families differ between the H_2-producing consortia derived from anaerobic sludge, compost, and wastewater treatment plants (Figure 7.4). However, they share five bacterial families: Clostridiaceae, Enterobacteriaceae, Lactobacillaceae, Thermoanaerobacteraceae, and Bifidobacteriaceae. The members

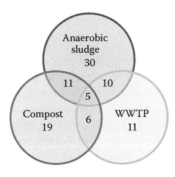

Figure 7.4 Venn diagram showing how many bacterial families are shared by the H_2-producing consortia derived from anaerobic digesters, compost, and wastewater treatment plants (WWTP).

of these families have been involved in ecological interactions such as commensalism, amenalism, synergism, and competition.

As positive interactions, commensalism is the relation between two species, for example, a polymer degrader and a nonhydrolytic H_2-producing bacterium (the commensal, Figure 7.5a): the commensal obtains monomers from the other without either harming or benefiting it (Valdez-Vazquez et al., 2015a). Another commensal relationship is established between a facultative anaerobe and a strict anaerobe (the commensal, Figure 7.5b): the commensal obtains an anoxic environment from the facultative anaerobe without either harming or benefiting it (Chang et al., 2008). Synergism is when the resources are divided (Figure 7.5c); two different species coferment glucose and xylose, improving the H_2 production potential (Zeidan and Van Niel, 2009). Amensalism occurs when lactic acid bacteria, such as *Lactobacillus* and *Enterococcus*, produce bacteriocins inhibiting the growth of *Clostridium* or other microorganisms without any costs or benefits received (Figure 7.5d) (Jo et al., 2007; Valdez-Vazquez et al., 2015a). The most important type of negative interaction reported in H_2-producing consortia is competition for nutrients. Competition occurs when different populations compete for substrate and nutrients, being a mutually detrimental interaction between populations (Figure 7.5e): the H_2 production potential decreases when non-H_2-producing bacteria are involved in the interaction, such as members of the families Bifidobacteriaceae and Lactobacillaceae (D.-H. Kim et al., 2008; Ohnishi et al., 2010; Valdez-Vazquez et al., 2015b). Finally, there is a type of commensal interaction that is not desired in the H_2-producing consortia (Figure 7.5f): methanogens, acetogens, and

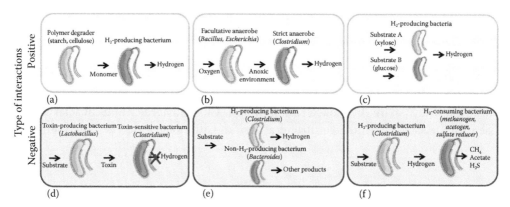

Figure 7.5   Illustrated examples of the microbial interactions established in the H_2-producing consortia (schemes based on Bernstein and Carlson, 2012). Positive interactions: (a) Commensalism, a hydrolytic bacterium provides monomers to a H_2-producing bacterium; (b) commensalism, a facultative bacterium provides the anaerobic environment required by a strict H_2-producing bacterium; and (c) synergism by division of resources, different groups of microorganisms coferment a mixture of substrates improving the H_2 production. Negative interactions: (d) Amensalism occurs when toxin-producing bacteria cause harm to other microorganisms without any costs or benefits received by themselves, (e) competition for substrate and nutrients between non-H_2-producing and H_2-producing bacteria reduces the H_2 production potential, and (f) H_2-consuming bacteria (the commensals) benefit from the H_2-producing bacteria without hurting or helping them.

sulfate reducers benefit from the H_2 produced by fermentative bacteria. In spite of fermentative bacteria being unaffected by the commensals, the H_2 production potential decreases.

Understanding how to avoid those negative interactions and how to strengthen those desired microbial interactions over long periods of time might be the main goal of such studies using engineered microbial consortia for H_2 production.

7.3.4 Case Study II: Total Production Cost of Lignocellulosic-Based Hydrogen by Consolidated Bioprocessing

Recently, we demonstrated the direct H_2 production from the wheat straw xylan using a synthetic microbial consortium with epiphytic strains of *Enterococcus*. The H_2 production stage continues with a solvent production stage from the remaining cellulose (Valdez-Vazquez et al., 2015a). Two microbial consortia perform the hydrolysis and fermentation of xylan and cellulose through two successive consolidated bioprocesses. Figure 7.6 shows a process block diagram of this two-stage biorefinery for H_2 and solvent production.

This configuration separates the consumption of five-carbon sugars in the first stage and the consumption of six-carbon sugars in the second stage using bacterial strains with specific hydrolytic capacities. These epiphytic enterococci were selected due to their capabilities to degrade and ferment xylan only. Thus, the lignocellulosic biomass, wheat straw, is enriched in cellulose. In addition, the first enterococcal consortium reduces the lignin content, which increases the cellulose availability for the second clostridial consortia. In the solvent stage, two clostridia species establish a mutualistic interaction. *C. cellulovorans* provides monomers to *C. beijerinckii*. Since *C. beijerinckii* quickly ferments glucose into H_2, it eliminates the catabolic repression in *C. cellulovorans*.

This dual consolidated bioprocess eliminates the biomass pretreatment and saccharification stages. Therefore, we might expect a significant reduction of investment and operating costs. Figure 7.7 shows the block diagram for the H_2 production consisting of only three stages: (1) particle reduction until 2–3 mm (Sánchez et al., 2013); (2) hydration stage, which takes place for 4 h to increase the substrate biodegradability

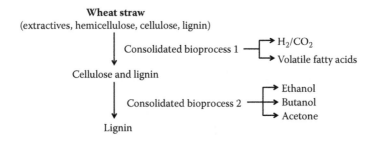

Figure 7.6 Process block diagram of a two-stage biorefinery for H_2 and solvent production.

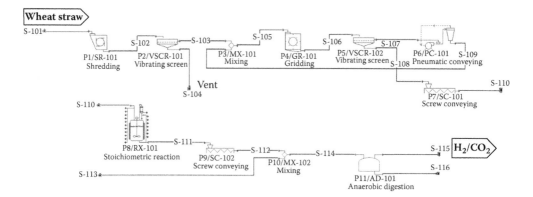

Figure 7.7 Block diagram of the milling stage (in red), hydration stage (in purple), and H_2 production stage (in blue). This facility receives 1000 tons/day of wheat straw whose composition has 16.5% (w/w) of xylan. In the milling stage, the particle size is reduced to 2–3 mm. Then, the feedstock biodegradability is improved by hydrating the wheat straw fibers with water for 4 h. After that, an enterococcal consortium converts the xylan fraction only into H_2/CO_2 and soluble metabolites. The process simulator (SuperPro Designer) serves to resolve the mass and energy balances.

(Lara-Vázquez et al., 2014a, 2014b); and (3) H_2 production from the wheat straw xylan by the enterococcal consortium with a residence time of 3 days. The facility receives 1000 tons of dry wheat straw per day and operates 330 days per year. The process simulator (SuperPro Designer®) resolves the mass and energy balances, as well as equipment parameters for the analyzed scenario. The nominal compositions of wheat straw used in the simulation were (mass%) 45.5% cellulose, 16.5% xylan, 6% lignin, 20.9% other solids, and 4% ashes. For the model presented, a wheat straw price of US$0.075/kg was used. A series of four shredders or grinders grind the feedstock (23 MT/h each). Then, a 3000 m³ tank receives the grind feedstock, along with 27 m³/h of water to hydrate and swell the feedstock fibers for 4 h without temperature control. After that, the hydrated feedstock is sent to one 3000 m³ anaerobic digester for H_2 production that operates at 37°C. In this anaerobic digester, just the wheat straw xylan is converted into H_2/CO_2 and soluble end products (acetic and butyric acids). The simulation results indicate that the facility produces 5850 tons of H_2 per year, with a total production cost of approximately US$7.0/kg H_2. It is important to highlight that in this model, all cellulose remains in the digestate to be fermented into solvents in the next bioprocess.

7.4 Renewable Resources for Dark Fermentative Hydrogen Production

The renewable biomass resources that can be used for bioH$_2$ production include crop residues, waste from processing of agricultural products, forest residues, livestock residues, energy crops, microalgae biomass, pulp and paper waste, food waste, and municipal solid waste. Agricultural residues include a wide range of plant material produced along with the main product of the crop. These include bagasse residue from sugarcane, rice husks, cotton stalks, orchard prunings, cottonseed hulls, palm, coconut, ground nut, cashew, wheat, coffee processing waste, barley, and oat straw (Saratale et al., 2013). Forestry residues comprise materials left behind after trees are felled or trimmed, sawdust, bark, and waste generated at sawmills. The total estimated global generating capacity from all these forest residues is around 10,000 MW, a figure that probably significantly underestimates the gross potential of these residues (Berndes et al., 2003). Moreover, livestock wastes, such as wet animal manure, and dry manure, such as poultry litter, can be used in biogas production. However, livestock waste is protein rich; thus, the yield of H_2 production is limited, but its addition to culture media seems to be feasible for $bioH_2$ production (Kapdan and Kargi, 2006). In the current scenario, energy crops also gain attention as feedstock for bioenergy production. Several energy crops, including perennial grasses, such as switchgrass, *Miscanthus*, and Napier grass, as well as some short-rotation forest species (e.g., *Eucalyptus*, poplars, and *Robinia*), are also being considered specifically for the purposes of accumulating biomass having high sugar and low lignin content (Somerville et al., 2010). Moreover, energy crops used for fermentative H_2 production can be divided in sugar-based crops (e.g., sweet sorghum and sugarcane); starch-based crops (e.g., corn and wheat); lignocellulose-based crops, including herbaceous biomass (e.g., switch grass and fodder grass); and woody crops (e.g., *Miscanthus* and poplar) (Saratale et al., 2013).

Microalgal biomass cultivation is regarded as highly sustainable with a number of advantages over first- and second-generation feedstock, including their rapid growth rate, their being a nonfood resource, the possibility of using infertile lands, their high capturing ability for CO_2, and their very short harvesting cycle (1–10 days) (Hwang et al., 2014). The carbohydrate from green microalgae comes from cellulose (on cell walls) and starch (in cytoplasm). Enzymatic saccharification of microalgal polysaccharides is necessary prior to fermentation, but in this case, the harsh pretreatment used for agricultural wastes may not be necessary due to the absence of lignin and hemicellulose in the microalgae biomass. Thus, algal biomass is very suitable as feedstock to produce various biofuels, for example, biodiesel, H_2, methane, and bioethanol. Demirbas (2010) demonstrated the production of H_2 from *Cladophora fracta* and *Chlorella prototohecoid* through pyrolysis and steam gasification. In addition, the residual microalgae biomass (e.g., *Scenedesmus*) derived from oil extraction processes was also used to produce H_2 (Yang et al., 2010).

Thus, these biomass-rich resources could be considered a very promising feedstock for $bioH_2$ production and could contribute to the displacement of fossil fuels and the reduction of greenhouse gas (GHG) emissions (Lo et al., 2009; Kumar et al., 2015). Various raw materials, including corn stover, wheat bran, wheat straw, rice straw, rice bran, sweet sorghum, potato steam peels, cassava stillage, sugarcane bagasse, and beer lees, have been utilized for H_2 production (Pattra et al., 2008; Lo et al., 2009; Saratale et al., 2010, 2013, 2015; Kumar et al., 2015). Numerous studies reported effective H_2 production using agricultural waste: *Ruminococcus albus* (sorghum extract) (Ntaikou et al., 2008), heat-treated anaerobically digested sludge (fodder maize, chicory fructooligosaccharides, and perennial ryegrass) (Kyazze et al., 2008),

C. butyricum CGS5 (rice husk hydrolyzate) (Lo et al., 2009), and *C. paraputrificum* M-2 (chitinous wastes) (Evvyernie et al., 2001). However, most H_2-producing micro-organisms cannot directly utilize cellulose or hemicellulose as a carbon source to grow and produce H_2. Therefore, pretreatment and hydrolysis of the lignocellulosic feedstock are often required prior to fermentative H_2 production. The cellulosic bio-mass can be treated by either physicochemical pretreatment or biological pretreat-ment processes, which are discussed in detail further.

7.4.1 Pretreatment of Lignocellulose-Based Biomass and Wastes

Lignocellulosic material possesses three major constituents: cellulose, hemicellulose, and lignin. The conversion to H_2 includes two processes: hydrolysis of cellulosic mate-rials to fermentable sugars and production of H_2 via dark fermentation. The factors that affect the hydrolysis of cellulose include the porosity (accessible surface area) of the waste materials, crystallinity of cellulose fiber, and lignin and hemicellulose con-tent (Kumar et al., 2008). The features of an effective pretreatment strategy include breaking the lignocellulosic complex, decreasing the cellulose crystallinity, preserv-ing the hemicellulose sugars, limiting the formation of degradation products that are inhibitory to hydrolysis and fermentation, and minimizing energy inputs and the use of extraneous chemical generation of toxic wastes (Saratale et al., 2008). Up to now, several pretreatment options have been used to treat the various biomass resources (Kumar et al., 2008; Saratale and Oh, 2015; Saratale et al., 2015).

7.4.1.1 Physicochemical Methods

Reduction of particle size by chipping, milling, or grinding is often needed to make material handling easier and to increase the surface-to-volume ratio, as well as to break the cellulose crystallinity. However, capital costs, operating costs, scale-up possibili-ties, and depreciation of equipment are important for mechanical pretreatment factors (Kumar et al., 2008). Chemical pretreatment methods, such as acid hydrolysis, alka-line hydrolysis, oxidative delignification, and solvent extraction, are purely initiated by chemical reactions for the disruption of the biomass structure (Behera et al., 2014). During dilute (mostly sulfuric) acid pretreatment, hydrolysis of hemicellulose occurs, which increases the porosity and improves the enzymatic digestibility, with release of monomeric sugars and soluble oligomers into the hydrolyzate (Saratale et al., 2015). In addition to this, different organic acids (e.g., maleic acid and fumaric acid) have been used for dilute acid pretreatment (Behera et al., 2014). Moreover, the major effect of alkaline pretreatment is the removal of lignin and various uronic acid substitutions from the hemicelluloses of the biomass, leading to an increase in the internal surface area and a decrease in the degree of polymerization, which improves the enzymatic hydrolysis of the remaining polysaccharides (Lo et al., 2009; Saratale and Oh, 2015; Saratale et al., 2015). Steam explosion is recognized as one of the most cost-effective pretreatment processes for hardwoods and agricultural residues, but has limitations due to the incomplete disruption of the lignin–carbohydrate matrix, and it generates compounds that may be inhibitory to microorganisms used in downstream processes

(Behera et al., 2014). Ammonia fiber explosion (AFEX) pretreatment for various lignocellulosic materials shows better performance (Kumar et al., 2008), but ammonia makes the process expensive and also causes secondary pollution problems. Recently, utilization of green liquor chemicals for the pretreatment of biomass was found to be an effective pretreatment option with less inhibitor formation (Saratale et al., 2016). To remove the inhibitory products, pretreated biomass needs to be washed by water, but this decreases the overall saccharification yields. Fan et al. (2006) compared the efficiency of H_2 production from raw wheat straw with that from HCl-pretreated wheat straw, by cow dung compost in batch tests. The experimental results in the acid-pretreated substrate showed the maximum cumulative H_2 yield of 68.1 mL H_2/g total volatile solids (TVS), and the value was about 136-fold compared with that of raw wheat straw wastes. In our recent study using sorghum husk hydrolyzates, compared with untreated ones, a sharp induction in H_2 production and H_2 yield by *C. beijerinckii* was observed in 0.2% H_2SO_4-pretreated hydrolyzates (Saratale et al., 2015).

However, in most cases, pretreatment induces the formation of technologically unfavorable compounds, known as fermentation inhibitors, after lignin and hemicellulose degradation, such as phenolic components, furfural and 5-hydroxymethylfurfural, and weak acids, for example, acetic acid and formic acid (Silva et al., 2013). Inhibitory substances might be present in the prehydrolyzate and are considered hydrolysis and fermentation barriers, severely depressing microbiological H_2 production and overall process performance. However, the effect of inhibitors mainly relies on their concentration and the specific tolerance of microorganisms toward them. To overcome these bottlenecks, some investigators suggested detoxification processes such as neutralization, washing, and removal by subjecting the prehydrolyzate to ion exchange chromatography, and applying advanced oxidation processes (Silva et al., 2013; Kumar et al., 2015).

7.4.1.2 Biological (Microbial and Enzymatic) Methods

Conventional physicochemical methods for cellulose hydrolysis not only require large inputs of energy but also make secondary pollution. Compared with conventional physicochemical methods, biological pretreatment of cellulosic materials can be achieved by using cellulolytic enzyme directly or using cellulolytic microorganisms. Filamentous fungi and actinomycetes, as well as some bacteria, are found to be effective in the hydrolysis of various lignocellulosic biomass by secreting cellulolytic enzymes and producing fermentable sugars (Saratale et al., 2010, 2012, 2014, 2015). Using an efficient fungal or bacterial system to hydrolyze biomass can be considered cost-effective, but the major obstacle in this process is the consumption of produced sugars during hydrolysis for their own growth. Enzymatic hydrolysis of cellulose is carried out by cellulose-hydrolyzing enzymes (cellulases), a mixture of several enzymes that hydrolyze crystalline or amorphous cellulose to fermentable sugar (Saratale et al., 2008, 2011; Kumar et al., 2008). There are several advantages of enzymatic hydrolysis, including little energy requirement and mild reaction conditions, high substrate specificity, high yield of sugars, and high hydrolysis efficiency. However, compared with chemical processes, enzymatic hydrolysis has certain disadvantages, including a low hydrolysis rate and high cost.

7.4.2 Bioconversion of Lignocellulosic Biomass to Hydrogen

Lignocellulosic bioH$_2$ production can proceed by three approaches. The first is based on a two-stage process, where pretreated cellulose is hydrolyzed in the first stage, followed by dark fermentation for H$_2$ production in the second stage. This process is named separated hydrolysis and fermentation (SHF). The other is to use a direct process in which cellulose hydrolysis and H$_2$ production take place in one reactor, called simultaneous saccharification and fermentation (SSF), or by the same microorganism, it is called CBP.

7.4.2.1 Separate Hydrolysis and Fermentation

The two-stage process of separate hydrolysis and fermentation is the most widely used for H$_2$ production from renewable biomass resources. The whole reaction is carried out in two apparatuses; hence, saccharification and fermentation can be carried out at the respective optimum conditions, leading to increased H$_2$ production efficiency. The major obstacle of SHF and even SSF is that the lignocellulosic biomass hydrolyzate consists of mainly hexose and pentose, which need to be further processed collectively into H$_2$. Although hexose is easy for bacteria to utilize, pentose utilization is harder. In order to make full utilization of lignocellulosic biomass, microorganisms that could ferment both hexoses and pentoses, especially pentoses for H$_2$ production, are essential. The known pentose-fermenting microorganisms, the extreme thermophilic *Thermoanaerobacter mathranii* A3N and the mesophilic *C. beijerinckii* YA001, revealed the highest production yields of 2.5 and 2.3 mol H$_2$ mol/mol xylose, respectively (Jayasinghearachchi et al., 2012; An et al., 2014). Recently, *Thermoanaerobacterium thermosaccharolyticum* W16 also metabolized pentoses efficiently, producing 2.19 mol H$_2$/mol xylose (Ren et al., 2008b).

In the SHF process, when lignocellulosic biomass is solubilized to reducing sugars, H$_2$ fermentation is easier to carry out. Therefore, an efficient pretreatment and hydrolysis process determines the best results regarding bioH$_2$ production. When size-reduced wheat stalks were used as substrate for H$_2$ production, a maximum H$_2$ yield of 17.6 mL/g substrate was achieved (Yuan et al., 2011). The application of acid and alkaline treatment to grass enabled the production of 72.2 and 39.5 mL H$_2$/g substrate, respectively (Cui and Shen, 2012). Additionally, biological pretreatment and saccharification processes are considered an alternative, with the advantages of low energy consumption and no inhibitor production. The white rot fungus *Phanerochaete chrysosporium* was used to treat cornstalk, and the results showed a 20.3% increase of the enzymatic saccharification. Using this hydrolyzate to produce H$_2$, the maximum H$_2$ yield was 80.3 mL/g pretreated cornstalk (Zhao et al., 2012). To avoid enzyme input during saccharification, on-site cellulases produced by *Trichoderma viride* were used to saccharify cornstalk. After optimization of the enzymatic saccharification process, a sugar yield of 38.5 g/L was obtained, and the corresponding H$_2$ yield reached 90.6 mL/g pretreated cornstalk (Zhao et al., 2013). Recently, we have also utilized the enzymes produced from *Cellulomonas biazotea* and *P. chrysosporium* by applying submerged and solid-state fermentation approaches. During submerged fermentation, enzymes were collected by centrifugation, whereas in solid-state fermentation the enzymes were squeezed through muslin cloth and collected through centrifugation and gave better saccharification of sugarcane bagasse and rice husk. The efficiency of fermentative conversion of the hydrolyzed sugarcane bagasse and rice husk for bioH$_2$ production by *C. butyricum* CGS5 and *C. beijerinckii* was checked. The maximum cumulative H$_2$ production and H$_2$ yield were 72.61 and 237.97 mL and 2.13 and 2.93 mmoL H$_2$/g reducing sugar, respectively (Saratale et al., 2010, 2015).

7.4.2.2 Simultaneous Saccharification and Fermentation

SSF has been reported as a promising alternative to improve bacterial H$_2$ evolution. SSF can shorten the operation time, reduce the equipment cost, and improve the H$_2$ production yield by eliminating end product inhibition during cellulose hydrolysis. Although the SSF process is more commercially feasible, its research is still in its primary stage. Ibrahim et al. (2015) found that the enzymatic saccharification conditions were similar to the conditions required for H$_2$ fermentation, so an SSF for H$_2$ production from pretreated oil palm empty fruit bunch (OPEFB) by *C. acetobutylicum* ATCC 824 was conducted. A maximum H$_2$ yield of 14.68 mL/g OPEFB was attained, which was 21.2% higher than that from the comparative SHF process. Li and Chen (2007) reported that steam-exploded corn straw (SECS) could improve the H$_2$-producing efficiency, and the maximum cumulative H$_2$ yield of 68 mL/g SECS and highest specific H$_2$ production rate of 126 mL/g volatile suspended solids (VSS)/day were achieved after 72 h fermentation through the SSF process conducted by *C. butyricum* AS1.209. Xu et al. (2010) used acetic acid steam explosion to enhance H$_2$ production from corn straw by SSF with the strain *Ethanoligenes harbinense* B49, which presented a high H$_2$ production potential. A maximum H$_2$ yield and maximum specific H$_2$ production rate of 72 mL/g acetic acid steam-exploded corn straw (ASCS) and 103 mL/g VSS/day were obtained, respectively. Zhao et al. (2013) employed fungal-pretreated cornstalk for H$_2$ production by SSF using crude enzymes from *T. viride* and *T. thermosaccharolyticum* W16, which produced a hydrogen yield of 89.3 mL/g cornstalk with an initial pH of 6.5, substrate concentration of 0.75%, and 34 FPU/g cellulose achieved after 72 h fermentation. Recently, Saratale et al. (2014) used crude enzymes produced from *P. chrysosporium* for the hydrolysis of rice husk. The maximum cumulative H$_2$ production of 237.97 mL/L was obtained using *C. beijerinckii* KCTC 1785 from rice husk enzymatic hydrolyzates.

7.4.2.3 Consolidated Bioprocessing of Lignocellulosic Biomass to Hydrogen

CBP, which has also been mentioned in direct microbial conversion, is an ideal option for H$_2$ production from lignocellulosic biomass. During this process, specific microorganisms (consortia) generate hemicellulolytic and cellulolytic enzyme cocktails, accomplish the hydrolysis of pretreated lignocellulosic biomass, and in situ convert the pentoses and hexoses to H$_2$. Compared with the SHF and SSF processes, CBP presents a highly integrated, one-stage design for the utilization of lignocellulosic biomass (Lynd et al., 2005). Direct H$_2$ production processes, that is, CBP from lignocellulosic biomass, have been found to be more cost-effective and commercially feasible (Taguchi et al.,

1996; Kapdan and Kargi, 2006; Lo et al., 2009; Saratale et al., 2013). Consequently, this process could effectively reduce the investment in installation, and lead to more competitive and economically feasible technology for bioH$_2$ production. During a CBP process, the presence of acclimatized consortia or purposefully chosen pure and functional cultures with cellulose- and hemicellulose-degrading and H$_2$-producing capability was regarded as the most crucial criterion. Some thermophilic H$_2$-producing bacteria, such as *T. kodakaraensis* KOD1, *C. thermolacticum*, *C. thermocellum* JN4, and *C. thermocellum* ATCC 27405, could be useful for the cellulosic biohydrogen production (Liu et al., 2008; Magnusson et al., 2008). Moreover, CBP of lignocellulosic biomass for H$_2$ production using mesophiles such as *Clostridium* sp. (Zhang et al., 2015) and *Thermotoga* sp. (Nguyen et al., 2010a), thermophiles such as *Caldicellulosiruptor* sp. (Talluri et al., 2013), *C. thermocellum* (Magnusson et al., 2008), and *Thermoanaerobacterium* sp. (Sheng et al., 2015) has also been reported. Recently, genetically modified *Caldicellulosiruptor bescii*, by deleting the L-lactate dehydrogenase gene (*ldh*), showed a 34% increase in the production of H$_2$ compared with the wild-type and parent strains (Cha et al., 2013). These results suggest that application of metabolic engineering is a powerful tool for the generation of desired end products by redirecting the metabolic fluxes via modification of the organism's genome. However, according to current research results, it is difficult to make an in-depth evaluation, mostly because experiments were usually carried out under different temperature and pH conditions and, more importantly, made use of different pretreatment methods that resulted in different features of the starting materials. As a result, to make an assessment of the different H$_2$ production processes, the use of a single substrate and similar pretreatment methods may be essential. Moreover, CBP was found to be overall less efficient because the preferred conditions for cellulose degradation and dark H$_2$ fermentation are significantly different from each other, and the microorganisms were inefficient utilizers of cellulose for H$_2$ production (Magnusson et al., 2008).

7.4.3 Different Wastes and Wastewater for Biohydrogen Production

The conversion of organic wastes into H$_2$ is attractive from both pollution control and energy recovery points of view. The waste or wastewater would be characterized as efficient feedstock for H$_2$ generation because of its high concentration of degradable organic compounds, high proportion of readily fermentable sugars, and low concentration of inhibitors to microbiological growth and activity. In the current scenario, research was carried out on fermentative H$_2$ production using solid waste (Lay et al., 1999). However, in addition to carbohydrates, such wastes usually contain a high concentration of proteins and fats, and thus their conversion efficiencies to H$_2$ are comparatively lower than those obtained from carbohydrate-based wastewaters. Literature studies using wastewaters from rice winery, noodle, sugar, sugar beet, and molasses manufacturing; food processing; and filtered leachate of municipal solid wastes have been reported (Kapdan and Kargi et al., 2006; Vijayaraghavan et al., 2006; Saratale et al., 2008; Kumar et al., 2015). Some studies used swine waste for H$_2$ production; however, very low H$_2$ yields of 3 mL H$_2$/g COD were achieved, equivalent to only

0.2% of energy recovery (Kotsopoulos et al., 2009). The highest H_2 yield of 321 mL H_2/g COD was demonstrated for the treatment of a sugar factory wastewater in a continuous stirred tank reactor (CSTR) with 63% glucose conversion efficiency. Hafez et al. (2009), using a novel bioH$_2$ system, later elaborated on utilizing corn syrup and achieved a maximum H_2 yield of 430 mL H_2/g COD, with an overall average of 385 mL H_2/g COD. Some scientists have conducted experiments to study the possibility of H_2 production from crude cheese whey using *C. saccharoperbutylacetonicum* (Ferchichi et al., 2005). Similarly, using solid waste consisting of jackfruit peel waste and sucrose-rich wastewater to H_2 with microflora isolated from cow dung and using mixed culture, sucrose-rich wastewater under upflow anaerobic sludge bed (UASB) operation gives 16.2 mmol/g VS destroyed and 1.61 mol H_2/mol glucose H_2 yield, respectively (Vijayaraghavan et al., 2006).

Molasses are another carbohydrate-rich substrate and a good source of sucrose. A maximum H_2 production rate of 5.57 m^3 H_2/m^3 reactor/day, with a specific H_2 production rate of 0.75 m^3 H_2/kg molasses volatile suspended solids (MLVSS)/day, was obtained using the molasses-based wastewater of a sugar refinery. The H_2 yield reached 26.13 mol H_2/kg COD removed within an OLR range of 35–55 kg COD/m^3 reactor/day (Ren et al., 2007).

Glycerol is generated in large amounts (about 10%) during the production of biodiesel. Estimation suggests that after transesterification of vegetable oils or animal fats, along with biodiesel, crude glycerol is generated. Recently, some studies reported the value of 1 mol/mol as the maximum theoretical H_2 yield from glycerol, based on the reaction of glycerol conversion to H_2, CO_2, and ethanol (Ito et al., 2005). Using a pure culture of *Klebsiella pneumoniae* for the fermentative H_2 production produced a H_2 yield of 0.53 mol H_2/mol glycerol, and the maximum H_2 production rate was 17.8 mmol/l/h, and 1,3-propanediol was the main by-product, reported earlier (Liu and Fang, 2007). Some investigators also studied the possibility of H_2 production from raw POME. An optimized thermophilic condition was developed with the prediction of H_2 gas at 6.5 L H_2/L POME (Thong et al., 2007). Moreover, some complex substrates (e.g., waste activated sludge, primary sludge, and hog manure) are not ideal for fermentative H_2 production due to their complex structures. However, after appropriate pretreatments, such as ultrasonication, acidification, freezing and thawing, sterilization, methanogenic inhibition, and microwave, the ability of H_2-producing bacteria to produce H_2 from waste activated sludge can be improved (Kapdan and Kargi, 2006; Guo et al., 2008; Kumar et al., 2015). Freezing and thawing (for *C. bifermentants*) and sterilization (for *Pseudomonas* sp. GZ1) (Guo et al., 2008) are superior pretreatment methods of wastewater sludge for fermentative H_2 production.

7.4.4 Parameters Affecting Dark Fermentative Hydrogen Production Using Biomass and Wastes

BioH$_2$ production is a complex process and is greatly influenced by many factors, including microorganism seeding, pH, temperature, OLR, metal ions, and oxidation–reduction potential. Optimization of these factors and controlling suitable environmental conditions could enhance dark fermentative H_2 production (J. K. Kim et al., 2008; Lee et al., 2014).

7.4.4.1 pH

During dark fermentative H_2 production, the formation of organic acids depletes the buffering capacity of the medium and results in a low final pH. It has been reported that low pH inhibits H_2 production because it affects hydrogenase activity and many aspects of microbial metabolism, including utilization of substrate, biogas content, and types of organic acids (Saratale et al., 2008; Lo et al., 2009). The optimal pH for maximum H_2 production varied within the range of 5.5 and 8.0 (Lay, 2001; Lin and Cheng, 2006; Saratale et al., 2010). A low pH of fermentation media usually results in lower intracellular ATP levels and causes inhibition of bacterial growth and changes in the metabolic pathways (acidogenesis to solventogenesis), resulting in low H_2 production (J. K. Kim et al., 2008; Saratale et al., 2010). In our recent study, the pH-controlled strategy (initial pH of 6.5, followed by pH control at 5.5) was found to be a better option for the efficient utilization of sorghum husk hydrolyzates and their conversion into H_2 by *C. beijerinckii* (Saratale et al., 2015). The foregoing results suggest that the proper control of pH appears to be essential for successful operation of bioH$_2$-producing reactors.

7.4.4.2 Temperature

Temperature is a vital factor that influences the growth rate and metabolic pathways, thus shifting the composition of the by-products, as well as the fermentative H_2 production (Lee et al., 2014). A range of operational temperatures, that is, mesophilic (35°C), thermophilic (55°C), and extreme thermophilic (>65°C), have been studied to evaluate their effect on bioH$_2$ production. Valdez-Vazquez et al. (2005) reported higher H_2 yields for thermophilic fermentation than in the mesophilic temperature range with acetate and butyrate as the dominant products, respectively. Conversely Fang et al. (2006b) compared H_2 production from rice slurry at 37°C and 55°C by an anaerobic digester sludge in batch experiments. The results showed that the mesophilic sludge was more effective than the thermophilic sludge for treating rice slurry. The maximum specific H_2 production rate and H_2 yield at 37°C were calculated as 2.1 L/g VSS/day and 346 mL/g carbohydrate. Gavala et al. (2006) found that the thermophilic conditions resulted in a low volumetric H_2 production rate with poor microbial mass production but a higher specific H_2 production rate, compared with mesophillic conditions at an HRT of 6–12 h. These studies have shown that the temperature can affect the metabolic pathways, as well as H_2 and metabolite production.

7.4.4.3 Inorganic Nutrients

The carbon-to-nitrogen (C/N) ratio is an important factor in the biological H_2 production process. Microorganisms require a proper nitrogen supplement for metabolism and for the production of essential enzymes during hydrogen fermentation (Bisaillon et al., 2006). Moreover, some studies suggest that the optimum phosphate concentration could enhance H_2 production (Lay et al., 2005; Bisaillon et al., 2006). Thus, appropriate C/N and C/P ratios are essential for better fermentative H_2 production.

7.4.4.4 Trace Metal Elements

$BioH_2$ production requires essential micronutrients for bacterial metabolism during fermentation. Sodium, magnesium, zinc, and iron are important trace metals that can affect H_2 production. Among them, iron is an important cofactor for hydrogenases activity, and therefore widely investigated in fermentative H_2 production (Wang and Wan, 2009). Zhang and Shen (2006) explored the effects of temperature and iron concentration on H_2 production from sucrose by a mixed culture where the organisms responsible for H_2 production were dominated by *C. pasteurianum*, in batch experiments. Magnesium, sodium, zinc, and iron were also found to be important trace metals affecting H_2 production, with magnesium being the most significant. However, at higher concentrations, metal ions may inhibit the activity of H_2-producing bacteria, so only trace levels are required for fermentative H_2 production (Li and Fang, 2007).

7.4.5 Case Study III: Biohydrogen Production from Lignocellulosic Biomass

The capacity of isolated actinomycetes *Saccharomyces* sp. MDS and *Nocardiopsis* sp. KNU and fungi *P. chrysosporium* for cellulase and xylanase production using a solid-state fermentation approach was investigated (Saratale et al., 2015). The produced crude enzymes could effectively hydrolyze sorghum husk biomass and release maximum sugar with higher glucose yield. Mild acidic pretreatment (0.2% H_2SO_4) was found to be the preferred method for increasing the accessibility of sorghum husk for enzymatic hydrolysis. The resulting sorghum husk hydrolyzates used for H_2 production by *C. beijerinckii* exhibited a maximum H_2 yield. The optimal conditions for H_2 production were 35°C and a sorghum husk hydrolyzate loading of 5.0 g reducing sugars/L. During dark fermentative H_2 production acidogenic metabolites decreased the pH of media, which hampered H_2 production. To overcome this, a pH control strategy (initial pH of 6.5, followed by pH control at 5.0, 5.5, and 6.0) was employed to improve the performance of $bioH_2$ production. Incubation at pH 5.5 showed the highest H_2 yield (1.05 mol H_2/ mol reducing sugars) with a higher cumulative H_2 production (1117 mL/L) and production rate (46.54 mL/L/h) (Table 7.1). Without the pH control set, a high level of acids and ethanol production during fermentation was observed, and these metabolites could act as inhibitors for H_2 production (Table 7.1), whereas in the set of controlled pH at 5.5, a higher production of acetate than of lactate, formate, and butyrate, and lower ethanol production were observed (Table 7.1). Utilization of xylose and arabinose by *C. beijerinckii* without pH control was approximately 22% and 14%, respectively, while in the case of pH control, the efficiencies increased around 54% and 45%, respectively, at pH 5.5 (Table 7.1). Thus, higher utilization of all sugars present in sorghum husk hydrolyzates and lower soluble metabolite production in the pH control approach may contribute to the increase in H_2 production and production rate. These results demonstrate that mild acid pretreatment, followed by enzymatic hydrolysis, is a promising technology for the enhancement of $bioH_2$ production from lignocellulosic biomass. This work may provide the promise of large-scale $bioH_2$ production using different abundant lignocellulosic feedstocks, such as rice waste biomass, which is abundant in South Korea.

TABLE 7.1 Effect of pH Control Strategy on H₂ Production, Soluble Metabolite Production, and Sugar Utilization by C. beijerinckii KCTC 1785 Using 0.2% Acid-Pretreated Sorghum Husk Hydrolyzates (RS 5.0 g/L at 35°C)

Parameters	SH Hydrolyzate Concentration (g/L)	H₂ Production			Soluble Metabolite Production (mg/L)					Sugar Utilization of SH Hydrolyzate (%)		
		H₂ Yield (mmol H₂/mol RS)	Maximum H₂ Production Rate (mL/L/h)	Cumulative H₂ Production (mL/L)	Acetate	Lactate	Butyrate	Formate	Ethanol	Glucose	Xylose	Arabinose
Without pH control	5.0	0.955 ± 0.01	952 ± 3.12	39.70 ± 1.54	47.2 ± 0.65	38.5 ± 0.54	19.1 ± 0.22	14.0 ± 0.18	54.0 ± 0.88	100 ± 1.25	22.0 ± 0.45	14.2 ± 0.18
Controlled pH 5.0	5.0	0.752 ± 0.01	650 ± 2.65	27.08 ± 1.65	38.5 ± 0.45	34.8 ± 0.35	19.0 ± 0.32	19.0 ± 0.14	25.1 ± 0.68	100 ± 1.0	25.2 ± 0.65	20.2 ± 0.22
Controlled pH 5.5	5.0	1.050 ± 0.02	1117 ± 4.15	46.54 ± 1.78	40.4 ± 0.35	32.4 ± 0.32	18.1 ± 0.35	11.2 ± 0.16	10.2 ± 0.45	100 ± 1.0	54.1 ± 0.54	45.4 ± 0.45
Controlled pH 6.0	5.0	0.689 ± 0.01	718 ± 3.45	29.91 ± 1.66	20.0 ± 0.18	19.1 ± 0.15	14.0 ± 0.21	10.2 ± 0.11	12.2 ± 0.45	100 ± 1.0	22.0 ± 0.32	18.2 ± 0.25

Source: Saratale, G. D. et al., *Biotechnol. Bioprocess Eng.*, 20(4), 733–43, 2015.

Note: Values are the mean of three experiments ± standard error of the mean (SEM). Statistics were determined by one-way ANOVA with Tukey–Kramer multiple comparisons test. SH, sorghum husk; RS, reducing sugars.

7.5 Life Cycle Analysis

BioH$_2$ can be potentially used in stationary or mobile applications. In the domain of mobile applications, passenger transportation is seen as an option. Many demonstration projects regarding bus fleets have already been deployed: the Clean Urban Transport for Europe (CUTE), the Global Hydrogen Bus Platform (HyFLEET:CUTE), the Sustainable Transport Energy Programme (STEP), and the Ecological City Transport System (ECTOS) (Ally and Pryor, 2007).

In Europe and the United States, there are already directives that biofuels (up to now, ethanol and biodiesel) must meet in order to be qualified as a biofuel usable in the transportation sector: the Renewable Energy Directive (REN) (EU, 2009) and the Renewable Fuel Standard (RFS) (Sissine, 2010). In both directives, the life cycle assessment (LCA) methodology (ISO, 2006) must be followed to compute the GHG savings in relation to the reference fossil system. In Europe, since April 2013, the classification of a fuel as biofuel requires at least a 35% GHG reduction potential. This value will increase to 50% by January 2017 and to 60% in 2018. In the U.S. directive, biofuels derived from feedstocks other than corn starch (kernels) must reduce GHG emissions by 50%, whereas cellulosic and agricultural waste feedstocks must achieve a reduction of at least 60%. Biomass-based biodiesel is an exception, requiring a GHG emission decrease of 50%.

The comparison between the use of H$_2$-powered vehicles and fossil fuel reference vehicles (gasoline or diesel) is usually performed through life cycle thinking. This method comprises an inventory of the input and output flows of CO$_2$, N$_2$O, and CH$_4$ emissions (whose totality is denominated GHG emissions) throughout the fuel chain, feedstock production, transportation to plant, plant transformation, and gate-to-pump transportation. This is classified as well-to-pump analysis. It can be furthered by the introduction of the analysis of materials extraction, production, assembling, dismantling, and recycling, giving origin to the cradle-to-grave (CTG) analysis. LCA studies, on the other hand, usually include more impact categories than well-to-wheels (WTW) and CTG studies, such as the analysis of acidification and eutrophication processes, ozone layer depletion, and carcinogenics. There are no standards for WTW or CTG studies. LCA studies might conform to the ISO 14040 series of standards, although many LCA studies do not make this claim.

The Ecoinvent database (Frischknecht et al., 2004) existent in commercial LCA software like Sigmapro is commonly used for fuel analysis. Specific software, for example, GREET (Wang, 2008), have databases of fuel life cycles for the United States. They include data as varied as H$_2$ production from electrolysis, and fossil and renewable natural gas resulting from landfill and anaerobic digestion of animal waste. The WTW energy and emission benefits of fuel cell vehicles were assessed, and the results reported up to a 79% reduction in GHG emissions through the replacement of fossil natural gas production systems with renewable natural gas. Stationary applications, such as the production of electricity from biofuels, can also be compared with their fossil counterparts in the same life cycle thinking: definition of boundaries, functional unit, reference fossil system, inventory of mass and energy flows, aggregations of inventoried values to produce impact categories, and interpretation of results.

Usually, biological processes for $bioH_2$ production are nonexistent in commercial databases, as well as data related to biomass feedstocks, such as olive mill wastes, carob pulp, and microalgae, which brings a clear limitation to the use of such programs for a complete LCA analysis.

Although commonly used for comparison purposes, LCA has several methodological issues, which are described in the next section.

7.5.1 Methodological Issues

$BioH_2$ production from several feedstocks can be compared through LCA with each other and/or with a reference fossil system in terms of carbon footprint, water footprint, and fossil energy use. However, results might not be comparable due to methodological differences. In case of microalgae feedstock for biofuels, a very interesting review that points out such methodological differences can be found in Collet et al. (2015). The results of LCA are highly variable due to different assumptions regarding indirect land use change (iLUC), allocation method, boundaries assumed, geographical differences, and disregarded items in the feedstock processing chain due to limited information. Also, a distinction can be made in the type of approach to the analysis: consequential LCA (CLCA) and attributional LCA (ALCA) (Brander et al., 2008). ALCA is based on the examination of the processes and material flows directly used in the production, consumption, and disposal of the product in order to clearly quantify the total emissions. On the other side, CLCA uses processes and material flows that are directly or indirectly affected by a marginal change in the output of a product (e.g., through market effects, substitution, and use of constrained resources), which makes it more affected by presumed economic scenarios.

Land use change (LUC) impact results from giving a different use to a portion of land previously with a certain carbon storage potential, which will cause a carbon debt (Djomo and Ceulemans, 2012). Soil organic carbon includes plant, animal, and microbial residues in all stages of decomposition. The loss of soil organic carbon by conversion of natural vegetation to cultivated use is discussed in Post and Kwon (2000). Direct LUC can be determined from comparison of the carbon balances of the previous land use with those obtained after the land has been used to produce biomass crops. iLUC adds more uncertainty to the analysis by assuming that, for example, the used land will cause forestland or wetlands somewhere else in the world to be converted to cropland. According to Searchinger et al. (2008), the additional carbon emissions occur as farmers worldwide respond to higher prices, converting forest and grassland to new cropland to replace the grain (or cropland) diverted to biofuels. The iLUC use is part of a CLCA analysis.

Allocation is inherent to a process with several products, by-products, and coproducts. The choice of the allocation method (mass, energy content, economic, system expansion, preferences of the LCA practitioner, or requests from the study's commissioner) can introduce differences in results by several orders of magnitude, therefore affecting decision making (Sandin et al., 2015). ALCA allocates emissions to coproducts based on economic value, energy content, or mass. CLCA uses system expansion to quantify the effect of coproducts on the overall GHG emissions. In a biorefinery, the environmental impact of the selected main product can be calculated as the emissions from the main production system minus the avoided emissions from the use of the by-products. This is called system expansion or substitution.

The functional unit is the function of the system under study and serves as the basis for the calculations. It can be difficult to identify in a biorefinery system, especially one that produces multiple outputs with different functions (Ahlgren et al., 2013), consequently making the process of narrowing it to a single main product or function rather difficult. A functional unit that avoids allocation between coproducts could be, for example, 1 tonne of biomass input, one biorefinery, or a combination of all outputs (i.e., production of 1 kg of product A, 2 MJ of product B, and 100 kg of product C). Cherubini (2010) uses the amount of biomass treated per year to compare a biorefinery concept that produces bioethanol, electricity, heat, and phenols from switchgrass, with a fossil reference system that delivers the same amount of products.

Direct carbon mass flows are usually not fully accounted for. It is usually assumed that the combustion of a biofuel burning emits the amount of carbon that was consumed during cultivation of the feedstock, that is, that the process is carbon neutral. Carbon gains in soil due to harvesting leftovers, direct carbon emissions during fermentation or within leftovers, are usually disregarded. This carbon flow accounting is defended in the annual basis carbon (ABC) analysis as possessing a less uncertain method of comparing two systems with the same function (DeCicco and Krishnan, 2015). In the report, it is also claimed that the biofuel carbon neutrality assumption built into LCA models does not hold up for real-world biofuel production. How to treat the timing of sequestration and emission of biogenic carbon may be an issue (Ahlgren et al., 2013).

As the carbon in soil may be measured (Slepetiene et al., 2008; Degerickx et al., 2015), and the carbon in cultivated biomass and processing leftovers may be quantified through elemental analysis, there is little reason to not include a complete carbon analysis of the production process. For example, for biofuel production from *Jatropha curcas*, soil organic carbon (Degerickx et al., 2015), husk, seed shell, tree branch (Murata et al., 2012), leaves (Méndez et al., 2014), whole seed, and seed pressed cake (Jourabchi et al., 2014) were measured and included in the analysis.

The carbon footprint is expressed in GHG emissions and is composed of the quantification of CO_2, N_2O, and CH_4 emissions, combined by using IPCC conversion factors for 100 years (IPCC, 2001). Some studies may include other gases, such as part of

TABLE 7.2   Differences of LCA in Biofuel Directives

	RED—Europe	EPA RFS—United States
Boundary	No infrastructures, no LUC	Infrastructures, LUC, and iLUC
GHG	Without biogenic carbon	With biogenic carbon
Allocation	Energy	System expansion

Note: EPA, Environmental Protection Agency.

the carbon footprint inventory, the GHG protocol corporate standard (Schmitz et al., 2000). It was launched in 1998 and covers the following GHG emissions: carbon dioxide (CO_2), methane (CH_4), nitrous oxide (N_2O), hydrofluorocarbons (HFCs), perfluorocarbons (PFCs), and sulfur hexafluoride (SF_6). As of May 2013, nitrogen trifluoride (NF_3) was included.

Significant methodological differences regarding GHG accounting are found between the REN European directive and the U.S. RFS, as seen in Table 7.2.

The other main issue in LCA uncertainty is the inventory data itself. For example, usually the electricity consumption of equipment is assumed based on equipment manufacturer specifications, instead of being measured through a portable oscilloscope. The sources of uncertainty in input data are further explored in Huijbregts et al. (2001).

7.5.2 Integrated Hydrogen Biorefinery and Scale-Up Examples

The Portuguese research and development (R&D) project ESIBITS (Evaluation of the Sustainability of Industrial Biohydrogen production by microalgae and Integration taxi/bus Transport Systems—EXPL/EMS-ENE/1078/2012) explored the use of dark fermentation technology for the production of bioH_2 (Silva et al., 2015). It focused on the conversion of microalgal biomass, analyzing several bench-scale experiments prior to the design of optimized scenarios for scale-up. A high-energy input-to-output ratio was registered. In order to achieve negative emissions, CO_2 uptake for microalgae growth was combined with the use of renewable electricity (see Table 7.3).

No allocation criteria were used (all impacts were imputed to the bioH_2 product), and the CO_2 uptake by all species was considered the same and equal to 1.83 kg/kg algae. This assumption may lead to uncertainty due to the variable carbon content of the different species. Proximate analysis of each species should be performed to decrease uncertainty. It is also interesting to note that in some studies, direct electricity consumption by each piece of equipment was taken into account, while in other cases, only working hours and equipment specifications were considered. This fact may affect results by almost 10 orders of magnitude. Further details on the considered processes and system boundaries can be found in Ferreira et al. (2012, 2013a, 2013b) and Pacheco et al. (2014). Also, it should be registered that some LCA studies are based on direct measurements instead of theoretical assumptions, causing yet another source of uncertainty and additional methodological issues.

7.5.3 Economics of Biohydrogen Production and Perspectives

BioH_2 as an energy source for stationary and mobile applications may only be considered competitive if its production cost does not surpass that of conventional production systems, such as wind-based water electrolysis (Ursua et al., 2012) and natural gas steam reforming (Blok et al., 1997). The cost may be expressed in terms of cost per mass unit or cost per energy unit, based on the lower heating value (LHV) of H_2

TABLE 7.3   Energy Consumption and CO_2 Emissions for Each $bioH_2$ Production Process Life Cycle Inventory (LCI)

$BioH_2$ Chain	Energy (MJ/MJ_{H2})	Minimum	Maximum	CO_2 (g/MJ_{H2})	Minimum	Maximum
$BioH_2$ from dried *S. obliquus*—best scenario	7.20	5.72	8.23	−659	−716	−613
$BioH_2$ from dried *S. obliquus*—optimized scenario	12.48	9.80	14.60	−311	−419	−209
$BioH_2$ as by-product from dried *S. obliquus*	33.28	27.28	37.26	2,172.0	1,935.1	2,331.6
$BioH_2$ from dried *S. obliquus*	88.00	70.31	99.49	5,776.0	5,118.4	6,268.0
$BioH_2$ as by-product from bioreference 1 *Nannochloropsis* sp.	147.00	119.00	164.00	9,665.0	8,645.0	10,369.0
$BioH_2$ as by-product from bioreference 2 *Nannochloropsis* sp.	168.00	136.00	187.00	11,020.0	9,858.0	11,820.0
$BioH_2$ as by product of *Spirogyra* sp. biorefinery	264.22	196.62	296.70	21,700.0	19,700.0	22,500.0
$BioH_2$ as by-product from *S. obliquus* hydrolyzate	364.30	281.20	404.90	27,198.0	24,149.0	29,218.0
$BioH_2$ by *Anabaena* sp. (photoautotrophic)	1,538.00	1,184.00	1,715.00	114,641.0	101,476.0	123,587.0
$BioH_2$ by *Anabaena* sp. cyanobacteria (photoautotrophic + fermentation)	1,723.00	1,327.00	1,919.00	128,502.0	113,825.0	266,887.0
$BioH_2$ from *S. obliquus* hydrolyzate	8,884.14	7,614.14	9,866.14	758,743.8	673,892.4	813,576.4
$BioH_2$ from *Nannochloropsis* sp.	9,058.00	7,285.00	10,123.00	591,112.0	527,022.0	634,402.0

(120 MJ/kg). Hydrogen sales should cover all costs plus profit: raw materials costs (e.g., natural gas or biomass), electricity, other operating expenses (e.g., consumables and staff), and recovery of capital invested. The U.S. Department of Energy (Wang, 2008) has defined as a target for 2015 \$3.10/kg for central H_2 plants and \$3.70/kg for distributed plants (Saur and Ainscough, 2011). The European Union cost targets for H_2 generation are 9.90 €/kg (0.08 €/MJ) in 2015 and 5.50 €/kg (0.05 €/MJ) in 2025 (McKinsey, 2010). After the quantification of costs for compression, storage, or dispensing, and including taxation, the final price will vary tremendously from country to country, as already seen with gasoline or diesel fuel, due to crude oil variations, labor variations, and so forth. Each country should set thresholds for the final price in order to allow elaboration in advance of a scheme of incentives by policy makers.

In Portugal, for example, a final price below 0.07 €/MJ would be attractive comparatively to gasoline use. A further decrease below 0.05 €/MJ would be attractive comparatively with diesel use, and below 0.03 €/MJ, H_2 use would be competitive even when compared with gasoline hybrid plug-in users and pure electric vehicle users (Baptista et al., 2010).

In view of all these considerations, it is important when designing a bioH$_2$ biorefinery facility for a variety of products, including H_2, that its final price can become competitive. A multiobjective procedure may be adopted (e.g., Wang et al., 2013) in order to theoretically predict the best biorefinery configuration that would be able to maximize profit and simultaneously satisfy a certain carbon footprint threshold. In this context, the LCA previously discussed is a valuable tool (Gebreslassie et al., 2013).

7.6 Future Prospects for Biohydrogen

Despite numerous reports on microorganisms capable of producing H_2, the efficiency of the bioconversion process is still very low. One reason for this is the fact that a microorganism that is robust enough or industrially able has not yet been found, which can be used as a platform for the genetic modification of metabolic pathways, to enable the production of more 4 mol H_2/mol glucose. According to Kalia and Purohit (2008), such an organism should be able to assimilate and ferment various sugars, withstand adverse environmental conditions, tolerate toxic compounds, and produce other compounds of economic interest. Potential candidates were listed in this chapter and are currently the object of intensive research efforts, pursuing the objective of placing bioH$_2$ on the map of hydrogen economy.

Acknowledgments

I.V.-V. would like to acknowledge the partial support of the CONACYT project (grant CB-2011/168921). Also, I.V.-V. thanks Dr. Arturo Sanchez, CINVESTAV, Guadalajara, for help with the SuperPro Designer software and data analysis. C.A.M.S. acknowledges the project IDL-UID/GEO/50019/2013.

References

Abdel-Banat, B., H. Hoshida, A. Ano, S. Nonklang, and R. Akada. 2010. High-temperature fermentation: How can processes for ethanol production at high temperatures become superior to the traditional process using mesophilic yeast? *Appl Microbiol Biotechnol* 85(4):861–7.

Adav, S. S., D.-J. Lee, A. Wang, and N. Ren. 2009. Functional consortium for hydrogen production from cellobiose: Concentration-to-extinction approach. *Bioresour Technol* 100(9):2546–50.

Ahlgren, S., A. Björklund, A. Ekman, H. Karlsson, J. Berlin, P. Börjesson, T. Ekvall, G. Finnveden, M. Janssen, and I. Strid. 2013. LCA of biorefineries identification of key issues and methodological recommendations. Göteborg, Sweden: Swedish Knowledge Centre for Renewable Transportation Fuels (f3 Centre).

Akutsu, Y., Y.-Y. Li, M. Tandukar, K. Kubota, and H. Harada. 2008. Effects of seed sludge on fermentative characteristics and microbial community structures in thermophilic hydrogen fermentation of starch. *Int J Hydrogen Energy* 33(22):6541–8.

Akutsu, Y., Y. Y. Li, H. Harada, and H. Q. Yu. 2009. Effects of temperature and substrate concentration on biological hydrogen production from starch. *Int J Hydrogen Energy* 34:2558–66.

Alamilla-Ortiz, Z. L., A. E. Escalante-Hernández, G. Hernández-Guzmán, and I. Valdez-Vazquez. 2015. Hydrogen production and diversity of two microbial communities using crystalline cellulose as sole substrate. Presented at the *HYPOTHESIS Proceedings*, Madrid, September 6–9.

Ally, J., and T. Pryor. 2007. Life-cycle assessment of diesel, natural gas and hydrogen fuel cell bus transportation systems. *J Power Sources* 170(2):401–11.

Alvarado-Cuevas, Z. D., A. M. López-Hidalgo, L. G. Ordonez, E. Oceregua-Contreras, J. T. Ornelas-Salas, and A. De León-Rodríguez. 2015. Biohydrogen production using psychrophilic bacteria isolated from Antarctica. *Int J Hydrogen Energy* 40(24):7586–92.

An, D., Q. Li, X. Wang, H. Yang, and L. Guo. 2014. Characterization on hydrogen production performance of a newly isolated *Clostridium beijerinckii* YA001 using xylose. *Int J Hydrogen Energy* 39(35):19928–36.

Argun, H., and F. Kargi. 2011. Bio-hydrogen production by different operational modes of dark and photo-fermentation: An overview. *Int J Hydrogen Energy* 36(13):7443–59.

Baghchehsaraee, B., G. Nakhla, D. Karamanev, and A. Margaritis. 2010. Fermentative hydrogen production by diverse microflora. *Int J Hydrogen Energy* 35(10):5021–7.

Baghchehsaraee, B., G. Nakhla, D. Karamanev, A. Margaritis, and G. Reid. 2008. The effect of heat pretreatment temperature on fermentative hydrogen production using mixed cultures. *Int J Hydrogen Energy* 33(15):4064–73.

Baptista, P., M. Tomás, and C. Silva. 2010. Plug-in hybrid fuel cell vehicles market penetration scenarios. *Int J Hydrogen Energy* 35(18):10024–30.

Basile, M. A., C. Carfagna, P. Cerruti, G. G. d'Ayala, A. Fontana, A. Gambacorta, M. Malinconico, and L. Dipasquale. 2012. Continuous hydrogen production by immobilized cultures of *Thermotoga neapolitana* on an acrylic hydrogel with pH-buffering properties. *RSC Adv* 2(9):3611.

Batista, A. P., L. Ambrosano, S. Graça, C. Sousa, P. A. S. S. Marques, B. Ribeiro, E. P. Botrel, P. Castro Neto, and L. Gouveia. 2015. Combining urban wastewater treatment with biohydrogen production—An integrated microalgae-based approach. *Bioresour Technol* 184:230–5.

Becker, E. W. 2007. Micro-algae as a source of protein. *Biotechnol Adv* 25(2):207–10.

Beckers, L., S. Hiligsmann, S. D. Lambert, B. Heinrichs, and P. Thonart. 2013. Improving effect of metal and oxide nanoparticles encapsulated in porous silica on fermentative biohydrogen production by *Clostridium butyricum*. *Bioresour Technol* 133:109–17.

Beckers, L., J. Masset, C. Hamilton, F. Delvigne, D. Toye, M. Crine, P. Thonart, and S. Hiligsmann. 2015. Investigation of the links between mass transfer conditions, dissolved hydrogen concentration and biohydrogen production by the pure strain *Clostridium butyricum* CWBI1009. *Biochem Eng J* 98:18–28.

Behera, R., N. Arora, S. Nandhagopal, and S. Kumar. 2014. Importance of chemical pretreatment for bioconversion of lignocellulosic biomass. *Renew Sustain Energy Rev* 36:91–106.

Berndes, G., M. Hoogwijk, and R. van den Broek. 2003. The contribution of biomass in the future global energy supply: A review of 17 studies. *Biomass Bioenergy* 25(1):1–28.

Bernstein, H. C., and R. P. Carlson. 2012. Microbial consortia engineering for cellular factories: In vitro to in silico systems. *Comput Struct Biotechnol J* 3(4):1–8.

Bisaillon, A., J. Turcot., and P. C. Hallenbeck. 2006. The effect of nutrient limitation on hydrogen production by batch cultures of *Escherichia coli*. *Int J Hydrogen Energy* 31(11):1504–8.

Blok, K., R. H. Williams, R. E. Katofsky, and C. A. Hendriks. 1997. Hydrogen production from natural gas, sequestration of recovered CO_2 in depleted gas wells and enhanced natural gas recovery. *Energy* 22(2):161–8.

Brander, M., R. Tipper, C. Hutchison, and G. Davis. 2008. Consequential and attributional approaches to LCA: A guide to policy makers with specific reference to greenhouse gas LCA of biofuels. Boston: Ecometrica Press.

Burgess, B. K., S. Wherland, W. E. Newton, and E. I. Stiefel. 1981. Nitrogenase reactivity: Insight into the nitrogen-fixing process through hydrogen-inhibition and HD-forming reactions. *Biochemistry* 20(18):5140–6.

Cai, G., B. Jin, P. Monis, and C. Saint. 2013. A genetic and metabolic approach to redirection of biochemical pathways of *Clostridium butyricum* for enhancing hydrogen production. *Biotechnol Bioeng* 110(1):338–42.

Cai, G., B. Jin, C. Saint, and P. Monis. 2011. Genetic manipulation of butyrate formation pathways in *Clostridium butyricum*. *J Biotechnol* 155(3):269–74.

Calusinska, M., C. Hamilton, P. Monsieurs, G. Mathy, N. Leys, F. Franck, B. Joris, P. Thonart, S. Hiligsmann, and A. Wilmotte. 2015. Genome-wide transcriptional analysis suggests hydrogenase- and nitrogenase-mediated hydrogen production in *Clostridium butyricum* CWBI 1009. *Biotechnol Biofuels* 8(1):1.

Calusinska, M., T. Happe, B. Joris, and A. Wilmotte. 2010. The surprising diversity of clostridial hydrogenases: A comparative genomic perspective. *Microbiology* 156(6):1575–88.

Cha, M., D. Chung, J. G. Elkins, A. M. Guss, and J. Westpheling. 2013. Metabolic engineering of *Caldicellulosiruptor bescii* yields increased hydrogen production from lignocellulosic biomass. *Biotechnol Biofuels* 6(1):1.

Chaganti, S. R., J. A. Lalman, and D. D. Heath. 2012. 16S rRNA gene based analysis of the microbial diversity and hydrogen production in three mixed anaerobic cultures. *Int J Hydrogen Energy* 37(11):9002–17.

Chang, J. J., C. H. Chou, C. Y. Ho, W. E. Chen, J. J. Lay, and C. C. Huang. 2008. Syntrophic co-culture of aerobic *Bacillus* and anaerobic *Clostridium* for bio-fuels and bio-hydrogen production. *Int J Hydrogen Energy* 33(19):5137–46.

Chen, C., H. Chang, and J. Chang. 2015. Producing carbohydrate-rich microalgal biomass grown under mixotrophic conditions as feedstock for biohydrogen production. *Int J Hydrogen Energy* 41(7):4413–20.

Chen, C. C., Y. S. Chuang, C. Y. Lin, C. H. Lay, and B. Sen. 2012. Thermophilic dark fermentation of untreated rice straw using mixed cultures for hydrogen production. *Int J Hydrogen Energy* 37(20):15540–6.

Chen, C. Y., G. D. Saratale, C. M. Lee, P. C. Chen, and J. S. Chang. 2008. Phototrophic hydrogen production in photobioreactors coupled with solar-energy-excited optical fiber. *Int J Hydrogen Energy* 33(23):6886–95.

Chen, J. S., J. Toth, and M. Kasap. 2001. Nitrogen-fixation genes and nitrogenase activity in *Clostridium acetobutylicum* and *Clostridium beijerinckii*. *J Ind Microb Biotechnol* 27(5):281–6.

Chen, W., Z. Tseng, K. Lee, and J. Chang. 2005. Fermentative hydrogen production with CGS5 isolated from anaerobic sewage sludge. *Int J Hydrogen Energy* 30(10):1063–70.

Cherubini, F. 2010. The biorefinery concept: Using biomass instead of oil for producing energy and chemicals. *Energy Convers Manag* 51(7):1412–21.

Chong, M., R. Rahim, Y. Shirai, and M. Hassan. 2009. Biohydrogen production by *Clostridium butyricum* EB6 from palm oil mill effluent. *Int J Hydrogen Energy* 34(2):764–71.

Chou, C., F. E. Jenney, M. W. W. Adams, and R. M. Kelly. 2008. Hydrogenesis in hyperthermophilic microorganisms: Implications for biofuels. *Metab Eng* 10(6):394–404.

Collet, P., A. Hélias, L. Lardon, J. P. Steyer, and O. Bernard. 2015. Recommendations for life cycle assessment of algal fuels. *Appl Energy* 154:1089–102.

Cui, M., and J. Shen. 2012. Effects of acid and alkaline pretreatments on the biohydrogen production from grass by anaerobic dark fermentation. *Int J Hydrogen Energy* 37:1120–24.

Das, D., and T. N. Veziroğlu. 2001. Hydrogen production by biological processes: A survey of literature. *Int J Hydrogen Energy* 26(1):13–28.

Davila-Vazquez, G., C. B. Cota-Navarro, L. M. Rosales-Colunga, A. León-Rodríguez, and E. Razo-Flores. 2009. Continuous biohydrogen production using cheese whey: Improving the hydrogen production rate. *Int J Hydrogen Energy* 34(10):4296–304.

DeCicco, J., and R. Krishnan. 2015. Annual basis carbon (ABC) analysis of biofuel production at the facility level. Ann Arbor: University of Michigan Energy Institute.

Degerickx, J., J. Almeida, P. C. J. Moonen, L. Vervoort, B. Muys, and W. M. J. Achten. 2015. Impact of land-use change to *Jatropha* bioenergy plantations on biomass and soil carbon stocks: A field study in Mali. *GCB Bioenergy* 8(2):443–55.

Demirbas, A. 2010. Hydrogen from mosses and algae via pyrolysis and steam gasification. *Energy Sources Part A-Recovery Util Environ Eff* 32:172–79.

de Vrije, T., R. R. Bakker, M. A. Budde, M. H. Lai, A. E. Mars, and P. A. Claassen. 2009. Efficient hydrogen production from the lignocellulosic energy crop *Miscanthus* by the extreme thermophilic bacteria *Caldicellulosiruptor saccharolyticus* and *Thermotoga neapolitana*. *Biotechnol Biofuels* 2(1):1.

Ding, J., B. F. Liu, N. Q. Ren, D. F. Xing, W. Q. Guo, J. F. Xu, and G. J. Xie. 2009. Hydrogen production from glucose by co-culture of *Clostridium butyricum* and immobilized *Rhodopseudomonas faecalis* RLD-53. *Int J Hydrogen Energy* 34 (9):3647–52.

Djomo, S. N., and R. Ceulemans. 2012. A comparative analysis of the carbon intensity of biofuels caused by land use changes. *GCB Bioenergy* 4(4):392–407.

Etchebehere, C., E. Castelló, J. Wenzel, M. P. Anzola-Rojas, L. Borzacconi, B. Buitrón et al. 2016. Microbial communities from 20 different hydrogen-producing reactors studied by 454 pyrosequencing. *Appl Microbiol Biotechnol* 100(7):3371–84.

Evvyernie, D., K. Morimoto, S. Karita, T. Kimura, K. Sakka, and K. Ohmiya. 2001. Conversion of chitinous wastes to hydrogen gas by *Clostridium paraputrificum* M-21 J. *Biosci Bioeng* 91(4):339–43.

EU (European Union). 2009. Directive 2009/28/EC of the European Parliament and of the Council of 23 April 2009 on the promotion of the use of energy from renewable sources and amending and subsequently repealing Directives 2001/77/EC and 2003/30/EC. *Off J Eur Union* 140:16–62.

Fan, Y., G. Zhang, X. Guo, Y. Xing, and M. Fan. 2006. Biohydrogen-production from beer lees biomass by cow dung compost. *Biomass Bioenergy* 30(5):493–6.

Fang, H. H. P., C. L. Li, and T. Zhang. 2006b. Acidophilic biohydrogen production from rice slurry. *Int J Hydrogen Energy* 31(6):683–92.

Fang, H. H. P., H. Zhu, and T. Zhang. 2006a. Phototrophic hydrogen production from glucose by pure and co-cultures of *Clostridium butyricum* and *Rhodobacter sphaeroides*. *Int J Hydrogen Energy* 31(15):2223–30.

Ferchichi, M., V. Crabbe, G. H. Gil, W. Hintz, and A. Almadidy. 2005. Influence of initial pH on hydrogen production from cheese whey. *J Biotechnol* 120(4):402–9.

Ferreira, A. F., A. C. Marques, A. P. Batista, P. A. S. S. Marques, L. Gouveia, and C.M Silva. 2012. Biological hydrogen production by *Anabaena* sp.—Yield, energy and CO_2 analysis including fermentative biomass recovery. *Int J Hydrogen Energy* 37(1):179–90.

Ferreira, A. F., J. Ortigueira, L. Alves, L. Gouveia, P. Moura, and C. Silva. 2013a. Biohydrogen production from microalgal biomass: Energy requirement, CO_2 emissions and scale-up scenarios. *Bioresour Technol* 144:156–64.

Ferreira, A. F., L. A. Ribeiro, A. P. Batista, P. A. S. S. Marques, B. P. Nobre, A. M. F. Palavra, P. P. da Silva, L. Gouveia, and C. Silva. 2013b. A biorefinery from *Nannochloropsis* sp. microalga—Energy and CO_2 emission and economic analyses. *Bioresour Technol* 138:235–44.

Foglia, D., M. Ljunggren, W. Wukovits, A. Friedl, G. Zacchi, K. Urbaniec, and M. Markowski. 2010. Integration studies on a two-stage fermentation process for the production of bio-hydrogen. *J Clean Prod* 18:72–80.

Frischknecht, R., N. Jungbluth, H. J. Althaus, G. Doka, R. Dones, T. Heck, S. Hellweg, H. Roland, T. Nemecek, G. Rebitzer, and M. Spielmann. 2004. The Ecoinvent database: Overview and methodological framework. *Int J Life Cycle Assess* 10(1):3–9.

Frock, A. D., J. S. Notey, and R. M. Kelly. 2010. The genus *Thermotoga*: Recent developments. *Environ Technol* 31(10):1169–81.

Gadow, S. I., H. Jiang, T. Hojo, and Y.-Y. Li. 2013. Cellulosic hydrogen production and microbial community characterization in hyper-thermophilic continuous bioreactor. *Int J Hydrogen Energy* 38(18):7259–67.

Gavala, H. N., I. V. Skiadas, and B. K. Ahring. 2006. Biological hydrogen production in suspended and attached growth anaerobic reactor systems. *Int J Hydrogen Energy* 31(9):1164–75.

Gebreslassie, B. H., M. Slivinsky, B. Wang, and F. You. 2013. Life cycle optimization for sustainable design and operations of hydrocarbon biorefinery via fast pyrolysis, hydrotreating and hydrocracking. *Comput Chem Eng* 50:71–91.

Gottshalk, G., J. R. Andreesen, and H. Hippe. 1981. The genus *Clostridium* (nonmedical aspects). In *The Prokaryotes, a Handbook on Habitats, Isolation and Identification of Bacteria*, ed. M. P. Starr, H. Stolp, H. G. Truper, A. Balows, and H. G. Schlegel, 1767–803. Vol. II. Berlin: Springer-Verlag.

Gouveia, L., and A. C. Oliveira. 2009. Microalgae as a raw material for biofuels production. *J Ind Microb Biotechnol* 36(2):269–74.

Goyal, Y., M. Kumar, and K. Gayen. 2013. Metabolic engineering for enhanced hydrogen production: A review. *Can J Microbiol* 59(2):59–78.

Guo, L., X. M. Li, X. Bo, Q. Yang, G. M. Zeng, D. X. Liao, and J. J. Liu. 2008. Impacts of sterilization, microwave and ultrasonication pretreatment on hydrogen producing using waste sludge. *Bioresour Technol* 99(9):3651–8.

Hafez, H., G. Nakhla, and H. El Naggar. 2009. Biological hydrogen production from corn-syrup waste using a novel system. *Energies* 2(2):445–55.

Hallenbeck, P. C., and D. Ghosh. 2009. Advances in fermentative biohydrogen production: The way forward? *Trends Biotechnol* 27(5):287–97.

Hiligsmann, S., L. Beckers, J. Masset, C. Hamilton, and P. Thonart. 2014. Improvement of fermentative biohydrogen production by *Clostridium butyricum* CWBI1009 in sequenced-batch, horizontal fixed bed and biodisc-like anaerobic reactors with biomass retention. *Int J Hydrogen Energy* 39(13):6899–911.

Huang, Y., W. Zong, X. Yan, R. Wang, C. L. Hemme, J. Zhou, and Z. Zhou. 2010. Succession of the bacterial community and dynamics of hydrogen producers in a hydrogen-producing bioreactor. *Appl Environ Microbiol* 76(10):3387–90.

Huijbregts, M. A. J., G. Norris, R. Bretz, A. Ciroth, B. Maurice, B. Bahr, B. Weidema, and A. S. H. Beaufort. 2001. Framework for modelling data uncertainty in life cycle inventories. *Int J Life Cycle Assess* 6(3):127–32.

Hwang, J. H., H. C. Kim, J. A. Choi, R. A. I. Abou-Shanab, B. A. Dempsey, J. M. Regan et al. 2014. Photoautotrophic hydrogen production by eukaryotic microalgae under aerobic conditions. *Nat Commun* 5:3234.

Ibrahim, M. F., S. Abd-Aziz, M. E. M. Yusoff, P. L. Yee, and M. A. Hassan. 2015. Simultaneous enzymatic saccharification and ABE fermentation using pretreated oil palm empty fruit bunch as substrate to produce butanol and hydrogen as biofuel. *Renew Energy* 77:447–55.

IPCC (Intergovernmental Panel on Climate Change). 2001. IPCC third assessment report: Climate change 2001. Geneva: IPCC.

Islam, R., N. Cicek, R. Sparling, and D. Levin. 2009. Influence of initial cellulose concentration on the carbon flow distribution during batch fermentation by *Clostridium thermocellum* ATCC 27405. *Appl Microbiol Biotechnol* 82:141–8.

ISO (International Organization for Standardization). 2006. Environmental management—Life cycle assessment—Requirements and guidelines. ISO 14044. Geneva: ISO.

Ito, Y., K. Nakashimada, T. Senba, N. Matsui, and N. Nishio. 2005. Hydrogen and ethanol production from glycerol-containing wastes discharged after biodiesel manufacturing process. *J Biosci Bioeng* 100(3):260–5.

Jayasinghearachchi, H. S., P. M. Sarma, and B. Lal. 2012. Biological hydrogen production by extremely thermophilic novel bacterium *Thermoanaerobacter mathranii* A3n isolated from oil producing cell. *Int J Hydrogen Energy* 37(7):5569–78.

Jeong, D.-Y., S.-K. Cho, H.-S. Shin, and K.-W. Jung. 2013. Application of an electric field for pretreatment of a seeding source for dark fermentative hydrogen production. *Bioresour Technol* 139:393–6.

Jiang, Z. Q., W. Deng, Y. P. Zhu, L. T. Li, Y. J. Sheng, and K. Hayashi. 2004. The recombinant xylanase B of *Thermotoga maritima* is highly xylan specific and produces exclusively xylobiose from xylans, a unique character for industrial applications. *J Mol Catal B Enzym* 27(4–6):207–13.

Jo, J. H., C. O. Jeo, D. S. Lee, and J. M. Park. 2007. Process stability and microbial community structure in anaerobic hydrogen-producing microflora from food waste containing kimchi. *Int J Hydrogen Energy* 131(3):300–8.

Jones, D. T., M. Shirley, X. Wu, and S. Keis. 2000. Bacteriophage infections in the industrial acetone butanol (ab) fermentation process. *J Mol Microbiol Biotechnol* 2(1):21–6.

Jourabchi, S. A., S. Gan, and H. K. Ng. 2014. Pyrolysis of *Jatropha curcas* pressed cake for bio-oil production in a fixed-bed system. *Energy Convers Manag* 78:518–26.

Junghare, M., S. Subudhi, and B. Lal. 2012. Improvement of hydrogen production under decreased partial pressure by newly isolated alkaline tolerant anaerobe, *Clostridium butyricum* TM-9A: Optimization of process parameters. *Int J Hydrogen Energy* 37(4):3160–8.

Kalia, V. C., and H. J. Purohit. 2008. Microbial diversity and genomics in aid of bioenergy. *J Ind Microbiol Biotechnol* 35(5):403–19.

Kapdan, I. K., and F. Kargi. 2006. Biohydrogen production from waste materials. *Enzyme Microb Technol* 38(5):569–82.

Kim, B. H., and G. M. Gadd. 2008. *Bacterial Physiology and Metabolism*. Cambridge: Cambridge University Press.

Kim, D.-H., S.-H. Kim, I.-B. Ko, C.-Y. Lee, and H.-S. Shin. 2008. Start-up strategy for continuous fermentative hydrogen production: Early switchover from batch to continuous operation. *Int J Hydrogen Energy* 33(5):1532–41.

Kim, D.-H., S.-H. Kim, and H.-S. Shin. 2009. Hydrogen fermentation of food waste without inoculum addition. *Enzym Microbial Technol* 45(3):181–7.

Kim, J. K., L. Nhat, Y. N. Chun, and S. W. Kim. 2008. Hydrogen production conditions from food waste by dark fermentation with *Clostridium beijerinckii* KCTC 1785. *Biotechnol Bioprocess Eng* 13(4):499–504.

Kim, M. S., J. S. Baek, Y. S. Yun, S. J. Sim, S. Park, and S. C. Kim. 2006. Hydrogen production from *Chlamydomonas reinhardtii* biomass using a two-step conversion process: Anaerobic conversion and photosynthetic fermentation. *Int J Hydrogen Energy* 31(6):812–6.

Kim, S.-H., and H.-S. Shin. 2008. Effects of base-pretreatment on continuous enriched culture for hydrogen production from food waste. *Int J Hydrogen Energy* 33(19):5266–74.

Kivisto, A., V. Santala, and M. Karp. 2010. Hydrogen production from glycerol using halophilic fermentative bacteria. *Bioresour Technol* 101(22):8671–7.

Klein, M., M. B. Ansorge-Schumacher, M. Fritsch, and W. Hartmeier. 2010. Influence of hydrogenase overexpression on hydrogen production of *Clostridium acetobutylicum* DSM 792. *Enzyme Microb Technol* 46(5):384–90.

Koskinen, P. E. P., A. H. Kaksonen, and J. A. Puhakka. 2007. The relationship between instability of H_2 production and compositions of bacterial communities within a dark fermentation fluidized-bed bioreactor. *Biotechnol Bioeng* 97(4):742–58.

Kotsopoulos, T. A., R. J. Zeng, and I. Angelidaki. 2006. Biohydrogen production in granular up-flow anaerobic sludge blanket (UASB) reactors with mixed cultures under hyper-thermophilic temperature (70°C). *Biotechnol Bioeng* 94(2):296–302.

Kumar, G., P. Bakonyi, S. Periyasamy, S. H. Kim, N. Nemestóthy, and K. Bélafi-Bakó. 2015. Lignocellulose biohydrogen: Practical challenges and recent progress. *Renew Sustain Energy Rev* 44:728–37.

Kumar, K., S. Roy, and D. Das. 2013. Continuous mode of carbon dioxide sequestration by *C. sorokiniana* and subsequent use of its biomass for hydrogen production by *E. cloacae* IIt-BT 08. *Bioresour Technol* 145:116–22.

Kumar, P., S. K. S. Patel, J. Lee, and V. C. Kalia. 2013. Extending the limits of *Bacillus* for novel biotechnological applications. *Biotechnol Adv* 31(8):1543–61.

Kumar, R., S. Singh, and O. V. Singh. 2008. Bioconversion of lignocellulosic biomass: Biochemical and molecular perspectives. *J Ind Microbiol Biotechnol* 35(5):377–91.

Kyazze, G., R. Dinsdale, F. R. Hawkes, A. J. Guwy, G. C. Premier, and I. S. Donnison. 2008. Direct fermentation of fodder maize, chicory fructans and perennial ryegrass to hydrogen using mixed microflora. *Bioresour Technol* 99(18):8833–9.

Lara-Vázquez, A. R., F. R. Quiroz-Figueroa, A. Sánchez, and I. Valdez-Vazquez I. 2014a. Particle size and hydration medium effects on hydration properties and sugar release of wheat straw fibers. *Biomass Bioenergy* 68:67–74.

Lara-Vázquez, A. R., A. Sánchez, and I. Valdez-Vazquez. 2014b. Hydration treatments increase the biodegradability of native wheat straw for hydrogen production by a microbial consortium. *Int J Hydrogen Energy* 39(35):19899–904.

Laurinavichene, T., and A. Tsygankov. 2015. Hydrogen photoproduction by co-culture *Clostridium butyricum* and *Rhodobacter sphaeroides*. *Int J Hydrogen Energy* 40(41): 14116–23.

Lay, C.-H., H.-C. Lin, B. Sen, C.-Y. Chu, and C.-Y. Lin. 2012. Simultaneous hydrogen and ethanol production from sweet potato via dark fermentation. *J Clean Prod* 27:155–64.

Lay, J. J. 2001. Biohydrogen generation by mesophilic anaerobic fermentation of microcrystalline cellulose. *Biotechnol Bioeng* 74(4):280–87.

Lay, J. J., K. S. Fan, J. I. Hwang, J. I. Chang, and P. C. Hsu. 2005. Factors affecting hydrogen production from food wastes by *Clostridium*-rich composts. *J Environ Eng* 131(4):595–602.

Lay, J. J., Y. J. Lee, and T. Noike. 1999. Feasibility of biological hydrogen production from organic fraction of municipal solid waste. *Water Res* 33(11):2579–86.

Lee, J., E. Seo, D. H. Kweon, K. Park, and Y. S. Jin. 2009. Fermentation of rice bran and defatted rice bran for butanol 5 production using *Clostridium beijerinckii* NCIMB 8052. *J Microbiol Biotechnol* 19(5):482–90.

Lee, K. S., L. M. Whang, G. D. Saratale, S. D. Chen, J. S. Chang, H. Hafez et al. 2014. Biological hydrogen production via dark fermentation. In *Hydrogen Energy Handbook* ed. S. A. Sherif, 181–250. Vol. 7. Boca Raton, FL: CRC Press/Taylor & Francis.

Lee, R., and J. M. Lavoie. 2013. From first-to-third-generation biofuels: Challenges of producing a commodity from a biomass of increasing complexity. *Animal Front* 3(2):6–11.

Lee, S. Y., J. H. Park, S. H. Jang, L. K. Nielsen, J. Kim, and K. S. Jung. 2008. Fermentative butanol production by *Clostridia*. *Biotechnol Bioeng* 101(2):209–28.

Levin, D. B., C. R. Carere, N. Cicek, and R. Sparling. 2009. Challenges for biohydrogen production via direct lignocellulose fermentation. *Int J Hydrogen Energy* 34(17):7390–403.

Levin, D. B., R. Islam, N. Cicek, and R. Sparling. 2006. Hydrogen production by *Clostridium thermocellum* 27405 from cellulosic biomass substrates. *Int J Hydrogen Energy* 31(11):1496–503.

Levin, D. B., L. Pitt, and M. Love. 2004. Biohydrogen production: Prospects and limitations to practical application. *Int J Hydrogen Energy* 29(2):173–85.

Li, C. L., and H. H. P. Fang. 2007. Inhibition of heavy metals on fermentative hydrogen production by granular sludge. *Chemosphere* 67(4):668–73.

Li, D. M., and H. Z. Chen. 2007. Biological hydrogen production from steam-exploded straw by simultaneous saccharification and fermentation. *Int J Hydrogen Energy* 32(12):1742–8.

Li, Y.-C., C.-Y. Chu, S.-Y. Wu, C.-Y. Tsai, C.-C. Wang, C.-H. Hung, and C.-Y. Lin. 2012a. Feasible pretreatment of textile wastewater for dark fermentative hydrogen production. *Int J Hydrogen Energy* 37(20):15511–7.

Li, Y.-C., M. M. Nissilä, S.-Y. Wu, C.-Y. Lin, and J.-A. Puhakka. 2012b. Silage as source of bacteria and electrons for dark fermentative hydrogen production. *Int J Hydrogen Energy* 37(20):15518–24.

Lin, C. Y., and C. H. Cheng. 2006. Fermentative hydrogen production from xylose using anaerobic mixed microflora. *Int J Hydrogen Energy* 31(7):832–40.

Lin, C.-Y., and W.-C. Hung. 2008. Enhancement of fermentative hydrogen/ethanol production from cellulose using mixed anaerobic cultures. *Int J Hydrogen Energy* 33(14):3660–7.

Liu, B. F., N. Q. Ren, J. Tang, J. Ding, W. Z. Liu, J. F. Xu et al. 2010. Bio-hydrogen production by mixed culture of photo- and dark-fermentation bacteria. *Int J Hydrogen Energy* 35(7):2858–62.

Liu C.-H., C.-Y. Chang, C.-L. Cheng, D.-J. Lee, J.-S. Chang. 2012. Fermentative hydrogen production by *Clostridium butyricum* CGS5 using carbohydrate-rich microalgal biomass as feedstock. *Int J Hydrogen Energy* 37:15458–64.

Liu, C. H., C. Y. Chang, Q. Liao, X. Zhu, C. F. Liao, and J. S. Chang. 2013. Biohydrogen production by a novel integration of dark fermentation and mixotrophic microalgae cultivation. *Int J Hydrogen Energy* 38(35):15807–14.

Liu, F., and B. Fang. 2007. Optimization of bio-hydrogen production from biodiesel wastes by *Klebsiella pneumonia*. *Biotechnol J* 2(3):374–80.

Liu, H., and G. Wang. 2012. Hydrogen production of a salt tolerant strain *Bacillus* sp. B2 from marine intertidal sludge. *World J Microbiol Biotechnol* 28(1):31–7.

Liu, X., Y. Zhu, and S. T. Yang. 2006. Butyric acid and hydrogen production by *Clostridium tyrobutyricum* ATCC 25755 and mutants. *Enzyme Microb Technol* 38(3):521–8.

Liu, Y., P. Yu, X. Song, and Y. Qu. 2008. Hydrogen production from cellulose by coculture of *Clostridium thermocellum* JN4 and *Thermoanaerobacterium thermosaccharolyticum* GD17. *Int J Hydrogen Energy* 33:2927–33.

Liu, Z., Y. Ying, F. Li, C. Ma, and P. Xu. 2010. Butanol production by *Clostridium beijerinckii* ATCC 55025 from wheat bran. *J Ind Microbiol Biotechnol* 37(5):495–501.

Lo, Y., W. Chen, C. Hung, S. Chen, and J. Chang. 2008. Dark H_2 fermentation from sucrose and xylose using H_2-producing indigenous bacteria: Feasibility and kinetic studies. *Water Res* 42(4):827–42.

Lo, Y., W. Lu, C. Chen, and J. Chang. 2010. Dark fermentative hydrogen production from enzymatic hydrolysate of xylan and pretreated rice straw by *Clostridium butyricum* CGS5. *Bioresour Technol* 101(15):5885–91.

Lo, Y.-C., G. D. Saratale, W. M. Chen, M. D. Bai, and J. Chang. 2009. Isolation of cellulose-utilizing bacteria for cellulosic biohydrogen production. *Enzyme Microb Technol* 44(6–7):417–25.

Louis, P., S. H. Duncan, S. I. Mccrae, M. S. Jackson, H. J. Flint, and J. Millar. 2004. Restricted distribution of the butyrate kinase pathway among butyrate-producing bacteria from the human colon. *J Bacteriol* 186(7):2099–106.

Lynd, L. R., W. H. Van Zyl, J. E. McBride, and M. Laser. 2005. Consolidated bioprocessing of cellulosic biomass: An update. *Curr Opin Biotechnol* 16(5):577–83.

Ma, Y., A. Huang, D. Zhu, G. Pan, and G. Wang. 2015. Biohydrogen production via the interaction of nitrogenase and anaerobic mixed-acid fermentation in marine bacteria. *Int J Hydrogen Energy* 40(1):176–83.

Magnusson, L., R. Islam, and R. Sparling. 2008. Direct hydrogen production from cellulosic waste materials with a single-step dark fermentation process. *Int J Hydrogen Energy* 33(20):5398–403.

Mariakakis, I., P. Bischoff, J. Krampe, C. Meyer, and H. Steinmetz. 2011. Effect of organic loading rate and solids retention time on microbial population during bio-hydrogen production by dark fermentation in large lab-scale. *Int J Hydrogen Energy* 36(17):10690–700.

Maru, B. T., M. Constanti, A. M. Stchigel, F. Medina, and J. E. Sueiras. 2012. Biohydrogen production by dark fermentation of glycerol using *Enterobacter* and *Citrobacter* sp. *Biotechnol Prog* 29(1):31–8.

Masset, J., M. Calusinska, C. Hamilton, S. Hiligsmann, B. Joris, A. Wilmotte, and P. Thonart. 2012. Fermentative hydrogen production from glucose and starch using pure strains and artificial co-cultures of *Clostridium* sp. *Biotechnol Biofuels* 5(1):1–15.

Masset, J., S. Hiligsmann, C. Hamilton, L. Beckers, F. Franck, and P. Thonart. 2010. Effect of pH on glucose and starch fermentation in batch and sequenced-batch mode with a recently isolated strain of hydrogen-producing *Clostridium butyricum* CWBI1009. *Int J Hydrogen Energy* 35(8):3371–8.

McKinsey, A. 2010. A portfolio of power-trains for Europe: A fact-based analysis. The role of battery electric vehicles, plug-in hybrids and fuel cell electric vehicles. Brussels: Fuel Cells and Hydrogen Joint Undertaking.

Méndez, L., J. Rojas, C. Izaguirre, B. Contreras, and R. Gómez. 2014. *Jatropha curcas* leaves analysis, reveals it as mineral source for low sodium diets. *Food Chem* 165:575–7.

Mertens, R., and A. Liese. 2004. Biotechnological applications of hydrogenases. *Curr Opin Biotechnol* 15(4):343–8.

Minnan, L., H. Jinli, W. Xiaobin, X. Huijuan, C. Jinzao, L. Chuannan et al. 2005. Isolation and characterization of a high H_2-producing strain *Klebsiella oxytoca* HP1 from a hot spring. *Res Microbiol* 156(1):76–81.

Moreno-Andrade, I., and B. Buitrón. 2015. Evaluation of particle size and initial concentration of total solids on biohydrogen production from food waste. *Fresenius Environ Bull* 24(7):2289–95.

Moreno-Andrade, I., G. Moreno, G. Kumar, and G. Buitrón G. 2015. Biohydrogen production from industrial wastewaters. *Water Sci Technol* 71:105–10.

Morimoto, K., T. Kimura, K. Sakka, and K. Ohmiya. 2005. Overexpression of a hydrogenase gene in *Clostridium paraputrificum* to enhance hydrogen gas production. *FEMS Microbiol Lett* 246(2):229–34.

Murata, K., Y. Liu, M. Inaba, and I. Takahara. 2012. Catalytic fast pyrolysis of *Jatropha* wastes. *J Anal Appl Pyrolysis* 94:75–82.

Nakayama, S., T. Kosaka, H. Hirakawa, K. Matsuura, S. Yoshino, and K. Furukawa. 2008. Metabolic engineering for solvent productivity by downregulation of the hydrogenase gene cluster hupCBA in *Clostridium saccharoperbutylacetonicum* strain N1-4. *Appl Genet Mol Biotechnol* 78(3):483–93.

Nasr, N., E. Elbeshbishy, H. Hafez, G. Nakhla, and M. H. El Naggar. 2011. Bio-hydrogen production from thin stillage using conventional and acclimatized anaerobic digester sludge. *Int J Hydrogen Energy* 36(20):12761–9.

Nath, K., and D. Das. 2004. Biohydrogen production as a potential energy resource—Present state-of-art. *J Sci Ind Res* 63:729–38.

Ngo, T. A., and H. T. V. Bui. 2013. Biohydrogen production using immobilized cells of hyperthermophilic eubacterium *Thermotoga neapolitana* on porous glass beads. *J Technol Innov Renew Energy* 2(3):231–8.

Nguyen, T. A. D., S. J. Han, J. P. Kim, M. S. Kim, and S. J. Sim. 2010. Hydrogen production of the hyperthermophilic eubacterium *Thermotoga neapolitana* under N_2 sparging condition. *Bioresour Technol* 101(1):S38–41.

Nguyen, T. A. D., K. R. Kim, M. T. Nguyen, M. S. Kim, D. Kim, and S. J. Sim. 2010. Enhancement of fermentative hydrogen production from green algal biomass of *Thermotoga neapolitana* by various pretreatment methods. *Int J Hydrogen Energy* 35(23):13035–40.

Ning, Y.-Y., D.-W. Jin, G.-P. Sheng, H. Harada, and X.-Y. Shi. 2012. Evaluation of the stability of hydrogen production and microbial diversity by anaerobic sludge with chloroform treatment. *Renew Energy* 38(1):253–7.

Nissilä, M. E., A. P. Tahti, J. A. Rintala, and J. A. Puhakka. 2011a. Thermophilic hydrogen production from cellulose with rumen fluid enrichment cultures: Effects of different heat treatments. *Int J Hydrogen Energy* 36(2):1482–90.

Nissilä, M. E., H. P. Tähti, J. A. Rintala, and J. A. Puhakka. 2011b. Effects of heat treatment on hydrogen production potential and microbial community of thermophilic compost enrichment cultures. *Bioresour Technol* 102(6):4501–6.

Ntaikou, I., H. N. Gavala, M. Kornaros, and G. Lyberatos. 2008. Hydrogen production from sugars and sweet sorghum biomass using *Ruminococcus albus*. *Int J Hydrogen Energy* 33(4):1153–63.

Oh, Y., S. M. Raj, G. Y. Jung, and S. Park. 2011. Current status of the metabolic engineering of microorganisms for biohydrogen production. *Bioresour Technol* 102(18):8357–67.

Ohnishi, A., Y. Bando, N. Fujimoto, and M. Suzuki. 2010. Development of a simple bio-hydrogen production system through dark fermentation by using unique microflora. *Int J Hydrogen Energy* 35(16):8544–53.

Ortigueira, J., T. Pinto, L. Gouveia, and P. Moura. 2015. Production and storage of biohy-drogen during sequential batch fermentation of *Spirogyra* hydrolyzate by *Clostridium butyricum*. *Energy* 88:528–36.

O-Thong, S., A. Hniman, P. Prasertsan, and I. Tsuyoshi. 2011. Biohydrogen production from cassava starch processing wastewater by thermophilic mixed cultures. *Int J Hydrogen Energy* 36(5):3409–16.

O-Thong, S., P. Prasertsan, and N.-K. Birkeland. 2009. Evaluation of methods for prepar-ing hydrogen-producing seed inocula under thermophilic condition by process perfor-mance and microbial community analysis. *Bioresour Technol* 100(2):909–18.

Pacheco, R., A. F. Ferreira, T. Pinto, B. P. Nobre, D. Loureiro, P. Moura et al. 2014. The pro-duction of pigments & hydrogen through a *Spirogyra* sp. biorefinery. *Energy Convers Manag* 89:789–97.

Pattra S., S. Sangyoka, M. Boonmee, and A. Reungsang. 2008. Bio-hydrogen production from the fermentation of sugarcane bagasse hydrolysate by *Clostridium butyricum*. *Int J Hydrogen Energy* 33(19):5256–65.

Pawar, S. S., and E. W. J. van Niel. 2013. Thermophilic biohydrogen production: How far are we? *Appl Microbiol Biotechnol* 97(18):7999–8009.

Pérez-Rangel, M., F. R. Quiroz-Figueroa, J. González-Casteñeda, and I. Valdez-Vazquez. 2015. Microscopic analysis of wheat straw cell wall degradation by microbial consortia for hydrogen production. *Int J Hydrogen Energy* 40(1):151–60.

Post, W. M., and K. C. Kwon. 2000. Soil carbon sequestration and land-use change: Processes and potential. *Glob Chang Biol* 6(3):317–27.

Prasertsan, P., S. O-Thong, and N.-R. Birkeland. 2009. Optimization and microbial commu-nity analysis for production of biohydrogen from palm oil mill effluent by thermophilic fermentative process. *Int J Hydrogen Energy* 34(17):7448–59.

Puhulwella, R. G., L. Beckers, F. Delvigne, A. S. Grigorescu, P. Thonart, and S. Hiligsmann. 2014. Mesophilic biohydrogen production by *Clostridium butyricum* CWBI1009 in trickling biofilter reactor. *Int J Hydrogen Energy* 39(30):16902–13.

Ren, N., W. Guo, B. Liu, G. Cao, and J. Ding. 2011. Biological hydrogen production by dark fermentation: Challenges and prospects towards scaled-up production. *Curr Opin Biotechnol* 22(3):365–70.

Ren, N. Q., G. L. Cao, and A. J. Wang. 2008b. Dark fermentation of xylose and glucose mix using isolated *Thermoanaerobacterium thermosaccharolyticum* W16. *Int J Hydrogen Energy* 33(21):6124–32.

Ren, N.-Q., W.-Q. Guo, X.-J. Wang, W.-S. Xiang, B.-F. Liu, X.-Z. Wang et al. 2008a. Effects of different pretreatment methods on fermentation types and dominant bacteria for hydrogen production. *Int J Hydrogen Energy* 33(16):4318–24.

Ren, Z., T. E. Ward, B. E. Logan, and J. M. Regan. 2007. Characterization of the cellulo-lytic and hydrogen-producing activities of six mesophilic *Clostridium* species. *J Appl Microbiol* 103(6):2258–66.

Rossi, D. M., J. B. da Costa, E. A. de Souza, M. C. Ruaro Peralba, D. Samios, and M. A. Záchia Ayub. 2011. Comparison of different pretreatment methods for hydrogen pro-duction using environmental microbial consortia on residual glycerol from biodiesel. *Int J Hydrogen Energy* 36(8):4814–9.

Rupprecht, J., B. Hankamer, J. H. Mussgnug, G. Ananyev, C. Dismukes, and O. Kruse. 2006. Perspectives and advances of biological H_2 production in microorganisms. *Appl Microbiol Biotechnol* 72(3):442–9.

Saint-Amans, S., L. Girbal, J. Andrade, K. Ahrens, and P. Soucaille. 2001. Regulation of car-bon and electron flow in *Clostridium butyricum* VPI 3266 grown on glucose-glycerol mixtures. *J Bacteriol* 183(5):1748–54.

Sánchez, A., V. Sevilla-Güitrón, G. Magaña, and L. Gutierrez. 2013. Parametric analysis of total costs and energy efficiency of 2G enzymatic ethanol production. *Fuel* 113:165–79.

Sandin, G., F. Røyne, J. Berlin, G. M. Peters, and M. Svanström. 2015. Allocation in LCAs of biorefinery products: Implications for results and decision-making. *J Clean Prod* 93:213–21.

Saratale, G. D., S. D. Chen, Y. C. Lo, R. G. Saratale, and J. S. Chang. 2008. Outlook of biohy-drogen production from lignocellulosic feedstock using dark fermentation—A review. *J Sci Ind Res* 67(11):962–79.

Saratale, G. D., I. J. Chien, and J. S. Chang. 2011. Enzymatic pretreatment of cellulosic wastes for anaerobic treatment and bioenergy production. In *Environmental Anaerobic Technology Applications and New Developments* ed. H.H.-P. Fang, 279–308. Vol. 13. London: Imperial College Press.

Saratale, G. D., S. D. Kshirsagar, V. T. Sampange, R. G. Saratale, S. E. Oh, S. P. Govindwar, and M. K. Oh. 2014. Cellulolytic enzymes production by utilizing agricultural wastes under solid state fermentation and its application for biohydrogen production. *Appl Biochem Biotechnol* 174(8):2801–17.

Saratale, G. D., S. D. Kshirsagar, R. G. Saratale, S. P. Govindwar, and M. K. Oh. 2015. Fermentative hydrogen production using sorghum husk as a biomass feedstock and process optimization. *Biotechnol Bioprocess Eng* 20(4):733–43.

Saratale, G. D., M. Y. Jung, and M. K. Oh. 2016. Reutilization of green liquor chemicals for pretreatment of whole rice waste biomass and its application to 2,3-butanediol produc-tion. *Bioresour Technol* 205:90–6.

Saratale, G. D., and M. K. Oh. 2015. Improving alkaline pretreatment method for preparation of whole rice waste biomass feedstock and bioethanol production. *RSC Adv* 5(118):91171–9.

Saratale, G. D., R. G. Saratale, and J. S. Chang. 2013. Biohydrogen from renewable resources. *Biohydrogen* 9:185–221.

Saratale, G. D., R. G. Saratale, Y. C. Lo, and J. S. Chang. 2010. Multicomponent cellulase pro-duction by *Cellulomonas biazotea* NCIM-2550 and its applications for cellulosic biohy-drogen production. *Biotechnol Prog* 26(2):406–16.

Saratale, G. D., R. G. Saratale, and S. E. Oh. 2012. Production and characterization of multiple cellulolytic enzymes by isolated *Streptomyces* sp. MDS. *Biomass Bioenergy* 47:302–15.

Sarma, S. J., V. Pachapur, S. K. Brar, Y. Le Bihan, and G. Buelna. 2015. Hydrogen biorefinery: Potential utilization of the liquid waste from fermentative hydrogen production. *Renew Sustain Energy Rev* 50:942–51.

Saur, G., and C. Ainscough. 2011. U.S. Geographic analysis of the cost of hydrogen from elec-trolysis. *Contract* 303:275–3000.

Schmitz, S., B. Dawson, M. Spannagle, F. Thomson, J. Koch, and R. Eaton. 2000. The Greenhouse Gas Protocol—A Corporate Accounting and Reporting Standard, Revised Edition. *GHG Protoc Corp Account Report Stand* 9:116.

Schut, G. J., and M. W. W. Adams. 2009. The iron-hydrogenase of *Thermotoga maritima* utilizes ferredoxin and NADH synergistically: A new perspective on anaerobic hydrogen production. *J Bacteriol* 191(13):4451–7.

Searchinger, T., R. Heimlich, R. A. Houghton, F. Dong, A. Elobeid, J. Fabiosa et al. 2008. Use of U.S. croplands for biofuels increases greenhouse gases through emissions from land-use change. *Science* 319(5867):1238–40.

Sheng, T., L. F. Gao, L. Zhao, W. Z. Liu, and A. J. Wang. 2015. Direct hydrogen production from lignocellulose by the newly isolated *Thermoanaerobacterium thermosaccharolyticum* strain DD32. *RSC Adv* 5(121):99781–8.

Siebers, B., and P. Schonheit. 2005. Unusual pathways and enzymes of central carbohydrate metabolism in Archaea. *Curr Opin Microbiol* 8(6):695–705.

Silva, C. M., R. C. Pacheco, T. M. Batista, A. F. Ferreira, and J. P. Ribau. 2015. Roadmap to decarburization and energy consumption minimization of the road transport sector: Biohydrogen production from several microalgae species and integration in optimized bus configurations. In *Energy Science and Technology*, vol. 11, *Hydrogen and Other Technologies*. New Delhi: Studium Press.

Silva, J. P. A., L. M. Carneiro, and I. C. Roberto. 2013. Treatment of rice straw hemicellulosic hydrolysates with advanced oxidative processes: A new and promising detoxification method to improve the bioconversion process. *Biotechnol Biofuels* 6(1):1.

Singh, L., Z. A. Wahid, M. F. Siddiqui, A. Ahmad, M. H. Ab Rahim, and M. Sakinah. 2013. Biohydrogen production from palm oil mill effluent using immobilized *Clostridium butyricum* EB6 in polyethylene glycol. *Proc Biochem* 48(2):294–8.

Sissine, F. 2010. Renewable Fuel Standard Program (RFS2) regulatory impact analysis. Technical Report EPA-420-R-10-006. Washington, DC: Assessment and Standards Division, Office of Transportation and Air Quality.

Sivagurunathan, P., B. Sen, and C.-Y. Lin. 2013. Batch fermentative hydrogen production by enriched mixed culture: Combination strategy and their microbial composition. *J Biosci Bioeng* 117(2):222–8.

Skonieczny, M. T., and V. Yargeau. 2009. Biohydrogen production from wastewaster by *Clostridium beijerinckii*: Effect of pH and substrate concentration. *Int J Hydrogen Energy* 34(8):3288–94.

Slepetiene, A., J. Slepetys, and I. Liaudanskiene. 2008. Standard and modified methods for soil organic carbon determination in agricultural soils. *Agro Res* 6(2):543–54.

Somerville C., H. Youngs, S. C. Taylor, S. C. Davis, and S. P. Long. 2010. Feedstocks for lignocellulosic biofuels. *Science* 329(5993):790–2.

Sreela-or, C., T. Imai, P. Plangklang, and A. Reungsang. 2011. Optimization of key factors affecting hydrogen production from food waste by anaerobic mixed cultures. *Int J Hydrogen Energy* 36(21):14120–33.

Taguchi, F., K. Yamada, K. Hasegawa, T. Takisaito, and K. Hara. 1996. Continuous hydrogen production by *Clostridium* sp. strain no. 2 from cellulose hydrolysate in aqueous two phase system. *J Ferment Bioeng* 82:80–3.

Talluri, S., S. M. Raj, and L. P. Christopher. 2013. Consolidated bioprocessing of untreated switchgrass to hydrogen by the extreme thermophile *Caldicellulosiruptor saccharolyticus* DSM 8903. *Bioresour Technol* 139:272–9.

Thauer, R. K., K. Jungermann, and K. Decker. 1977. Energy conservation in chemotrophic anaerobic bacteria. *Bacteriological Rev* 41(1):100–80.

Thong, S. O., P. Prasertsan, N. Intrasungkha, S. Dhamwichukorn, and N. K. Birkeland. 2007. Improvement of biohydrogen production and treatment efficiency on palm oil mill effluent with nutrient supplementation at thermophilic condition using an anaerobic sequencing batch reactor. *Enzyme Microb Technol* 41:583–90.

Tracy, B. P., S. W. Jones, A. G. Fast, D. C. Indurthi, and E. T. Papoutsakis. 2012. Clostridia: The importance of their exceptional substrate and metabolite diversity for biofuel and biorefinery applications. *Curr Opin Biotechnol* 23(3):364–81.

Ursua, A., L. M. Gandia, and P. Sanchis. 2012. Hydrogen production from water electrolysis: Current status and future trends. *Proc IEEE* 100:410–26.

Valdez-Vazquez, I., A. E. Escalante-Hernández, Z. L. Alamilla-Ortiz, A. J. García-Pérez, M. R. Mercado-Ireta, J. González-Castañeda et al. 2015b. Pyrosequencing of two H₂-producing consortia from native wheat straw. Presented at the *HYPOTHESIS Proceedings*, Madrid, September 6–9.

Valdez-Vazquez, I., M. Pérez-Rangel, A. Tapia, G. Buitrón, C. Molina, G. Hernández et al. 2015a. Hydrogen and butanol production from native wheat straw by synthetic microbial consortia integrated by species of *Enterococcus* and *Clostridium*. *Fuel* 159:214–22.

Valdez-Vazquez, I., and H.-M. Poggi-Varaldo. 2009. Hydrogen production by fermentative consortia. *Renew Sustain Energy Rev* 13(5):1000–13.

Valdez-Vazquez, I., E. Ríos-Leal, F. Esparza-García, F. Cecchi, and H. M. Poggi-Varaldo. 2005. Semi-continuous solid substrate anaerobic reactors for H₂ production from organic waste: Mesophilic versus thermophilic regime. *Int J Hydrogen Energy* 30(13):1383–91.

Van De Werken, H. J. G., M. R. A. Verhaart, A. L. VanFossen, K. Willquist, D. L. Lewis, J. D. Nichols et al. 2008. Hydrogenomics of the extremely thermophilic bacterium *Caldicellulosiruptor saccharolyticus*. *Appl Environ Microbiol* 74(21):6720–9.

Varrone, C., S. Rosa, F. Fiocchetti, B. Giussani, G. Izzo, G. Massini, et al. 2013. Enrichment of activated sludge for enhanced hydrogen production from crude glycerol. *Int J Hydrogen Energy* 38(3):1319–31.

Vignais, P. M., B. Billoud, and J. Meyer. 2001. Classification and phylogeny of hydrogenases 1. *FEMS Microbiol Rev* 25(4):455–501.

Vijayaraghavan, K., D. Ahmad, and M. K. B. Ibrahim. 2006. Biohydrogen generation from jackfruit peel using anaerobic contact filter. *Int J Hydrogen Energy* 31(5):569–79.

Wan, J., Y. Ning, X. Shi, D. Jin, S. Li, and Y. Chen. 2013. Microbial community structure of anaerobic sludge for hydrogen production under different acid pretreatment conditions. *J Renew Sustain Energy* 5(2):023126.

Wang, B., B. H. Gebreslassie, and F. You. 2013. Sustainable design and synthesis of hydrocarbon biorefinery via gasification pathway: Integrated life cycle assessment and technoeconomic analysis with multiobjective superstructure optimization. *Comput Chem Eng* 52:55–76.

Wang, C. C., C. W. Chang, C. P. Chu, D. J. Lee, B. V. Chang, and C. S. Liao. 2003. Producing hydrogen from wastewater sludge by *Clostridium bifermentans*. *J Biotechnol* 102(1):83–92.

Wang, J. L., and W. Wan. 2009. Factors influencing fermentative hydrogen production: A review. *Int J Hydrogen Energy* 34(2):799–811.

Wang, M. 2008. The greenhouse gases, regulated emissions, and energy use in transportation (GREET) model: Version 1.5. Chicago: Center for Transportation Research, Argonne National Laboratory.

Wang, M., B. H. Olson, J. Chang. 2008b. Relationship among growth parameters for *Clostridium butyricum*, hydA gene expression, and biohydrogen production in a sucrose-supplemented batch reactor. *Appl Microb Cell Physiol* 78(3):525–32.

Wang, M. Y., Y. L. Tsai, B. H. Olson, and J. S. Chang. 2008a. Monitoring dark hydrogen fermentation performance of indigenous *Clostridium butyricum* by hydrogenase gene expression using RT-PCR and qPCR. *Int J Hydrogen Energy* 33(18):4730–8.

Wang, S., J. Chen, and J. L. Johnson. 1990. A nitrogen-fixation gene (*nifC*) in *Clostridium pasteurianum* with sequence similarity to chlJ of *Escerichia coli*. *Biochem Biophys Res Commun* 169(3):1122–8.

Wang, X., D. Hoefel, C. P. Saint, P. T. Monis, and B. Jin. 2007. The isolation and microbial community analysis of hydrogen producing bacteria from activated sludge. *J. Appl Microbiol* 103(5):1415–23.

Wang, X., and B. Jin. 2009. Process optimization of biological hydrogen production from molasses by a newly isolated *Clostridium butyricum* W5. *J Biosci Bioeng* 107(2):138–44.

Wang, X., P. T. Monis, C. P. Saint, and B. Jin. 2009. Biochemical kinetics of fermentative hydrogen production by *Clostridium butyricum* W5. *Int J Hydrogen Energy* 34(2):791–8.

Xu, J. F., N. Q. Ren, D. X. Su, and J. Qiu. 2010. Bio-hydrogen production from acetic acid steam-exploded corn straws by simultaneous saccharification and fermentation with *Ethanoligenens harbinense* B49. *Int J Energy Res* 34(5):381–6.

Yang, Z. M., R. B. Guo, X. H. Xu, X. L. Fan, and X. P. Li. 2010. Enhanced hydrogen production from lipid-extracted microalgal biomass residues through pretreatment. *Int J Hydrogen Energy* 35(18):9618–23.

Yokoi, H., A. Saitsu, H. I. Uchida, J. Hirose, S. Hayashi, and Y. Takasaki. 2001. Microbial hydrogen production from sweet potato starch residue. *J Biosci Bioeng* 91(1):58–63.

Yokoi, H., T. Tokushige, J. Hirose, S. Hayashi, and Y. Takasaki. 1998. H$_2$ production from starch by a mixed culture of *Clostridium butyricum* and *Enterobacter aerogenes*. *Biotechnol Lett* 20(2):143–7.

Yokoyama, H., N. Moriya, H. Ohmori, M. Waki, A. Ogino, and Y. Tanaka. 2007. Community analysis of hydrogen-producing extreme thermophilic anaerobic microflora enriched from cow manure with five substrates. *Appl Microbiol Biotechnol* 77(1):213–22.

Yossan, S., S. O-Thong, and P. Prasertsan. 2012. Effect of initial pH, nutrients and temperature on hydrogen production from palm oil mill effluent using thermotolerant consortia and corresponding microbial communities. *Int J Hydrogen Energy* 37(18):13806–14.

Yuan, X., X. Shi, P. Zhang, Y. Wei, R. Guo, and L. Wang. 2011. Anaerobic biohydrogen production from wheat stalk by mixed microflora: Kinetic model and particle size influence. *Bioresour Technol* 102(19):9007–12.

Zeidan, A. A., and E. W. J. Van Niel. 2009. Developing a thermophilic hydrogen-producing co- culture for efficient utilization of mixed sugars. *Int J Hydrogen Energy* 34(10):4524–8.

Zhang, C., F. X. Lu, and X. H. Xing. 2011. Bioengineering of the *Enterobacter aerogenes* strain for biohydrogen production. *Bioresour Technol* 102(18):8344–9.

Zhang, C., H. Yang, F. Yang, and Y. Ma. 2009. Current progress on butyric acid production by fermentation. *Curr Microbiol* 59(6):656–63.

Zhang, J. N., Y. H. Zheng, H. Q. Fan, Y. T. Hou, and H. Wei, H. 2015. Direct degradation of cellulosic biomass to bio-hydrogen from a newly isolated strain *Clostridium sartagoforme* FZ11. *Bioresour Technol* 192:60–7.

Zhang, Y., and J. Shen. 2006. Effect of temperature and iron concentration on the growth and hydrogen production of mixed bacteria. *Int J Hydrogen Energy* 31(4):441–6.

Zhao, H., Y. Lu, L. Wang, C. Zhang, C. Yang, and X. Xing. 2015. Disruption of lactate dehydrogenase and alcohol dehydrogenase for increased hydrogen production and its effect on metabolic flux in *Enterobacter aerogenes*. *Bioresour Technol* 194:99–107.

Zhao, L., G. L. Cao, A. J. Wang, H. Y. Ren, D. Dong, Z. N. Liu et al. 2012. Fungal pretreatment of cornstalk with *Phanerochaete chrysosporium* for enhancing enzymatic saccharification and hydrogen production. *Bioresour Technol* 114:365–9.

Zhao, L., G. L. Cao, A. J. Wang, H. Y. Ren, C. J. Xu, and N. Q. Ren. 2013. Enzymatic saccharification of cornstalk by onsite cellulases produced by *Trichoderma viride* for enhanced biohydrogen production. *GCB Bioenergy* 5(5):591–8.

Zheng, Y., J. Kahnt, I. H. Kwon, R. I. Mackie, and R. K. Thauer. 2014. Hydrogen formation and its regulation in *Ruminococcus albus*: Involvement of an electron-bifurcating [FeFe]-hydrogenase, of a non electron-bifurcating [FeFe]-hydrogenase and of a putative hydrogen-sensing [FeFe]-hgydrogenase. *J Bacteriol* 196(22):3840–52.

8

Engineering Strategies for Enhancing Photofermentative Biohydrogen Production by Purple Nonsulfur Bacteria Using Dark Fermentation Effluents

Anish Ghimire, Giovanni Esposito, Vincenzo Luongo, Francesco Pirozzi, Luigi Frunzo, and Piet N.L. Lens

Contents

8.1 Introduction

8.1.1 H_2 Production Processes

Most energy fuels, chemicals, and raw materials in our daily lives are derived from petroleum-based refineries. However, depleting fossil fuel reserves and increasing greenhouse gas emissions and severe pollution problems as the consequences of by-products from fossil fuel utilization are driving interests toward biorefineries for the production of energy and useful chemicals (Cherubini 2010; Menon and Rao 2012). In the energy and environmental sector, hydrogen (H_2) has gained considerable interest owing to its higher specific energy content (122 MJ/kg), as well as water and energy being the sole oxidative reaction by-products (Balat and Kırtay 2010). At present, H_2 production for industrial applications is mainly derived from thermocatalytic and gasification processes, which are highly dependent on fossil fuels. In comparison with the energy-intensive physicochemical routes for H_2 production, biological processes can be operated at ambient conditions and are advantageous, as they can utilize renewable biomass (Das and Veziroglu 2001; Ghimire et al. 2015).

Based on light dependency as an energy source for biochemical reactions, biological H_2 production pathways can be broadly categorized into light-dependent and -independent processes (Das and Veziroglu 2008; Hallenbeck and Ghosh 2009). The light-dependent photohydrogen production systems can be further classified into (1) direct photolysis, where water is broken down into H_2 and O_2 gas by algae and cyanobacteria; (2) indirect photolysis, in which cyanobacteria or cyanophytes synthesize H_2 in the presence of light and inorganic carbon; and (3) photofermentation (PF), carried out by photosynthetic bacteria where photodecomposition of organic compounds occurs. The light-independent processes include (1) dark fermentation (DF), which involves fermentative hydrogen production from carbohydrate-rich organic biomass, and (2) H_2 from bioelectrochemical systems or microbial electrolysis cells.

8.1.2 Light-Independent H_2 Production

DF is a well-studied biological route for the production of hydrogen from organic biomass, including waste, owing to its higher H_2 production rates than light-dependent processes (Ghimire et al. 2015). However, due to the thermodynamic constraints, dark fermentative conversion of carbohydrate-rich organic biomass offers lower H_2 yields and gives incomplete conversion of organic biomass; that is, organic acids and alcohols remain as major fermentation by-products. On the brighter side, the PF processes can convert these dark fermentative by-products to biohydrogen. Moreover, PF processes have higher H_2 yields and generate less residues than DF processes (Li and Fang 2009; Lo et al. 2010). A dual system can integrate the conversion of carbohydrates to organic acids in the first stage (DF) and the utilization of its by-products in the second stage (PF) (Redwood et al. 2008).

8.1.3 Light-Dependent H_2 Production

Using light as a source of energy, purple nonsulfur bacteria (PNSB) synthesize H_2 by carrying out an anaerobic photosynthesis. In PNSB, this takes place in the presence of the nitrogenase enzyme and light, with reduced carbon sources such as organic acids. In addition, under certain operating conditions, PNSB also synthesize cell reserve materials or biopolymers, that is, polyhydroxybutyrate (PHB) molecules (Khatipov et al. 1998; De Philippis et al. 1992).

An example of PF is the conversion of acetic acid to biohydrogen and/or biopolymers. It can be expressed by the following equations (Equations 8.1 and 8.2):

$$2CH_3COOH + 4H_2O \xrightarrow{\text{Light energy}} 8H_2 + 4CO_2 \tag{8.1}$$

$$2CH_3COOH + 2[H] \xrightarrow{\text{Light energy}} PBH\text{-monomer} + H_2O \tag{8.2}$$

$$C_6H_{12}O_6 + 2H_2O \rightarrow 2CH_3COOH + 2CO_2 + 4H_2 \tag{8.3}$$

Photofermentative H_2 production systems are attractive because of their higher H_2 yield potential, that is, 66.67 mmol H_2/g chemical oxygen demand (COD) (Equation 8.1) from PF systems, compared with only 22.22 mmol H_2/g COD from the DF process, with acetate as the sole by-product (Equation 8.3). Moreover, the biopolymer production can add an economic value to the PF process. However, H_2 and PHB photofermentative production are competing processes (Khatipov et al. 1998; Wu et al. 2012). Nonetheless, a concomitant production of H_2 and PHB is also possible, as shown in a study by Montiel-Corona et al. (2015). The photofermentative H_2 and PHB production depends on several operating conditions, such as nutrient availability (carbon-to-nitrogen [C/N] ratio), PNSB strain (mixed or pure culture), pH, light intensity, and the presence of physical-chemical stress, for example, the presence of inhibitors of H_2 formation, such as ammonium, in the culture medium (Li and Fang 2009; Chen et al. 2011; Adessi and De Philippis 2014).

The ability of PNSB to convert reduced carbon sources such as organic acids and alcohols to H_2 and PHB makes PNSB-based PF a good posttreatment process for dark fermentation effluents (DFEs) (Rai et al. 2014a; Cheng et al. 2015; Chookaew et al. 2015; Dipasquale et al. 2015; Nasr et al. 2015). Moreover, the potential of the PF process to be operated as a stand-alone system for wastewater treatment has also been reported (Eroğlu et al. 2008; Li and Fang 2009; Hülsen et al. 2014). With the increasing application of DF processes for H_2 production, the integrated DF-PF process can enhance H_2 yields, thus providing sustainability to scaled-up biohydrogen production processes. Likewise, the potential of PF processes for the production of biopolymers can give further economic gain.

8.1.4 Scope of This Chapter

This chapter aims to summarize the state of the art of PF processes for H_2 production by overviewing existing understanding of the microbiology of the PF process, different photobioreactor (PBR) designs, conversion efficiencies of different PNSB strains, and process operational parameters such as pH, temperature, and nutrient requirements. This competence can be applied for the valorization of DFEs and wastewater through H_2 and PHB production. In addition, this work presents current approaches of the mathematical modeling of PF, as well as highlights the economics of the process.

8.2 Microbiology and Phototrophic Metabolism of PNSB

8.2.1 Bacterial Photosynthesis

Bacterial photosynthesis can be divided into two types depending on the presence or absence of oxygen for the metabolism of bacteriochlorophyll, a bacterial photosynthetic pigment. Oxygenic photosynthesis is carried out by cyanobacteria and prochlorophytes, whereas anoxygenic photosynthesis can be generally mediated by purple bacteria, green sulfur bacteria, heliobacteria, and others (Kim and Gadd 2008). Photosynthetic anoxygenic bacteria are a very diverse group of bacteria that carry out bacteriochlorophyll-dependent photosynthesis as a metabolic process (McEwan 1994). The anoxygenic phototrophic bacteria can be broadly grouped into different classes (Figure 8.1), based on their photosynthetic pigments and electron donors (Kim and Gadd 2008; McEwan 1994). Depending on the electron donors used, purple bacteria can be further divided into purple sulfur bacteria (use sulfur compounds as electron donors) and nonsulfur bacteria (use organic substances as electron donors).

Some drawbacks of this photofermentative system, as pointed by Hallenbeck and Benemann (2002), include the inherent high energy demand associated with

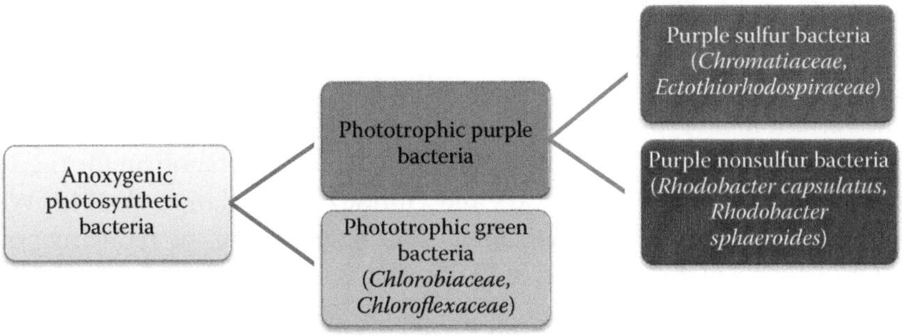

Figure 8.1   Classification of anoxygenic photosynthetic bacteria.

the nitrogenase enzyme, lower photoconversion efficiencies, and economic issues of anaerobic PBRs covering large areas. These drawbacks can be overcome by effective design and operation of the PBRs and selecting proper strains or enrichment of PNSB for an efficient conversion to photo-H_2.

8.2.2 Purple Nonsulfur Bacteria

Among the anoxygenic bacteria, the PNSB exhibit very diverse morphological, biochemical, and metabolic properties (Imhoff et al. 1984). PNSB are gram-negative photoheterotrophs, which normally carry out photosynthesis under anaerobic conditions. Although PNSB are facultative anaerobes, they can also grow chemotropically under oxygenic conditions using oxygen as an electron acceptor (McEwan 1994). Different from purple sulfur bacteria, which use elemental sulfur as the electron donor, PNSB typically use organic electron donors, such as organic acids; however, they can also use hydrogen gas as an electron donor (Kim and Gadd 2008).

PNSB can utilize various types of carbon sources, such as short-chain organic acids and glucose. The theoretical photofermentative conversion of different organic acids, typically present in DFEs, to H_2 can be expressed by the following reactions (Equations 8.4 through 8.7) (De Philippis et al. 1992; Barbosa et al. 2001; Han et al. 2012):

$$\text{Lactate:}\quad C_3H_6O_3 + 3H_2O \rightarrow 6H_2 + 3CO_2 \tag{8.4}$$

$$\text{Acetate:}\quad CH_3COOH + 2H_2O \rightarrow 4H_2 + 2CO_2 \tag{8.5}$$

$$\text{Propionate:}\quad C_3H_6O_2 + 4H_2O \rightarrow 7H_2 + 3CO_2 \tag{8.6}$$

$$\text{Butyrate:}\quad C_4H_8O_2 + 6H_2O \rightarrow 10H_2 + 4CO_2 \tag{8.7}$$

However, the conversion ability of different PNSB for different substrates varies (Barbosa et al. 2001; Bianchi et al. 2010). Some species prefer a certain sole carbon source, while H_2 yields seem to be higher with mixed sources of carbon (Han et al. 2012). The variation in H_2 production from different carbon sources can be explained by differences in their reduction states and the associated metabolism for the assimilation of different carbon sources (Kars and Gündüz 2010; Han et al. 2012; Wang et al. 2014). Similarly, when the carbon source is acetate, most of the reducing power of the PNSB is utilized for the synthesis of PHB rather than H_2 (Hustede et al. 1993; Kars and Gündüz 2010).

8.2.3 Photosystem of PNSB

The photosynthetic apparatus of PNSB is simple, as it contains only one photosystem (PS), unlike the two PSs in algae and cyanobacteria. PNS bacterial cells contain bacteriochlorophyll α or β located on the cytoplasmic membrane. The PS of PNSB contains the light harvesting complexes that absorb photons, initiating a charge (electron–hole) separation through excitation (Figure 8.2). Electrons that are liberated from organic acids are transported around through a number of electron carriers, that is, the cytochrome C_2 complex, cytochrome bc_1 complex (Cyt bc_1), and quinone Q (Figure 8.2). The transfer of electrons across the membranes creates a large proton

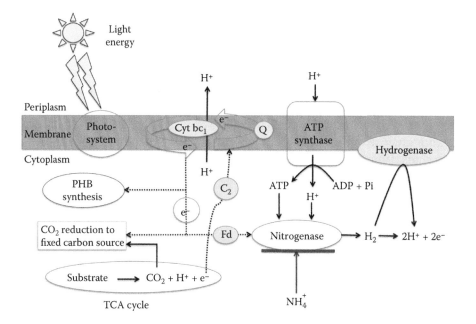

Figure 8.2 Schematic representation of mechanisms of photofermentative H_2 and PHB production in PNSB. (Adapted and modified from Adessi, A., and De Philippis, R., *Int. J. Hydrogen Energy*, 39(7), 3127–41, 2014; Akkerman, I. et al., *Int. Assoc. Hydrogen Energy*, 27, 1195–208, 2002; Kars, G., and Gündüz, U., *Int. J. Hydrogen Energy*, 35(13), 6646–56, 2010.)

gradient, which drives the synthesis of ATP from ADP by ATP synthase (Figure 8.2) (Akkerman et al. 2002; Hu et al. 2002). The extra energy in the form of ATP will be used to reduce ferredoxin (fd). Then, the ATP and reduced ferredoxin drives the proton reduction to hydrogen by nitrogenase (Hallenbeck and Ghosh 2009). Thus, as a result of anoxygenic photosynthesis, conversion of organic substances into H_2 takes place.

Nitrogenase and hydrogenase are the two enzymes that strongly influence hydrogen production: nitrogenase promotes its production, whereas hydrogenase consumes hydrogen (Figure 8.2). Besides the light conditions, the PF culture medium should be under nitrogen limitation and oxygen should be absent, as their presence inhibits the nitrogenase activity (Koku et al. 2002; Li and Fang 2009; Kars and Gündüz 2010). The activity of the nitrogenase enzyme is of fundamental importance for efficient photo-H_2 production (Hallenbeck and Benemann 2002). Equations 8.8 and 8.9 explain the effect of N_2 on the metabolism of PNSB (Das and Veziroglu 2001):

$$\text{With dinitrogen:} \quad N_2 + 8H^+ + 8e^- + 16ATP \rightarrow 2NH_3 + H_2 + 16ADP + 16P_i \quad (8.8)$$

$$\text{Without dinitrogen:} \quad 8H^+ + 8e^- + 16ATP \rightarrow 4H_2 + 16ADP + 16P_i \quad (8.9)$$

The presence of nitrogen, either in gaseous form or in the culture medium, can thus inhibit the activity of the nitrogenase enzyme that synthesizes molecular H_2. Therefore, substrates with a high C/N ratio are more suitable for H_2 conversion in these systems. Thus, the nitrogen content in the substrate (e.g., DFEs) should be considered when coupling PF to DF.

8.2.4 PHB Accumulation by PNSB

PNSB accumulates poly-β-hydroxybutyrate (PHB), an intracellular storage of carbon and energy formed under physiological stress, particularly at high C/N ratios, higher ammonia concentrations, or sulfur-deprived conditions (Khatipov et al. 1998; Waligórska et al. 2009; Eroglu and Melis 2011). The production of PHB and H_2 functions as the way to dissipating the excess reducing power, and the PHB synthesis competes with the H_2 production (Figure 8.2). Thus, depending on the aim of the process, the PF can be directed toward H_2 production by suppressing the PHB synthesis through genetic engineering of the PNSB (Kim et al. 2011). Kars and Gündüz (2010) reviewed the different genetic manipulation strategies to improve photofermentative biohydrogen production. They proposed to modify the acetate assimilation pathways that share the common biosynthetic route of PHB.

After deletion of the PHB-producing gene from *Rhodobacter sphaeroides* KD131, the H_2 production rate was increased from 36. to 43.8 1 mL H_2/L/h (Kim et al. 2011), in accordance with the study of Hustede et al. (1993), who observed an increase in cell growth and H_2 production when eliminating the gene for PHB synthesis in *R. sphaeroides*.

In addition, PNSB produce light harvesting bacterial pigments (bacteriochloro-phylls and carotenoids) that can be of commercial interests (Venil et al. 2013). This ability of PNSB has been highlighted in a few older studies and needs to be explored again (Cohen-Bazire et al. 1965; Schmidt 1971).

8.3 Photohydrogen Conversion Efficiencies

Akkerman et al. (2002) suggested three parameters to evaluate the photo-H_2 pro-duction process: H_2 production yield, the yield coefficient of H_2 produced relative to the carbon source consumed, and the photochemical efficiency (PE). Table 8.1 compares PF and DF systems in terms of H_2 yields from substrate conversion and production rate. PF systems are superior in terms of substrate-to-H_2 conversion, while they have slower H_2 production kinetics than DF systems. Considering the theoretical conversion of substrate to H_2 from Equations 8.4 through 8.7 and the experimental results reported in past studies (Table 8.1), PNSB have a very ver-satile metabolism and high substrate-to-H_2 conversion efficiency (McEwan 1994; Rupprecht et al. 2006; Bianchi et al. 2010). Their PF system lacks the oxygen sen-sitivity issues that are encountered in biophotolysis. Moreover, their light utiliza-tion proficiency is high, as PNSB can absorb and utilize both visible (400–700 nm) and near-infrared (700–900 nm) light. Also, PNSB use a wide variety of substrates (Eroglu and Melis 2011). The application of PNSB can be promising for PF sys-tems, as they not only give a higher substrate-to-product conversion and higher H_2 yield, but also benefit in their capability to reduce pollution loads, for example, the

TABLE 8.1 Comparison of Photo and Fermentation Dark Systems for Biohydrogen Production

Bio H_2 Systems (Microorganisms)	Carbon Source	H_2 Production Rate (mL H_2/L/h)	H_2 Yield (mL H_2/g COD)[a]	Reference
Photofermentation				
R. palustris WP3-5	DFEs	25.2	235.1	Chen et al. 2010
R. sphaeroides RV	Succinate	16.5	158.7	Han et al. 2012
Mixed culture	DFEs	5.7	568.5	Montiel-Corona et al. 2015
Dark Fermentation				
Kitchen waste compost	Vegetable waste	1000	38	Lee et al. 2010
Clostridium thermocellum 7072	Corn stalks	740	140	Cheng and Liu 2011
Klebsiella sp. TR17	Glycerol	48	128.6[b]	Chookaew et al. 2015

[a] mL H_2/g COD is calculated from the data provided in the publications.
[b] mL H_2/g COD consumed.

treatment of effluents (organic acids) from DF, with an added economic benefit in the form of PHB production, a valuable biopolymer.

The photofermentative H_2 production efficiency can also be measured as PE, which is an efficiency parameter with which the light is utilized to produce energy stored as hydrogen in a PF process. The PE depends on the photosynthetically active radiation (PAR) range, which determines the light energy absorbed by the photofermentative species. For example, green algae have a PAR range of 400–700 nm, while the range for PNSB is 400–950 nm (Figure 8.3). Akkerman et al. (2002) reported that the PE values vary between 3% and 10% in green algae. Redwood et al. (2012a) achieved a 71% increase in combined photosynthetic activity by illuminating both *R. sphaeroides* and *Arthrospira (Spirulina) platensis* by dividing a single beam of simulated sunlight using a dichroic mirror.

In addition to H_2 yield and the other parameters mentioned before, the performance of a PF process can be evaluated by the light conversion efficiency. Light or solar energy conversion efficiency can be calculated as the ratio of total energy produced, that is, stored in the form of H_2, to the total energy input to the bioreactor (energy as photons in the case of solar conversion efficiencies). The light conversion efficiency (η) can be calculated with an empirical formula, that is, the ratio of the total

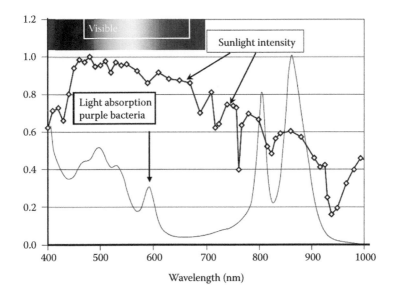

Figure 8.3 Sunlight and light absorption by purple bacteria. (Adapted and modified from Akkerman, I. et al., in *Bio-Methane and Bio-Hydrogen: Status and Perspectives of Biological Methane and Hydrogen Production*, ed. J.H. Reith et al., Dutch Biological Hydrogen Foundation: Hague, the Netherlands, 2003, pp. 124–45. Reprinted with permission of Prof. Dr. Rene Wijffels, Bio-methane and bio-hydrogen: Status and perspectives of biological methane and hydrogen production, Dutch Biological Hydrogen Foundation.)

energy (heat of combustion) value of hydrogen to energy input to the PBR by solar radiation (Koku et al. 2002). The η can be evaluated as

$$\eta(\%) = \frac{[33.61 \cdot \rho_{H_2} \cdot V_{H_2}]}{[I \cdot A \cdot t]} \cdot 100 \tag{8.10}$$

where V_{H_2} is the volume of produced H_2 (L), ρ_{H_2} is the density of the produced hydrogen gas (g/L), I is the light intensity (W/m^2), A is the irradiated area (m^2), and t is the duration of hydrogen production (hours).

Koku et al. (2002) reported a varying range of light conversion efficiencies between 1% and 5%, on average, for different strains of *R. sphaeroides*. According to the best knowledge of the authors, a PE higher than 10% has not been reported so far. However, a wide range of approaches to increase the PE has been suggested in the literature, including the design of efficient PBR and improving lighting conditions (Chen et al. 2011; Adessi and De Philippis 2014), genetic modifications of PNSB for enhancing nitrogenase activity (Ozturk et al. 2006), reduction of the pigment content for higher light uptake (Kondo et al. 2002), deletion or inactivation of the genes responsible for PHB synthesis (Franchi et al. 2005; Kim et al. 2011), and development of hydrogenase deficient (*hup*-) mutant PNSB strains responsible for H_2 uptake (Franchi et al. 2005; Uyar et al. 2015).

8.4 Operating Conditions of Photobioreactors for H_2 Production

8.4.1 PNSB Inoculum

PNSB are widely distributed in nature and prefer aquatic environments with low oxygen concentrations, significant amounts of soluble organic matter, moderate temperatures, and weak as well as stronger light conditions (Imhoff et al. 2005). Besides freshwater, members of the PNSB group can also be found in marine and hypersaline environments and even in sediments that are exposed to light. These organisms can also thrive in thermal springs and alkaline soda lakes (Imhoff et al. 2005). A eutrophic lake is an example of a favorable habitat for members of these genera (Imhoff et al. 2005; Bianchi et al. 2010).

Rhodopseudomonas palustris, *R. sphaeroides*, *Rhodobacter capsulatus*, and others are the most studied PNSB stains. However, PNSB strains capable of utilizing the substrates and light at higher conversion efficiencies are of research interest. Many studies have successfully isolated H_2-producing PNSB strains from different mixed consortia. Some examples of isolated H_2-producing PNSB and their H_2 yields and production rates are presented in Table 8.2.

Afsar et al. (2011) carried out PF studies using different PNSB strains, which showed that the PF efficiency highly depends on the effluent composition and bacterial strain used. The PF was carried out using the effluents from the thermophilic DF of glucose and potato steam peel hydrolysate as a carbon source under indoor

TABLE 8.2   Comparison of Photo-H₂ Production by Different Isolated and Mixed PNSB Strains from Various Inoculum Sources

Microbial Inoculum Sources	Isolated PNSB Members	Highest H₂-Producing Strain	Main Carbon Source	Temperature (°C)	pH	Light Intensity (lux)	Maximum H₂ Yield (mL H₂/g COD)[a]	Maximum H₂ Production Rate (mL H₂/L/h)	Reference
Pig dung	Not reported	*R. palustris*	Acetate	30	7	5,000	171.8	–	Yanling et al. 2008
Wastewater ponds	*Rhodobacter* sp.	*R. sphaeroides* ZX-5	Butyrate	30	6.0–9.0	4,000 (tungsten lamps)	–	118	Tao et al. 2008
Water and lake bed samples	Not reported	Unidentified PNSB strain TN1	Acetate	30	–	3,000	647.5	43	Suwansaard et al. 2009
Freshwater pond sludge	Not reported	*R. faecalis* strain RLD-53	Malate	35	7	4,000 (incandescent lamp)	1,242.5	25	Ren et al. 2009
Lake water and sediment samples	*R. capsulatus*, *Rhodospirillum rubrum*, *R. sphaeroides*, *R. palustris* strain AV33	*R. palustris* strain AV33	Lactate	30	6.8	200 μmol (photons) m²/s (incandescent lamp)	–	50.7	Bianchi et al. 2010
Activated sludge	*R. palustris*	Unidentified PNSB mixed culture	DFEs of starch wastewater	31	5.5	48[b] (equivalent to 190 W/m² of tungsten lamp)	970.0	120.8 ± 7	Tawfik et al. 2014
Activated sludge	Not reported	Unidentified enriched PNSB	DFEs	30	7.0	3,000 (LEDs and halogen lamps)	321.3 ± 3.7	5.7	Montiel-Corona et al. 2015
Silt sewage, pig manure, and cow dung	Not reported	Unidentified PNSB mixed culture	Enzymatic hydralysate of corncob	30	7.0	4,000 (LED lamps)	–	165	Z. Zhang et al. 2015

[a]COD values are calculated from theoretical oxygen demand of main carbon sources.
[b]Calculated using the available data and conversion values from http://www.egc.com/useful_info_lighting.php.

batch conditions. The PNS strains, such as *R. capsulatus* (DSM1710), *R. capsulatus* hup⁻ (YO3), *R. sphaeroides* O.U.001 (DSM5864), *R. sphaeroides* O.U.001 hup⁻, and *R. palustris*, were used in the study. The results showed that *R. sphaeroides* gave the highest amount of hydrogen from PF of glucose DFEs, while *R. capsulatus* produced better results on effluents from the DF of potato steam peels' hydrolysate.

However, the use of pure cultures of bacterial strains demands maintenance of sterile conditions in the bioreactors. The varying PF efficiencies of different PNS bacterial strains on different substrates suggest, for substrates such as DFEs, which contain mixed organic acids, the use of mixed consortia of PNSB in order to exploit the substrate utilization capacity of different PNS bacterial strains. In a study by Montiel-Corona et al. (2015), the H_2 yields from enriched mixed PNSB cultures were higher (1478 ± 17 mL H_2/L) than those from pure *R. capsulatus* cultures (1252 ± 20 mL H_2/L).

8.4.1.1 Inoculum Age

The selection of inoculum culture age can be critical to obtain a higher performance of PF systems. It has been found that the PNSB inoculum from the exponential phase of the growth curve is suitable for better performance of PBRs for biohydrogen production (Basak and Das 2007). Koku et al. (2003) found vast differences in total H_2 production, H_2 production rates, and the overall substrate conversion rates when *R. sphaeroides* O.U.001 of two different inoculum ages was used in the PF of malic acid. The inoculum harvested from the midexponential phase gave a higher total gas production (357 mL H_2), gas production rate (0.009 mL H_2/L/h), and overall substrate conversion rate (35%) than the inoculum harvested at the stationary phase, which gave a lower total gas production (236 mL H_2), gas production rate (0.003 mL H_2/L/h), and overall substrate conversion rate (24%).

In a study by Sasikala et al. (1991) on the effect of culture age on the photoproduction of hydrogen by *R. sphaeroides* O.U.001, the inoculum with a 20 h culture period gave the highest H_2 evolution (60 mL H_2/L reactor), while it was lower for a short (4 h) or long (38 h) culture period. A range of optimal inoculum ages has been reported in the literature. Akroum-Amrouche et al. (2011) reported an optimum inoculum age of 36–48 h in PF using *R. sphaeroides* CIP 60.6, while Liu et al. (2011) reported an inoculum age of 24 h for *Rhodopseudomonas faecalis* RLD-53 as optimum.

The aged inoculum can give poor performance in terms of H_2 production, and large retention times may shift the metabolic pathways to accumulation of PHB (Koku et al. 2003). They also reported that a repeated culture of PNSB might lead to loss of H_2 production capacity due to a decline in the activity of the electron carrier ferredoxin.

8.4.1.2 Cell Immobilization

Studies have used different cell immobilization techniques in order to have the advantage of operating the PF process in the exponential growth phase for an infinite period of time and protect the culture strains from the inhibitory effects of chemicals that might be present in influent (Zhu et al. 1999a, 1999b; Chen and Chang 2006; Liu et al. 2011). However, a major limitation in cell-immobilized PF systems is the penetration and transmission of light through the immobilization media. Also, the cell immobilization technology might not be practical when the PNSB cells are required to be harvested for PHB production.

Zhu et al. (1999a) used cationic polyelectrolytes, such as chitosan, poly-L-lysine (PLL), polyethyleneimine (PEI), and trimethylammonium glycol chitosan iodide (TGCI), to entrap *R. sphaeroides* in order to prevent the inhibitory effect of NH_4^+ on H_2 production. In another study by Chen and Chang (2006), a small amount of solid carrier, for example, activated carbon, silica gel, or clay, was used for immobilization of *R. palustris* WP3-5 cells. The results of the study showed 67.2%–50.9% and 37.2%–32.5% increases in the H_2 production rate and H_2 yield, respectively, when clay and silica gel were used. Similarly, Zhu et al. (1999b) demonstrated that the immobilization in agar gels could protect the PNS strains from inhibitory effects of the ammonium ion in photofermentative hydrogen production from tofu wastewater using *R. sphaeroides*.

8.4.2 Carbon and Nutrient Sources

The substrate types and their concentration used in PF can influence the H_2 production rates and yields. Han et al. (2012) studied the effect of different carbon sources and their concentrations on the photo-H_2 production using a batch culture of *R. sphaeroides* RV. The substrates used were either individual substrates, such as acetate, propionate, butyrate, lactate, malate, succinate, ethanol, glucose, citrate, or sodium carbonate, or mixed carbon sources, such as malate and succinate, or lactate and succinate. The results of the study showed that the H_2 production for the mixed substrates is higher (794 mmol H_2/mol substrate for 2.02 g/L lactate and 2.0 g/L succinate) than for a single substrate (424 mmol H_2/mol substrate for 0.8 g/L sodium propionate). This makes PF prominent for the application in the treatment of DFEs that typically contain more than one organic acid (Rai et al. 2014b; Nasr et al. 2015).

8.4.2.1 Effect of Organic Loading Rate and Hydraulic Retention Time

The organic loading rate (OLR) and the hydraulic retention time (HRT) strongly affect the performance of PBRs as they determine the substrate degradation efficiency and the hydrogen production rate. Venkata Mohan et al. (2008) studied the effect of different OLRs on photo-H_2 production and substrate degradation efficiency. The synthetic wastewater gave the maximum substrate degradation efficiency (1.4 kg COD/m³/day) at an OLR of 2.45 kg COD/m³/day, while higher specific H_2 production (19.29 mol H_2/kg COD$_{removed}$) was achieved at an OLR of 1.4 kg COD/m³/day with 45% COD removal. In another study, Tawfik et al. (2014) studied the effect of varying OLR (3.2–16 kg COD/m³/day) using mixed PNSB cultures, which resulted in maximum H_2 production at an OLR of 6.4 kg COD/m³/day. Increasing OLR caused volatile fatty acid (VFA) accumulation, which might inhibit the PNSB. Therefore, inhibition of the nitrogenase activity resulted in decreasing H_2 production when the OLR was higher than 6.4 kg COD/m³/day (Tawfik et al. 2014). This is supported by another PF study carried out with acid-hydrolyzed wheat starch and a pure culture of *Rhodobacter* sp. (Kapdan et al. 2009). The results of the study showed that upon increasing the initial sugar concentration from 2.2 to 13.0 g/L, the H_2 yield (H_2Y) increased, with a maximum H_2Y achieved at 5 g/L (143.5 mL H_2/g COD).

A range of optimum HRTs, varying from 2.5 h (Tawfik et al. 2014) to 3 days (Ozmihci and Kargi 2010), have been reported in the literature for achieving higher photo-H_2 production in a continuous reactor. Tawfik et al. (2014) found an optimum HRT at 2.5 h (0.97 ± 0.12 LH_2/gCOD$_{removed}$/day) when studying a range of HRTs from 0.9 to 4.0 h. They also observed the improvement in removal efficiency of butyrate and lactate when the HRT was increased. Similarly, another study carried out with mixed PNSB by Z. Zhang et al. (2015) showed that varying HRTs from 12 to 72 h significantly affected the H_2Y, with the highest H_2Y of 482.4 mmol H_2/L obtained at an HRT of 36 h. In contrast, Ozmihci and Kargi (2010) obtained the highest H_2Y and production rate at an HRT of 72 h during PF of DFEs using *R. sphaeroides*. The differences in optimum HRT may be attributed to differences in PNSB strains, substrate concentration, C/N ratio, and other operating conditions, such as pH, temperature, and light intensity.

8.4.2.2 *Effect of C/N Ratio*

The C/N ratio plays an important role in the growth of PNSB and photo-H_2 and PHB production. However, higher levels of nitrogen inhibit H_2 production, while higher C/N ratios enhance the production of PHB (Eroglu et al. 1999; Koku et al. 2003; Argun et al. 2008; Waligórska et al. 2009). A low C/N ratio can result in the accumulation of ammonia, which inhibits the nitrogenase and thus the H_2 production process. Therefore, it is always desirable to have nitrogen-limited conditions in the PBR. Due to the nitrogen requirements for bacterial photosynthetic metabolism and inhibition of nitrogenase at higher ammonium concentrations, there is a trade-off between the minimum amount of nitrogen for bacterial growth and noninhibiting levels.

A range of C/N ratios has been reported in the literature, that is, from as low as 8 to as high as 120. Eroglu et al. (1999) reported the optimum C/N ratio of 30 with 15 mM to 2 mM (malic acid to glutamic acid) for the maximum hydrogen production rate. In another study, Boran et al. (2010) reported a C/N ratio of 38.6 with 40 mM acetic acid and 2 mM sodium glutamate in PF by *R. capsulatus* in a solar tubular PBR under outdoor conditions. Similarly, Argun et al. (2008) reported the optimum total VFA and NH_4^+-N concentrations of 2350 and 47 mg/L, respectively, for increasing the H_2 production by *R. sphaeroides* strains. In another study (Eroğlu et al. 2009), the highest H_2 production potential of 19.9 m^3 H_2/m^3 was obtained from olive mill wastewater with the highest C/N molar ratio of 73.8.

Waligórska et al. (2009) found that accumulation of PHB increased 30-fold when the C/N ratio increased from 6 to 120 in *R. sphaeroides*. However, the amount of PHB accumulation mainly depends on the PNSB strains and the other process operational conditions (De Philippis et al. 1992; Montiel-Corona et al. 2015). As PHB biosynthesis is a H_2-competing pathway, its concomitant production with H_2 could raise future interests, as PHB possesses economic value as a biodegradable polymer (Koku et al. 2002). Some of the results from previous studies on H_2 and PHB production in PF processes are summarized in Table 8.3.

TABLE 8.3   Comparison of Hydrogen and PHB Production by Different Isolated Strains and Enriched Mixed Cultures of PNS via Photofermentation of Various Carbon Sources

Microbial Inoculum Sources	Main Carbon and Nitrogen Source (COD in g/L)[a]	C/N Ratio	Light Intensity (Lux)	PHB (% Dry Cell Weight)	Volumetric H$_2$ Yield (mL H$_2$/L)	Reference
R. sphaeroides 17023 (wild type)	30 mM acetate and 7 mM glutamic acid (1.9)	8.6	1,500 (incandescent light)	70.0	0	Hustede et al. 1993
R. sphaeroides 17023 (wild type)	30 mM lactate and 7 mM glutamic acid (2.8)	12.86	1,500 (incandescent light)	24.0	2,310	Hustede et al. 1993
Enriched photoheterotrophic culture	DFEs (11.61 g/L butyric, 1.76 g/L propionic, and 1.01 g/L acetic acids and 0.78 g/L total ammonia) (24.8)	10.63	3,000 (LED and halogen lamps)	5.0	1,478 ± 17	Montiel-Corona et al. 2015
R. capsulatus				29.0	1,252 ± 20	Montiel-Corona et al. 2015
R. sphaeroides O.U.001 (DSM5648)	0.54 g/L malic acid and 0.8 g/L sodium glutamate in sugar refinery wastewater (30% v/v in medium) (NA[b])	–	200 W/m²	70.4	648	Yiğit et al. 1999
R. sphaeroides AV1a	DFEs of food waste (3.5)	–	40,000	32.5 ± 3	914.1 ± 8	Ghimire et al. 2016
R. sphaeroides strain RV	40 mM acetate only with nitrogen limitation (1.9)	–	5,000 (incandescent light)	38.0	0	Khatipov et al. 1998

Note: –, data not available.

[a]COD values are calculated from the theoretical oxygen demand of main carbon sources.

8.4.2.3 Micronutrients

Microorganisms need different micronutrients, such as iron and nickel, for their metabolism and growth. The PF process relies on the photosynthetic electron transport systems from which bacteria obtain their energy (Figure 8.2). The constituents of the electron transport systems, such as cytochromes, are Fe protein complexes, and PNSB strains have 24 Fe atoms in each nitrogenase (Zhu et al. 2007). Another electron carrier, ferrodoxin, also contains Fe. Thus, Fe limitation can influence the metabolism of PNSB and the production of H_2.

Uyar et al. (2009) found that the hydrogen yield increases from 0.3 to 1.0 $L/L_{culture}$ when iron is added to micronutrient. They suggested 0.1 mM ferric citrate as an optimum concentration for hydrogen production. Similarly, Zhu et al. (2007) studied the effect of ferrous ion (0–3.2 mg/L) on PF using *R. sphaeroides* and found that the photo-H_2 production was significantly suppressed when Fe^{2+} was limited. The H_2 production increased when increasing the Fe^{2+} concentration and reached a maximum at the concentration of 2.4 mg/L. In another study, Rai et al. (2014b) studied the effects of Ni^{2+}, Fe^{2+}, and Mg^{2+} on the PF of cheese whey for H_2 production, and showed significant effects of Ni^{2+} and Fe^{2+} supplementation on H_2 yields. However, the presence of nickel might also enhance the hydrogenase activity, which takes up the H_2 produced by the nitrogenase activity, thus decreasing the net H_2 production yield (Li and Fang 2009).

8.4.2.4 Presence of Bicarbonate

Some studies have shown that the addition of bicarbonate and carbonate ions enhances the H_2 production in PF (Takabatake et al. 2004; Montiel-Corona et al. 2015). Bicarbonate and carbonate function as electron acceptors and enhance the utilization of butyric and propionic acids, while their absence unbalances the oxidation–reduction potential, resulting in decreased H_2 production. Takabatake et al. (2004) reported that the presence of carbonate improves the assimilation of ammonium (NH_4^+) and VFAs. They also observed that the uptake of 1 mM acetate releases 0.37–0.40 carbonate. However, this internal production of carbonate was not enough to promote butyrate and propionate consumption, which are more reductive than bacterial cells. For PNSB growth on butyrate, each mole of butyrate requires 0.7 mol CO_2 (Montiel-Corona et al. 2015).

8.4.3 Environmental Conditions

8.4.3.1 Effect of Light Intensity and Wavelength

The light conversion efficiency (η) varies for different PNSB strains because of their different light harvesting antenna pigments; thus, they have a different PAR range. However, η also depends on the light intensity, illuminated area of the PBR, reactor design, and other operational conditions of the PF process. Generally, the intensity of light has a positive influence on the H_2 production. There are some studies dedicated to assess the effect of the light intensity on the growth and H_2 production by PNSB (Koku et al. 2002; Uyar et al. 2007; Sevinç et al. 2012; Androga et al. 2014; Akman et al. 2015).

Uyar et al. (2007) studied the effect of light intensity, light wavelength, and the illumination protocol on the growth and H_2 production by *R. sphaeroides* O.U.001 in PBRs (Figure 8.4). The hydrogen production increased with increasing light intensity, and the highest production was reached at 270 W/m^2. The results also showed a decrease in photoproduction of hydrogen by 39% when there was a lack of infrared light (750–950 nm wavelength). The substrate conversion efficiency was increased and hydrogen production was stimulated when the light was illuminated after inoculation, and no hydrogen was produced during the dark periods.

Sevinç et al. (2012) studied the effects of temperature (20°C, 30°C, and 38°C) and light intensity (1500, 2000, 3000, 4000, and 5000 lux) on the kinetic parameters and hydrogen production in PF of acetic and lactic acids using *R. capsulatus*. The results of the study reported the maximum hydrogen production at 5000 lux for 20°C and 3000 lux for 30°C and 38°C. In a more recent study, Androga et al. (2014) established an optimal light intensity and temperature of 287 W/m^2 (4247.6 lux) and 27.5°C, respectively, in PF tests carried out using *R. capsulatus* DSM1710 in a medium containing acetate, lactate, and glutamate. In another recent study, Akman et al. (2015) reported an optimum light intensity of 263.6 W/m^2 (3955 lux) in a PF study carried out with acetate as the carbon source and *R. capsulatus*, which is in accordance with the study from Androga et al. (2014).

Future development of PF systems requires an economical solution to provide the sources of light, so that outdoor systems utilizing natural sunlight become a practical option. Therefore, research interests have been growing to exploit the natural sunlight in PF processes (Androga et al. 2011, 2012a; Avcioglu et al. 2011; Montiel-Corona et al. 2015). Even though sunlight cannot ensure continuous light conditions, there are some studies that have shown that the dark and light cycles might not have significant effects on photo-H_2 production (Li et al. 2011) or have positive effects on H_2 production, depending on the exposure duration of the light and dark conditions

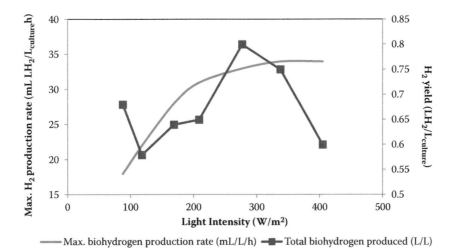

Figure 8.4   Effect of light intensity on biohydrogen production by *R. sphaeroides* O.U.001. (Based on the data obtained from Uyar, B. et al., *Int. J. Hydrogen Energy*, 32(18), 4670–7, 2007.)

(Sargsyan et al. 2015). Montiel-Corona et al. (2015) reported a 40.25% reduction in H_2 yields during PF using mixed PNSB in comparison with indoor conditions. However, H_2 yields obtained from outdoor reactors can be comparable to those under indoor conditions (Androga et al. 2011). In addition to the type of light source, photofermentative H_2 production also depends on other operating conditions of the PBRs, such as mixing conditions that affect the distribution of light, culture temperature, and pH. Furthermore, harnessing the natural light in upscale applications of PF might reduce the cost of long-term PBR operation.

8.4.3.2 Culture Temperature and pH

The operating temperature of a culture is one of the important parameters that affects the bacterial metabolism or metabolic pathways, as well as substrate conversion efficiency and thus H_2 production. Basak and Das (2007) reported 31°C–36°C as the optimum temperature range for *Rhodobacter* sp., while Androga et al. (2014) reported 26.8°C as the optimum culture temperature for a higher H_2 yield. Moreover, culture pH affects the biochemical reactions, as it determines the ionic form of the active sites for enzymatic activity (Chen et al. 2011). PF studies have been carried out in the pH range varying between 5.5 and 7.5 (Tables 8.2 and 8.4). Akroum-Amrouche et al. (2011) reported an optimum pH of 7.5 (± 0.1) for the H_2 production by *R. sphaeroides*, while Nath and Das (2009) reported an optimum H_2 production at pH 6.5 for the same PNSB species. This difference of change in optimum pH can be attributed to the difference in substrate type used in PF experiments, as lactate was used as a sole carbon source in the former, while DF spent medium was used in the latter study. In another study, Koku et al. (2002) reported an optimum pH of 7.1–7.3 for the activity of the nitrogenase enzyme, while the range of 6.5–7.5 is optimum for the activity of the hydrogenase enzyme.

During most of the PF tests, the pH has shown an increasing trend, which could be due to PHB production (Khatipov et al. 1998; Nath and Das 2009). Eroglu et al. (1999) reported a slight decrease in pH during the bacterial growth phase and a pH increase during H_2 production. The effluents from DF are generally in the acidic pH range (Ghimire et al. 2015), and are required to be adjusted to a pH range of 6.5–7.5 to ensure the optimum operating conditions in the PF process. However, the range of optimum pH seems to be dependent on the PNSB species. Some studies by Tawfik et al. (2014) and Tao et al. (2008) have shown the feasibility of H_2 production by mixed PNSB at pH 5.5–6.0, which is generally an ideal pH range of DFEs obtained from DF processes.

8.4.3.3 Effect of Mixing

Mixing is required in PBRs to keep the PNSB biomass suspended and to uniformly distribute the substrates and nutrients in the culture medium. Moreover, since the light source is nonmobile, mixing would only ensure the uniform distribution of light for the suspended microorganisms throughout the PBRs, thus avoiding light gradients. It also helps to maintain sufficient mass transfer, which generally includes the exchange of gases, that is, H_2 and CO_2. Akroum-Amrouche et al. (2011) found unstable H_2 production with a 13.0% and 60.8% reduction of the average and maximum

TABLE 8.4   Variation of Different Operational Parameters in PF Studies

PNS Strains	Carbon (and Nitrogen) Source	Culture Type (Reactor Type)	Culture Temperature (°C)	pH	Light Intensity (lux)	Maximum H_2 Yield (mL H_2/g COD)[b]	Maximum H_2 Production Rate (mL H_2/L/h)	Reference
R. sphaeroides O.U.001 (DSM586)	DFEs of glucose	Batch	30	6.4	38–50[a] (tungsten lamp)	–	26.4	Afsar et al. 2011
R. capsulatus (DSM1710)	DFEs of potato steam peels' hydrolysate	Batch	30	6.4	38–50[a] (tungsten lamp)	–	12.3	Afsar et al. 2011
R. capsulatus (hup⁻)	Acetic acid (glutamate)	Continuous Tubular PBR	<40	<8.0	Natural sunlight (outdoor conditions)	122.5	8.9	Boran et al. 2012
R. sphaeroides CIP 60.6	Lactate (glutamate)	Batch	30	7.0	4,500–8,500 (tungsten lamp)	–	39.8	Akroum-Amrouche et al. 2011
R. capsulatus YO3 (hup⁻)	Acetate (glutamate)	Fed-batch panel PBR	35	7.0	Natural sunlight (outdoor conditions)	–	11.4	Androga et al. 2011
R. palustris WP3-5	Formic, acetic, butyric, and lactic acids (glutamate)	Continuous-column PBR	28–35	6.8	4,000–7,000	–	13.2	Lee et al. 2011
R. sphaeroides O.U.001	Malate (glutamate)	Batch annular PBR	32	6.8	5.1[a] (equivalent to 15 W/m² internal illumination)	1,050	6.5	Basak and Das 2009

(Continued)

TABLE 8.4 (CONTINUED)　Variation of Different Operational Parameters in PF Studies

PNS Strains	Carbon (and Nitrogen) Source	Culture Type (Reactor Type)	Culture Temperature (°C)	pH	Light Intensity (lux)	Maximum H_2 Yield (mL H_2/g COD)[b]	Maximum H_2 Production Rate (mL H_2/L/h)	Reference
Mixed culture	Acetate and glucose as cosubstrate (glutamate)	Batch	34	6–7	4,000 (fluorescent light)	78.6[c]	–	Srikanth et al. 2009
	Butyrate and glucose as cosubstrate (glutamate)					74.5[c]	–	
R. capsulatus (DSM155)	DFEs of Miscanthus hydrolysate (with iron addition)	Batch	30–33	6.6–6.8	4,000	–	19.0	Uyar et al. 2009
R. sphaeroides O.U.001 (DSM5864)	Malate	Flat-panel PBR	32	6.8	50[a] (tungsten lamp)	1,073.3	10.0	Eroglu et al. 2008
R. palustris WP3-5	Butyrate (glutamic acid)	Batch	32	7.1	10,000 (tungsten lamp)	803.6	24.9	Chen et al. 2007

Note: –, data not available.

[a] Calculated using the available data and conversion values from http://www.egc.com/useful_info_lighting.php.

[b] COD values are calculated from the theoretical oxygen demand of the main carbon sources.

[c] mL H_2/g COD/day.

H_2 production rates when mixing was stopped during the exponential phase of PF. In another study, Li et al. (2011) reported that mixing during the H_2 production phase of the PNSB stationary growth phase is vital for higher H_2 yields during the exponential cell growth phase. Moreover, the type of mixing system may also affect the photo-H_2 production performance. Z. Zhang et al. (2015) showed that baffled PBRs can outperform magnetic-stirred PBRs, as supported by higher H_2 yields, as well as faster cell growth and substrate conversion. This higher H_2 production can be attributed to enhanced gas transfer and distribution of light in the PBRs due to good mixing conditions.

8.4.3.4 Inhibition of Photo-H_2 Production

Nitrogenase plays an important role in hydrogen generation. Thus, the presence of chemical substances that disrupt the nitrogenase activity decreases the photo-H_2 production. Koku et al. (2002) reported that the presence of N_2 and NH_4^+ inhibits the H_2 production. Also, CO, EDTA, and O_2 are likely to inhibit the nitrogenase activities. Similarly, an elevated level of CO_2 inside the reactor inhibits photo-H_2 production, while lower levels (4%–18% w/v) favor the growth phase of PNSB, and thus H_2 production (see Section 8.4.2). Furthermore, a lower C/N ratio does not favor photo-H_2 production, as it could result in the accumulation of ammonium and inhibition of nitrogenase in a PF process for H_2 production.

8.5 PBR Systems

8.5.1 PBR Configurations

The design considerations of PBRs for photo-H_2 production are similar to those of PBRs for algal biomass production. However, anaerobic conditions are required for the PF process using PNSB. Most of the published reviews on the design of PBRs for biohydrogen production are based on bioreactors for algal biomass production (Akkerman et al. 2002; Carvalho et al. 2006; Dasgupta et al. 2010). In some more recent works, Adessi and De Philippis (2014) and Chen et al. (2011) have summarized the knowledge on the design, illumination, and culture strategies of PBR systems aimed at enhancing photo-H_2 production with PNSB.

The most common reactor types reported in the literature are presented in Figure 8.5. More insight has been provided on the performance of different reactors, with more elaboration on tubular and flat-panel reactors, as these reactor configurations have been the subject of major interest because of their practicality in scaled-up PF processes.

8.5.1.1 Plate Reactors

Plate reactors are flat panels that consist of a rectangular transparent box with a depth varying between 1 and 5 cm (Akkerman et al. 2002). These reactors have received research attention for photo-H_2 production because of their large illumination area, possibilities of scaling up, and suitability in outdoor conditions. Flat-plate PBRs are

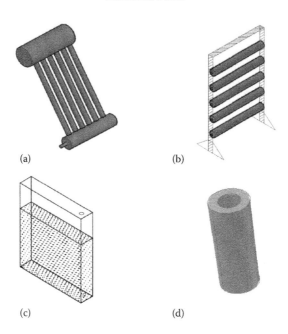

(a)   (b)

(c)   (d)

Figure 8.5 Schematic representation of PBR systems for PF: (a) nearly horizontal tubular PBR, (b) tubular (fence-type) PBR, (c) flat-panel PBR, and (d) anular-type PBR.

constructed with cheap materials, which are generally transparent to achieve the maximum utilization and conversion of solar energy.

Eroglu et al. (2008) investigated the performance of an 8 L flat-plate PBR under outdoor operating conditions using a culture of *R. sphaeroides* O.U.001. Among the different carbon sources, such as malate, lactate, acetate, and olive mill wastewater, used in the PF, the highest hydrogen production rate (10 mLH$_2$/L/h) was reached with malate as the carbon source, and formate was found to be the dominant end product. Ugwu et al. (2008) reported the following potential problems that flat-plate systems can face during scale-up:

- Requirement of many compartments and support materials
- Difficulty in operational temperature control
- Wall growth resulting in reduced light penetration

8.5.1.2 Tubular Reactors

Tubular PBRs contain a long transparent tube with a length ranging from 10 to 100 m and diameters ranging from 3 to 6 cm (Akkerman et al. 2002). These PBRs are one of the most suitable reactors for outdoor conditions. Generally, tubular PBRs are constructed with transparent glass or plastic tubes. The culture is recirculated with a mixing system (such as a pump) to provide efficient mass transfer and equal light distribution.

Boran et al. (2010) successfully developed and demonstrated a pilot-scale (80 L) tubular PBR for PF of acetate using *R. capsulatus* in outdoor operating conditions (during winter seasons) in Ankara (Turkey). The PBR gave an average molar productivity of 0.31 mol $H_2/m^3/h$ during daylight hours, and the gas contained 99% hydrogen and 1% carbon dioxide by volume. The system provided an overall hydrogen yield of 0.6 mol H_2/mol acetate, and the H_2 production with respect to the total illuminated surface area amounted to 0.112 mol H_2/m^2/day.

Ugwu et al. (2008) reported some limitations that tubular PBRs face during scale-up:

- Difficulty in operational temperature control
- Fouling and growth on the walls of the tubes
- Large space requirements

One of the major problems during the scaling up of tubular PBRs is the decrease in the ratio of the illumination surface to volume because of the increase in the tube diameter. This causes a decrease in light intensity (light shading effect) for the cells at the lower part of the tube, which negatively affects the cell growth. However, a good mixing system also provides an efficient light distribution (Ugwu et al. 2003).

8.5.1.3 Vertical-Column Reactors

Vertical-column reactors have been the subject of research for algal biomass production, as they are compact, have low cost, and are easy to operate (Ugwu et al. 2008). Bubble columns, airlift reactors, and annular column reactors are common vertical-column PBR configurations (Posten 2009). Bubble column reactors have a larger diameter than tubular reactors and are frequently used indoors (at a larger lab scale) or outdoors. Because of the larger diameters in these reactors, darker zones are created at the center of the column, which might be disadvantageous for photosynthetic bacterial growth. Besides these three major reactor types, laboratory-scale PF research has been carried out in internally illuminated reactors (Chen et al. 2010). Dasgupta et al. (2010) have described the possibility of using different configurations, such as torus-shaped and helical reactors.

The concept of an annular column reactor aims to overcome the problem associated with the central darker zones in bubble column reactors (Posten 2009). The major advantages of this reactor configuration are high mass transfer rate, good mixing conditions with less shear on bacterial cells, low energy consumption, and the potential for industrial application. However, the small illumination surface makes this configuration less competitive than other counterparts.

8.5.1.4 Comparison between Panel and Tubular PBRs

Table 8.5 compares studies done in various configurations of PBRs. Flat-panel and tubular reactors have the highest theoretical efficiencies and have been used at the pilot scale under outdoor conditions (Eroglu et al. 2008; Boran et al. 2010; Gebicki et al. 2010). These studies opened perspectives for scaling up these two promising PBRs for photo-H_2 production using PNSB cultures. In some lab-scale studies, higher H_2 productivities were obtained with flat-panel PBRs, while some studies with tubular

TABLE 8.5 Comparison of Tubular and Plate PBRs under Outdoor Conditions

PNS Strains	Carbon Source	PBR Type	Volume (L) (Dimension)	Operations Conditions	Maximum H_2 Yield	Maximum H_2 Production Rate (L H_2/m³/h)	Productivity per Illuminated Surface Area (L H_2/(m²-day)	Productivity per Ground Area (L H_2/[m²-day])	Light Conversion Efficiency[a]	Reference
R. capsulatus DSM155	Acetate, sodium lactate, and glutamate	Flat panel	4 × 25	Summer (Aachen, Germany)	–	12.3	3.69	29.52	0.20%	Gebicki et al. 2010
R. capsulatus DSM156	Acetate, sodium lactate, and glutamate	Tubular	60 (0.12 m diameter and 0.65 m length)	Summer (Aachen, Germany)	–	6.3	3.35	3.35	0.19%	Gebicki et al. 2010
R. capsulatus DSM1710	Acetate, lactate, and glutamate	Tubular	80	Winter (Ankara, Turkey)	15%	6.9	2.46	1.74	1%	Boran et al. 2010
R. capsulatus YO3 (hup⁻)	Acetate and glutamate	Flat panel	4	Summer (Ankara, Turkey)	53%	11.4	1.5	4.93	–	Androga et al. 2011
R. capsulatus YO3 (hup⁻)	Acetate and glutamate	Tubular	90	Outdoor conditions (Ankara, Turkey)	35%	8.9	9.6	6.7	0.20%	Boran et al. 2012
A. platensis M2 (cyanobacteria)	CO_2	Tubular	34	Summer (Florence, Italy)	–	1,176.0	738.0	–	5.6%[b]	Tredici and Zittelli 1998
A. platensis M2 (cyanobacteria)	CO_2	Flat panel	5.4	Summer (Florence, Italy)	–	1,017.3	686.6	–	4.8%[b]	Tredici and Zittelli 1998

[a] The light conversion efficiency was calculated using Equation 8.10.
[b] The efficiency of the cultures was calculated by multiplying the reactor productivity by the mean enthalpy value of the biomass of *A. platensis* M2 cultivated outdoors (21.56 kJ/g) and dividing by the mean visible solar energy input on the PBR surface (14.08 MJ/day).

PBRs have shown good performance under outdoor light conditions. Moreover, tubular PBRs are easier to manage and scale up.

Photo-H_2 production through PF can be a promising technology for clean energy recovery. In addition, recovery of PHB can be of further interest. To establish PF as the posttreatment of DFEs, more research needs to be performed for improving the system efficiency through optimization of different operating parameters. The system efficiency can be improved by providing optimum culture conditions and bioreactor design. The PF systems have been presented as the bottlenecks in the integrated DF-PF process because of their higher production cost. Thus, innovative low-cost mixing, heating, and cooling systems need to be explored, and PBR designs for improving the ratio of the surface area to volume (A/V) require future research. Moreover, PHB can add economic value to the PF process. Using mixed PNSB to utilize the conversion efficiencies of different microbial consortia can give an economic advantage by reducing the cost of H_2 production.

8.5.2 Design Considerations for PBRs

In addition to the physical parameters, such as quantity of light penetrating into the bioreactor, a good PBR design should consider various physicochemical parameters, such as pH, temperature, dissolved oxygen and CO_2, shear due to agitation, C/N ratio, carbon sources, and availability of nutrients. As mentioned earlier, these parameters influence various biochemical pathways and ultimately the H_2 production in PBRs.

A general consideration to achieve a good design of PBRs as reported by Dasgupta et al. (2010) includes the following physicochemical parameters, which affect the performance of PBRs:

- High light penetration into PBRs
- High A/V ratio (higher illumination area)
- Temperature and pH control
- Good mixing system
- Better gas exchange or mass transfer
- Transparency and durability of the materials

8.5.2.1 Ratio of Surface Area to Volume
The amount of light absorbed by a reactor system is a limiting factor in PBR systems. The A/V ratio is one of the important parameters to be considered during the design of PBRs, as it determines the amount of light entering into the system. The higher the A/V ratio, the larger will be the surface area for receiving light for growth and metabolism. Therefore, the A/V ratio can be directly correlated with cell concentration and the volumetric productivity of the system (Dasgupta et al. 2010).

Gebicki et al. (2009) compared hydrogen productivities of a flat-panel (A/V ratio of 20 m^{-1}) and an inclined horizontal tubular (A/V ratio of 15.38 m^{-1}) PBR with respect to an illuminated surface area and ground area occupied by the reactor. The mean hydrogen productivity of the flat-panel reactor was 1250 mL $H_2/(m^2_{illuminated\ surface}/day)$, while that of the tubular reactor was 1100/L $H_2/(m^2_{illuminated\ surface}/day)$. The illuminated area per unit ground area occupied by the panel reactor was 8.9 times higher than that of the tubular reactor, which gives the economic edge of the comparison. However, a fenced-type tubular PBR (Figure 8.5) could be of research interest in the future, as this reactor configuration occupies less space than inclined horizontal tubular PBRs.

8.5.2.2 Mixing Systems

Mixing systems in PBRs could include pumping, mechanical stirring, and airlift mixers. Ugwu et al. (2003) proposed a static mixer for tubular bioreactors. The selection of the type of mixing system is important, as the pumps used for mixing or recirculation exert shear forces that might be harmful to PNSB. Another disadvantage of the mixing system is the additional cost due to the required energy for its operation.

8.5.2.3 Construction Materials

The selection of materials during the construction of PBRs determines not only the economy, but also the performance of the system. Several factors should be considered while selecting the construction materials. PBRs can be constructed from glass, polyvinyl chloride (PVC) material, low-density polyethylene (LDPE), poly-methyl methacrylate (PMMA), and fiberglass. Dasgupta et al. (2010) reported the following considerations for the selection of the construction material for PBRs:

- High transparency
- Durable and low cost
- Nontoxic to PNS strains and resistant to chemicals and metabolites produced by the PNS strains
- High weathering resistance and easiness in cleaning

The results of the net energy analysis of three different materials—glass, LDPE, and PMMA—done by Gebicki et al. (2010) suggest the use of LDPE for the construction of tubular and panel PBRs.

8.6 Mathematical Modeling of Growth and Product Kinetics of PNSB

Knowledge on the kinetics of the biological process becomes vital to have a better design and control of the process. The strong influence of operational parameters, such as light intensity and substrate concentrations, on photofermentative H_2 and PHB synthesis has been demonstrated (Hustede et al. 1993; Uyar et al. 2007; Han et al. 2012; Wu et al. 2012; Androga et al. 2014). However, very limited work has been done on the kinetic analysis of the PF process (Koku et al. 2002; Gadhamshetty et al. 2008; D. Zhang et al. 2015).

8.6.1 Biomass Growth

Few mathematical models have been proposed to study growth kinetics of PNSB cultures. A theoretical cell growth rate can be expressed as

$$\frac{dX}{dt} = \mu X - mX \tag{8.11}$$

where X is the cell dry weight concentration (g/L), m is maintenance coefficient for biomass (decay rate), and μ is the specific growth rate (h^{-1}). Gadhamshetty et al. (2008) proposed the Monod equation to provide the expression for μ to model the growth curve in a batch of PBRs with the assumption that sufficient light and an optimal C/N ratio are available under stressful nitrogen concentrations. The proposed model simulates the biomass growth under substrate-limited conditions as

$$\frac{dX}{dt} = \mu X = \left(\frac{\mu_m S}{K_s + S} \right) X \tag{8.12}$$

where the specific growth rate μ (h^{-1}) depends on both the maximum specific growth rate μ_m (h^{-1}) and the half saturation constant K_S (mg/L).

However, the growth curve obtained for *R. sphaeroides* O.U.001 deviated from the Monod model (Koku et al. 2003). Equation 8.12 needs to include the substrate inhibition and inhibition due to higher biomass concentration. Moreover, the inhibition from higher substrate levels could be due to osmotic stress and/or the presence of one or more unknown inhibitors, such as pigments (Gadhamshetty et al. 2008). Besides, a higher biomass concentration reduces the light intensity inside the PBR, causes self-shading effects, and limits the substrate diffusion, which in turn affects the hydrogen production. Thus, the specific growth rate (μ) in Equation 8.12 is modified in Equation 8.13 to include the two inhibitory effects:

$$\mu = \frac{\mu_m S}{K_s + S + \dfrac{S^2}{K_{Xi}}} \left(1 - \frac{X}{X_m} \right) \tag{8.13}$$

The inhibitory effect due to biomass concentration is provided by a logistic model. The term X_m is the maximum cell dry mass concentration at which growth will cease. The specific growth rate in Equation 8.13 is further modified to include the effect of the light exposure on PNSB. The modification included the declining effect of excess light on biomass growth, as the surplus absorbed light energy may result in damage and degradation of the reaction center involved in the photosynthetic process. The final equation is expressed as

$$\mu = \frac{\mu_m S}{K_s + S + \dfrac{S^2}{K_{Xi}}}\left(1 - \frac{X}{X_m}\right)\left(\frac{I}{K_{XI} + I + K_I I^2}\right) \tag{8.14}$$

where K_I is the light inhibition constant of cell formation (m^2/W), K_{XI} is the light saturation constant of cell formation (W/m^2), and K_{Xi} is the substrate inhibition constant of cell formation (g/L), which depends on the type of substrate and PNSB strains or culture. The smaller the value of K_I, the larger is the inhibition effect of light on PNSB growth.

8.6.2 Consumption of Substrate

The Contois model can be used to describe the consumption of the substrate:

$$\frac{dS}{dt} = -\frac{\mu_m S}{Y(S + K_S X)} X \tag{8.15}$$

or

$$\frac{dS}{dt} = \frac{1}{Y_{\frac{X}{S}}} \mu X \tag{8.16}$$

8.6.3 Relation between Biomass Growth and Product Formation

Mu et al. (2006) used the modified Luedeking–Piret model to establish the relationship between product (P_i) formation, substrate (S) degradation, and biomass (X) growth for DF hydrogen production by mixed anaerobic cultures. The following Luedeking–Piret model can be used to describe the relationship between three parameters. The Luedeking–Piret model and its modified form can also describe the relationship between the formation of H_2 and PHB as products and biomass:

$$\frac{dP_i}{dt} = -Y_{\frac{P_i}{X}}\frac{dX}{dt} + \beta X \tag{8.17}$$

$$\frac{dP_i}{dt} = -Y_{\frac{P_i}{X}}\frac{dX}{dt} \tag{8.18}$$

where P_i is the concentration of product i and $Y_{pi/X}$ is the yield of product i with respect to biomass X.

Similarly, the formation of products with respect to consumption of substrate can be written as

$$\frac{dP_i}{dt} = -Y_{\frac{Pi}{S}} \frac{dS}{dt} \qquad (8.19)$$

where P_i is the concentration of product i and Y_{pi}/s is the yield of product i with respect to substrate S.

The growth of biomass can be expressed in relation to the substrate consumption as

$$\frac{dX}{dt} = -Y_{\frac{X}{S}} \frac{dS}{dt} \qquad (8.20)$$

where X is the concentration of the biomass and Y_X/s is the yield of biomass with respect to the substrate S.

On integrating Equation 8.20 from the initial concentration (S_0) to the final substrate concentration (S) and product (from initial concentration of 0 to final product concentration P_i), it is possible to write the following equations:

$$dP_i = -Y_{P_i} dS$$
$$\int_0^{P_i} dP_i = -Y_{P_i} \int_{S_0}^S dS \qquad (8.21)$$

$$P_i = -Y_{P_i}(S_0 - S) \qquad (8.22)$$

with $i = H_2$ and PHB.

These relationships can be applied to model the kinetics of substrate consumption, PNSB growth, and product formation (H_2 and PHB) in the PF process. Some reported values of kinetic parameters established in different PF processes are presented in Table 8.6.

TABLE 8.6   Some Reported Values of Kinetic Model Parameters

S.N.	Kinetic Parameters	Values	Reference
1.	μ_m (h^{-1})	0.17	Gadhamshetty et al. 2008
2.	K_S (g/L)	0.0912	Gadhamshetty et al. 2008
3.	$Y_{\frac{X}{S}}$ (g/g)	3.7	Eroglu et al. 1999
4.	K_I (m^2/W)	0.0975	Sasikala et al. 1991
5.	K_{XI} (W/m^2)	22	Gadhamshetty et al. 2008
6.	K_{Xi} (g/L)	7	Gadhamshetty et al. 2008

Note: S.N., serial number.

8.7 Future Perspectives

8.7.1 Economics

There are very few studies aimed at determining the economics of photo-H_2 production (Benemann 1997; HYVOLUTION 2011). Benemann (1997) presented an economic analysis of a conceptual two-stage process where microalgae are used to produce a carbohydrate-rich biomass cultivated in large open ponds and hydrogen will be produced in tubular PBRs. The paper reported the estimated overall total hydrogen production costs of \$9.5/GJ.

HYVOLUTION (2011) conducted a cost analysis of integrated thermophilic DF, followed by the PF process, with a 60 kg H_2/h (2 MW thermal power equivalent) production capacity, operating for 8000 h/year (PF operating for 10 h/day and 3000 h annually). In this integrated concept, the total cost per kilogram of biohydrogen is €56 (~\$75.15), with €47/kg H_2 for the tubular PBR-based PF process. This cost of production is much higher than that of hydrogen from coal (\$0.36–\$1.83/kg H_2) and natural gas (\$2.48–\$3.17/kg H_2) (Bartels et al. 2010; HYVOLUTION 2011). However, HYVOLUTION (2011) reported a cost of €21/kg H_2 for a stand-alone thermophilic DF system, which is lower than the production cost of a combined system (i.e., €56/kg H_2); thus, the PF process needs improvement to lower the unit production cost of H_2.

The light conversion efficiencies of the PF play an important role in determining the economics of photohydrogen production. In addition, the substrates and the PNSB strains are also crucial factors. The selection of PBRs also influences the capital and operational cost and, in the end, the unit cost of the photohydrogen production (HYVOLUTION 2011).

8.7.2 Integration with Dark Fermentation Process

PF can be applied as a posttreatment stage on DFEs, which mostly contains organic acids and alcohols (Figure 8.6). The integrated DF-PF process has been demonstrated by several studies (Rai et al. 2014a; Tawfik et al. 2014; Yang et al. 2015). DF has the unique capability to utilize a wide range of complex waste biomass that can ensure the future supply of feedstock, and combining the two processes (DF + PF) can provide the complete conversion of organic substrate in addition to enhanced H_2 yields. Typical chemical reactions of conversion of organic acids produced in mixed-type fermentation to photo-H_2 are presented in Equations 8.4 through 8.7.

Redwood et al. (2008) reviewed different possible integration strategies for coupling DF-PF processes. In general, DF-PF systems can be integrated in three possible ways: (1) utilizing DFEs produced in PF systems, (2) cultivating dark and photofermentative microorganisms in one reactor system (Liu et al. 2010; Chandra et al. 2015), or (3) separating the two systems by a physical barrier, such as a membrane (Redwood et al. 2011; Liu et al. 2015). DF followed by photo-H_2 production is well studied by many researchers (Ghimire et al. 2015).

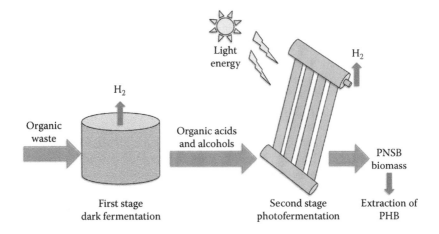

Figure 8.6   Sequential DF-PF processes for production of H_2 and PHB.

Depending on the process operating parameters, such as pH, substrate loading, and substrate type, DFE generally has an acidic pH (<6.0) and inhibiting levels of ammonia and organic acids. Therefore, the DFE requires pretreatment, such as pH adjustment, dilution, and removal of ammonia before feeding into a PF process. Ammonia concentrations exceeding 2–5 mM inhibit the photo-H_2 production (Argun et al. 2008; Lee et al. 2011). Therefore, substrates with a higher C/N ratio are usually preferred for PF. Depending on the DFE requirements, several ammonia removal strategies, such as stripping, treatment with natural zeolites, and membrane processes, can be applied (Androga et al. 2012b; Redwood et al. 2012b). However, most continuous dark fermentative processes lack high ammonia levels due to incomplete conversion of proteins or amino acids present in the substrates, making them ideal substrates for the PF processes.

Acknowledgment

The authors would like to acknowledge the Erasmus Mundus Joint Doctorate Programme ETeCoS[3] (Environmental Technologies for Contaminated Solids, Soils and Sediments) under the EU grant agreement FPA no. 2010–0009. This research was also supported by the project "Modular Photo-Biologic Reactor for Bio-Hydrogen: Application to Dairy Waste—RE-MIDA" from the Agriculture Department of the Campania Region in the context of the Programme of Rural Development 2007–2013, Measure 124.

References

Adessi, A., and R. De Philippis. 2014. Photobioreactor Design and Illumination Systems for H_2 Production with Anoxygenic Photosynthetic Bacteria: A Review. *International Journal of Hydrogen Energy* 39 (7): 3127–41.

Afsar, N., E. Özgür, M. Gürgan, S. et al. 2011. Hydrogen Productivity of Photosynthetic Bacteria on Dark Fermenter Effluent of Potato Steam Peels Hydrolysate. *International Journal of Hydrogen Energy* 36 (1): 432–38.

Akkerman, I., M. Janssen, J. Rocha, and R. H. Wijffels. 2002. Photobiological Hydrogen Production: Photochemical Efficiency and Bioreactor Design. *International Journal of Hydrogen Energy* 27: 1195–208.

Akkerman, I., M. Janssen, J. Rocha, and R. H. Wijffels. 2003. Photobiological Hydrogen Production: Photochemical Efficiency and Bioreactor Design. In Bio-methane and bio-hydrogen: Status and perspectives of biological methane and hydrogen production, ed. J.H. Reith, R.H. Wijffels and H. Barten, 124–145. Dutch Biological Hydrogen Foundation, Hague, the Netherlands.

Akman, M. C., T. H. Erguder, U. Gündüz, and İ. Eroğlu. 2015. Investigation of the Effects of Initial Substrate and Biomass Concentrations and Light Intensity on Photofermentative Hydrogen Gas Production by Response Surface Methodology. *International Journal of Hydrogen Energy* 40(15): 5042–49.

Akroum-Amrouche, D., N. Abdi, H. Lounici, and N. Mameri. 2011. Effect of Physico-Chemical Parameters on Biohydrogen Production and Growth Characteristics by Batch Culture of *Rhodobacter Sphaeroides* CIP 60.6. *Applied Energy* 88 (6): 2130–35.

Androga, D. D., E. Özgür, I. Eroglu, U. Gündüz, and M. Yücel. 2012a. Photofermentative Hydrogen Production in Outdoor Conditions. In Hydrogen Energy—Challenges and Perspectives, ed. D. Minic, 77–120. InTech.

Androga, D. D., E. Özgür, I. Eroglu, U. Gündüz, and M. Yücel. 2012b. Amelioration of Photofermentative Hydrogen Production from Molasses Dark Fermenter Effluent by Zeolite-Based Removal of Ammonium Ion. *International Journal of Hydrogen Energy* 37 (21): 16421–29.

Androga, D. D., E. Ozgur, U. Gunduz, M. Yucel, and I. Eroglu. 2011. Factors Affecting the Longterm Stability of Biomass and Hydrogen Productivity in Outdoor Photofermentation. *International Journal of Hydrogen Energy* 36 (17): 11369–78.

Androga, D. D., P. Sevinç, H. Koku, M. Yücel, U. Gündüz, and I. Eroglu. 2014. Optimization of Temperature and Light Intensity for Improved Photofermentative Hydrogen Production Using *Rhodobacter Capsulatus* DSM 1710. *International Journal of Hydrogen Energy* 39 (6): 2472–80.

Argun, H., F. Kargi, and I. Kapdan. 2008. Light Fermentation of Dark Fermentation Effluent for Bio-Hydrogen Production by Different *Rhodobacter* Species at Different Initial Volatile Fatty Acid (VFA) Concentrations. *International Journal of Hydrogen Energy* 33 (24): 7405–12.

Avcioglu, S. G., E. Ozgur, I. Eroglu, M. Yucel, and U. Gunduz. 2011. Biohydrogen Production in an Outdoor Panel Photobioreactor on Dark Fermentation Effluent of Molasses. *International Journal of Hydrogen Energy* 36 (17): 11360–68.

Balat, H., and E. Kırtay. 2010. Hydrogen from Biomass—Present Scenario and Future Prospects. *International Journal of Hydrogen Energy* 35 (14): 7416–26.

Barbosa, M. J., J. M. Rocha, J. Tramper, and R. H. Wijffels. 2001. Acetate as a Carbon Source for Hydrogen Production by Photosynthetic Bacteria. *Journal of Biotechnology* 85 (1): 25–33.

Bartels, J. R., M. B. Pate, and N. K. Olson. 2010. An Economic Survey of Hydrogen Production from Conventional and Alternative Energy Sources. *International Journal of Hydrogen Energy* 35 (16): 8371–84.

Basak, N., and D. Das. 2007. The Prospect of Purple Non-Sulfur (PNS) Photosynthetic Bacteria for Hydrogen Production: The Present State of the Art. *World Journal of Microbiology and Biotechnology* 23 (1): 31–42.

Basak, N., and D. Das. 2009. Photofermentative Hydrogen Production Using Purple Non-Sulfur Bacteria *Rhodobacter Sphaeroides* O.U.001 in an Annular Photobioreactor: A Case Study. *Biomass and Bioenergy* 33 (6–7): 911–19.

Benemann, J. R. 1997. Feasibility Analysis of Photobiological Hydrogen Production. *International Association of Hydrogen Energy* 22: 979–87.

Bianchi, L., F. Mannelli, C. Viti, A. Adessi, and R. De Philippis. 2010. Hydrogen-Producing Purple Non-Sulfur Bacteria Isolated from the Trophic Lake Averno (Naples, Italy). *International Journal of Hydrogen Energy* 35 (22): 12216–23.

Boran, E., E. Özgür, J. Van Der Burg, M. Yücel, U. Gündüz, and Inci Eroglu. 2010. Biological Hydrogen Production by *Rhodobacter Capsulatus* in Solar Tubular Photo Bioreactor. *Journal of Cleaner Production* 18: S29–35.

Boran, E., E. Ozgur, M. Yucel, U. Gunduz, and I. Eroglu. 2012. Biohydrogen Production by *Rhodobacter Capsulatus* Hup⁻ Mutant in Pilot Solar Tubular Photobioreactor. *International Journal of Hydrogen Energy* 37: 16437–45.

Carvalho, A. P., A. Meireles, F. X. Malcata, L. A. Meireles, and F. X. Malcata. 2006. Microalgal Reactors: A Review of Enclosed System Designs and Performances. *Biotechnology Progress* 22 (6): 1490–1506.

Chandra, R., G. Nikhil, and S. Mohan. 2015. Single-Stage Operation of Hybrid Dark-Photo Fermentation to Enhance Biohydrogen Production through Regulation of System Redox Condition: Evaluation with Real-Field Wastewater. *International Journal of Molecular Sciences* 16 (5): 9540–56.

Chen, C., W. Lu, J. Wu, and J. Chang. 2007. Enhancing Phototrophic Hydrogen Production of *Rhodopseudomonas Palustris* via Statistical Experimental Design. *International Journal of Hydrogen Energy* 32 (8): 940–49.

Chen, C.-Y., and J.-S. Chang. 2006. Enhancing Phototropic Hydrogen Production by Solid-Carrier Assisted Fermentation and Internal Optical-Fiber Illumination. *Process Biochemistry* 41 (9): 2041–49.

Chen, C.-Y., C.-H. Liu, Y.-Ch. Lo, and J.-S. Chang. 2011. Perspectives on Cultivation Strategies and Photobioreactor Designs for Photo-Fermentative Hydrogen Production. *Bioresource Technology* 102 (18): 8484–92.

Chen, C.-Y., K.-L. Yeh, Y.-C. Lo, H.-M. Wang, and J.-S. Chang. 2010. Engineering Strategies for the Enhanced Photo-H$_2$ Production Using Effluents of Dark Fermentation Processes as Substrate. *International Journal of Hydrogen Energy* 35 (24): 13356–64.

Cheng, J., L. Ding, A. Xia, R. Lin, Y. Li, J. Zhou, and K. Cen. 2015. Hydrogen Production Using Amino Acids Obtained by Protein Degradation in Waste Biomass by Combined Dark- and Photo-Fermentation. *Bioresource Technology* 179: 13–19.

Cheng, X.-Y., and C.-Z. Liu. 2011. Hydrogen Production via Thermophilic Fermentation of Cornstalk by Clostridium Thermocellum. *Energy & Fuels* 25 (4): 1714–20.

Cherubini, F. 2010. The Biorefinery Concept: Using Biomass instead of Oil for Producing Energy and Chemicals. *Energy Conversion and Management* 51 (7): 1412–21.

Chookaew, T., S. O-thong, and P. Prasertsan. 2015. Biohydrogen Production from Crude Glycerol by Two Stage of Dark and Photo Fermentation. *International Journal of Hydrogen Energy* 40 (24): 2–7.

Cohen-Bazire, G., W. R. Sistrom, and R. Y. Stanier. 1965. Kinetic Studies of Pigment Synthesis by Non-Sulfur Purple Bacteria. *Journal of Cellular and Comparative Physiology* 49 (1): 25–68.

Das, D., and T. N. Veziroglu. 2008. Advances in Biological Hydrogen Production Processes. *International Journal of Hydrogen Energy* 33 (21): 6046–57.

Das, D., and T. N. Veziroglu. 2001. Hydrogen Production by Biological Processes: A Survey of Literature 26: 13–28.

Dasgupta, C. N., J. J. Gilbert, P. Lindblad et al. 2010. Recent Trends on the Development of Photobiological Processes and Photobioreactors for the Improvement of Hydrogen Production. *International Journal of Hydrogen Energy* 35 (19): 10218–38.

De Philippis, R., A. Ena, M. Guastini, C. Sili, and M. Vincenzini. 1992. Factors Affecting Poly-β-Hydroxybutyrate Accumulation in *Cyanobacteria* and in Purple Non-Sulfur Bacteria.pdf. *FEMS Microbiological Reviews* 103: 187–94.

Dipasquale, L., A. Adessi, G. d'Ippolito, F. Rossi, A. Fontana, and R. De Philippis. 2015. Introducing Capnophilic Lactic Fermentation in a Combined Dark-Photo Fermentation Process: A Route to Unparalleled H_2 Yields. *Applied Microbiology and Biotechnology* 99 (2): 1001–10.

Eroğlu, E., I. Eroğlu, U. Gündüz, and M.Yücel. 2008. Effect of Clay Pretreatment on Photofermentative Hydrogen Production from Olive Mill Wastewater. *Bioresource Technology* 99 (15): 6799–6808.

Eroğlu, E., İ. Eroğlu, U. Gündüz, and M. Yücel. 2009. Comparison of Physicochemical Characteristics and Photofermentative Hydrogen Production Potential of Wastewaters Produced from Different Olive Oil Mills in Western-Anatolia, Turkey. *Biomass and Bioenergy* 33 (4): 706–11.

Eroglu, E., and A. Melis. 2011. Photobiological Hydrogen Production: Recent Advances and State of the Art. *Bioresource Technology* 102 (18): 8403–13.

Eroglu, I., K. Aslan, and U. Gu. 1999. Substrate Consumption Rates for Hydrogen Production by *Rhodobacter Sphaeroides* in a Column Photobioreactor 70: 103–13.

Eroglu, I., A. Tabanoglu, U. Gunduz, E. Eroglu, and M. Yucel. 2008. Hydrogen Production by *Rhodobacter Sphaeroides* O.U.001 in a Flat Plate Solar Bioreactor. *International Journal of Hydrogen Energy* 33 (2): 531–41.

Franchi, E., C. Tosi, G. Scolla, G. D. Penna, F. Rodriguez, and P. M. Pedroni. 2005. Metabolically Engineered *Rhodobacter Sphaeroides* RV Strains for Improved Biohydrogen Photoproduction Combined with Disposal of Food Wastes. *Marine Biotechnology (New York, N.Y.)* 6 (6): 552–65.

Gadhamshetty, V., A. Sukumaran, N. Nirmalakhandan, and M. Theinmyint. 2008. Photofermentation of Malate for Biohydrogen Production—A Modeling Approach. *International Journal of Hydrogen Energy* 33 (9): 2138–46.

Gebicki, J., M. Modigell, M. Schumacher, Job van der Burg, and Eugenè Roebroeck. 2010. Comparison of Two Reactor Concepts for Anoxygenic H_2 Production by *Rhodobacter Capsulatus*. *Journal of Cleaner Production* 18: S36–42.

Gebicki, J., M. Modigell, M. Schumacher, J. V. D. Burg, and E. Roebroeck. 2009. Development of Photobioreactors for Anoxygenic Production of Hydrogen by Purple Bacteria. *Chemical Engineering Transactions* 18: 363–66.

Ghimire, A., L. Frunzo, F. Pirozzi, E. Trably, R. Escudie, P. N. L. Lens, and G. Esposito. 2015. A Review on Dark Fermentative Biohydrogen Production from Organic Biomass: Process Parameters and Use of by-Products. *Applied Energy* 144: 73–95.

Ghimire, A., S. Valentino, L. Frunzo, F. Pirozzi, P. N. L. Lens, and G. Esposito. 2016. Concomitant Biohydrogen and Poly-β-Hydroxybutyrate Production from Dark Fermentation Effluents by Adapted Rhodobacter Sphaeroides and Mixed Photofermentative Cultures. *Bioresource Technology* 217: 157–64.

Hallenbeck, P. C., and J. R. Benemann. 2002. Biological Hydrogen Production; Fundamentals and Limiting Processes 27: 1185–93.

Hallenbeck, P. C., and D. Ghosh. 2009. Advances in Fermentative Biohydrogen Production: The Way Forward? *Trends in Biotechnology* 27 (5): 287–97.

Han, H., B. Liu, H. Yang, and J. Shen. 2012. Effect of Carbon Sources on the Photobiological Production of Hydrogen Using Rhodobacter Sphaeroides RV. *International Journal of Hydrogen Energy* 37 (17): 12167–74.

Hu, X., T. Ritz, A. Damjanović, F. Autenrieth, and K. Schulten. 2002. Photosynthetic Apparatus of Purple Bacteria. *Quarterly Reviews of Biophysics*. 35 (1): 1–62.

Hülsen, T., D. J. Batstone, and J. Keller. 2014. Phototrophic Bacteria for Nutrient Recovery from Domestic Wastewater. *Water Research* 50: 18–26.

Hustede, E., A. Steinbiichel, and H. G. Schlegel. 1993. Relationship between the Photoproduction of Hydrogen and the Accumulation of PHB in Non-Sulphur Purple Bacteria. *Applied Microbiology and Biotechnology* 39: 87–93.

HYVOLUTION. 2011. Non-Thermal Production of Pure Hydrogen-Final Activity Report. Wageningen, The Netherlands. https://www.google.com/url?sa=t&rct=j&q=&esrc=s&source=web&cd=1&ved=0ahUKEwit54ylyYDUAhVBLo8KHRszD8cQFggiMAA&url=http%3A%2F%2Fwww.biohydrogen.nl%2Fdownloadattachment%2F22636%2F26107%2FPES_P5_Final_Report_08092011.pdf&usg=AFQjCNFru4elHuVu3REZnE3HAmPNSJp8EA&sig2=_PtQWpZWgNQXn3xmJPRRHw (accessed May 20, 2017).

Imhoff, J. F., H. G. Truper, and N. Pfennig. 1984. Rearrangement of the Species and Genera of the Phototrophic Purple Nonsulfur Bacteria. *International Journal of Systematic Biology* 34 (3): 340–43.

Imhoff, J. F., A. Hiraishi, and J. Suling. 2005. Anoxygenic Phototrophic Purple Bacteria. In *Bergey's Manual® of Systematic Bacteriology of Systematic Bacteriology*, 119–32. Springer US.

Kapdan, I. K., F. Kargi, R. Oztekin, and H. Argun. 2009. Bio-Hydrogen Production from Acid Hydrolyzed Wheat Starch by Photo-Fermentation Using Different Rhodobacter Sp. *International Journal of Hydrogen Energy* 34 (5): 2201–7.

Kars, G., and U. Gündüz. 2010. Towards a Super H_2 Producer: Improvements in Photofermentative Biohydrogen Production by Genetic Manipulations. *International Journal of Hydrogen Energy* 35 (13): 6646–56.

Khatipov, E., M. Miyake, J. Miyake, and Y. Asada. 1998. Accumulation of Poly-Beta-Hydroxybutyrate by *Rhodobacter Sphaeroides* on Various Carbon and Nitrogen Substrates. *FEMS Microbiology Letter* 162 (1): 39–45.

Kim, B. Hong, and G. M. Gadd. 2008. *Bacterial Physiology and Metabolism*. Cambridge University Press.

Kim, M.-S., D.-H. Kim, H.-N. Son, L. N. Ten, and J. K. Lee. 2011. Enhancing Photo-Fermentative Hydrogen Production by *Rhodobacter Sphaeroides* KD131 and Its PHB Synthase Deleted-Mutant from Acetate and Butyrate. *International Journal of Hydrogen Energy* 36 (21): 13964–71.

Koku, H., I. Eroglu, U. Gunduz, M. Yucel, and L. Turker. 2002. Aspects of the Metabolism of Hydrogen Production by *Rhodobacter Sphaeroides*. *International Journal of Hydrogen Energy* 27 (11–12): 1315–29.

Koku, H., I. Eroglu, U. Gunduz, M. Yucel, and L. Turker. 2003. Kinetics of Biological Hydrogen Production by the Photosynthetic Bacterium *Rhodobacter Sphaeroides* O.U. 001. *International Journal of Hydrogen Energy* 28 (4): 381–88.

Kondo, T., M. Arakawa, T. Hirai, T. Wakayama, M. Hara, and J. Miyake. 2002. Enhancement of Hydrogen Production by a Photosynthetic Bacterium Mutant with Reduced Pigment. *Journal of Bioscience and Bioengineering* 93 (2): 145–50.

Lee, C.-M., G.-J. Hung, and C.-F. Yang. 2011. Hydrogen Production by *Rhodopseudomonas Palustris* WP 3-5 in a Serial Photobioreactor Fed with Hydrogen Fermentation Effluent. *Bioresource Technology* 102 (18): 8350–56.

Lee, Z.-K., S.-L. Li, P.-C. Kuo et al. 2010. Thermophilic Bio-Energy Process Study on Hydrogen Fermentation with Vegetable Kitchen Waste. *International Journal of Hydrogen Energy* 35 (24): 13458–66.

Li, R. Y., and H. H. P. Fang. 2009. Heterotrophic Photo Fermentative Hydrogen Production. *Critical Reviews in Environmental Science and Technology* 39 (12): 1081–108.

Li, X., Y. Wang, S. Zhang, J. Chu, M. Zhang, M. Huang, and Y. Zhuang. 2011. Effects of Light/dark Cycle, Mixing Pattern and Partial Pressure of H_2 on Biohydrogen Production by *Rhodobacter Sphaeroides* ZX-5. *Bioresource Technology* 102 (2): 1142–48.

Liu, B.-F., N.-Q. Ren, J. Tang et al. 2010. Bio-Hydrogen Production by Mixed Culture of Photo- and Dark-Fermentation Bacteria. *International Journal of Hydrogen Energy* 35 (7): 2858–62.

Liu, B.-F., G.-J. Xie, J. D., and N.-Q. Ren. 2011. Optimization of Photo-Hydrogen Production by Immobilized *Rhodopseudomonas Faecalis* RLD-53. *Natural Resources* 02 (01): 1–7.

Liu, B.-F., G.-J. Xie, R.-Q. Wang et al. 2015. Simultaneous Hydrogen and Ethanol Production from Cascade Utilization of Mono-Substrate in Integrated Dark and Photo-Fermentative Reactor. *Biotechnology for Biofuels* 8: 8.

Lo, Y.-C., C.-Y. Chen, C.-M. Lee, and J.-S. Chang. 2010. Sequential Dark–photo Fermentation and Autotrophic Microalgal Growth for High-Yield and CO_2-Free Biohydrogen Production. *International Journal of Hydrogen Energy* 35 (20): 10944–53.

McEwan, A. G. 1994. Photosynthetic Electron Transport and Anaerobic Metabolism in Purple Non-Sulfur Phototrophic Bacteria. *Antonie van Leeuwenhoek* 66 (1–3): 151–64.

Menon, V., and M. Rao. 2012. Trends in Bioconversion of Lignocellulose: Biofuels, Platform Chemicals & Biorefinery Concept. *Progress in Energy and Combustion Science* 38 (4): 522–50.

Montiel-Corona, V., S. Revah, and M. Morales. 2015. Hydrogen Production by an Enriched Photoheterotrophic Culture Using Dark Fermentation Effluent as Substrate: Effect of Flushing Method, Bicarbonate Addition, and Outdoor–indoor Conditions. *International Journal of Hydrogen Energy* 40 (30): 9096–105.

Mu, Y., G. Wang, and H.-Q. Yu. 2006. Kinetic Modeling of Batch Hydrogen Production Process by Mixed Anaerobic Cultures. *Bioresource Technology* 97 (11): 1302–7.

Nasr, M., Ahmed T., Shinichi O., M. Suzuki, S. Kumari, and F. Bux. 2014. Continuous Biohydrogen Production from Starch Wastewater via Sequential Dark-Photo Fermentation with Emphasize on Maghemite Nanoparticles. *Journal of Industrial and Engineering Chemistry*, March. The Korean Society of Industrial and Engineering Chemistry.

Nath, K., and D. Das. 2009. Effect of Light Intensity and Initial pH during Hydrogen Production by an Integrated Dark and Photofermentation Process. *International Journal of Hydrogen Energy* 34 (17): 7497–501.

Ozmihci, S., and F. Kargi. 2010. Bio-Hydrogen Production by Photo-Fermentation of Dark Fermentation Effluent with Intermittent Feeding and Effluent Removal. *International Journal of Hydrogen Energy* 35 (13): 6674–80.

Ozturk, Y., M. Yucel, F. Daldal, S. Mandaci, U. Gunduz, L. Turker, and I. Eroglu. 2006. Hydrogen Production by Using *Rhodobacter Capsulatus* Mutants with Genetically Modified Electron Transfer Chains. *International Journal of Hydrogen Energy* 31 (11): 1545–52.

Posten, C. 2009. Design Principles of Photo-Bioreactors for Cultivation of Microalgae. *Engineering in Life Sciences* 9 (3): 165–77.

Rai, P. K., S. P. Singh, and R. K. Asthana. 2014a. Biohydrogen Production from Sugarcane Bagasse by Integrating Dark- and Photo-Fermentation. *Bioresource Technology* 152: 140–46.

Rai, P. K., R. K. Asthana, and S.P. Singh. 2014b. Optimization of Photo-Hydrogen Production Based on Cheese Whey Spent Medium. *International Journal of Hydrogen Energy* 39 (14): 7597–603.

Redwood, M. D, R. Dhillon, R. L. Orozco et al. 2012a. Enhanced Photosynthetic Output via Dichroic Beam-Sharing. *Biotechnology Letters* 34 (12): 2229–34.

Redwood, M. D., R. L. Orozco, A. J. Majewski, and L. E. Macaskie. 2012b. An Integrated Biohydrogen Refinery: Synergy of Photofermentation, Extractive Fermentation and Hydrothermal Hydrolysis of Food Wastes. *Bioresource Technology* 119: 384–92.

Redwood, M. D., R. L. Orozco, A. Majewski, and L. E. Macaskie. 2011. Biohydrogen Production by Extractive Fermentation and Photo fermentation. In 4th World Hydrogen Technologies Convention, 14–16 Sept 2011, Glasgow, U.K.

Redwood, M. D., M. Paterson-Beedle, and L. E. Macaskie. 2008. Integrating Dark and Light Bio-Hydrogen Production Strategies: Towards the Hydrogen Economy. *Reviews in Environmental Science and Bio/Technology* 8 (2): 149–85.

Ren, N.-Q., B.-F. Liu, J. Ding, and G.-J. Xie. 2009. Hydrogen Production with R. Faecalis RLD-53 Isolated from Freshwater Pond Sludge. *Bioresource Technology* 100 (1): 484–87.

Rupprecht, J., B. Hankamer, J. H. Mussgnug, G. Ananyev, C. Dismukes, and O. Kruse. 2006. Perspectives and Advances of Biological H_2 Production in Microorganisms. *Applied Microbiology and Biotechnology* 72 (3): 442–49.

Sargsyan, H., L. Gabrielyan, L. Hakobyan, and A. Trchounian. 2015. Light–dark Duration Alternation Effects on *Rhodobacter Sphaeroides* Growth, Membrane Properties and Bio-Hydrogen Production in Batch Culture. *International Journal of Hydrogen Energy* 40 (11): 4084–91.

Sasikala, K., C.V. Ramana, and P.R. Roa. 1991. Environmental Regulation for Optimal Biomass Yield and Photoreduction of Hydrogen by *Rhodobacter Sphaeroides* O.U.001. pdf. *International Journal of Hydrogen Energy* 16 (9): 597–601.

Schmidt, K. 1971. Carotenoids of Purple Nonsulfur Bacteria. *Archives of Microbiology* 77 (3): 231–38.

Sevinç, P., U. Gündüz, I. Eroglu, and M. Yücel. 2012. Kinetic Analysis of Photosynthetic Growth, Hydrogen Production and Dual Substrate Utilization by *Rhodobacter Capsulatus*. *International Journal of Hydrogen Energy* 37 (21): 16430–36.

Srikanth, S., S. Venkata Mohan, M. Prathima Devi, D. Peri, and P. N. Sarma. 2009. Acetate and Butyrate as Substrates for Hydrogen Production through Photo-Fermentation: Process Optimization and Combined Performance Evaluation. *International Journal of Hydrogen Energy* 34 (17): 7513–22.

Suwansaard, M., W. Choorit, J. H. Zeilstra-Ryalls, and P. Prasertsan. 2009. Isolation of Anoxygenic Photosynthetic Bacteria from Songkhla Lake for Use in a Two-Staged Biohydrogen Production Process from Palm Oil Mill Effluent. *International Journal of Hydrogen Energy* 34 (17): 7523–29.

Takabatake, H., K. Suzuki, I. Beom Ko, and T. Noike. 2004. Characteristics of Anaerobic Ammonia Removal by a Mixed Culture of Hydrogen Producing Photosynthetic Bacteria. *Bioresource Technology* 95 (2): 151–58.

Tao, Y., H. Yangling, Y. Wu, F. Liu, X. Li, W. Zong, and Z. Zhou. 2008. Characteristics of a New Photosynthetic Bacterial Strain for Hydrogen Production and Its Application in Wastewater Treatment. *International Journal of Hydrogen Energy* 33 (3): 963–73.

Tawfik, A., H. El-Bery, S. Kumari, and F. Bux. 2014. Use of Mixed Culture Bacteria for Photofermentive Hydrogen of Dark Fermentation Effluent. *Bioresource Technology* 168: 119–26.

Tredici, M. R., and G. Ch. Zittelli. 1998. Efficiency of Sunlight Utilization: Tubular Versus Flat Photobioreactors. Biotechnology and Bioengineering 57: 187–97.

Ugwu, C. U., H. Aoyagi, and H. Uchiyama. 2008. Photobioreactors for Mass Cultivation of Algae. *Bioresource Technology* 99 (10): 4021–28.

Ugwu, C. U., J. C. Ogbonna, and H. Tanaka. 2003. Design of Static Mixers for Inclined Tubular Photobioreactors. *Journal of Applied Phycology* 15 (2/3): 217–23.

Uyar, B., I. Eroglu, M. Yucel, U. Gunduz, and L. Turker. 2007. Effect of Light Intensity, Wavelength and Illumination Protocol on Hydrogen Production in Photobioreactors. *International Journal of Hydrogen Energy* 32 (18): 4670–77.

Uyar, B., M. Gürgan, E. Özgür, U. Gündüz, M.Yücel, and I. Eroglu. 2015. Hydrogen Production by Hup– Mutant and Wild-Type Strains of *Rhodobacter Capsulatus* from Dark Fermentation Effluent of Sugar Beet Thick Juice in Batch and Continuous Photobioreactors. *Bioprocess and Biosystems Engineering.*

Uyar, B., M. Schumacher, J. Gebicki, and M. Modigell. 2009. Photoproduction of Hydrogen by *Rhodobacter Capsulatus* from Thermophilic Fermentation Effluent. *Bioprocess and Biosystems Engineering* 32 (5): 603–6.

Venil, C. K., Z. A. Zakaria, and W. A. Ahmad. 2013. Bacterial Pigments and Their Applications. *Process Biochemistry* 48 (7): 1065–79.

Venkata Mohan, S., S. Srikanth, P. Dinakar, and P. N. Sarma. 2008. Photo-Biological Hydrogen Production by the Adopted Mixed Culture: Data Enveloping Analysis. *International Journal of Hydrogen Energy* 33 (2): 559–69.

Waligórska, M., K. Seifert, K. Górecki, M. Moritz, and M. Laniecki. 2009. Kinetic Model of Hydrogen Generation by *Rhodobacter Sphaeroides* in the Presence of NH4+ Ions. *Journal of Applied Microbiology* 107 (4): 1308–18.

Wang, R., C. Cui, Y. Jin et al. 2014. Photo-Fermentative Hydrogen Production from Mixed Substrate by Mixed Bacteria. *International Journal of Hydrogen Energy* 39 (25): 13396–400.

Wu, S. C., S. Z. Liou, and C. M. Lee. 2012. Correlation between Bio-Hydrogen Production and Polyhydroxybutyrate (PHB) Synthesis by *Rhodopseudomonas Palustris* WP3-5. *Bioresource Technology* 113: 44–50.

Yang, H., B. Shi, H. Ma, and L. Guo. 2015. Enhanced Hydrogen Production from Cornstalk by Dark- and Photo-Fermentation with Diluted Alkali-Cellulase Two-Step Hydrolysis. *International Journal of Hydrogen Energy* 40 (36): 1–8.

Yanling, Y., L. V. Zhenmei, M. Hang, and C. Jun. 2008. Dynamic Changes of Microbial Community Diversity in a Photohydrogen Producing Reactor Monitored by PCR-DGGE. *Journal of Environmental Science* 20: 1118–25.

Yiğit, D. Ö., U. Gündüz, L. Türker et al. 1999. Identification of by-Products in Hydrogen Producing Bacteria; *Rhodobacter Sphaeroides* O.U. 001 Grown in the Waste Water of a Sugar Refinery. *Progress in Industrial Microbiology* 35 (C): 125–31.

Zhang, D., P. Dechatiwongse, E.A. del Rio-Chanona, G.C. Maitland, K. Hellgardt, and V.S. Vassiliadis. 2015. Modelling of Light and Temperature Influences on *Cyanobacterial* Growth and Biohydrogen Production. *Algal Research* 9: 263–74.

Zhang, Z., Y. Wang, J. Hu, Q. Wu, and Q. Zhang. 2015. Influence of Mixing Method and Hydraulic Retention Time on Hydrogen Production through Photo-Fermentation with Mixed Strains. *International Journal of Hydrogen Energy* 40 (20): 6521–29.

Zhu, H., H. Fang, T. Zhang, and L. Beaudette. 2007. Effect of Ferrous Ion on Photo Heterotrophic Hydrogen Production by *Rhodobacter Sphaeroides*. *International Journal of Hydrogen Energy* 32 (17): 4112–18.

Zhu, H., T. Wakayama, T. Suzuki, Y. Asada, and J. Miyake. 1999. Entrapment of *Rhodobacter Sphaeroides* RV in Cationic Polymer/agar Gels for Hydrogen Production in the Presence of NH4+. *Journal of Bioscience and Bioengineering* 88 (5): 507–12. http://www.ncbi.nlm.nih.gov/pubmed/16232653.

Zhu, H., T. Suzuki, A. Tsygankov, Y. Asada, and J. Miyake. 1999. Hydrogen Production from Tofu Wastewater by *Rhodobacter Sphaeroides* Immobilized in Agar Gels. *International Journal of Hydrogen Energy* 24 (4): 305–10.

9

Hydrogen Photoproduction by Oxygenic Photosynthetic Microorganisms

Fabrice Franck, Bart Ghysels, and Damien Godaux

Contents

9.1 Introduction

More than 70 years has passed now since the first detailed report on H_2 photoevolution by a green microalga (Gaffron and Rubin, 1942). This initiated an ongoing era of research on biological H_2 photoproduction, motivated by the ultimate ideal of establishing clean biofuel technologies on the basis of the photosynthetic process for the future. Oxygenic photosynthesis uses light energy absorbed by photosystem II (PSII) to split water into oxygen, protons, and electrons. While oxygen is lost as a waste product, electrons are channeled through an electron transport chain to reduce acceptors of low redox potential after a second step of light activation in photosystem I (PSI). Once reduced, these acceptors will be used in turn to reduce the CO_2 fixed by the RubisCO enzyme (for a general description of photosynthesis, see Walker 1992). The general principle of photosynthesis-based H_2 photoproduction is to derive strong reductants generated through photosynthesis to provide electrons that will be combined with protons to generate H_2. This can be achieved directly with the terminal electron acceptor ferredoxin (Fd) of the electron transport chain, thereby bypassing the CO_2 reducing reactions. Alternatively, carbohydrates resulting from photosynthetic CO_2 reduction can be reoxidized and provide the necessary reducing equivalents. In both cases, the formation of H_2 from protons and electron donors must be catalyzed by appropriate enzymes. It appears that the only known natural enzymes with such ability are hydrogenases and nitrogenases, and that both of them are inactivated by dioxygen. This is an unfortunate fatality if photosynthetic water splitting has to be combined one way or another with continuous H_2 production. However, local anoxia can be maintained in the vicinity of hydrogenases or nitrogenases by various means, either by natural compartmentation or temporal separation, or by manipulating cellular metabolisms in such a way that oxygen levels remain low. This was the general frame of the large body of research on H_2 photoproduction by photosynthetic microorganisms in the last 70 years. Even though noticeable success was obtained, H_2 evolution rates and light conversion efficiencies remained well below values expected for biotechnological applications. But research is ongoing, and new ideas are constantly being expressed and tested, especially among highly motivated young researchers.

In this chapter, we review the diversity, function, and physiological roles of H_2-producing enzymes. Then we address the principles and results of various approaches by which cyanobacterial and microalgal metabolisms were oriented toward a continuous photoproduction of H_2. Finally, through the case study of a mutant, we discuss some tracks for future improvements of H_2 photoproduction by green microalgae.

9.2 Enzymes Involved in H_2 Production in Cyanobacteria and Green Microalgae

9.2.1 Hydrogenases

Hydrogenase enzymes catalyze the very simple chemical reaction $2H^+ + 2e^- \leftrightarrow H_2$. The reaction is reversible, and its direction depends on the redox potential of the components able to interact with the enzyme. In the presence of H_2 and an electron acceptor, hydrogenase will act as a hydrogen uptake enzyme, but in the presence of an electron donor of low potential, it may use protons as electron acceptors and release H_2.

Many living organisms use H_2 as an energy source or carrier. Hydrogenases are widely spread among the microorganisms' diversity with numerous representatives in the domain of bacteria and archaea, and are also present in some unicellular eukaryotic organisms. This group of enzymes has been shown to be involved in a large variety of bioenergetic processes leading to ATP generation or redox power dissipation, allowing the growth of host microorganism under various conditions.

A serious issue when considering hydrogenases for hydrogen biogeneration is the extreme sensitivity of these enzymes to oxygen *in vitro* and *in vivo*. O_2 binds as a ligand to the active site, accepts electrons, and is reduced to reactive oxygen species (ROS) trapped in the enzyme. This may lead to permanent damage when ROS survives long enough to attack the vulnerable catalytic center.

Despite their diversity in many respects (size, quaternary structure, electron donors, or acceptors), hydrogenases consist of three phylogenetically distinct classes: the [Fe]-hydrogenases, the [NiFe]-hydrogenases, and the [FeFe]-hydrogenases, generally presented as the result of a convergent evolution (for review, see Vignais and Billoud, 2007). Each class is characterized by a distinctive functional core.

9.2.1.1 [Fe]-Hydrogenases

The catalytic center of [Fe]-hydrogenases does not contain any Fe-S or Ni clusters and was therefore named iron-sulfur cluster-free hydrogenase, or [Fe]-hydrogenase (Shima and Thauer, 2007). [Fe]-hydrogenases are restricted to some methanogens (Pilak et al., 2006) where they are essential for growth under nickel limitation (Thauer, 1998). Due to their low distribution, this class of hydrogenases has been poorly studied. The vast majority of known hydrogenases belong to two other classes.

9.2.1.2 [NiFe]-Hydrogenases

These enzymes are common in prokaryotes with numerous representatives in bacteria and archaea. [NiFe]-Hydrogenase forms globular heteromultimers (Volbeda et al., 1995) where the bimetallic NiFe active site is coordinated by four cysteines and three nonprotein ligands, one carbon monoxide (CO) molecule, and two cyanide ions (CN), bound to the Fe atom (Happe et al., 1997; Pierik et al., 1999). The NiFe cluster is located in the large subunit and deeply buried in the protein. The small subunit contains several Fe-S clusters that conduct electrons from the active site to the physiological electron acceptor or donor. The [4Fe-4S] cluster proximal to the active site is regarded as essential for activity (Albracht, 1994). In additional to structural genes, there are accessory genes for maturation and the insertion of Ni, Fe, CO, and CN at the active site of the heterodimer. The maturation of hydrogenases follows a complex pathway, which involves at least seven auxiliary proteins (HypA-F and an endopeptidase). The corresponding genes are clustered in a majority of host organisms, and transcriptional control is regulated by several factors, including hydrogen, oxygen, carbon monoxide, and redox state of the cell.

[NiFe]-hydrogenases have a high affinity (low apparent Km) for H_2, indicating that they act mostly in consuming H_2 in the different organisms, even if some enzymes display good H_2-producing capacity. They are described as more resistant to O_2 damages than [FeFe]-hydrogenases, as catalytic activity is reversibly inactivated by O_2. Some microorganisms, such as *Ralstonia* sp., have even evolved oxygen-tolerant [NiFe]-hydrogenases able to oxidize hydrogen in the presence of air.

[NiFe]-hydrogenase classification is based on subunit sequence alignments and divides the [NiFe]-hydrogenases into four groups. Remarkably, this classification based on amino acid sequences is in good agreement with the groups derived from cellular functions.

9.2.1.2.1 Membrane-Bound Respiratory Uptake Hydrogenases (Group 1).

Present in bacteria as well as in archaea, membrane-bound respiratory uptake hydrogenases are the most numerous and best documented. They transfer electrons from H_2 to a terminal electron acceptor (such as O_2 [aerobic respiration], NO_3^-, SO_4^-, fumarate, or CO_2 [anaerobic respiration]) through a membrane-located complex. Hydrogen uptake through hydrogenase activity is coupled to an energy-conserving electron transport chain that generates a proton motive force (Mitchell, 1961, 1966). Oxidative phosphorylation provides ATP to the cell as a means of energy conservation. These enzymes are thus capable of supporting growth with hydrogen as an energy source (for review, see Vignais and Billoud, 2007).

9.2.1.2.2 Cyanobacterial Uptake Hydrogenases (Group 2a).

Cyanobacterial uptake hydrogenases have been found in all nitrogen-fixing cyanobacteria. These proteins are soluble in the cytoplasm and function in association with the oxygen-sensitive nitrogenase that performs reduction of nitrogen to ammonia in heterocyst cells (see below). Such a reaction is highly endergonic and requires ATP, as well as reducing power. Hydrogen evolution concomitant to nitrogen fixation by nitrogenase might be considered a loss of energy so that nitrogen-fixing cyanobacteria develop uptake hydrogenases to recycle energy stored in H_2. Such recycling is very efficient as nitrogen-fixing cyanobacteria show no or little net hydrogen production. It is suggested to have different beneficial functions for nitrogen-fixing cyanobacteria, that is, reinjecting electrons in the respiratory chain to generate ATP and reducing oxygen to prevent nitrogenase inhibition (for review, see Bothe et al., 2010).

9.2.1.2.3 Hydrogen Sensors (Group 2b).

Mainly found in bacteria, hydrogen sensors are located in the cytoplasm. Together with kinases, they act as hydrogen signal transducers to control the phosphorylation state of transcription activators, inducing the synthesis of hydrogenases belonging to group 1 (for review, see Lenz et al., 2002).

9.2.1.2.4 Bidirectional Heteromultimeric Cytoplasmic Hydrogenases (Group 3).

The [NiFe]-hydrogenases of group 3 are able to bind a soluble cofactor, for example, F420, ferredoxin, or $NADP^+$, and function reversibly. Many members of this group are found in archaea but are also present in bacteria, including cyanobacteria. Particularly in cyanobacteria, bidirectional hydrogenases are common enzymes in both diazotrophic and nondiazotrophic cyanobacteria (Serebriakova et al., 1994). They are tetrameric, soluble, or loosely attached to the membrane. Physiological function is still a matter of debate, but the enzyme might function as an electron security valve to dissipate excess electrons in unfavorable redox conditions, therefore maintaining an adequate oxidation–reduction state in the cell (for review, see Carrieri et al., 2011). Bidirectional hydrogenases have been particularly well studied in *Synechocystis* PCC 6803, where they might function to regenerate $NAD(P)^+$ from $NAD(P)H$ (Dutta et al., 2005) or Fd_{ox} from Fd_{red} (Gutekunst et al., 2014) during fermentation to maintain ATP synthesis and cell homeostatis. Hydrogenase activity also allows reoxidation of the photosynthetic electron transport chain in the light after dark anaerobic incubation. It is therefore involved in activation of photosynthesis in such conditions (just like [Fe-Fe]-hydrogenase in green microalgae see Section 9.2.1.3). After the onset of light, hydrogen photoproduction is followed after a few minutes by photosynthetic oxygen evolution (Cournac et al., 2002). Bidirectional hydrogenase activity is also important to regulate the photosynthetic electron flow under high or changing light intensity (Appel et al., 2000).

9.2.1.2.5 Hydrogen-Evolving and Energy-Conserving Membrane-Associated Hydrogenases (Group 4). These multimeric hydrogenases, present in both bacteria and archaea, dispose of excess reducing power from anaerobic fermentation to generate a proton gradient for ATP synthesis; this process results in hydrogen evolution. Two (or more) subunits are integral membrane proteins sharing similarities with mitochondrial complex I subunits and playing a crucial role in energy conservation (for review, see Hedderich, 2004).

9.2.1.3 [FeFe]-Hydrogenases
Biodiversity of [FeFe]-hydrogenases is poor compared with that of [NiFe]-hydrogenases. These enzymes are present in anaerobic prokaryotes (genera *Clostridium* and *Desulfovibrio*), but also in a few lower eukaryotes, such as anaerobic fungi (*Neocallimastix frontalis*), parasitic protozoas (*Trichomonas vaginalis*), and unicellular microalgae (*Scenedesmus obliquus*, *Chlorella fusca*, and *Chlamydomonas reinhardtii*). They are the only type of hydrogenase found in eukaryotes, with genes being located in the nucleus but enzymes exclusively expressed in organelles.

[FeFe]-hydrogenases are mainly monomeric enzymes. Their catalytic center is highly conserved and consists of a binuclear iron site bound to a [4Fe-4S] by a cysteine bridge. Nonprotein ligands CN– and CO are attached to the iron atoms of the binuclear Fe center (Nicolet et al., 2000). The Fe atoms also share two bridging sulfur ligands of a small five-atom molecule (Nicolet et al., 2002). Additional domains accommodate the Fe-S center and relay the electron transfer from the external electron source to the H cluster buried inside this monomeric protein. Moreover, a hydrophobic channel from the surface to the active site provides access for protons and egress for H_2. This unique active center leads to about a 100-fold higher activity than for the other types of hydrogenases (Frey, 2002; Vogt et al., 2007). Three accessory proteins, named HydE, HydF, and HydG, are required for the proper biosynthesis of [FeFe]-hydrogenases. HydE and HydG belong to the radical S-adenosylmethionine (radical SAM) superfamily, whereas the HydF maturation factor belongs to the GTPase protein family (for review, see Vignais et al., 2001).

Whereas [NiFe]-hydrogenases tend to be involved in hydrogen uptake, [FeFe]-hydrogenases thermodynamically allow ferredoxin–NADH reoxidation, and thus generate hydrogen using protons as terminal electron acceptors. Yet, [FeFe]-hydrogenases present severe sensitivity to O_2, as the enzyme is irreversibly damaged after exposure to small concentrations of O_2 (Ghirardi et al., 1997). This represents a major limitation to achieve a profitable hydrogen biogeneration based on [FeFe]-hydrogenases. Cohen and coworkers demonstrated that oxygen diffusion is clearly limited to specific regions located within the conserved active site domain, offering important clues on how to overcome the oxygen sensitivity of [FeFe]-hydrogenases (Cohen et al., 2005).

[FeFe]-hydrogenases are usually involved in the disposal of excess reductant generated during fermentation under anaerobic conditions, resulting in fermentative hydrogen evolution. This step is fundamental to prevent fermentation cessation. Although hydrogenases are involved in anaerobic fermentation, they are also capable of interaction with the photosynthetic electron transport chain in unicellular photosynthetic microalgae, such as *C. reinhardtii*. Hydrogenases accept electrons from reduced ferredoxins downstream of PSI (Florin et al., 2001), resulting in a light-dependent production of H_2. Like in cyanobacteria, this process enables algae to oxidize and thereby activate their photosynthetic electron transport system after anoxic incubation.

9.2.2 Nitrogenases

Nitrogenases catalyze the reduction of molecular nitrogen N_2 into ammonium NH_3 in diazotrophic bacteria. Since they are inactivated by oxygen, these enzymes are found only in bacterial cells in which an anaerobic environment can be kept. For filamentous cyanobacteria, this condition is achieved in heterocyst cells of diazotrophic species, such as *Anabaena* sp., which served as experimental species in the first studies on cyanobacterial hydrogen photoproduction (Benemann and Weare, 1974; Weissman and Benemann, 1977). For diazotrophic unicellular cyanobacteria, nitrogenase activity generally occurs during the night period, when oxygenic photosynthesis is inactive. An example of this kind is *Cyanothece* sp., which has been used more recently to achieve high rates of nitrogenase-based H_2 photoproduction (Bandyopadhyay et al., 2010).

Due to the stability of the N_2 molecule, its reduction is highly energy demanding. Chemical reduction by H_2 through the Haber–Bosch reaction, used for industrial ammonium production, operates efficiently only at high pressures (200 bars) and high temperatures (450°C) in the presence of catalysts. The nitrogenase-catalyzed reaction uses chemical energy in two forms: redox energy and phosphate-bound energy. The complex catalytic process needs the cooperation of two enzyme subcomplexes:

dinitrogenase reductase and dinitrogenase. Dinitrogenase reductase uses reduced ferredoxin and ATP (in the presence of Mg^{2+}) as substrates to transfer electrons to the dinitrogenase moiety, which uses these electrons, as well as protons, to reduce N_2 to NH_3. The latter process first involves H_2 formation, followed by its substitution by N_2 on the catalytic site of dinitrogenase for further reduction. H_2 is therefore a side product of N_2 reduction when the N_2 substrate is available. The most studied nitrogenase (Mo-nitrogenase), containing a Mo-Fe cofactor, catalyzes the reaction according to the stoichiometry

$$N_2 + 8H^+ + 8Fd^- + 16ATP \rightarrow 2NH_3 + H_2 + 8Fd + 16ADP + 16Pi$$

Nitrogenase also reduces C_2H_2 (acetylene) to C_2H_4 (ethylene), a property commonly used to estimate enzyme activity using gas chromatography. In the absence of reducible substrate other than H^+, nitrogenase combines protons with electrons from ferredoxin in an ATP-dependent manner, with the stoichiometry

$$2H^+ + 2Fd^- + 4ATP \rightarrow H_2 + 2Fd + 4ADP + 4Pi$$

At equal rates of electron transfer from dinitrogenase reductase to dinitrogenase, the rate of H_2 formation is then increased fourfold compared with rates in the presence of N_2. Many experiments on nitrogenase-dependent H_2 photoevolution in cyanobacteria were therefore conducted under an argon atmosphere. On the other hand, carbon monoxide is a potent inhibitor of N_2 and C_2H_2 reduction but not of H_2 formation (Hwang et al., 1973; Smith et al., 1976), whereas C_2H_2 inhibits hydrogenases that may consume hydrogen. Complex gas mixtures, including inhibitors (such as Ar + CO + C_2H_2), were therefore sometimes used to maximize H_2 evolution (reviewed in Lopes Pinto et al., 2002).

The consumption of ATP (used during electron transfer to dinitrogenase) makes nitrogenase-catalyzed H_2 evolution more energy demanding than hydrogenase-catalyzed processes. On the other hand, nitrogenase is a "slow" enzyme (6.4 s^{-1} [Hallenbeck, 2012]). Although not very appealing from thermodynamic and kinetic considerations, nitrogenase-catalyzed H_2 evolution has continued to attract interest, essentially because long-term production (days or weeks) is easily achieved using diazotrophic cyanobacteria.

The dinitrogenase part of Mo-nitrogenase is a protein heterodimer made of the *nifD* and *nifK* gene products. The heterodimer contains the Fe-Mo cofactor at the catalytic site of N_2 reduction, and an 8Fe-7S P-cluster used to transport electrons from dinitrogenase reductase. The latter (sometimes called Fe-protein) is a homodimer of *nifH* gene products and contains a 4Fe-4S cofactor for electron acceptance from Fd^-. The need for 2ATP at each electron transfer stems from the association–dissociation cycle between dinitrogenase reductase and dinitrogenase.

Besides Mo-nitrogenases, V-nitrogenases (with a V-Fe cofactor) or Fe-nitrogenases (with a Fe-only cofactor) also exist. Cyanobacteria possessing Mo- or V-nitrogenase, or both, occur, but Fe-nitrogenases are generally not detected in cyanobacteria (reviewed in Bothe et al., 2010; Bothe and Newton, 2014). Compared with Mo-nitrogenases, V- and Fe-nitrogenases show higher H_2/N_2 stoichiometric ratios during N_2 reduction, but probably keep an equal ATP/e$^-$ ratio of 2 because the mechanism of electron transfer from dinitrogenase reductase to dinitrogenase is a general feature for all nitrogenases.

From a biological perspective, the requirement of a micro-oxic environment for nitrogenase activity makes N_2-based nitrogen assimilation incompatible with carbon assimilation through photosynthesis. Therefore, the two processes must be either spatially or temporally separated. However, the nitrogenase function requires a continuous supply of reducing equivalents and ATP, which for photoautotrophic organisms are provided directly or indirectly as a consequence of photosynthesis.

Spatial separation occurs in heterocyst-forming filamentous cyanobacteria, such as *Anabaena*, which served as a useful model. Heterocysts only differentiate when combined nitrogen (nitrate or ammonium) is missing. Their formation is under the control of complex gene expression patterns (reviewed in Kumar et al., 2010), most probably triggered by oxoglutarate (the substrate for ammonium incorporation into glutamate), which accumulates when combined nitrogen is missing. This results, among others, in expression of *nif* genes, breakdown of PSII, stimulation of respiration, and formation of the thick cell wall that helps maintain micro-oxic conditions. Reductants for nitrogenase activity are supplied by oxidative breakdown (oxidative pentose pathway) of sucrose synthesized in photosynthetically active, nonheterocyst cells. Continuous ATP renewal is maintained through photophosphorylation supported by PSI-dependent cyclic electron flow (CEF). The ammonium resulting from N_2 reduction is fixed in glutamine, which is further used for glutamate formation in nonheterocyst cells.

The interdependency between nonheterocyst and heterocyst cells has consequences when nitrogenase is used as the main device for H_2 photobiological production. Limitations can occur due to a limited supply of reduced carbon from nonheterocyst cells. Also, when H_2 production is optimized by use of a N_2-free gas phase, N deficiency can have general adverse effects on the metabolism of both cell types in the long term, an effect that was eventually alleviated by providing brief ammonium pulses (Cournac et al., 2004).

Temporal separation is observed in single-cell diazotrophic cyanobacteria, such as fast-growing *Cyanothece* strains (Reddy et al., 1993). In this case, the circadian oscillator controls large-scale photoperiodic changes in gene expression (reviewed in Sherman et al., 2010). The cell shifts between carbon and nitrogen assimilation as a function of day and night, a repetitive pattern that can be maintained for some time after cells are transferred to constant conditions, as is typical for circadian oscillations. *Cyanothece* also shows aerobic nitrogenase activity when adapted for long periods to constant light or dark conditions, which indicates that intracellular compartmentation, possibly associated with glycogen granules, ensures oxygen-protecting mechanisms for nitrogenase activity (Reddy et al., 1993; Toepel et al., 2008). In recent years, *Cyanothece* (mainly strain ATCC 51142, among several sequenced strains) has become the preferred model system for experimentation on single-cell nitrogenase-dependent H_2 photoproduction.

9.3 H$_2$ Photoproduction Processes Using Cyanobacteria

Although there is evidence that significant fermentative hydrogen evolution in darkness can take place in some cyanobacterial species (Schütz et al., 2004), we will limit this review to photoproduction processes. Protocols, rates, and metabolic aspects to be considered for optimization vary considerably depending on the category of cyanobacteria. It is therefore convenient to discuss separately the results of experiments made with diazotrophic filamentous cyanobacteria, with diazotrophic unicellular cyanobacteria, and with nondiazotrophic cyanobacteria.

9.3.1 Specific Activities versus Volumetric Rates

In most instances, rates are expressed as specific activities on a chlorophyll basis (μmol H$_2$·mg Chl^{-1}·h^{-1}). Specific activity values are of biological relevance when comparing the properties of different strains, or of a given strain in different conditions (Lopes Pinto et al., 2002). But even after this normalization on Chl, comparisons remain only indicative because results can vary largely depending on the measuring conditions, especially incident light intensity, vessel geometry, and cell concentration. At the level of each cell, H$_2$ photoproduction has a nonlinear dependency on light intensity, and photosynthetic pigments (Chl, phycobilins, and carotenoids) exert a strong shading effect on short distances even at relatively low cell concentrations (typically for Chl concentrations higher than 1 μg·mL^{-1}). Light scattering also influences the light profile inside the vessel. Ideally, measurements should be done at very low cell densities and using a light path short enough to avoid light attenuation. The light intensity should be such that activity values close to maximal ones are obtained. For practical reasons (mainly linked to the sensitivity of the equipment), these conditions are rarely met.

From a biotechnological standpoint, specific activities are of little use, and volumetric productivities (usually expressed in mL H$_2$·L^{-1}·h^{-1}) are preferred because they give an idea of the volume of H$_2$ gas that can be produced by a given volume of illuminated cell suspension. From this perspective, the highest values will be obtained when the cell density is high enough for complete absorption of incident light. The problem of cell density effects has been extensively discussed in the context of microalgal cultivation in order to optimize biomass production rates (Tredici and Zittelli, 1998; Formighieri et al., 2012). In this context, too high cell concentrations can also be detrimental due to the appearance of a "dark zone" in the vessel, where respiratory losses can occur. Relationships between cell density, light path, and light intensity effects are also relevant for optimizing H$_2$ photoproduction rates by cyanobacteria or microalgae (Kosourov et al., 2011), but in this case, the consequences of the appearance of a dark zone in the vessel at very high cell densities are not known.

In the following discussion, we focus more on volumetric productivity achievements than on specific activities for given processes. This is convenient if cyanobacteria and green microalgae are to be compared since cyanobacteria and green microalgae (having different pigment compositions) cannot easily be compared on a pigment basis.

9.3.2 Nitrogenase-Based Photoproduction by Filamentous Diazotrophic Cyanobacteria

H_2 photoproduction by heterocystous filamentous cyanobacteria such as *Anabaena* sp. or *Nostoc* sp. is the most largely documented cyanobacterial H_2 photoproduction process. It was reported more than 40 years ago (Benemann and Weare, 1974), with some indications of its natural occurrence already in 1896 (see Benemann and Weare, 1974). In this process, water-splitting photosynthesis takes place in vegetative cells and produces reduced carbon compounds (sucrose and maltose) that are transferred to the heterocysts, where they enter the oxidative pentose pathway. This provides reducing equivalents that can be used to reduce ferredoxin for nitrogenase function, while ATP will be produced in heterocysts through PSI-dependent cyclic photophosphorylation (Figure 9.1a). Under the N_2-depleted gas phase (e.g., Ar + CO_2) and nitrate-free medium, nitrogenase in heterocysts will produce H_2 for long periods of time (days), while cells become N depleted and slowly lose their photosynthetic ability (and phycobilin pigments), and filaments tend to become more fragile. In the Benemann and Weare (1974) paper, stable H_2 photoproduction (at rates equivalent to around 2 mL·L^{-1}·h^{-1} for rather diluted algal suspension) was reported for 3 h and was shown resistant to oxygen in the gas phase, but inhibited largely by N_2. Subsequent work (Weissman and Benemann, 1977) investigated the effects of light intensity, cell concentration, and repeated ammonium additions (to slow N deprivation-induced degradation). High H_2 photoproduction rates were obtained during several days (30 mL·L^{-1}·h^{-1}, slowly declining to 12 mL·L^{-1}·h^{-1} after 8 days). These works also showed that the PSII inhibitor 3-(3,4-dichlorophenyl)-1,1-dimethylurea (DCMU) (which inhibits water splitting in vegetative cells) was progressively more efficient in inhibiting H_2 photoproduction, meaning that the dependence of H_2 photoproduction on photosynthetic water splitting increased with time as carbohydrate reserves progressively exhausted.

From these early works, it was understood that the presence of an uptake hydrogenase (*hup* genes) in the heterocysts had adverse effects on the rate of H_2 photoproduction (Figure 9.1a and b). Later on, several mutants of *Anabaena* sp. or *Nostoc* sp. deprived of uptake hydrogenase activity were obtained, either by random mutagenesis or by genetic engineering (Mikheeva et al., 1995; Lindberg et al., 2002; Lindblad et al., 2002; Khetkorn et al., 2012). Improvement of H_2 photoproduction was generally observed with these strains, in particular when the experimental devices did not allow fast removal of produced gases. H_2 photoproduction by such mutants seemed sufficiently robust to justify the first outdoor experiments in photobioreactors (Lindblad et al., 2002; Tsygankov et al., 2002).

In the London outdoor experiments by Tsygankov et al. (2002), the PK84 mutant of *Anabaena variabilis* (devoid of Hup activity [Mikheeva et al., 1995]) was used for cultivation in batch or chemostat culture mode in a 4.35 L tubular photobioreactor (described in Tsygankov et al., 1998) under CO_2-enriched air. N_2 reduction by nitrogenase therefore occurred and cells were actively growing while evolving H_2 as the side reaction of nitrogenase. In such conditions, solar energy was used for biomass

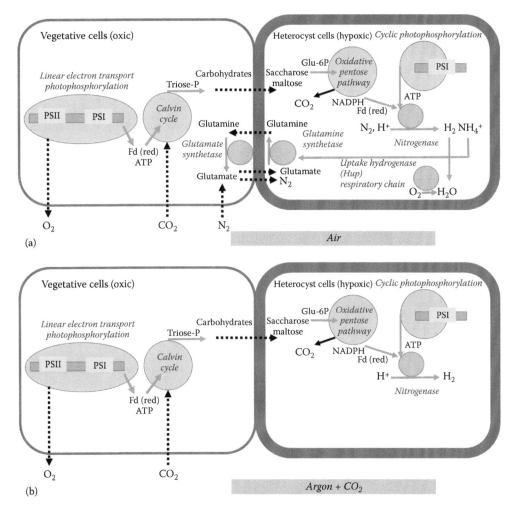

Figure 9.1 Schematic representation of the metabolic pathways involved in H_2 photoproduction by filamentous, heterocystous cyanobacteria. (a) Natural conditions in which H_2 is only produced as a by-product of nitrogenase-mediated N_2 fixation. (b) Optimized conditions in which N_2 fixation is suppressed by changing the gas phase composition and uptake hydrogenase (Hup) activity is suppressed after mutagenesis. Dotted black line, diffusion or transport of metabolites or gases.

accumulation and for H_2 photoproduction. In these nonoptimal conditions for H_2 photoproduction, rather large volumetric rates were observed in the daytime (with some production also during the night), with peaks of 23 mL·L^{-1}·h^{-1} and average values around 7 mL·L^{-1}·h^{-1} during a 5-day experiment (Table 9.1). Photoproduction was found to increase with the density of actively growing cells and be essentially limited by temperature (between 24°C and 30°C), with moderate effects of light intensity variations. H_2 photoproduction could be observed during long-term experiments (40 days) with no need for refreshing the culture, and contamination by eukaryotic

TABLE 9.1 Typical Specific Activities and Volumetric Rates of Hydrogen Photoproduction by Cyanobacteria, Taken from Three Representative Studies

Strain	Reference	Conditions/Comments	Specific Activity (µmol $H_2 \cdot$mg Chl$^{-1} \cdot$h^{-1})	Volumetric Productivity (mL $H_2 \cdot$L$^{-1} \cdot$h^{-1})
Nitrogenase-Based, Filamentous Diazotrophic Cyanobacteria				
A. variabilis PK84 mutant (impaired in Hup uptake hydrogenase)	Tsygankov et al., 2002	Outdoor PBR (4.35 l) in London (natural light) operated during several weeks. Batch under N-fixing conditions. Air/ CO_2 (98:02). V-nitrogenase (induced by replacement of Mo by V in medium).	Maximum: ±200 *µmol $H_2 \cdot$mg Chl$^{-1} \cdot$h^{-1}* Average: ±150 *µmol $H_2 \cdot$mg Chl$^{-1} \cdot$h^{-1}*	Maximum: *23 mL $H_2 \cdot$L$^{-1} \cdot$h^{-1}* Average: *±7 mL $H_2 \cdot$L$^{-1} \cdot$h^{-1}*
Nitrogenase-Based, Unicellular Diazotrophic Cyanobacteria				
Cyanothece sp. ATCC 51142	Bandyopadhyay et al., 2010	Vials with 20–25 mL cultures. Batch under N-fixing conditions. Transfer from 12 h photoperiod to continuous light (100 µmol·m^{-2}·s^{-1}) at the start of subjective night (measurements during 60 h). Glycerol was used for mixotrophic cultivation.	Average: ±125 *µmol $H_2 \cdot$mg Chl$^{-1} \cdot$h^{-1}* (photoautotrophy in air) ±175 *µmol $H_2 \cdot$mg Chl$^{-1} \cdot$h^{-1}* (photoautotrophy in CO_2-enriched air) ±200 *µmol $H_2 \cdot$mg Chl$^{-1} \cdot$h^{-1}* (mixotrophy in air) Maximum: 467 µmol $H_2 \cdot$mg Chl$^{-1} \cdot$h^{-1} (mixotrophy in argon)	Maximum: *±30 mL $H_2 \cdot$L$^{-1} \cdot$h^{-1}* (photoautotrophy in air)
Hydrogenase-Based, Nondiazotrophic Cyanobacteria				
Synechocystis sp. PCC 6803 M55 mutant deficient for type I NADPH-dehydrogenase	Cournac et al., 2004	1.5 mL measuring chamber. Preadaptation to anaerobiosis (Ar sparging) for 1 h. Experiment duration: 25 min. Light intensity: 300 µmol·m^{-2}·s^{-1}. Sustained production only in the presence of glucose.	4.8 *µmol $H_2 \cdot$mg Chl$^{-1} \cdot$h^{-1}* (glucose) 6 µmol $H_2 \cdot$mg Chl$^{-1} \cdot$h^{-1} (glucose + glucose oxidase)	*1.6 mL $H_2 \cdot$L$^{-1} \cdot$h^{-1}* (glucose) 2 mL $H_2 \cdot$L$^{-1} \cdot$h^{-1} (glucose + glucose oxidase)

Note: When needed (values shown in italics), specific activities and/or volumetric rates were calculated from information contained in the papers, assuming standard conditions with 1 µmol H_2 = 22.4 µL. PBR, photobioreactor.

microalgae was avoided by cycloheximide addition. Lindblad et al. (2002) used the same experimental facility for about 1 week to test another *hup* mutant of *Anabaena* PCC 7120. They found peak values of 15 $mL \cdot L^{-1} \cdot h^{-1}$ and confirmed temperature effects.

In these studies, low values of light energy conversion into H_2 were always reported (up to 0.9%), which leaves much space for improvement since the maximum photosynthetic energy conversion yield is 12%. In *hup* mutants under optimized photobioreactor conditions indoors, energy efficiencies of up to 4% could be obtained (Nyberg et al., 2015). An evident limitation in using heterocystous cyanobacteria for both long-term and efficient H_2 photoproduction is also the competition between N_2 and proton reduction by nitrogenase. An approach to overcome this problem consists of engineering the nitrogenase to limit its affinity for N_2. In this line, *Anabaena* PCC 7120 mutants with amino acid substitutions close to the Fe-Mo cofactor were found to photoproduce H_2 in a N_2-insensitive manner, while keeping a low N_2-reducing activity useful for growth (Masukawa et al., 2014).

9.3.3 Nitrogenase-Based Photoproduction by Unicellular Diazotrophic Cyanobacteria

Several single-cell cyanobacteria are known to possess the ability to perform N_2 assimilation at night, when a suboxic environment for nitrogenase activity can be established. A remarkable potential for H_2 photoproduction was demonstrated for one of them, *Cyanothece* ATCC 51142, when it was continuously illuminated in N_2-fixing conditions (the absence of combined nitrogen in the medium) (Bandyopadhyay et al., 2010). High rates of H_2 photoproduction were found during subjective nights[*] when cells were illuminated continuously during several days after being grown in photoperiodic conditions (Melnicki et al., 2012). In such conditions, oscillations in H_2 photoproduction rates were observed. Substantial H_2 evolution could be observed in air (Bandyopadhyay et al., 2010), but was largely enhanced under an Ar atmosphere (to prevent N_2 reduction by nitrogenase). Under continuous sparging with Ar + CO_2, rhythmic O_2 and CO_2 evolution occurred, with O_2 evolution increasing during the subjective days, whereas CO_2 evolution increased during subjective nights, concomitantly with the increase in H_2 evolution (Melnicki et al., 2012). H_2 evolution was therefore to some extent resistant to the presence of dissolved oxygen in the medium.

Evidence of this type of behavior in a *Cyanothece* strain was first reported by Mitsui and Suda (1995).[†] This was followed by the study of Bandyopadhyay et al. (2010), who reported the effects of various factors on the H_2 photoevolution performances of *Cyanothece* ATCC 51142 in airtight vials. Their study emphasized the high production rates of this strain when compared with other well-known cyanobacterial H_2 producers (Table 9.1), with the highest rates being obtained under Ar atmosphere and

---

[*] When a continuous illumination regime is applied after several days of photoperiodic growth, circadian oscillations continue for several days and cells go through a succession of subjective nights and days.

[†] With a strain named *Synechococcus* sp. Miami BG 043511, later classified as *Cyanothece* Miami BG 043511 (Zehr et al., 2001).

with the addition of glycerol, the organic carbon substrate for mixotrophic growth in this species. Remarkably, H_2 accumulation proceeded at substantial rates in air without an added carbon source (around 100 µmol $H_2 \cdot$mg Chl$^{-1} \cdot$h^{-1}, leading to the accumulation of 900 mL·L^{-1} H_2 within 5 days in the experimental conditions used in this work), albeit about four times slower than in optimal conditions (Table 9.1).

From the inhibition of H_2 photoproduction by combined nitrogen (Bandyopadhyay et al., 2010), it was clear that H_2 formation was driven essentially by nitrogenase, possibly with a minor contribution of reversible hydrogenase (Min and Sherman, 2010). Hence, two essential questions arose from these observations: (1) What are the pathways for fueling the nitrogenase with reducing equivalents in these cells, and (2) how, in the presence of oxygen, can a suboxic cellular environment be maintained in order to avoid nitrogenase inhibition?

On the basis of a detailed analysis of gas exchange, inhibitor effects, and thermodynamic considerations, it was concluded by Melnicki et al. (2012) that PSII-driven water photolysis was an important contributor to generate electrons for nitrogenase. It is not clear however, if this water photolysis-dependent H_2 photoproduction operates by using reduced ferredoxin directly or if temporary storage of photosynthetically reduced carbon, followed by reoxidation, occurs. In any case, it is likely that the oxidation of glycogen (and glycerol, when added) provides reducing power for an additional indirect pathway, which would necessitate PSI excitation, and whose extent may vary depending on conditions. This explains the slow, rather than immediate, inhibition of H_2 photoproduction after addition of the PSII inhibitor DCMU (Melnicki et al., 2012). A recent transcriptomic study (Bernstein et al., 2015) revealed that expression of PSII core genes correlated with that of nitrogenase during the time course of H_2 photoproduction after N depletion in nonsynchronized cultures, which (together with electron transport analysis) led the authors to suggest a direct use of photogenerated electrons by nitrogenase. It is therefore possible, although not firmly established, that *Cyanothece* sp. is the only known case of a cyanobacterium capable of operating a nitrogenase-based H_2 production driven directly by water photolysis.

The concomitant production of H_2 by nitrogenase and O_2 by PSII in single cells such as *Cyanothece* sp. is amazing, since one would expect fast nitrogenase inhibition. Two processes likely contribute to decreasing the local O_2 levels: oxidative phosphorylation with carbohydrates (glycogen) as substrates, and O_2 consumption driven by Hup uptake dehydrogenase. The role of the latter was highlighted in a study of a knockdown *hupL* mutant, which was found unable to accumulate nitrogenase under N_2-fixing conditions in aerobic conditions (Zhang et al., 2014). Western blot analysis in the same study further indicated that the uptake hydrogenase, as well as nitrogenase, was thylakoid associated.

In recent years, the potential of *Cyanothece* sp. for future biotechnological H_2 production processes, added to its fascinating rhythmic patterns, has prompted proteomic and transcriptomic studies of *Cyanothece* strains in order to create robust models at the system level (Stöckel et al., 2008; McDermott et al., 2011; Aryal et al., 2013; Welkie et al., 2014; Bernstein et al., 2015). By this approach, it should become easier to identify targets for genetic manipulations and to optimize experimental conditions for improved *Cyanothece* performances.

9.3.4 Hydrogenase-Based Photoproduction by Nondiazotrophic Cyanobacteria

Synechocystis PCC 6803 is a strain of unicellular, nondiazotrophic cyanobacteria that has been largely used as a model for cyanobacterial photosynthetic studies. It contains a bidirectional [NiFe]-hydrogenase (Hox), which is able to use electrons from the photosynthetic electron transport chain via ferredoxin or flavodoxin (Gutekunst et al., 2014) to generate H_2, and which is less sensitive to O_2 than other hydrogenases (Appel et al., 2000). In wild-type (WT) cells adapted to anoxia in the dark, H_2 is only transiently evolved after the onset of light until full activation of photosynthesis has proceeded, similarly to what is observed in the green microalga *Chlamydomonas* (see Section 9.4). The reversibility of the hydrogenase reaction and enzyme inhibition by evolved O_2 contribute to the transient nature of the H_2 burst. In addition, other consumers of reducing equivalents, such as the respiratory chain and the nitrate assimilation pathway, compete with hydrogenase, thereby affecting H_2 photoproduction (Gutthann et al., 2007). H_2 photoevolution on a much longer timescale than in the WT was described and analyzed for the M5 mutant strain, which is deficient for type I NADPH-dehydrogenase (Cournac et al., 2004). It is believed that the absence of this dehydrogenase allows the maintenance of a high NADPH/NADP$^+$ ratio, which is necessary to thermodynamically maintain hydrogenase activity in the direction of proton reduction. In this mutant, H_2 photoproduction could be extended to some minutes (instead of some seconds in the WT) until O_2 inhibition stops the process. The source of electrons for hydrogenase was the photosynthetic electron chain, in a direct pathway from water to H_2 (direct biophotolysis) involving PSII and PSI in series. An indirect pathway (involving only PSI) could be promoted by glucose addition to fuel the electron transport chain at the level of plastoquinones (probably through a succinate dehydrogenase or a type II dehydrogenase [Gutthann et al., 2007]). Then O_2 levels remained low, and a stable photoproduction at around 2 mL $H_2 \cdot L^{-1} \cdot h^{-1}$ was observed during the observation time of 25 min (Table 9.1).

Due to the difficulty of maintaining substantial H_2 production in the light with *Synechocystis*, more recent research has focused on two-stage processes, with a first phase of reduced carbon compound (glycogen) accumulation through photosynthesis (light phase), followed by dark fermentative H_2 production. This was found to be relatively efficient with *Synechocystis* PCC 6803, as well as with other cyanobacterial strains, such as strains of the *Gloeocapsa*, *Lyngbya*, or *Microcoleus* genera, or with the easily cultured *Arthrospira maxima* (Antal and Lindblad, 2005; Ananyev et al., 2012; Kothari et al., 2012, 2014). Optimization of such a two-phase process relies on manipulating experimental conditions in order to favor glycogen accumulation in the light phase (e.g., using S or N deprivation) and limit competing sinks and reversibility during the dark phase.

9.4 Hydrogen Photoproduction Using the Green Microalga *C. reinhardtii*

C. reinhardtii is a unicellular eukaryotic green alga that belongs to Chlorophyceae. It is widely regarded as a model organism for bioenergetics studies on mitochondrial respiration and photosynthesis due to the availability of genetic and molecular tools (sequenced genomes, insertional mutagenesis, ability to easily control its life cycle in the laboratory, etc.). It is also the model organism for the study of H_2 photoproduction by eukaryotic microalgae.

C. reinhardtii possesses two genomic sequences of [FeFe]-hydrogenases that are targeted to the chloroplast of the cell, named *HydA1* and *HydA2* (Happe et al., 2002). HydA1 and HydA2 proteins are closely related (68% identity and 74% similarity at the amino acid level), with molecular masses of about 49 kDa (Melis et al., 2004). These smallest [FeFe]-hydrogenases consist of the catalytic subunit (H cluster) only, being directly reduced by electrons from ferredoxin (Happe and Naber, 1993; Florin et al., 2001). Because of this particular feature, these [FeFe]-hydrogenases possess the highest hydrogen production capacity recorded to date. HydA1 has been shown to be responsible for the main hydrogenase activity, as HydA2's contribution is weak (Godman et al., 2010; Meuser et al., 2012). Hydrogenase activity is sensitive to molecular oxygen (Ghirardi et al., 1997). Inactivation occurs within minutes of exposure to atmospheric levels of O_2 (Cohen et al., 2005). The expression of enzymes and maturation factors also requires anoxic conditions, and is regulated by Crr1, a transcription regulator also induced during copper starvation (Pape et al., 2012).

Because expression and activity are sensitive to oxygen, the chloroplastic hydrogenases only operate under anoxic conditions. The enzymes then allow reoxidation of ferredoxin for continuous bacterial-type anaerobic fermentation, as well as for photosynthesis reactivation upon illumination.

9.4.1 Fermentation-Dependent H_2 Production

The microalga *C. reinhardtii* grows aerobically thanks to the oxygenic process of photosynthesis, but it can also be subject to a number of anoxic situations, especially in soil environments with high concentrations of microbes. As an example of its metabolic flexibility, *Chlamydomonas* possesses fermentative pathways typically found in facultative anaerobic bacteria of the soil. These pathways act to provide ATP from the oxidation of organic or inorganic compounds in the absence of oxygen (Müller and Gorrell, 1983). As part of the fermentation processes, hydrogenases play a crucial role in the reoxidation of reduced ferredoxin, allowing us to evacuate excess reducing power in H_2 as a safety valve to sustain fermentation (Happe et al., 2002). Anaerobically maintained *Chlamydomonas* cells generate H_2 in the dark, but the rate is about 10 times slower than that of light-dependent hydrogen evolution (Gaffron and Rubin, 1942).

9.4.2 H_2 Photoproduction by C. reinhardtii

An intriguing question to address is the meaning of the presence of such oxygen-sensitive hydrogenases, together with oxygenic photosynthesis. Apart from their obvious function in fermentation, their possible function in the light has not yet been clearly established in microalgae. Typically, H_2 evolution in *C. reinhardtii* can be observed when shifting an anoxic cell suspension of *Chlamydomonas* cells from dark to light. In laboratory conditions, anoxia is achieved by cellular respiration in airtight containers in the dark. The process can also be sped up by bubbling the culture with an inert gas like argon or nitrogen, or by addition of an oxygen-consuming enzyme (i.e., glucose oxidase). Anoxia then drives the expression of hydrogenases. A relatively high but transient H_2 photoproduction can be observed after a sudden dark-to-light shift of the anoxic algal cells. PSII activity is the main source of electrons to the hydrogenase in this condition since the addition of the PSII inhibitor DCMU strongly decreases H_2 evolution (Cournac et al., 2002). Hydrogenases accept electrons from Fd reduced by the photosynthetic pathway until the enzyme is inhibited by O_2 evolved by PSII (Hemschemeier et al., 2009; Stripp et al., 2009) and/or due to the lack of reduced Fd substrate due to Calvin cycle reactivation (Godaux et al., 2015).

Given the above observations, it seems rather straightforward to suspect a possible role of hydrogenases in photosynthetic induction, as was done already in the seventies (Kessler, 1973; Schreiber and Vidaver, 1974) and further supported by Cournac and coworkers (2002). Recently, our group investigated the involvement of hydrogenase activity in the reactivation of the Calvin cycle after adaptation to dark anoxic conditions in *C. reinhardtii* (Ghysels et al., 2013; Godaux et al., 2013, 2015). We have shown that hydrogenase activity is responsible for the entire photosynthetic electron transport flux during the first seconds of illumination, by serving as the final electron acceptor following short- or long-term acclimation to dark and anoxic conditions. Hydrogenases can therefore be viewed as safety valves through which excess photosynthetic electrons are dissipated when other PSI acceptors cannot be regenerated. Additionally, these studies showed that the absence of hydrogenase activity could be unfavorable for the growth of *Chlamydomonas* in its natural habitat (Godaux et al., 2015). Indeed, hydrogenase activity shortens the delay for Calvin cycle reactivation because the hydrogenase-dependent alternative electron flow participates in ATP synthesis for carbon fixation.

H_2 evolution by the above-described dark anoxic incubation protocol is unsustainable, making this approach useless from a biotechnological point of view. Actual application-oriented research on H_2 photoproduction with *Chlamydomonas* started when Melis and coworkers developed a procedure for sustainable H_2 evolution less than 20 years ago. It is based on a two-stage protocol that temporally separates photosynthetic O_2 evolution from the H_2 production phase, enabling a relatively high rate of H_2 photoproduction for several days (Melis et al., 2000). In the first, aerobic stage, cells are grown in the light in a complete medium. Cells are then transferred to a sulfur-deprived medium. Sulfur deprivation causes a rapid decline of the photosynthesis rate to a level below that of respiration (Wykoff et al., 1998). This is due to prompt but reversible inhibition of PSII activity, caused by the chloroplast inability to sustain the high turnover of the D1 polypeptide of the PSII reaction center (Mattoo and Edelman, 1987; Wykoff et al., 1998). Under these conditions, anoxia can be reached in the light thanks to mitochondrial respiration, thereby permitting expression and sustained activity of the hydrogenase (Melis et al., 2000). In addition to the deleterious effect on PSII efficiency, sulfur deprivation induces a massive degradation of the RubisCO enzyme (Zhang et al., 2002) so that the Calvin cycle ability for CO_2 reduction is lost. Thus, the major photosynthetic electron sink is no longer operative, and hydrogenase becomes the main electron acceptor downstream of PSI. This light-driven electron transport to hydrogenase generates a proton gradient across the thylakoid membrane for ATP synthesis (for review, see Ghysels and Franck, 2010).

Regarding the sustained photoproduction of hydrogen in *C. reinhardtii* during S deprivation, two electron pathways to the hydrogenase are considered, both light-dependent and engaging PSI and ferredoxin. These pathways, named the direct and indirect pathways, are defined by the origin of the electrons for proton reduction. The direct pathway involves the PSII-dependent water oxidation to feed hydrogenase with electrons after their transfer through the whole electron transport chain. The indirect pathway involves the nonphotochemical reduction of the photosynthetic electron transport chain at the level of plastoquinones in a PSII-independent manner (Figure 9.2). Therefore, the direct pathway is often referred to as the PSII-dependent pathway, and the indirect pathway is often named the PSII-independent pathway. The indirect pathway relies on the activity of Nda2, a type II NAD(P)H dehydrogenase (Jans et al., 2008) that catalyzes plastoquinone reduction by NAD(P)H derived from

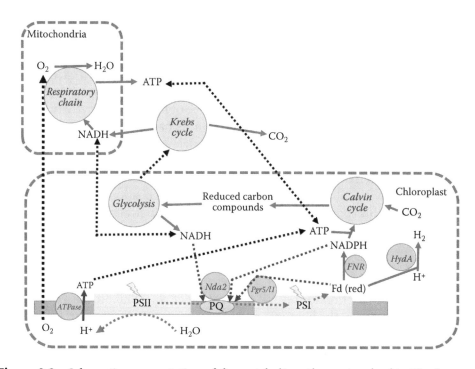

Figure 9.2 Schematic representation of the metabolic pathways involved in H_2 photoproduction by the unicellular eukaryotic green alga *C. reinhardtii*. Dotted blue line, photosynthetic linear electron transport from water splitting to ferredoxin reduction; dotted red line, photosynthetic cyclic electron transports, including Pgr5/l1 pathway and Nda2 pathway; continuous red line, ATP production by ATPase activity. Dotted black line: diffusion or transport of metabolites or gases within the cell.

the degradation of carbon reserves (mainly starch, accumulated in an earlier stage in response to S deprivation) (Fouchard et al., 2005; Chochois et al., 2009).

Contributions of the direct and indirect pathway to H_2 photoproduction by *Chlamydomonas* under sulfur deprivation are still largely debated. Several studies showed a strong reduction in the H_2 production rate when PSII is inhibited by DCMU addition (Chochois et al., 2009; Mignolet et al., 2012; Baltz et al., 2014). This indicates that the PSII-dependent pathway prevails in the absence of inhibitors. However, a mutant for Nda2 displays a lower rate of H_2 production compared with the WT, suggesting an important role for the PSII-independent pathway (Jans et al., 2008; Mignolet et al., 2012). Addition of a hexose symporter in *Chlamydomonas* allows glucose utilization by the PSII-independent pathway and leads to improved H_2 photoproduction (Doebbe et al., 2007). This result contrasts with the observation that a *sta6* mutant lacking starch produces similar hydrogen amounts as the parental strain (Chochois et al., 2009). Moreover, overexpression of Nda2 did not significantly improve the amount of hydrogen produced (Baltz et al., 2014). The implications of one or the other pathway may have varied among different studies due to different experimental conditions, and a systematic study should be carried out to close the debate.

Typical maximal volumetric H_2 production rates obtained by S deprivation vary from 1 to 2 $mL \cdot h^{-1} \cdot L^{-1}$ of cell suspension in WT strains (Melis et al., 2000), and can be improved to about 10 $mL \cdot h^{-1} \cdot L^{-1}$ in mutants with increased H_2 production ability (Kruse et al., 2005; Tolleter et al., 2011; Volgusheva et al., 2013; Steinbeck et al., 2015). Generally, the highest rates are maintained for about 70 h before the H_2 production starts to slowly decline. The evolved gas is almost pure H_2, with traces of CO_2 (Melis et al., 2000). It was shown (Kosourov et al., 2012; Mignolet et al., 2012) that product (H_2) inhibition of H_2 photoproduction might be partly responsible for the observed decline and has to be considered when discussing kinetics of H_2 evolution. A systematic removal of H_2 (every 24 h) was shown to double the overall rate of H_2 evolution in WT *Chlamydomonas* in S deprivation. Release of H_2 partial pressures also stimulated H_2 production in low-light hypoxic S-replete cultures (Jurado-Oller et al., 2015).

When making the cost–energy efficiency balance of H_2 photoproduction by S deprivation, it should be taken into account that most studies dealing with the subject, including the ones cited in this chapter, make use of acetate, a reduced carbon source, in the culture medium. This means that H_2 is, per definition, produced mixotrophically and not photoautotrophically. Acetate is used for boosting mitochondrial respiration in order to reach and sustain the anoxic conditions required for hydrogenase activity. Most of the acetate production today depends on chemical production processes requiring a substantial amount of energy, and its use also leads to nonnegligible CO_2 production in the culture, which somehow is incompatible with the principle of generating a clean renewable fuel. Attempts were made to omit acetate in the H_2 production protocol and replace it with inorganic carbon supplied as $NaHCO_3$ or CO_2 bubbling in the media, but reaching anoxia in autotrophy proved to be difficult (Fouchard et al., 2005). Progress was made by Tsygankov and coworkers (2006), who obtained sustainable H_2 evolution autotrophic conditions by attenuation of the incident light intensity, in combination with S deprivation. Still, the best obtained yields were rather disappointing (0.8 mL·h⁻¹·L⁻¹ during 43 h) when compared with mixotrophic H_2 production in S deprivation (2 mL·h⁻¹·L⁻¹ during 80 h) (Melis et al., 2000; Zhang and Melis, 2002). Mathematical modeling of the different culture parameters of the H_2 evolution process in photoautotrophy helped to determine culture conditions leading to higher H_2 evolution rates (up to 2.75 mL·h⁻¹·L⁻¹ during 70 h) in photobioreactors, comparable to those obtained in mixotrophic conditions (Fouchard et al., 2005; Degrenne et al., 2011b). A key parameter with this respect turned out to be the illuminated fraction in the photobioreactor, which depends on biomass concentration and incident light intensity at the surface of the reactor.

9.5 Case Study: *pgrl1*, a *Chlamydomonas* Photosynthetic Mutant with Enhanced H_2 Photoproduction under Anoxia

From the perspective of improving the H_2 photoproduction potential of *Chlamydomonas*, mainly two aspects have been investigated by mutagenetic approaches: electron supply to hydrogenase (through the PSII-dependent and/or PSII-independent pathways) and hydrogenase expression or activity. Because hydrogenase interacts with the photosynthetic machinery, modifications of the features of the photosynthetic electron transport chain are likely to influence hydrogenase activity, and vice versa. Chlorophyll fluorescence measurements have been largely used in order to reveal particular aspects of the functioning of the electron transport chain, as well as to isolate photosynthetic mutants and/or mutants of the hydrogenase function. Some of these mutants showed improved H_2 photoproduction potential. Here, we present the general principles of the chlorophyll fluorescence approach for mutant isolation, and we discuss the properties of one of them.

9.5.1 Fluorescence-Based Methodology for Isolation of Photosynthetic Mutants

During photosynthesis, light is absorbed by chlorophyll molecules, which undergo transitions from the ground state to an excited singlet state. Within nanoseconds, chlorophylls in their excited state return to their ground state either by transferring excitation energy to neighbor pigments for trapping in reaction centers (where photochemical charge separation occurs), or by losing excitation energy in dissipation processes. These dissipation processes include nonradiative internal conversion and radiative internal conversion. Chlorophyll fluorescence produced by radiative energy losses in intact photosynthetic organisms is emitted in the red and near-infrared spectral region and is easily recorded using fluorimeters specially designed for this purpose. It arises mainly from chlorophyll *a* (Chl *a*) associated to the PSII antenna bed, PSI showing only weak fluorescence (Lavorel, 1962; Franck et al., 2002).

Because the different deactivation processes of excited PSII Chl *a* compete with each other, changes in the fluorescence quantum yield (and therefore of the intensity of the recorded fluorescence light) indicate opposite changes either in the photochemical quantum yield or in the thermal dissipation quantum yield of PSII. This is why *in vivo* fluorescence recordings provide a rich but complex source of information on the functioning of the photosynthetic electron transport chain (reviewed by Baker, 2008). Chlorophyll fluorescence in continuously illuminated samples is very sensitive to the saturation level of the photosynthetic electron transport chain, because saturation leads to a reduction in PSII photochemical quantum yield. In the context of H_2 photoproduction by *Chlamydomonas*, this relationship was used to show that hydrogenase is very efficient as a transient sink for photogenerated electrons shortly after the onset of light under anoxia, and that this favors the reactivation of the Calvin cycle (Ghysels et al., 2013; Godaux et al., 2015). Under oxic conditions, the fluorescence signal will also be sensitive to alterations of many photosynthetic functions, such as alternative electron pathways (reviewed in Cardol et al., 2011) or processes that control thermal dissipation of absorbed energy. This sensitivity of the chlorophyll fluorescence approach was therefore exploited in order to identify and analyze mutants that were generated in order to better understand photosynthetic regulation or, more specifically, hydrogenase-dependent photosynthetic electron flow.

Two molecular biology techniques are commonly used to generate mutations: site-directed mutagenesis and insertional mutagenesis. Site-directed mutagenesis is a direct genetic method used to obtain specific and targeted gene silencing, for example, by RNA interference. Phenotypic properties of the mutant are then used to understand the function of the gene product in the WT. This approach calls for prerequisites, such as knowledge of the targeted gene sequence. Insertional mutagenesis is an indirect genetic method used to randomly inactivate gene products, for example, by the insertion of a cassette coding for an antibiotic resistance. Transformants are screened for a specific phenotypic change in comparison with the WT. If a mutant exhibits the targeted phenotype, it can be assumed that the insertion has caused the inactivation of a gene related to the phenotypic change. The inactivated gene can be identified by molecular tools and the mutant characterized in depth. This approach requires fast and powerful screening methods to rapidly identify strains of interest among a large insertional mutant library.

Screening procedures based on chlorophyll fluorescence are widely used to spot mutants impaired in photosynthetic activity (Kruse et al., 1999; Houille-Vernes et al., 2011; Tolleter et al., 2011). The screening system requires sophisticated fluorescence imaging equipment, composed of a fast digital camera able to record the time course of chlorophyll fluorescence in response to light from a large number of algal colonies or cultures simultaneously (on a petri dish or in wells of multiwell plates) (Johnson et al., 2009). Such a system has been used for the isolation of algal insertional mutants deficient in H_2 photoproduction, with the aim of identifying novel genes involved in the process. Recently, a new fluorescence-based screening method was developed in our group to isolate mutants impaired in hydrogenase expression or activity, as well as mutants altered in related H_2 metabolism in anoxia (Godaux et al., 2013). The screen is based on the relationship between the PSII photochemical quantum yield (calculated from the fluorescence signal) and the hydrogenase capacity during the first seconds of illumination after a prolonged dark anoxic period. Such a relationship is expected because hydrogenase is the main electron sink for the whole electron transport chain for short illumination times under anoxia. Compared with previously described screens for mutants impaired in H_2 metabolism and hydrogenase function (Seibert et al., 2001; Wecker et al., 2011), the fluorescence-based screening method is remarkably fast (thousands of transformants analyzed per day) and sensitive (due to the linear relationship between hydrogenase activity and the fluorescence signal).

Mutants isolated in oxic conditions on the basis of specific photosynthetic phenotypes might also display modified potential for H_2 photoproduction processes, if the altered photosynthetic function has a positive or adverse effect on H_2 photoproduction under anoxia. Some years ago, Kruse and coworkers (2005) isolated the *stm6* mutant on the basis of a disrupted capacity to regulate the distribution of excitation energy between the two photosystems (a process known as "state transition"). The reported H_2 production rates of *stm6* were higher (5–13 times) than that of the control strain. Yet, the high H_2-producing phenotype of the mutant could not be explained on the basis of altered state transition only. It probably resulted from a complex and not well-understood remodeling of the cellular metabolism (increase in mitochondrial respiration, inhibition of photosynthetic cyclic electron pathway, large amount of starch stock, etc.). More recently, a mutant impaired in cyclic electron flow (CEF), named *pgrl1*, has been isolated on the basis of a specific alteration of the fluorescence signal (Tolleter et al., 2011). The *pgrl1* mutant also displays remarkable features in H_2 photoproduction processes. This will be discussed in more detail below.

9.5.2 Properties of the Chlamydomonas pgrl1 *Mutant in the Context of H_2 Photoproduction*

9.5.2.1 *Isolation and Main Characteristics of the* pgrl1 *Mutant, a Mutant of Cyclic Electron Flow*

The *pgrl1* mutant was isolated by a chlorophyll fluorescence-based screen in oxic conditions, designed for the identification of mutants showing altered Chl fluorescence signatures during dark-to-light and light-to-dark transitions (Tolleter et al., 2011). Dark–light and light–dark transitions are characterized by stronger activity of auxiliary pathways of electron transport, such as cyclic electron transport, and pathways depending on O_2 as the alternative electron sink (Peltier et al., 2010). The screen was designed with the conviction that mutants resulting in changes in these pathways would be revealed by their fluorescence signature during dark–light transitions. The underlying idea was that mutations affecting auxiliary electron flow pathways would also have an impact on the electron flow to the hydrogenase (Hemschemeier and Happe, 2011). The case of the *pgrl1* mutant nicely illustrates this.

The mutant did not show the typical transient fluorescence rise (15–20 s) observed in the WT upon the switch from dark to light. Molecular and genetic characterization showed that the mutant phenotype was the result of a single insertion of a paromomocyin resistance cassette disrupting the *pgrl1* gene. This gene was earlier identified in *Arabidopsis*, where it was shown to be involved in cyclic electron transport around PSI (DalCorso et al., 2008). By two independent spectroscopic methods

probing fast redox-state variations of thylakoid electron transporters, a similar deficiency in cyclic electron transport ability was detected in the *Chlamydomonas pgrl1*
mutant (Tolleter et al., 2011). The *pgrl1* mutant also displayed an increase of H$_2$ evolution capacity during S deprivation experiments, as well as during short-time H$_2$
photoevolution experiments upon dark anaerobic incubation (Tolleter et al., 2011;
Steinbeck et al., 2015).

9.5.2.2 Cyclic Electron Flow and Its Interactions with H$_2$ Evolution

Understanding the H$_2$ phenotype of the *pgrl1* mutant requires some reflection on
cyclic electron transport (CEF) and its role in photosynthesis. Photosynthesis is generally represented as a linear flow of electrons from PSII to PSI, leading to the reduction of NADP to NADPH downstream of PSI. The electron transfer reactions taking
place in the thylakoid membrane are coupled to a transfer of protons across the membrane leading to the formation of a proton gradient. The resulting proton motive
force drives the generation of ATP by means of the ATP-ase complex, spanning the
membrane. The linear electron transport and the associated proton gradient hence
provide ATP and NADPH required to drive the metabolic reactions of the Calvin
cycle by which CO$_2$ is reduced. However, for reasons of stoichiometry, the overall
cycle requires more ATP than is supplied by the linear electron flow operating from
PSII to PSI (Alric et al., 2010). Different mechanisms, such as a deviation of part of
the electron flow toward O$_2$-reducing reactions and cyclic electron transport around
PSI, have been proposed to participate in equilibrating the ATP/NADPH balance
(Cardol et al., 2011). CEF is a feedback loop reinjecting electrons derived from PSI-
generated reductants into the chain in between PSII and PSI (Figure 9.2). In this way,
an additional proton gradient is created to generate ATP without net production of
NADPH. Two CEF pathways have been identified (Alric et al., 2010; Peltier et al., 2010;
Cardol et al., 2011). One pathway relies on the presence of the Nda2 NAD(P)H dehydrogenase to reinject electrons into the chain at the expense of NADPH oxidation.
This pathway probably predominates in nonreducing metabolic conditions (Alric,
2014). The other pathway relies on PGR5/PGRL1 proteins and directly reinjects electrons derived from ferredoxin (the PSI acceptor) back into the plastoquinone (PQ)
pool between PSII and PSI. The actors of this second pathway seem to be conserved
throughout the green lineage (higher plants and algae). The PGR5/PGRL1-dependent
pathway, activated in reducing conditions in microalgae, outcompetes the NAD(P)H
dehydrogenase pathway in terms of electron transfer capacity (Alric, 2014).

Higher plant CEF mutants generally show more severe growth deficiencies than
algal mutants (Cardol et al., 2011). In microalgae, CEF is made redundant to some
extent with other mechanisms that help to balance ATP/NADPH ratios during
steady-state photosynthesis. The *pgrl1* mutant, for instance, displays increased rates
of mitochondrial respiration and chloroplastic O$_2$ photoreduction in order to compensate for its decreased CEF ability (Dang et al., 2014).

A particularly strong impact of CEF was shown on the ability of *Chlamydomonas*
cells to reactivate the Calvin cycle following a dark–light transition in an anoxic
environment (Godaux et al., 2015). In this very reducing metabolic context, CEF
is proposed to counterbalance overreduction of the PSI acceptor pool (Alric, 2014;

Johnson et al., 2014). A similar role was assigned here to hydrogenase-dependent electron flow (Ghysels et al., 2013; Godaux et al., 2015). Cyclic and hydrogenase-dependent electron flows seem to rescue the cells of a deadlock situation, caused by a combination of severe ATP depletion with strongly reducing metabolic conditions and by the absence of O_2 as an alternative sink, in which induction of photosynthetic electron transport and Calvin cycle activation would hardly be possible otherwise. As CEF and hydrogenase-dependent electron flows seem to cooperate at the onset of photosynthesis in anoxic conditions, it is not surprising that they influence each other.

The link between PGRL1/PGR5-dependent CEF and H_2 evolution was confirmed with other CEF-deficient *Chlamydomonas* mutants, such as *pgr5* (Johnson et al., 2014; Steinbeck et al., 2015) and a *pgrl1/pgr5* double mutant (Steinbeck et al., 2015). All these mutants display higher H_2 evolution ability. Can we ascribe this effect merely to competition between the two pathways for electrons, or are there other mechanisms that account for the increased H_2 evolution in CEF-deficient mutants? An indirect effect of CEF might be taken into account here: by modulating the transmembrane pH gradient, CEF indirectly controls electron transport at the level of the intersystem electron carrier, cytb6f (Joliot and Johnson, 2011). Tolleter et al. (2011) held the suppression of this control mechanism responsible for the increased evolution of H_2 in the *pgrl1* mutant. Other possible mechanisms were proposed recently and will be discussed below.

As earlier described, photoevolution of H_2 gas in *C. reinhardtii* can be obtained by two ways. One way is the preincubation of *Chlamydomonas* cells in a state of anoxia in the dark, a condition that leads to hydrogenase expression. This condition is also marked by the lack of electron sink capacity, which is the result of three parallel mechanisms:

1. The inactivity of the Calvin cycle upon dark acclimation and its slow reactivation (in the minute timescale after onset of light)
2. The very small pool of oxidized redox equivalents in the absence of oxidative phosphorylation
3. The absence of O_2 as an alternative sink when all redox equivalents are in a reduced state

Transiently, hydrogenase is then the only available electron sink that explains the strong H_2 evolution burst when the anoxic cell suspension is suddenly exposed to light. How does diminished CEF ability affect H_2 evolution in this condition? The difference does not lie in the rate but in the duration of H_2 evolution. Tolleter et al. (2011) showed that initial rates of light-driven H_2 evolution after dark anoxic incubation are the same in a *pgrl1* mutant and WT, but an earlier decline of H_2 evolution is observed in the WT compared within the mutant. In WT cells, the fast decline of the H_2 evolution rate is usually assigned to hydrogenase inhibition by O_2 released during the starting PSII activity. However, when comparing WT and *pgrl1* mutant cells for their electron flux distributions through CEF, hydrogenase, and carbon fixation, Godaux et al. (2015) concluded that the decline of the H_2 evolution rate is more likely the result of the rising electron flux directed to carbon fixation through the activation of the Calvin cycle. This explains the longer duration of the H_2 evolution in the mutant, where the lack of CEF is detrimental to Calvin cycle activation due to an altered ATP/NADPH ratio. Competition between hydrogenase-based proton reduction and Calvin cycle-dependent CO_2 reduction therefore seems critical in this context.

In long-term H_2 production assays by S deprivation, the three mutants impaired in PGRL1/PGR5-dependent CEF all display a similar H_2-overproducing phenotype: between five and eight times more than their respective WTs, with the highest H_2 production rates measured in the *pgr5* mutant (800 mL·L^{-1} of culture accumulated over 8 days, or an average of 100 mL·L^{-1}·day^{-1}) (Tolleter et al., 2011; Steinbeck et al., 2015). Steinbeck et al. (2015) proposed a model to explain the H_2-overproducing phenotype of the CEF mutants, which, in contrast to an earlier hypothesis (Tolleter et al., 2011), does not rely on a primary interaction of cyclic- and hydrogenase-dependent electron flows. The improved H_2 evolution ability of the CEF mutants during S deprivation can be explained almost entirely by their higher O_2 consumption capacity. The latter appears to be a structural adaptation of the mutant strains to their diminished ability for ATP production by CEF in the chloroplast (Dang et al., 2014). In response to S deprivation, this secondary effect of the CEF deficiency enables these mutants to reach anoxia faster, with a better-preserved PSII efficiency. Furthermore, the high rates of O_2 consumption during the H_2 production phase help to maintain the activity of the hydrogenase despite the higher PSII activity. This altogether leads to higher PSII-driven H_2 evolution. This model is supported by the observation that the indirect, PSII-independent pathway contributes very little to the improvement of H_2 production ability, as was shown in experiments in which PSII activity was blocked by addition of the inhibitor DCMU (Steinbeck et al., 2015).

9.5.3 Lessons from the pgrl1 *Mutant Phenotype for Future* H_2-Producing Strategies with Chlamydomonas

The *pgrl1* mutant combines several interesting phenotypic aspects related to H_2 evolution that may inspire future efforts for improving H_2 photoproduction by *C. reinhardtii* or other microalgae.

For instance, the studies with *pgrl1* have highlighted the potential of improving the H_2 evolution ability by increasing the O_2 uptake capacity and increasing the PSII stability during S deprivation. In this respect, the *pgr* mutants behave similar to another mutant, *stm6*, that shows increased H_2 evolution ability (Kruse et al., 2005; Volgusheva et al., 2013). The *Stm6* mutant carries an insertional knockout of a mitochondrial transcription factor-encoding gene (Wobbe and Nixon, 2013), but despite this entirely different genetic deficiency, the *stm6* mutant shares many features of its pleiotropic phenotype with the *pgrl1* and *pgr5* mutants. This includes the increased H_2-producing ability, enhanced respiration rates, and diminished CEF ability. Coupled to its increased O_2 uptake capacity, *stm6* also shows improved PSII stability under conditions of S deprivation (Volgusheva et al., 2013). Just like for the *pgr* mutants (Steinbeck et al., 2015), these features contribute to increasing the rate of PSII-dependent electron flow to hydrogenase. Independently, improved stability of the D1 subunit of PSII in conditions of S deprivation was also obtained by a site-directed mutagenesis approach (Torzillo et al., 2009, 2015; Scoma et al., 2012) and shown to lead to an increase of H_2 production under S deprivation. A similar effect was obtained by readdition of small concentrations of S during the H_2 production phase in order to reboost D1 repair and concomitant PSII activity (Laurinavichene et al., 2008). All these studies show that interventions favoring electron flow by the direct PSII-dependent pathway to hydrogenase bear a high potential for H_2 evolution improvement, whether they are obtained by genetic modification of the strains or by improvement of the culture parameters. Within this perspective, the screening for mutants with increased respiration rates is an approach that deserves further investigation (Rühle et al., 2008), in particular because the studies with *pgrl1* and *stm6* tend to show that during S deprivation, PSII stability may benefit from the shorter oxic lag phase caused by the increased O_2 uptake rates in the mutants (Kruse et al., 2005; Volgusheva et al., 2013).

Some aspects of the *pgrl1* mutant phenotype deserve attention with respect to the development of alternative procedures for H_2 production aiming to omit nutrient stress-based approaches, such as S deprivation. The S deprivation protocol is indisputably the most effective protocol today for microalgal-based H_2 photoproduction, but it has serious flaws. First, S deprivation stress affects cell viability in the long term. Second, H_2 evolution is based on a two-stage procedure, requiring a change of media to switch between biomass and H_2 production. This would be a serious drawback for its implementation on an industrial scale. Several attempts have been made to omit the S deprivation-based approach. Hypoxic hydrogen evolution in light was observed in nonnutrient stressed conditions based on light attenuation and concentration of biomass (Degrenne et al., 2011a). The obtained H_2 production yields with this approach remain low despite the presumably ideal combination of high PSII efficiencies in a low-O_2 environment. This is probably because only two out of three conditions for efficient H_2 evolution are fulfilled here. As shown by the works with *pgrl1* (Godaux et al., 2015), hydrogenase as an electron sink cannot compete with a fully operational CO_2-fixing metabolism. This illustrates that one of the key aspects of S deprivation-triggered H_2 photoevolution is the loss of the Calvin cycle as the primary sink for reduced ferredoxin (Zhang and Melis, 2002). No alternative approach for H_2 evolution will be able to match the efficiency of the S deprivation approach unless it also includes elimination of CO_2 fixation as a sink during the H_2 production phase. This aspect also constitutes one of the most important biological limitations for H_2 photoproduction in this microalga. Even if we were to succeed in making the hydrogenase insensitive to O_2, we would have to face the problem that biomass maintenance and accumulation and H_2 evolution rely on directly competing pathways, which means that they can hardly occur simultaneously. Therefore, a two-phase approach including a reversible inhibition of CO_2 fixation will probably remain the only option. As shown by the works with the *pgrl1* mutant, proton gradient formation driven by CEF (together with a hydrogenase-supported linear electron) is critical to meet the ATP demand for CO_2 fixation in anoxia. It shows that conditional elimination of CO_2 fixation as an electron sink in order to promote H_2 evolution can be obtained by creating a conditional ATP depletion in the chloroplast. To end this case study, we propose some strategies by which this might be achieved, some of them involving the *pgrl1* mutant:

1. Conditional uncoupling of photosynthetic electron transport to proton gradient formation by heterologous chloroplast expression or targeting of uncoupling proteins under the control of an inducible promotor.
2. Combining the *pgrl1* knockout with a mutation causing a deficiency of the carbon concentration mechanism (CCM). Due to low CO_2 solubility in water, microalgae strongly depend on CCM for sustaining the Calvin cycle in the absence of external CO_2 supply (Wang et al., 2015). Such a double mutant could be used in order to switch between biomass production when sufficient CO_2 is supplied and H_2 evolution under limiting CO_2 in anoxia. The CCM deficiency in combination with a local ATP depletion due to a lack of CEF may even strengthen the inhibition of CO_2-fixing reactions.
3. The extended duration of anoxic H_2 photoevolution in the *pgrl1* mutant (10 min instead of the typical 2–3 min for WT [Tolleter et al., 2011]) opens some perspectives for applying a procedure of alternating dark–light cycles in constant hypoxic conditions for H_2 evolution. It is at least a possibility that deserves investigation. There would of course be a substantial loss of efficiency of the process due to the dark periods, but that could be compensated by the far greater PSII activity than under S deprivation and would result in higher light-to-H_2 conversion efficiency during the light periods (three to four times higher than under S deprivation [Cournac et al., 2002]).

The first two approaches should be combined with a method to create a condition of hypoxia in the light and under nutrient-replete conditions, for instance, by control of the incident light and biomass density as described by Degrenne et al. (2011a), or by using a strain that carries a molecular switch to conditionally turn down PSII activity, such as the recently developed temperature-sensitive control of PSII function (Bayro-Kaiser and Nelson, 2016).

9.6 Conclusion

Efforts to use photosynthetic microorganisms as platforms for sustained H_2 photoproduction have largely contributed to revealing the metabolic flexibility of cyanobacteria and green microalgae. But the performances in terms of H_2 photoproduction rates have remained low, although they were progressively improved by exploiting biological diversity together with mutagenesis approaches. With the recent developments of systems biology and of synthetic biology, our capacity to understand cell metabolism at the system level and to engineer profound changes in pathway networks is rapidly growing. This will most probably lead to new improvements of biological H_2 photoproduction processes in the near future (Khetkorn et al., 2013). But as noted by the Nobel Prize biochemist Hartmut Michel (2012), even if the sunlight conversion efficiency of any photosynthesis-based biofuel production process were increased close to the one of photosynthetic carbon fixation (around 5%), it would remain significantly lower than the efficiency of already available photovoltaic cells.

This is due essentially to a poor utilization of the solar spectrum and to energy losses that are intrinsic to the functioning of photosystems. To overcome this barrier, a long-term and perhaps chimeric objective would be the reshaping of photosystems and their associated light harvesting systems (Blankenship and Chen, 2013). Besides research aimed at increasing rates and yields, it will also be important to analyze socioeconomics and the sustainability aspects of any photobiological H_2 photoproduction process emerging from research labs.

References

Albracht, S.P.J. 1994. Nickel hydrogenases: In search of the active site. *Biochim. Biophys. Acta* 1188, 167–204.

Alric, J. 2014. Redox and ATP control of photosynthetic cyclic electron flow in *Chlamydomonas reinhardtii*. II. Involvement of the PGR5–PGRL1 pathway under anaerobic conditions. *Biochim. Biophys. Acta* 1837, 825–834.

Alric, J., Lavergne, J., Rappaport, F. 2010. Redox and ATP control of photosynthetic cyclic electron flow in *Chlamydomonas reinhardtii*. I. Aerobic conditions. *Biochim. Biophys. Acta* 1797, 44–51.

Ananyev, G.M., Skizim, N.J., Dismukes, G.C. 2012. Enhancing biological hydrogen production from cyanobacteria by removal of excreted products. *J. Biotechnol.* 162, 97–104.

Antal, T.K., Lindblad, P. 2005. Production of H2 by sulphur-deprived cells of the unicellular cyanobacteria *Gloeocapsa alpicola* and *Synechocystis* sp. PCC 6803 during dark incubation with methane or at various extracellular pH. *J. Appl. Microbiol.* 98, 114–120.

Appel, J., Phunpruch, S., Steinmüller, K., Schulz, R. 2000. The bidirectional hydrogenase of *Synechocystis* sp. PCC 6803 works as an electron valve during photosynthesis. *Arch. Microbiol.* 173, 333–338.

Aryal, U.K., Callister, S.J., Mishra, S., Zhang, X., Shutthanandan, J.I., Angel, T.E., Shukla, A.K. et al. 2013. Proteome analyses of strains ATCC 51142 and PCC 7822 of the diazotrophic cyanobacterium *Cyanothece* sp. under culture conditions resulting in enhanced H2 production. *Appl. Environ. Microbiol.* 79, 1070–1077.

Baker, N.R. 2008. Chlorophyll fluorescence: A probe of photosynthesis in vivo. *Annu. Rev. Plant Biol.* 59, 89–113.

Baltz, A., Dang, K.-V., Beyly, A., Auroy, P., Richaud, P., Cournac, L., Peltier, G. 2014. Plastidial expression of type II NAD(P)H dehydrogenase increases the reducing state of plastoquinones and hydrogen photoproduction rate by the indirect pathway in *Chlamydomonas reinhardtii*. *Plant Physiol.* 165, 1344–1352.

Bandyopadhyay, A., Stöckel, J., Min, H., Sherman, L.A., Pakrasi, H.B. 2010. High rates of photobiological H2 production by a cyanobacterium under aerobic conditions. *Nat. Commun.* 1, 139.

Bayro-Kaiser, V., Nelson, N. 2016. Temperature-sensitive PSII: A novel approach for sustained photosynthetic hydrogen production. *Photosynth. Res.* 1–9.

Benemann, J.R., Weare, N.M. 1974. Hydrogen evolution by nitrogen-fixing *Anabaena cylindrica* cultures. *Science* 184, 174–175.

Bernstein, H.C., Charania, M.A., McClure, R.S., Sadler, N.C., Melnicki, M.R., Hill, E.A., Markillie, L.M., Nicora, C.D., Wright, A.T., Romine, M.F., Beliaev, A.S. 2015. Multiomic dynamics associate oxygenic photosynthesis with nitrogenase-mediated H2 production in *Cyanothece* sp. ATCC 51142. *Sci. Rep.* 5, 16004.

Blankenship, R.E., Chen, M. 2013. Spectral expansion and antenna reduction can enhance photosynthesis for energy production. *Curr. Opin. Chem. Biol.* 17, 457–461.

Bothe, H., Newton, W.E. 2014. Nitrogenase-dependent hydrogen production by cyanobacteria. In *Microbial Bioenergy: Hydrogen Production*, ed. D. Zannoni, R. De Philippis. Dordrecht, the Netherlands: Springer, pp. 137–153.

Bothe, H., Schmitz, O., Yates, M.G., Newton, W.E. 2010. Nitrogen fixation and hydrogen metabolism in cyanobacteria. *Microbiol. Mol. Biol. Rev.* 74, 529–551.

Cardol, P., Forti, G., Finazzi, G. 2011. Regulation of electron transport in microalgae. *Biochim. Biophys. Acta* 1807, 912–918.

Carrieri, D., Wawrousek, K., Eckert, C., Yu, J., Maness, P.-C. 2011. The role of the bidirectional hydrogenase in cyanobacteria. *Bioresour. Technol.* 102, 8368–8377.

Chochois, V., Dauvillée, D., Beyly, A., Tolleter, D., Cuiné, S., Timpano, H., Ball, S., Cournac, L., Peltier, G. 2009. Hydrogen production in *Chlamydomonas*: Photosystem II-dependent and -independent pathways differ in their requirement for starch metabolism. *Plant Physiol.* 151, 631–640.

Cohen, J., Kim, K., Posewitz, M., Ghirardi, M.L., Schulten, K., Seibert, M., King, P. 2005. Molecular dynamics and experimental investigation of H2 and O2 diffusion in [Fe]-hydrogenase. *Biochem. Soc. Trans.* 33, 80–82.

Cournac, L., Guedeney, G., Peltier, G., Vignais, P.M. 2004. Sustained photoevolution of molecular hydrogen in a mutant of *Synechocystis* sp. strain PCC 6803 deficient in the type I NADPH-dehydrogenase complex. *J. Bacteriol.* 186, 1737–1746.

Cournac, L., Mus, F., Bernard, L., Guedeney, G., Vignais, P., Peltier, G. 2002. Limiting steps of hydrogen production in *Chlamydomonas reinhardtii* and *Synechocystis* PCC 6803 as analysed by light-induced gas exchange transients. *Int. J. Hydrog. Energy* 27, 1229–1237.

DalCorso, G., Pesaresi, P., Masiero, S., Aseeva, E., Schünemann, D., Finazzi, G., Joliot, P., Barbato, R., Leister, D. 2008. A complex containing PGRL1 and PGR5 is involved in the switch between linear and cyclic electron flow in *Arabidopsis*. *Cell* 132, 273–285.

Dang, K.-V., Plet, J., Tolleter, D., Jokel, M., Cuiné, S., Carrier, P., Auroy, P. et al. 2014. Combined increases in mitochondrial cooperation and oxygen photoreduction compensate for deficiency in cyclic electron flow in *Chlamydomonas reinhardtii*. *Plant Cell* 26, 3036–3050.

Degrenne, B., Pruvost, J., Legrand, J. 2011a. Effect of prolonged hypoxia in autotrophic conditions in the hydrogen production by the green microalga *Chlamydomonas reinhardtii* in photobioreactor. *Bioresour. Technol.* 102, 1035–1043.

Degrenne, B., Pruvost, J., Titica, M., Takache, H., Legrand, J. 2011b. Kinetic modeling of light limitation and sulfur deprivation effects in the induction of hydrogen production with *Chlamydomonas reinhardtii*. Part II: Definition of model-based protocols and experimental validation. *Biotechnol. Bioeng.* 108, 2288–2299.

Doebbe, A., Rupprecht, J., Beckmann, J., Mussgnug, J.H., Hallmann, A., Hankamer, B., Kruse, O. 2007. Functional integration of the HUP1 hexose symporter gene into the genome of *C. reinhardtii*: Impacts on biological H2 production. *J. Biotechnol.* 131, 27–33.

Dutta, D., De, D., Chaudhuri, S., Bhattacharya, S.K. 2005. Hydrogen production by cyanobacteria. *Microb. Cell Factories* 4, 36.

Florin, L., Tsokoglou, A., Happe, T. 2001. A novel type of iron hydrogenase in the green alga *Scenedesmus obliquus* is linked to the photosynthetic electron transport chain. *J. Biol. Chem.* 276, 6125–6132.

Formighieri, C., Franck, F., Bassi, R. 2012. Regulation of the pigment optical density of an algal cell: Filling the gap between photosynthetic productivity in the laboratory and in mass culture. *J. Biotechnol.* 162, 115–123.

Fouchard, S., Hemschemeier, A., Caruana, A., Pruvost, J., Legrand, J., Happe, T., Peltier, G., Cournac, L. 2005. Autotrophic and mixotrophic hydrogen photoproduction in sulfur-deprived *Chlamydomonas* cells. *Appl. Environ. Microbiol.* 71, 6199–6205.

Franck, F., Juneau, P., Popovic, R. 2002. Resolution of the photosystem I and photosystem II contributions to chlorophyll fluorescence of intact leaves at room temperature. *Biochim. Biophys. Acta* 1556, 239–246.

Frey, M. 2002. Hydrogenases: Hydrogen-activating enzymes. *ChemBioChem* 3, 153–160.

Gaffron, H., Rubin, J. 1942. Fermentative and photochemical production of hydrogen in algae. *J. Gen. Physiol.* 26, 219–240.

Ghirardi, M.L., Togasaki, R.K., Seibert, M. 1997. Oxygen sensitivity of algal H2-production. In *Biotechnology for Fuels and Chemicals*, ed. B.H. Davison, C.E. Wyman, M. Finkelstein. New York: Humana Press, pp. 141–151.

Ghysels, B., Franck, F. 2010. Hydrogen photo-evolution upon S deprivation stepwise: An illustration of microalgal photosynthetic and metabolic flexibility and a step stone for future biotechnological methods of renewable H2 production. *Photosynth. Res.* 106, 145–154.

Ghysels, B., Godaux, D., Matagne, R.F., Cardol, P., Franck, F. 2013. Function of the chloroplast hydrogenase in the microalga *Chlamydomonas*: The role of hydrogenase and state transitions during photosynthetic activation in anaerobiosis. *PLoS One* 8, e64161.

Godaux, D., Bailleul, B., Berne, N., Cardol, P. 2015. Induction of photosynthetic carbon fixation in anoxia relies on hydrogenase activity and proton-gradient regulation-like1-mediated cyclic electron flow in *Chlamydomonas reinhardtii*. *Plant Physiol.* 168, 648–658.

Godaux, D., Emonds-Alt, B., Berne, N., Ghysels, B., Alric, J., Remacle, C., Cardol, P. 2013. A novel screening method for hydrogenase-deficient mutants in *Chlamydomonas reinhardtii* based on in vivo chlorophyll fluorescence and photosystem II quantum yield. *Int. J. Hydrog. Energy* 38, 1826–1836.

Godman, J.E., Molnár, A., Baulcombe, D.C., Balk, J. 2010. RNA silencing of hydrogenase(-like) genes and investigation of their physiological roles in the green alga *Chlamydomonas reinhardtii*. *Biochem. J.* 431, 345–352.

Gutekunst, K., Chen, X., Schreiber, K., Kaspar, U., Makam, S., Appel, J. 2014. The bidirectional NiFe-hydrogenase in *Synechocystis* sp. PCC 6803 is reduced by flavodoxin and ferredoxin and is essential under mixotrophic, nitrate-limiting conditions. *J. Biol. Chem.* 289, 1930–1937.

Gutthann, F., Egert, M., Marques, A., Appel, J. 2007. Inhibition of respiration and nitrate assimilation enhances photohydrogen evolution under low oxygen concentrations in *Synechocystis* sp. PCC 6803. *Biochim. Biophys. Acta* 1767, 161–169.

Hallenbeck, P.C. 2012. Hydrogen production by cyanobacteria. In *Microbial Technologies in Advanced Biofuels Production*, ed. P.C. Hallenbeck. Boston: Springer, pp. 15–28.

Happe, R.P., Roseboom, W., Pierik, A.J., Albracht, S.P.J., Bagley, K.A. 1997. Biological activation of hydrogen. *Nature* 385, 126–126.

Happe, T., Hemschemeier, A., Winkler, M., Kaminski, A. 2002. Hydrogenases in green algae: Do they save the algae's life and solve our energy problems? *Trends Plant Sci.* 7, 246–250.

Happe, T., Naber, J.D. 1993. Isolation, characterization and N-terminal amino acid sequence of hydrogenase from the green alga *Chlamydomonas reinhardtii*. *Eur. J. Biochem.* 214, 475–481.

Hedderich, R. 2004. Energy-converting [NiFe] hydrogenases from archaea and extremophiles: Ancestors of complex I. *J. Bioenerg. Biomembr.* 36, 65–75.

Hemschemeier, A., Happe, T. 2011. Alternative photosynthetic electron transport pathways during anaerobiosis in the green alga *Chlamydomonas reinhardtii*. *Biochim. Biophys. Acta* 1807, 919–926.

Hemschemeier, A., Melis, A., Happe, T. 2009. Analytical approaches to photobiological hydrogen production in unicellular green algae. *Photosynth. Res.* 102, 523–540.

Houille-Vernes, L., Rappaport, F., Wollman, F.-A., Alric, J., Johnson, X. 2011. Plastid terminal oxidase 2 (PTOX2) is the major oxidase involved in chlororespiration in *Chlamydomonas*. *Proc. Natl. Acad. Sci. U.S.A.* 108, 20820–20825.

Hwang, J.C., Chen, C.H., Burris, R.H. 1973. Inhibition of nitrogenase-catalyzed reductions. *Biochim. Biophys. Acta* 292, 256–270.

Jans, F., Mignolet, E., Houyoux, P.-A., Cardol, P., Ghysels, B., Cuiné, S., Cournac, L., Peltier, G., Remacle, C., Franck, F. 2008. A type II NAD(P)H dehydrogenase mediates light-independent plastoquinone reduction in the chloroplast of *Chlamydomonas*. *Proc. Natl. Acad. Sci. U.S.A.* 105, 20546–20551.

Johnson, X., Steinbeck, J., Dent, R.M., Takahashi, H., Richaud, P., Ozawa, S.-I., Houille-Vernes, L. et al. 2014. Proton gradient regulation 5-mediated cyclic electron flow under ATP- or redox-limited conditions: A study of ΔATPase pgr5 and ΔrbcL pgr5 mutants in the green alga *Chlamydomonas reinhardtii*. *Plant Physiol.* 165, 438–452.

Johnson, X., Vandystadt, G., Bujaldon, S., Wollman, F.-A., Dubois, R., Roussel, P., Alric, J., Béal, D. 2009. A new setup for in vivo fluorescence imaging of photosynthetic activity. *Photosynth. Res.* 102, 85–93.

Joliot, P., Johnson, G.N. 2011. Regulation of cyclic and linear electron flow in higher plants. *Proc. Natl. Acad. Sci. U.S.A.* 108, 13317–13322.

Jurado-Oller, J.L., Dubini, A., Galván, A., Fernández, E., González-Ballester, D. 2015. Low oxygen levels contribute to improve photohydrogen production in mixotrophic non-stressed *Chlamydomonas* cultures. *Biotechnol. Biofuels* 8, 149.

Kessler, E. 1973. Effect of anaerobiosis on photosynthetic reactions and nitrogen metabolism of algae with and without hydrogenase. *Arch. Mikrobiol.* 93, 91–100.

Khetkorn, W., Khanna, N., Incharoensakdi, A., Lindblad, P. 2013. Metabolic and genetic engineering of cyanobacteria for enhanced hydrogen production. *Biofuels* 4, 535–561.

Khetkorn, W., Lindblad, P., Incharoensakdi, A. 2012. Inactivation of uptake hydrogenase leads to enhanced and sustained hydrogen production with high nitrogenase activity under high light exposure in the cyanobacterium *Anabaena siamensis* TISTR 8012. *J. Biol. Eng.* 6, 19.

Kosourov, S.N., Batyrova, K.A., Petushkova, E.P., Tsygankov, A.A., Ghirardi, M.L., Seibert, M. 2012. Maximizing the hydrogen photoproduction yields in *Chlamydomonas reinhardtii* cultures: The effect of the H2 partial pressure. *Int. J. Hydrog. Energy* 37, 8850–8858.

Kosourov, S.N., Ghirardi, M.L., Seibert, M. 2011. A truncated antenna mutant of *Chlamydomonas reinhardtii* can produce more hydrogen than the parental strain. *Int. J. Hydrog. Energy* 36, 2044–2048.

Kothari, A., Parameswaran, P., Garcia-Pichel, F. 2014. Powerful fermentative hydrogen evolution of photosynthate in the cyanobacterium *Lyngbya aestuarii* BL J mediated by a bidirectional hydrogenase. *Front. Microbiol.* 5, 680.

Kothari, A., Potrafka, R., Garcia-Pichel, F. 2012. Diversity in hydrogen evolution from bidirectional hydrogenases in cyanobacteria from terrestrial, freshwater and marine intertidal environments. *J. Biotechnol.* 162, 105–114.

Kruse, O., Nixon, P.J., Schmid, G.H., Mullineaux, C.W. 1999. Isolation of state transition mutants of *Chlamydomonas reinhardtii* by fluorescence video imaging. *Photosynth. Res.* 61, 43–51.

Kruse, O., Rupprecht, J., Bader, K.-P., Thomas-Hall, S., Schenk, P.M., Finazzi, G., Hankamer, B. 2005. Improved photobiological H2 production in engineered green algal cells. *J. Biol. Chem.* 280, 34170–34177.

Kumar, K., Mella-Herrera, R.A., Golden, J.W. 2010. Cyanobacterial heterocysts. *Cold Spring Harb. Perspect. Biol.* 2, a000315.

Laurinavichene, T.V., Kosourov, S.N., Ghirardi, M.L., Seibert, M., Tsygankov, A.A. 2008. Prolongation of H2 photoproduction by immobilized, sulfur-limited *Chlamydomonas reinhardtii* cultures. *J. Biotechnol.* 134, 275–277.

Lavorel, J. 1962. Hétérogénéité de la chlorophylle in vivo I. Spectres d'émission de fluorescence. *Biochim. Biophys. Acta* 60, 510–523.

Lenz, O., Bernhard, M., Buhrke, T., Schwartz, E., Friedrich, B. 2002. The hydrogen-sensing apparatus in *Ralstonia eutropha*. *J. Mol. Microbiol. Biotechnol.* 4, 255–262.

Lindberg, P., Schütz, K., Happe, T., Lindblad, P. 2002. A hydrogen-producing, hydrogenase-free mutant strain of *Nostoc punctiforme* ATCC 29133. *Int. J. Hydrog. Energy* 27, 1291–1296.

Lindblad, P., Christensson, K., Lindberg, P., Fedorov, A., Pinto, F., Tsygankov, A. 2002. Photoproduction of H2 by wildtype *Anabaena* PCC 7120 and a hydrogen uptake deficient mutant: From laboratory experiments to outdoor culture. *Int. J. Hydrog. Energy* 27, 1271–1281.

Lopes Pinto, F.A., Troshina, O., Lindblad, P. 2002. A brief look at three decades of research on cyanobacterial hydrogen evolution. *Int. J. Hydrog. Energy* 27, 1209–1215.

Masukawa, H., Sakurai, H., Hausinger, R.P., Inoue, K. 2014. Sustained photobiological hydrogen production in the presence of N2 by nitrogenase mutants of the heterocyst-forming cyanobacterium *Anabaena*. *Int. J. Hydrog. Energy* 39, 19444–19451.

Mattoo, A.K., Edelman, M. 1987. Intramembrane translocation and posttranslational palmitoylation of the chloroplast 32-kDa herbicide-binding protein. *Proc. Natl. Acad. Sci. U.S.A.* 84, 1497–1501.

McDermott, J.E., Oehmen, C.S., McCue, L.A., Hill, E., Choi, D.M., Stöckel, J., Liberton, M., Pakrasi, H.B., Sherman, L.A. 2011. A model of cyclic transcriptomic behavior in the cyanobacterium *Cyanothece* sp. ATCC 51142. *Mol. Biosyst.* 7, 2407–2418.

Melis, A., Seibert, M., Happe, T. 2004. Genomics of green algal hydrogen research. *Photosynth. Res.* 82, 277–288.

Melis, A., Zhang, L., Forestier, M., Ghirardi, M.L., Seibert, M. 2000. Sustained photobiological hydrogen gas production upon reversible inactivation of oxygen evolution in the green alga *Chlamydomonas reinhardtii*. *Plant Physiol.* 122, 127–136.

Melnicki, M.R., Pinchuk, G.E., Hill, E.A., Kucek, L.A., Fredrickson, J.K., Konopka, A., Beliaev, A.S. 2012. Sustained H2 production driven by photosynthetic water splitting in a unicellular cyanobacterium. *mBio* 3, e00197-12.

Meuser, J.E., D'Adamo, S., Jinkerson, R.E., Mus, F., Yang, W., Ghirardi, M.L., Seibert, M., Grossman, A.R., Posewitz, M.C. 2012. Genetic disruption of both *Chlamydomonas reinhardtii* [FeFe]-hydrogenases: Insight into the role of HYDA2 in H2 production. *Biochem. Biophys. Res. Commun.* 417, 704–709.

Michel, H. 2012. Editorial: The nonsense of biofuels. *Angew. Chem. Int. Ed.* 51, 2516–2518.

Mignolet, E., Lecler, R., Ghysels, B., Remacle, C., Franck, F. 2012. Function of the chloroplastic NAD(P)H dehydrogenase Nda2 for H2 photoproduction in sulphur-deprived *Chlamydomonas reinhardtii*. *J. Biotechnol.* 162, 81–88.

Mikheeva, L.E., Schmitzh, O., Shestakov, S.V., Bothe, H. 1995. Mutants of the cyanobacterium *Anabaena variabilis* altered in hydrogenase activities. *Z. Naturforsch. C* 50, 505–510.

Min, H., Sherman, L.A. 2010. Hydrogen production by the unicellular, diazotrophic cyanobacterium *Cyanothece* sp. strain ATCC 51142 under conditions of continuous light. *Appl. Environ. Microbiol.* 76, 4293–4301.

Mitchell, P. 1961. Coupling of phosphorylation to electron and hydrogen transfer by a chemiosmotic type of mechanism. *Nature* 191, 144–148.

Mitchell, P. 1966. Chemiosmotic coupling in oxidative and photosynthetic phosphorylation. *Biol. Rev.* 41, 445–501.

Mitsui, A., Suda, S. 1995. Alternative and cyclic appearance of H2 and O2 photoproduction activities under non-growing conditions in an aerobic nitrogen-fixing unicellular cyanobacterium *Synechococcus* sp. *Curr. Microbiol.* 30, 1–6.

Müller, M., Gorrell, T.E. 1983. Metabolism and metronidazole uptake in *Trichomonas vaginalis* isolates with different metronidazole susceptibilities. *Antimicrob. Agents Chemother.* 24, 667–673.

Nicolet, Y., Cavazza, C., Fontecilla-Camps, J.C. 2002. Fe-only hydrogenases: Structure, function and evolution. *J. Inorg. Biochem.* 91, 1–8.

Nicolet, Y., Lemon, B.J., Fontecilla-Camps, J.C., Peters, J.W. 2000. A novel FeS cluster in Fe-only hydrogenases. *Trends Biochem. Sci.* 25, 138–143.

Nyberg, M., Heidorn, T., Lindblad, P. 2015. Hydrogen production by the engineered cyanobacterial strain Nostoc PCC 7120 ΔhupW examined in a flat panel photobioreactor system. *J. Biotechnol.* 215, 35–43.

Pape, M., Lambertz, C., Happe, T., Hemschemeier, A. 2012. Differential expression of the *Chlamydomonas* [FeFe]-hydrogenase-encoding HYDA1 gene is regulated by the copper response regulator. *Plant Physiol.* 159, 1700–1712.

Peltier, G., Tolleter, D., Billon, E., Cournac, L. 2010. Auxiliary electron transport pathways in chloroplasts of microalgae. *Photosynth. Res.* 106, 19–31.

Pierik, A.J., Roseboom, W., Happe, R.P., Bagley, K.A., Albracht, S.P.J. 1999. Carbon monoxide and cyanide as intrinsic ligands to iron in the active site of [NiFe]-hydrogenases NiFe(CN)2CO, biology's way to activate H2. *J. Biol. Chem.* 274, 3331–3337.

Pilak, O., Mamat, B., Vogt, S., Hagemeier, C.H., Thauer, R.K., Shima, S., Vonrhein, C., Warkentin, E., Ermler, U. 2006. The crystal structure of the apoenzyme of the iron–sulphur cluster-free hydrogenase. *J. Mol. Biol.* 358, 798–809.

Reddy, K.J., Haskell, J.B., Sherman, D.M., Sherman, L.A. 1993. Unicellular, aerobic nitrogen-fixing cyanobacteria of the genus *Cyanothece. J. Bacteriol.* 175, 1284–1292.

Rühle, T., Hemschemeier, A., Melis, A., Happe, T. 2008. A novel screening protocol for the isolation of hydrogen producing *Chlamydomonas reinhardtii* strains. *BMC Plant Biol.* 8, 107.

Schreiber, U., Vidaver, W. 1974. Chlorophyll fluorescence induction in anaerobic *Scenedesmus obliquus. Biochim. Biophys. Acta* 368, 97–112.

Schütz, K., Happe, T., Troshina, O., Lindblad, P., Leitão, E., Oliveira, P., Tamagnini, P. 2004. Cyanobacterial H(2) production—A comparative analysis. *Planta* 218, 350–359.

Scoma, A., Krawietz, D., Faraloni, C., Giannelli, L., Happe, T., Torzillo, G. 2012. Sustained H2 production in a *Chlamydomonas reinhardtii* D1 protein mutant. *J. Biotechnol.* 157, 613–619.

Seibert, M., Benson, D.K., Flynn, T.M. 2001. Method and apparatus for rapid biohydrogen phenotypic screening of microorganisms using a chemochromic sensor. U.S. Patent 6277589. Midwest Research Institute.

Serebriakova, L., Zorin, N.A., Lindblad, P. 1994. Reversible hydrogenase in *Anabaena variabilis* ATCC 29413. *Arch. Microbiol.* 161, 140–144.

Sherman, L.A., Min, H., Toepel, J., Pakrasi, H.B. 2010. Better living through *Cyanothece*—Unicellular diazotrophic cyanobacteria with highly versatile metabolic systems. In *Recent Advances in Phototrophic Prokaryotes*, ed. P.C. Hallenbeck. New York: Springer, pp. 275–290.

Shima, S., Thauer, R.K. 2007. A third type of hydrogenase catalyzing H2 activation. *Chem. Rec.* 7, 37–46.

Smith, L.A., Hill, S., Yates, M.G. 1976. Inhibition by acetylene of conventional hydrogenase in nitrogen-fixing bacteria. *Nature* 262, 209–210.

Steinbeck, J., Nikolova, D., Weingarten, R., Johnson, X., Richaud, P., Peltier, G., Hermann, M., Magneschi, L., Hippler, M. 2015. Deletion of proton gradient regulation 5 (PGR5) and PGR5-like 1 (PGRL1) proteins promote sustainable light-driven hydrogen production in *Chlamydomonas reinhardtii* due to increased PSII activity under sulfur deprivation. *Front. Plant Sci.* 6.

Stöckel, J., Welsh, E.A., Liberton, M., Kunnvakkam, R., Aurora, R., Pakrasi, H.B. 2008. Global transcriptomic analysis of *Cyanothece* 51142 reveals robust diurnal oscillation of central metabolic processes. *Proc. Natl. Acad. Sci. U.S.A.* 105, 6156–6161.

Stripp, S.T., Goldet, G., Brandmayr, C., Sanganas, O., Vincent, K.A., Haumann, M., Armstrong, F.A., Happe, T. 2009. How oxygen attacks [FeFe] hydrogenases from photosynthetic organisms. *Proc. Natl. Acad. Sci. U.S.A.* 106, 17331–17336.

Thauer, R.K. 1998. Biochemistry of methanogenesis: A tribute to Marjory Stephenson. 1998 Marjory Stephenson Prize Lecture. *Microbiol. Read. Engl.* 144 (Pt 9), 2377–2406.

Toepel, J., Welsh, E., Summerfield, T.C., Pakrasi, H.B., Sherman, L.A. 2008. Differential transcriptional analysis of the cyanobacterium *Cyanothece* sp. strain ATCC 51142 during light-dark and continuous-light growth. *J. Bacteriol.* 190, 3904–3913.

Tolleter, D., Ghysels, B., Alric, J., Petroutsos, D., Tolstygina, I., Krawietz, D., Happe, T. et al. 2011. Control of hydrogen photoproduction by the proton gradient generated by cyclic electron flow in *Chlamydomonas reinhardtii*. *Plant Cell* 23, 2619–2630.

Torzillo, G., Scoma, A., Faraloni, C., Ena, A., Johanningmeier, U. 2009. Increased hydrogen photoproduction by means of a sulfur-deprived *Chlamydomonas reinhardtii* D1 protein mutant. *Int. J. Hydrog. Energy* 34, 4529–4536.

Torzillo, G., Scoma, A., Faraloni, C., Giannelli, L. 2015. Advances in the biotechnology of hydrogen production with the microalga *Chlamydomonas reinhardtii*. *Crit. Rev. Biotechnol.* 35, 485–496.

Tredici, M.R., Zittelli, G.C. 1998. Efficiency of sunlight utilization: Tubular versus flat photobioreactors. *Biotechnol. Bioeng.* 57, 187–197.

Tsygankov, A., Kosourov, S., Tolstygina, I., Ghirardi, M., Seibert, M. 2006. Hydrogen production by sulfur-deprived *Chlamydomonas reinhardtii* under photoautotrophic conditions. *Int. J. Hydrog. Energy* 31, 1574–1584.

Tsygankov, A.A., Fedorov, A.S., Kosourov, S.N., Rao, K.K. 2002. Hydrogen production by cyanobacteria in an automated outdoor photobioreactor under aerobic conditions. *Biotechnol. Bioeng.* 80, 777–783.

Tsygankov, A.A., Hall, D.O., Liu, J., Rao, K.K. 1998. An automated helical photobioreactor incorporating cyanobacteria for continuous hydrogen production. In *Biohydrogen*. Berlin: Springer, pp. 431–440.

Vignais, P.M., Billoud, B. 2007. Occurrence, classification, and biological function of hydrogenases: An overview. *Chem. Rev.* 107, 4206–4272.

Vignais, P.M., Billoud, B., Meyer, J. 2001. Classification and phylogeny of hydrogenases. *FEMS Microbiol. Rev.* 25, 455–501.

Vogt, S., Lyon, E.J., Shima, S., Thauer, R.K. 2007. The exchange activities of [Fe] hydrogenase (iron–sulfur-cluster-free hydrogenase) from methanogenic archaea in comparison with the exchange activities of [FeFe] and [NiFe] hydrogenases. *JBIC J. Biol. Inorg. Chem.* 13, 97–106.

Volbeda, A., Charon, M.-H., Piras, C., Hatchikian, E.C., Frey, M., Fontecilla-Camps, J.C. 1995. Crystal structure of the nickel–iron hydrogenase from *Desulfovibrio gigas*. *Nature* 373, 580–587.

Volgusheva, A., Styring, S., Mamedov, F. 2013. Increased photosystem II stability promotes H2 production in sulfur-deprived *Chlamydomonas reinhardtii*. *Proc. Natl. Acad. Sci. U.S.A.* 110, 7223–7228.

Walker, D. 1992. Tansley review no. 36 excited leaves. *New Phytol.* 121, 325–345.

Wang, Y., Stessman, D.J., Spalding, M.H. 2015. The CO2 concentrating mechanism and photosynthetic carbon assimilation in limiting CO2: How *Chlamydomonas* works against the gradient. *Plant J.* 82, 429–448.

Wecker, M.S.A., Meuser, J.E., Posewitz, M.C., Ghirardi, M.L. 2011. Design of a new biosensor for algal H2 production based on the H2-sensing system of *Rhodobacter capsulatus*. *Int. J. Hydrog. Energy* 36, 11229–11237.

Weissman, J.C., Benemann, J.R. 1977. Hydrogen production by nitrogen-starved cultures of *Anabaena cylindrica*. *Appl. Environ. Microbiol.* 33, 123–131.

Welkie, D., Zhang, X., Markillie, M.L., Taylor, R., Orr, G., Jacobs, J., Bhide, K. et al. 2014. Transcriptomic and proteomic dynamics in the metabolism of a diazotrophic cyanobacterium, *Cyanothece* sp. PCC 7822 during a diurnal light–dark cycle. *BMC Genomics* 15, 1185.

Wobbe, L., Nixon, P.J. 2013. The mTERF protein MOC1 terminates mitochondrial DNA transcription in the unicellular green alga *Chlamydomonas reinhardtii*. *Nucleic Acids Res.* 41, 6553–6567.

Wykoff, D.D., Davies, J.P., Melis, A., Grossman, A.R. 1998. The regulation of photosynthetic electron transport during nutrient deprivation in *Chlamydomonas reinhardtii*. *Plant Physiol.* 117, 129–139.

Zehr, J.P., Waterbury, J.B., Turner, P.J., Montoya, J.P., Omoregie, E., Steward, G.F., Hansen, A., Karl, D.M. 2001. Unicellular cyanobacteria fix N2 in the subtropical North Pacific Ocean. *Nature* 412, 635–638.

Zhang, L., Happe, T., Melis, A. 2002. Biochemical and morphological characterization of sulfur-deprived and H2-producing *Chlamydomonas reinhardtii* (green alga). *Planta* 214, 552–561.

Zhang, L., Melis, A. 2002. Probing green algal hydrogen production. *Philos. Trans. R. Soc. B Biol. Sci.* 357, 1499–1511.

Zhang, X., Sherman, D.M., Sherman, L.A. 2014. The uptake hydrogenase in the unicellular diazotrophic cyanobacterium *Cyanothece* sp. strain PCC 7822 protects nitrogenase from oxygen toxicity. *J. Bacteriol.* 196, 840–849.

Section II

Liquid Biofuels

10

Microbial Production of Liquid Biofuels through Metabolic Engineering

Wei Ning Chen and Jiahua Shi

Contents

10.1 Introduction

10.1.1 General Context

The total reserve of fossil fuels is depleting gradually all over the world. According to the statistical review of world energy by British Petroleum (BP 2015), the total reserve of coal, crude oil, and natural gas in the world is estimated at 892 billion tonnes, 1688 billion barrels, and 186 trillion cubic meters, respectively. The numbers seem to be huge at first glance, but based on the current level of energy consumption, it is estimated that coal would be exhausted in 2128, crude oil used up by 2067, and the last cubic meter of natural gas by 2069. The fossil fuels (e.g., diesel and gasoline) play important roles in energy supply for transportation, agriculture, industries, and so forth (Demirbas 2009). Thus, the depletion of fossil fuel could lead to an energy crisis and is becoming a serious concern for governments. Moreover, the world is facing the potential threat of global climate change, which can be principally attributed to the rising greenhouse gas (GHG) emissions from fossil fuel usage, according to the statistics by the International Energy Agency (IEA) (Ugarte et al. 2003; IEA 2015).

Therefore, it is of great importance to develop alternative fuels with high energy content. Biofuel has been identified as nontoxic, biodegradable, and renewable fuel, which can contribute to the minimization of fossil fuel burning and GHG production (Li et al. 2008; Naik et al. 2010). Biofuel refers to the fuels produced from biomass predominantly, and it can be in gas, liquid, or solid form. Biofuel can be produced from waste oils, biological waste, and renewable resources, which makes it an economically alternative supply of energy compared with the traditional fossil fuels (Meher et al. 2006). A lot of biofuels can be produced from biomass, including biodiesels, ethanol, methanol, and hydrogen (Demirbas 2008). It is important to develop renewable biofuels for economic sustainability and environmental concerns.

10.1.2 Three Generations of Liquid Biofuels

The first generation, or conventional biofuels, such as biodiesel, bioethanol, and biogas, is mainly made from food crops, for example, cereals, oilseeds, and sugar crops. The need for first-generation biofuels has been observed to increase during the past few years, reaching 22.5 billion L in 2012 (REN21 2014; Tolmac et al. 2014). However, this increase of first-generation fuels also accelerates the competition with water and land used for food production.

The second generation of biofuels is derived from nonfood biomass, such as lignocellulosic biomass or woody crops, agricultural residues, and nonedible parts of plants. Second-generation biofuels could help solve the problems encountered with first-generation fuels and also provide a larger proportion of global fuel supply, with greater benefits for the environment. It is reported that by 2022, the renewable fuels estimated to be used in the United States will reach 36 billion gallons, and 16 billion gallons are supposed to be generated from cellulosic sources and residual nonfood

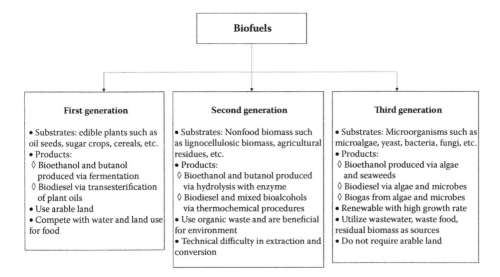

Figure 10.1   Classification of the first, second, and third generations of biofuels based on their substrates, products, advantages, and disadvantages.

parts of current crops (Schnepf and Yacobucci 2013). The problem that second-generation biofuel encounters is the difficulty of extracting useful feedstock from cellulosic material, and the technology to change it to biofuels is at an early stage of exploration.

The third generation of biofuels is produced from microorganisms, mainly micro-algae, bacteria, yeast, and fungi. The introduction of third-generation microbial-produced biofuels can rectify the problems of first- and second-generation ones, such as land needs and possible environmental pollution (Thornley et al. 2008). The expense of third-generation biofuels from microorganisms can be significantly decreased compared with that of traditional fossil fuels and other generations. Indeed, micro-bial biofuels can be produced from a variety of materials, such as wastewater, food waste, residual biomass, and many other by-products as the carbon source, which reduces the competition for resources with land and food production. Besides, the microbial biofuels are more easily renewable with a high growth rate and high lipid content.

Figure 10.1 shows the classification of first, second, and third generation of biofuels based on their substrates, products, advantages, and disadvantages.

10.1.3 Workflow of Industrial Liquid Biofuel Production

A flowsheet of the industrial biofuel production process of different generations is shown in Figure 10.2. The industrial production of bioethanol is used as an exam-ple to illustrate the process in first- and second-generation biofuel production, as in Figure 10.2a. Typically, the first-generation bioethanol production utilizes food crops that contain sugars or starch as substrates, whereas the second-generation

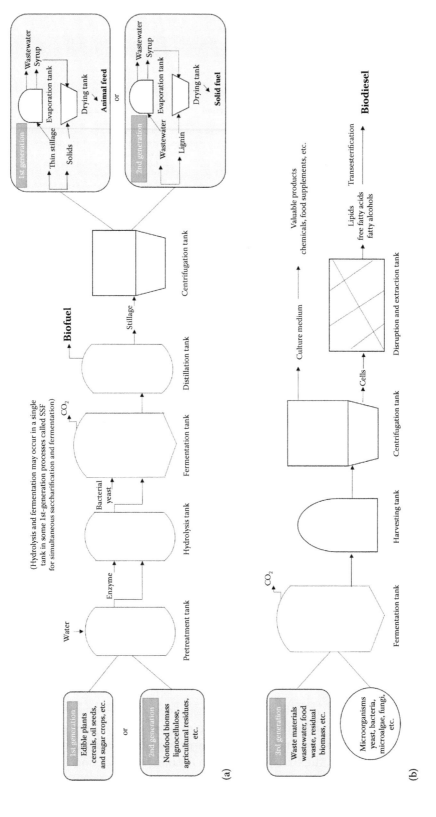

Figure 10.2 General flowsheet of industrial biofuel production processes: (a) first- and second-generation biofuel production and (b) third-generation biofuel production.

bioethanol production uses different types of lignocellulosic materials (Nigam and Singh 2011). The substrates first go through a pretreatment tank or equipment, where they can be milled and liquefied, facilitating the separation of complex carbohydrate molecules constituting cellulose, hemicellulose, and lignin. This is followed by the hydrolysis process catalyzed by enzymes or an acid or alkali to transform the substrates into simple sugars. The third step is the fermentation process carried out by yeasts such as *Saccharomyces cerevisiae* and yielding ethanol and carbon dioxide. The temperature is kept at 28°C–30°C to fit the optimum cultivation environment for yeast. The range of pH is adjusted between 5.0 and 6.0, which is crucial for the activity of enzymes and transport proteins in yeast and preventing bacteria contamination at the same time. The next step is distillation of the fermentation liquid to separate and purify the ethanol to a concentration above 99.7% for fuel application. The stillage from the distillation tank is centrifuged, consisting of residual substrate, yeast cells, by-products, and so forth. It is then treated by evaporation and a drying process and used as animal feed or solid fuels (Taherzadeh et al. 2013; Patrik et al. 2014).

Figure 10.2b exhibits the industrial process of third-generation biofuel production. Different from those of the first and second generations, the substrates utilized in the third generation can be waste products and solve the resource problems in first and second generations, such as competing land need with food and possible environmental pollution. The microorganisms can use these waste resources directly in fermentation at optimized conditions. However, different microorganisms have various requirements. Microalgae needs light and CO_2 to carry out photosynthesis, and the other microorganisms do not need these. The temperature range can be 20°C–35°C, and the pH value range can be 4–10, according to different species. Strain selection is critical and oleaginous microorganisms, which can naturally produce more than 20% lipid of their cell dry weight, have been hot research targets in the last few decades. Fast development of genetic engineering technologies has also improved the quality of strains, leading them to produce desired target products. The case study in the latter part of this chapter exhibits the application of genetic engineering in *S. cerevisiae* to produce desired products by changing the conventional pathways. After fermentation, the cells of microorganisms are harvested and centrifuged to separate them from the culture medium. The cells are disrupted and inclusions, mostly lipids, free fatty acids, or fatty alcohols, are extracted and turned into biofuels by transesterification. The culture medium contains secretions from cells and can be extracted and separated to become value-added chemicals, food supplements, and so forth.

10.1.4 Microorganisms for Microbial Liquid Biofuel Production

10.1.4.1 Microalgae
Microalgae refer to a diverse group of aquatic microorganisms that can be identified in many different environments, such as freshwater and sea water in saline conditions (Carlsson et al. 2007; Schenk et al. 2008). They can grow photoautotrophically or

heterotrophically and are able to use carbon dioxide and sunlight to accumulate oils through photosynthesis under specific conditions. Microalgae is a potential biofuel feedstock because they can produce polysaccharides or sugar and triacylglycerides or fats, depending on the species and cultivation conditions (e.g., nitrogen sources, pH, and salinity), which are the raw materials for bioethanol and biodiesel production. Among these conditions, the nitrogen limitation degree and light intensity are the key parameters in biofuel production (Solovchenko et al. 2008). Many microalgae have been reported with high oil accumulation abilities, such as *Botryococcus braunii*, *Chlorella* sp., *Nannochloropsis* sp., and *Phaeodactylum triornutum* (Largeau et al. 1980; Chisti 2007; Metzger et al. 2008). Although microalgae can produce different biofuels with high contents and sufficient diversities (Posten and Schaub 2009), they have some problems, such as the growth conditions being relatively expensive with light requirements and CO_2 and freshwater supply, the scale-up procedures being complicated, and the growth rate being low compared with that of other microorganisms.

10.1.4.2 Bacteria

Bacteria consist of a large domain of prokaryotic microorganisms and are found in a huge variety of habitats, such as soil, water, plants, animals and some extreme environments. They are capable of accumulating specialized lipids, such as fatty alcohols, wax esters, hydrocarbons, triacylglycerols (TAGs), and peptidolipids (Leman 2009). A lot of bacteria are considered to be potential oil producers, such as *Escherichia coli*, *Arthrobacter* sp., and *Rhodococcus opacus* (Alvarez and Steinbüchel 2002; Kalscheuer et al. 2006; Meng et al. 2009). Bacteria present advantages in producing biofuels when compared with microalgae. Their growth rate is much higher, and the cultivation methods are relatively simple. The technologies and tools for genetic engineering and metabolic engineering to modify bacteria performance to improve oil accumulation are very mature and widely used. However, the produced lipids are often complex, and few species are able to produce TAG.

10.1.4.3 Yeasts

Yeasts are single-cell eukaryotic microorganisms that are widely dispersed in nature, such as plant leaves, fruits, and soil. Some species can accumulate lipids, which mainly consist of TAGs and free fatty acids, under proper cultivation conditions with adjusted carbon-to-nitrogen (C/N) ratios, pH, oxygen, temperature, and inorganic salt (Li et al. 2008). Numerous yeasts have been reported for biofuel production, such as *S. cerevisiae*, *Rhodosporidium toruloides*, and *Yarrowia lipolytica* (Hassan et al. 1996; Papanikolaou and Aggelis 2009). Yeast exhibits considerable advantages in lipid production, such as high growth rate, high ability of lipid accumulation, easy culture methods, and ability to be grown in the conventional bioreactors. The variety of substrates that yeasts can utilize is also an enormous advantage. Carbohydrates, hydrocarbons, and waste materials can be effectively utilized by yeast and converted into lipids (Hu et al. 2008; Azócar et al. 2010; Papanikolaou and Aggelis 2011). The only problem is that yeast cultivation needs organic carbon sources to produce biofuel.

10.1.4.4 Fungi

Fungi constitute a eukaryotic organism kingdom that is usually identified in soil and dead matter. Although various species of fungi are reported to accumulate lipids, such as *Mucor circinelloides, Aspergillus oryzae, Humicola lanuginose,* and *Mortierella vinacea* (Vicente et al. 2009), few reports describe the use of these fungal species for direct biodiesel production. Most reports describe the use of these fungi for the production of special lipids, such as eicosapentaenoic acid, arachidonic acid, decosahexaenoic acid, and γ-linolenic acid (Papanikolaou et al. 2004). Although the growth rate is high and culture methods are simple, the drawback is that fungi need organic carbon sources.

Table 10.1 shows the comparison of major microorganisms based on their differences in lipid content and composition, and their advantages and disadvantages.

10.1.5 Types of Microbial Liquid Biofuels

10.1.5.1 Bioalcohols

Bioalcohols are produced mainly by direct and indirect fermentation. In direct fermentation, the starting carbonaceous raw materials (starches, sugarcane bagasses, cellulosic materials, etc.) are first reduced to simple sugar monomers by chemical or microbial methods. Then the simple sugar monomers are converted to biofuels by microbial fermentation, for example, by wild-type or genetically engineered bacteria or yeast strains (Panwar et al. 2013; Singh and Nigam 2014). The indirect fermentation involves feedstock pyrolysis first, and then it converts the produced gas into alcohol using acetogenic bacteria (Klasson et al. 1992; Elshahed 2010).

Generally, the substrates are first pretreated using various methods, such as steam explosion, ammonia fiber explosion, acid or alkali hydrolysis, and biological or enzymatic processes, which are introduced to facilitate the breakdown of cellulose and hemicellulose, the complex carbohydrate molecules in the substrates (Prasad et al. 2007). This step is then followed by the saccharification process, which releases the sugars from polysaccharides, and the fermentation process, which produces bioalcohols by microorganisms. The final step is the distillation process for the separation and collection of bioalcohols.

In the fermentation process, yeast such as *S. cerevisiae* (Liu et al. 2008) and bacteria such as *Zymomonas mobilis* (Joachimsthal and Rogers 2000) are the commonly used microorganisms in the bioalcohol production industries. The development of genetic engineering has enabled the expression of foreign genes in yeast and bacteria to efficiently take up various substrates and produce a high yield of bioalcohols.

One important bioalcohol is bioethanol, which is ethanol or ethyl alcohol produced from biomass by hydrolysis and fermentation. It can be used together with traditional gasoline as fuel in automobiles. The mixture of bioethanol and gasoline

TABLE 10.1 Comparison of Major Microorganisms as Feedstock for Biofuel Production

Microorganism	Lipid Content (% Dry Weight)	Main Lipid Composition	Advantages	Disadvantages
Microalgae	20–70	Fatty acid Free fatty acid Hydrocarbon Unusual lipid	High lipid accumulation yield Some species can produce hydrocarbons directly Possibility of heterotrophic growth Convert carbon dioxide to potential biofuel Can provide several different types of renewable biofuels	Expensive, complicated scale-up Necessity of light Low growth rate
Bacteria	18–70	Free fatty acid Unusual lipid Wax esters	High growth rate Easy culture method Relatively easy-to-use biological engineering technology, genetic engineering, and metabolic engineering to modify bacteria performance to improve oil accumulation Can be grown in conventional bioreactor	Production of complex lipids Few species able to accumulate TAG
Yeast	25–70	Triacylglycerols Polyhydroxyalkanoates Free fatty acid	High growth rate High lipid accumulation Easy culture method Can be grown in conventional bioreactor	Need organic carbon sources
Fungi	25	Unusual lipid Fatty acid	High growth rate Easy culture method Can be grown in conventional bioreactor	Need organic carbon sources

Source: From Azócar, L. et al., *Appl. Microbiol. Biotechnol.*, 88, 621–636, 2010.

has a lower sulfur content than gasoline alone, which contributes to a lower emission of sulfur oxide—the major component of acid rain. Another significant bioalcohol is biobutanol, which is a four-carbon alcohol containing more carbon and hydrogen than ethanol. Compared with bioethanol, biobutanol is easier to blend with gasoline, is less corrosive, and has a higher density and lower volatility. All these advantages make biobutanol an essential target in industries.

10.1.5.2 Biodiesels

Biodiesel refers to oil and diesel fuel and is made up of long-chain alkyl, methyl, ethyl, or propyl esters. Conventional biodiesels are mainly produced from vegetable oil or animal fat by chemical transesterification. However, the land yield of this traditional production is relatively low, and it competes for land for food production. The microbial production of biodiesels is of increasing interest to industries, and they can be produced from a wide range of raw materials, such as biomass, corn, plant lipids, animal fats, and even used cooking oils, by esterification or transesterification of triglycerides with alcohol (Fukuda et al. 2001; Shi et al. 2011). Microbial oils, also called single-cell oils, produced by oleaginous microorganisms, such as yeast, fungi, bacteria, and algae, can be used as potential feedstock for the production of biodiesels. The oleaginous microorganisms are rich in oil content, have a short life cycle, are easy to scale up, and can be produced from residual biomass as a carbon source, consequently reducing the competition for resources with land and food production.

Microbial oils are synthesized in all microorganisms, whereas only oleaginous microorganisms accumulate significant quantities of oils, to an extent of more than 20% of their cell dry biomass under certain cultivation conditions. Oil accumulation is favored when a high C/N ratio ranging from 120–140 is utilized in cultivation (Shi et al. 2013). Different organic nitrogen sources also affect oil production. Although both organic and inorganic nitrogen sources can be utilized to culture yeast, cell growth is favored over lipid production when inorganic nitrogen sources are used, whereas organic nitrogen sources favor oil production. Trace metal ions, such as Mn^{2+}, Mg^{2+}, Cu^{2+}, Zn^{2+}, Fe^{3+}, and Ca^{2+}, are also reported to influence biomass and lipid accumulation. An increase in oil content could be induced with optimized concentrations of trace metal ions (Hassan et al. 1996; Liu et al. 2000; Li et al. 2006; Huang et al. 2010). For different microorganisms and different culture conditions (such as temperature, pH, and culture time), the oil contents and compositions are also different. For microalgae, the suitable ranges for the temperature, pH, and light intensity (lux) are 16°C–27°C, 7–9, and 1,000–10,000 (depends on volume and density), respectively; for bacteria, the suitable temperature range is 20°C–37°C and the pH is 7; for yeast, the suitable ranges for temperature and pH are 20°C–30°C and 5.5–6.5, respectively; and for fungi, the suitable ranges for temperature and pH are 10°C–50°C and 5.0–7.0, respectively. The optimal culture time will vary with the size of containers, the inoculation amount, and so on. Table 10.2 lists some examples of oleaginous microorganisms and their oil contents and lipid compositions (Meng et al. 2009). The oil from these microorganisms can be transformed into (m)ethyl esters by esterification and provide sufficient resources for biodiesel production.

TABLE 10.2 Some Oleaginous Microorganisms and Their Oil Contents and Compositions

Microorganism	Oil Content (% Dry Weight)	Lipid Composition (% Total Lipid)					
		C16:0	C16:1	C18:0	C18:1	C18:2	C18:3
Microalgae		12–21	55–57	1–2	58–60	4–20	14–30
Botryococcus braunii	25–75						
Schizochytrium sp.	50–77						
Nitzschia sp.	45–47						
Yeast		11–37	1–6	1–10	28–66	3–24	1–3
Rhodosporidium toruloides	66						
Lipomyces starkeyi	64						
Cryptococcus albidus	65						
Bacteria		8–10	10–11	11–12	25–28	14–17	–
Bacillus alcalophilus	18–24						
Rhodococcus opacus	25						
Arthrobacter sp.	40						
Fungi		7–23	1–6	2–6	19–81	8–40	4–42
Mortierella vinacea	66						
Aspergillus oryzae	57						
Humicola lanuginosa	75						

10.2 Metabolic Engineering of Microorganisms

10.2.1 Overview

Microorganisms provide an alternative way to produce numerous important chemicals and biofuels from renewable biomass, which makes the development of sustainable processes important for industries (Peralta-Yahya et al. 2012). In the past, microorganisms were genetically modified by introducing chemical mutation to increase the yield of target metabolites. This procedure was generally a random mutagenesis that was followed by a screening method to choose the mutant strains that can overexpress the desired products (Stephanopoulos and Vallino 1991). However, the specific metabolic pathways are not clear using this method, leading to a lack of details on the limitations or enzymes involved in the process. Metabolic engineering is a new technique that emerged in the last two decades that can analyze the metabolic pathway of microorganisms and provide details on the constraints and their effects to facilitate a better production of the desired products. Based on the biological information, metabolic engineering can genetically engineer the cellular metabolism pathways and improve the microorganism properties for industrial use (Dai and Nielsen 2015). The rapid developments in DNA synthesis, synthetic biology, and systems biology also shed light on the metabolic engineering-related industries for the production of essential chemicals and metabolites. Compared with the traditional random mutagenesis and chemical synthesis methods, metabolic engineering of microorganisms is less expensive, effort saving, and environmentally friendly (Jullesson et al. 2015). Metabolic engineering includes the enhancement and redirection of metabolic flux in microorganisms by modifying the metabolic pathways through genetic modifications, which include the deletion or overexpression of genes, the introduction of recombinant DNA cassettes, and the replacement of gene expression signals (Ostergaard et al. 2000).

10.2.2 Metabolic Engineering Applications in Biofuel Production

Based on the successful development of tools and strategies for metabolic engineering in microorganisms, numerous chemicals, including biofuels, have been reported to be produced with increasing yield and productivity, as shown in Table 10.3. The microbial production of bioalcohols, such as 1-butanol and 1-propanol, and biodiesel feedstock, such as fatty acid ethyl esters (FAEEs), has been reported using metabolic engineering strategies. These strategies have been applied in the well-studied model microorganisms, such as *Escherichia coli* and *S. cerevisiae* (Zhang et al. 2011). They have been studied in depth, and their characterization has been extensively explored with plenty of the genetic manipulation tools available. Moreover, they can utilize a wide range of waste materials as carbon sources, grow at a rapid rate, and be potentially engineered to tolerate toxic biofuels; thus, they can be widely used for biofuel production (Alper et al. 2006; Trinh et al. 2008).

10.3 Case Study: Enhanced Production of Fatty Alcohols by Engineering the TAG Synthesis Pathway in *S. cerevisiae* (Tang and Chen 2014)

10.3.1 Metabolism of S. cerevisiae

S. cerevisiae is often used as a microbial production host for fatty acids and fatty acid derivatives because of its wealth of available genetic information and the versatile tools designed to manipulate its metabolic pathways (Nielsen 2009; Tang et al. 2013). In *S. cerevisiae*, the direct product of de novo fatty acid synthesis is acyl-coenzyme A (acyl-CoA). The fatty acyl-CoAs are generated and released to cytosol during fatty acid synthesis and then fed into a lipid storage pathway, where the TAGs and sterol esters are formed (Tehlivets et al. 2007). In *S. cerevisiae*, the acyl-CoA, diacylglycerol acyltransferase (DAGAT), encoded by *dga1*, has been demonstrated to be one of the essential enzymes in TAG synthesis (Mora et al. 2012). It is responsible for the terminal step of TAG formation by condensation of acyl-CoAs and diacylglycerol. Overexpression of the active DAGAT in *S. cerevisiae* resulted in a significant increase in lipid content, while the mutant strain lacking the *dga1* gene was found to synthesize a reduced level of TAGs (Oelkers et al. 2002; Kamisaka et al. 2013). Since DAGAT is the rate-limiting enzyme involved in the acyl-CoA-dependent TAG synthesis pathway, the cellular acyl-CoA production levels are expected to be altered through regulation of DAGAT activity. The product of *S. cerevisiae* fatty acid synthase (FAS) is already in the activated form of fatty acyl-CoAs; therefore, a modified pathway to

TABLE 10.3 Representative Fuels Produced Using Systems Metabolic Engineering

Biofuels	Production Performances			Systems Metabolic Engineering Strategies	References
	Titer	Yield (Biofuel/Substrate)	Productivity	Strains (Substrate)	
Ethanol	48 g/L	0.37 g/g	0.79 g/L/h	*S. cerevisiae* (xylose)	Engineering to ferment mixtures of xylose and cellobiose as carbon sources	Ha et al. (2011)
1-Propanol	10.8 g/L	0.11 g/g	0.14 g/L/h	*E. coli* (glucose/glycerol)	Releasing feedback inhibition of amino acid biosynthesis, and deleting competing pathway and stress response gene	Choi et al. (2012)
1-Butanol	585.3 g/L	0.31 g/g	1.32 g/L/h	*Clostridium acetobutylicum* (glucose)	Reinforcing the direct butanol route by deleting other acid-forming routes and overexpressing the mutant adhE1 gene	Jang et al. (2012)
Hydrocarbon	580 mg/L			*E. coli* (glucose)	Engineering fatty acid biosynthesis and degradation pathways	Choi and Lee (2013)
Fatty acid ethyl esters	25.4 mg/L			*S. cerevisiae* (glucose)	Engineering to increase cytosolic acyl-CoA pools and medium optimization and codon-optimized acyltransferase genes	Thompson and Trinh (2014)
Fatty acid short-chain esters (FASEs)	1008 mg/L	17.1 mg/g	21.9 mg/L/h	*E. coli* (glycerol)	Generating an artificial metabolic pathway to produce short-chain alcohols, and combining fatty acyl-CoA and short-chain alcohols to produce FASEs	Guo et al. (2014)

Source: From Cho, C. et al., *Biotechnol. Adv.*, 33, 1455–1466, 2015.

reduce the turnover of acyl-CoAs to lipids becomes an attractive approach to provide an increased amount of intermediates for producing fatty acid derivatives.

A series of enzymes have been found to have the ability to convert fatty acyl-CoAs to useful chemicals, such as fatty aldehydes, fatty alcohols, and even alkanes or alkenes (Metz et al. 2000; Wahlen et al. 2009; Schirmer et al. 2010). The fatty acyl-CoA reductase (FAR) is prominent as one of the most studied enzymes. It can be divided into aldehyde-forming and alcohol-forming classes (Metz et al. 2000). The aldehyde-forming FAR catalyzes a two-electron reduction of the active forms of fatty acids to fatty aldehydes, while the alcohol-forming FAR catalyzes a four-electron reduction of the active forms of fatty acids to fatty alcohols. In this case study, *S. cerevisiae* was metabolically modified to produce enhanced fatty acid derivatives. The gene disruption strategy, together with the heterologous gene expression strategy, was applied to alter the intracellular metabolic pathways in *S. cerevisiae*. The *S. cerevisiae* biosystem was designed to be a direct way to overproduce fatty acid derivatives, realizing the idea of applying biotechnologies in yeast cells for the production of engineering chemicals.

10.3.2 Materials and Methods

10.3.2.1 Metabolic Pathway Design
The strategy herein aims to supply sufficient activated forms of fatty acids for enhanced production of fatty acid derivatives. The *dga1* gene responsible for the rate-limiting step of the fatty acyl-CoA-dependent pathway of TAG synthesis is going to be deleted, to release more fatty acyl-CoAs from TAGs. The *FAR* gene will be cloned and overexpressed in the gene-disrupted strain, to convert the accumulated fatty acyl-CoAs to the corresponding fatty acid derivatives, as shown in Figure 10.3. The optimization of

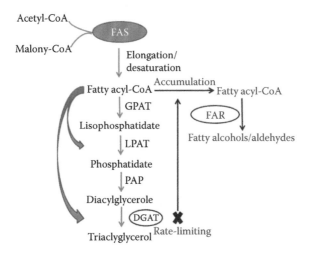

Figure 10.3   Workflow of the metabolic engineering pathway in *S. cerevisiae* for enhanced fatty acid derivative production. GPAT, glycerol-3-phosphate acyltransferase; LPAT, lysophosphatidate acyltransferase; PAP, phosphatidate phosphatase; DGAT, diglyceride acyltransferase; FAR, fatty acid reductase. The designed pathway is red.

culture medium will be carried out with the engineered strain for further enhancement of fatty acid derivative production.

10.3.2.2 Strains and Cultivations

The *S. cerevisiae* wild-type BY4741 (ΔMAT, *his3Δ*, *leu2Δ*, *met15Δ*, and *ura3Δ*) was obtained from the American Tissue Culture Collection (ATCC) and cultured in YPD medium (10 g/L yeast extract, 20 g/L peptone, and 20 g/L dextrose) at 30°C, 250 rpm. The Δ*dga1* strain, which is the strain with a deletion of the *dga1* gene, was obtained in this study, and the correct disrupted colonies were screened on minimal selective medium with the *his*+ marker: a 6.7 g/L yeast nitrogen base with 5 g/L ammonium sulfate, but without amino acids; 20 g/L dextrose; and a 0.65 g/L *his* dropout amino acid mixture that contains all the necessary amino acids for yeast except histidine at 30°C. The *FAR* gene from birds (accession JN638549) was obtained from Professor Frentzen's lab (Hellenbrand et al. 2011) and inserted into the pYES-DEST52 expression vector under the galactose-inducible promoter (GAL1). The recombinant vector was transformed into the *S. cerevisiae* wild-type strain to construct the "wt-FAR" and into the Δ*dga1*-disrupted strain to construct the "Δ*dga1*-FAR." Meanwhile, the empty pYES vector was transformed into the *S. cerevisiae* wild-type strain to construct the "wt-E" and into the Δ*dga1*-disrupted strain to construct the "Δ*dga1*-E" in order to eliminate the possible influence of the pYES vector to the strain and consider it as a control, compared with those with the pYES-DEST52 expression vector. All the recombinant strains were cultivated in minimal selective medium with the *ura*+ marker (0.77 g/L) and induced by 20 g/L galactose at 30°C, 250 rpm.

The optimization of fatty alcohol production in the engineered strain was carried out in culture medium with different carbon and nitrogen concentrations: the yeast nitrogen base without ammonium sulfate and amino acids was added to a concentration of 1.7 g/L; the *ura* dropout amino acid mixture was added to 0.77 g/L; the raffinose was added to 10 g/L; the galactose concentration was fixed at 20 and 60 g/L, respectively; and the ammonium was added to a reduced concentration of 2.5 and 0.1 g/L, respectively. All the cultivations were kept at 30°C, 250 rpm.

10.3.2.3 Target Gene Disruption of Δdga1

The polymerase chain reaction (PCR)-mediated one-step gene disruption method was applied to obtain the Δ*dga1* mutant (Hegemann et al. 2006). The pUG 27 plasmid containing the *his*+ marker was used for the generation of the disruption cassette. The primers designed for the PCR procedure were as follows: 5′ATACATAAGGAAACGCAGAGGCATACAGTTTGAACAGTCACATA<u>AC AGCTGAAGCTTCGTACGC</u>3′ for the forward strand, which contained 45 nucleotide stretches that are homologous to a sequence upstream of the start codon of *dga1*, and 5′CCTT ATTTATTCTAACATATTTTGTGTTTTCCAATGAATTCATTA<u>GCATA GGCCACTAGTGGATCTG</u>3′ for the reverse strand, which contained nucleotide stretches that are homologous to a sequence downstream of the stop codon of *dga1*. The PCR program was carried out as follows: 95°C for 30 s, 55°C for 30 s, and 72°C for 2 min in one cycle; a total of 35 cycles were performed. The PCR product was purified and transformed into *S. cerevisiae*-competent cells by the PEG-LiAc method, and the correct colonies were screened on a minimal *his*⁺-selective agar plate and determined by colony PCR.

10.3.2.4 Metabolite Extraction

Five milliliters of *S. cerevisiae* cells from the stationary phase was collected and centrifuged at 10,000*g* for 10 min, to remove the supernatants. The cell pellets were washed twice with distilled water and redissolved in 400 mL of ethyl acetate. Ten milliliters of heptadecanoic acid (10 mg/mL) was added as an internal control, and the same volume of glass beads (425–600 mm) was added. The sample tubes were vortexed in a bead mill for cell breakage according to the following procedure: speed of 4.0 m/s and vortex time for one cycle of 20 s; a total of six cycles were carried out. After cell lysis, the mixtures were vibrated at room temperature for 1 h and then centrifuged at 10,000 rpm for 10 min. The supernatants containing the intracellular metabolites were carefully collected. To extract the metabolites from culture broth, 5 mL of collected culture medium was added to 10 mL of heptadecanoic acid (10 mg/mL) as an internal control, 500 mL of acetic acid, 500 mL of saturated NaCl, and 2 mL of ethyl acetate. The mixture tubes were placed in a rotating analyzer at room temperature for metabolite extraction. After 1 h, the upper layers of ethyl acetate were carefully collected. The mixture tubes were added to another 2 mL of ethyl acetate and rotated for one more hour. The layers of ethyl acetate were collected again and mixed with the previous collections. All the collections were prepared for further analysis.

10.3.2.5 Derivatization of Fatty Alcohols and Lipids

The derivatization of fatty alcohols was carried out by adding N,O-bistrifluoroacetamide (BSTFA) to 1% chlorotrimethylsilane (TMCS): 60 mL of the collections from both the intracellular extracts and supernatant extracts was added to 40 mL of BSTFA, respectively, and kept at 70°C for 30 min. After reaction, the sample tubes were shaken at room temperature for 1 h. Through this operation, the trimethylsilyl group was added to the fatty alcohols at the position of the active hydrogen. The derivative samples were then transferred to gas chromatography–mass spectrometry (GC-MS) vials and analyzed by the GC-MS analysis method. The derivatization of lipid was carried out by the BF3/MeOH method. The collection samples from the intracellular extracts and supernatant extracts were evaporated to complete dryness. The derivatization was initiated by adding 500 mL of boron trifluoride solution to 14% methanol. The reaction was carried out at 95°C for 30 min, followed by adding 50 mL of saturated sodium chloride solution immediately. Through this operation, both the TAGs and free fatty acids were esterified to fatty acid methyl esters (FAMEs). After cooling to room temperature, the FAMEs were extracted by 300 mL of n-hexane. The derivative samples were then collected and transferred into GC-MS vials for the analysis. The lipid content was calculated as total fatty acids in the *S. cerevisiae* cells.

10.3.2.6 GC-MS Procedures

The derivative samples were analyzed on an Agilent 7890A and 5975C GC-MS system equipped with a DB-5 capillary column (30 m, 250 mm inner diameter, and 0.25 mm thickness; Agilent J&W Scientific, Folsom, California). One milliliter of each sample was injected into the GC-MS system by an Agilent autosampler and separated by a DB-5 capillary column in a splitless mode. Helium was used as the carrier gas with a flow rate of 1.1 mL/min. The inlet temperature was set at 250°C, and the MS source temperature was set at 230°C. The oven temperature was first set at 80°C for 1 min and then increased with a ramp-up rate of 7°C/min to 250°C, where it was held for 10 min.

10.3.2.7 Statistical Analysis

For each calculation, three biological replicates were performed. The FAME standard mix (C8-C24, Sigma, St. Louis, Missouri) and fatty alcohol standards were used as references for the sample calculation. The mean value and standard deviation calculated from three different batches of data were used to represent the production levels of the components. Student's t-test was applied in the analysis of statistical significance, and a $p < 0.05$ threshold was used to reflect the significant difference in this study.

10.3.3 Results

10.3.3.1 Effect of the Engineered Pathway and Lipid Content

The deletion of the $dga1$ gene slowed down the TAG synthesis pathway in S. cerevisiae, and the TAG content was supposed to be reduced. Herein, the lipid contents in the wild-type strain and the gene-disrupted strain were investigated. When the strains were grown in culture medium with raffinose and galactose as carbon sources, the majority of the detected fatty acids included in the lipids were myristic acid (C14:0), palmitoleic acid (C16:1), palmitic acid (C16:0), oleic acid (C18:1), and stearic acid (C18:0). As shown in Table 10.4, palmitoleic acid was found to be the most abundant in both the wild-type strain and the gene-disrupted strain, representing about 60% of the total fatty acids. The production of palmitic acid, oleic acid, and stearic acid accounted for about 10%–15% of the total fatty acids, and myristic acid constituted only 1%–2% of that.

The lipid contents in the $\Delta dga1$ strain were dramatically decreased compared with those in the wild-type strain, as shown in Figure 10.4. The total fatty acids reduced from 116 to 77 mg/L ($p < 0.01$). Among the fatty acid compositions, the production level of myristic acid decreased to 87% of that in the wild-type strain, from 1.5 to 1.3 mg/L ($p < 0.05$); the palmitoleic acid decreased to 64%, from 71 to 45 mg/L

TABLE 10.4 Fatty Acid Compositions Included in the Lipid Content in the Wild-Type (wt) Strain and *dga1* Gene-Disrupted (Δ*dga1*) Strain

Strain	Total Fatty Acid (mg/L)	Compositions				
		C14:0	C16:0	C16:1	C18:0	C18:1
wt	116	1.3%	11.7%	61.2%	12.8%	13.0%
Δ*dga1*	77	1.7%	12.1%	58.9%	16.3%	11.0%

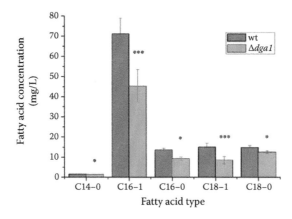

Figure 10.4 Fatty acid production levels between the wild-type strain and the *dga1* gene-disrupted strain. The standard deviation was calculated from three independent experimental results. The asterisk indicates the statistical significance of the fatty acid changes between the wild-type strain and the gene-disrupted strain: *, $p < 0.05$; **, $p < 0.01$; ***, $p < 0.005$.

($p < 0.005$); the palmitic acid decreased to 68%, from 13.5 to 9.3 mg/L ($p < 0.05$); the oleic acid decreased to 56%, from 15.0 to 8.0 mg/L ($p < 0.005$); and the stearic acid decreased to 85%, from 14.8 to 12.6 mg/L ($p < 0.05$). The obtained results reflected that the disruption of the *dga1* gene resulted in the reduction of lipid contents, which was regarded as a consequence of the significantly blocked TAG synthesis pathway, depending on the fatty acyl-CoAs as substrates. The gene-disrupted strain was then used for the enhanced production of fatty acid derivatives, in which the FAR was overexpressed.

10.3.3.2 Enhanced Production of Fatty Alcohols in Engineered S. cerevisiae

By overexpressing the *FAR* gene in the *S. cerevisiae* strains, the fatty alcohols were observed to be produced as shown in Figure 10.5, while no fatty aldehydes were detected. In both the wt-FAR and Δ*dga1*-FAR strains, the detected fatty alcohols in the cells were myristyl alcohol (C14:0), cetyl alcohol (C16:0), palmitoleyl alcohol (C16:1), and stearyl alcohol (C18:0). As shown in Table 10.5, cetyl alcohol was found to be the most abundant, representing about 80% of the total fatty alcohols, and stearyl alcohol represented more than 10% of the total fatty alcohols. Myristyl alcohol and palmitoleyl alcohol only accounted for a small percentage, about 2%–3% and 0.4%, respectively. The wt-E and Δ*dga1*-E strains did not produce any detectable levels of fatty alcohols, as there is no enzyme to convert fatty acyl-CoA to fatty alcohols.

Due to the blockage of the fatty acyl-CoA-dependent TAG synthesis pathway, the fatty alcohols produced in the Δ*dga1*-FAR cells increased drastically, as shown in Table 10.5. The total fatty alcohols increased to about 1.7 times that in wt-FAR ($p < 0.05$), and the maximum production level reached 45 mg/L, compared with 26 mg/L in the wt-FAR strain. The production level of myristyl alcohol increased by three times ($p < 0.005$), from 0.5 to 1.5 mg/L; the palmitoleyl alcohol yield increased by about two times ($p < 0.05$), from 0.1 to 0.2 mg/L; and the cetyl alcohol and stearyl

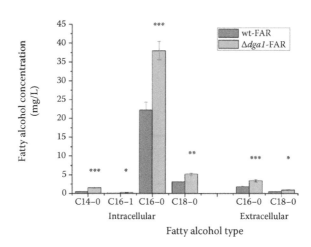

Figure 10.5 Intracellular and extracellular fatty alcohol production levels between the wt-FAR strain and the $\Delta dga1$-FAR strain. The standard deviation was calculated from three independent experimental results. The asterisk indicates the statistical significance of the fatty alcohol changes between the wt-FAR strain and the $\Delta dga1$-FAR strain: *, $p < 0.05$; **, $p < 0.01$; ***, $p < 0.005$.

TABLE 10.5   Fatty Alcohol Compositions in the wt-FAR Strain and $\Delta dga1$-FAR Strain

	Strain Type	Total Fatty Alcohols (mg/L)	Compositions			
			C14:0	C16:0	C16:1	C18:0
Intracellular	wt-FAR	26	2.0%	85.7%	0.4%	11.9%
	$\Delta dga1$-FAR	45	3.4%	84.7%	0.4%	11.5%
Extracellular	wt-FAR	2.2	–	78.7%	–	21.3%
	$\Delta dga1$-FAR	4.3	–	79.5%	–	20.5%

alcohol production levels increased by about 1.7 times, from 22 to 38 mg/L ($p < 0.005$) and 3.0 to 5.1 mg/L ($p < 0.01$) respectively. The results were consistent with the designed pathway for the enhanced production of fatty acid derivatives.

The culture broth collected from the stationary phase was extracted by ethyl acetate. The obtained metabolites were derived by the same BSTFA + TMCS method and analyzed by GC-MS. The results showed that the fatty alcohol contents secreted in the culture medium were much less than those in the cells. The total fatty alcohol yield was only 10% of the intracellular ones (Figure 10.5). The detectable fatty alcohols were cetyl alcohol (C16:0) and stearyl alcohol (C18:0). Cetyl alcohol represented about 80% of the total fatty alcohols in the supernatant, and stearyl alcohol represented about 20% (Table 10.4). The results reflect that the yields of both the intracellular and extracellular fatty alcohols were improved by applying the engineered pathway in *S. cerevisiae*.

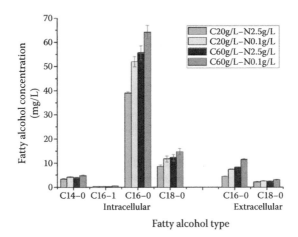

Figure 10.6   Intracellular and extracellular fatty alcohol production levels in Δ*dga1*-FAR strain grown in culture conditions with different carbon and nitrogen concentrations. C2%-N2.5 g/L, culture condition with 20 g/L galactose and 2.5 g/L ammonium sulfate; C2%-N0.1 g/L, 20 g/L galactose and 0.1 g/L ammonium sulfate; C6%-N2.5 g/L, 60 g/L galactose and 2.5 g/L ammonium sulfate; C6%-N0.1 g/L, 60 g/L galactose (wt/vol) and 0.1 g/L ammonium sulfate. The standard deviation was calculated from three independent experimental results.

10.3.3.3 Optimization of Fatty Alcohol Production in Engineered S. cerevisiae

The production of fatty alcohols in the engineered strain was further optimized in different culture media, and the final titer of fatty alcohols was investigated. With the increase of carbon concentration and decrease of nitrogen concentration, the yield of fatty alcohols in the Δ*dga1*-FAR strain was significantly increased. It was reflected in the intracellular production of myristyl alcohol, cetyl alcohol, and stearyl alcohol, and the secretion of cetyl alcohol and steary alcohol, as shown in Figure 10.6. When the galactose concentration was fixed at 20 g/L, the yield of total intracellular fatty alcohols increased to 51 mg/L in the culture condition with 2.5 g/L ammonium sulfate, and to 73 mg/L in that with 0.1 g/L ammonium sulfate. This yield was further increased to 68 and 84 mg/L, respectively, when the galactose concentration was increased to 60 g/L in the culture medium. The increasing phenomenon was also found in the secretion of fatty alcohols in the Δ*dga1*-FAR strain. The yield of extracellular fatty alcohols detected in the broth was finally increased to 14 mg/L in the condition with 60 g/L galactose and 0.1 g/L ammonium sulfate. It was concluded that the fatty alcohol production could be dramatically enhanced by optimizing the culture conditions with an increased carbon concentration and limited nitrogen concentration.

10.3.4 Discussion

The TAGs are the most important storage forms for energy and fatty acids in the yeast cells, and they are always regarded as potential fuel targets (Yu et al. 2012).

Three different pathways were discovered that contribute to the TAG biosynthesis in *S. cerevisiae*:

1. Acyl-CoA: DAGAT encoded by the *dga1* gene was responsible for the rate-limiting step of TAG formation by condensation of acyl-CoA and diacylglycerol (Oelkers et al. 2002; Sorger and Daum 2002).
2. Phospholipid: DAGAT encoded by the *lro1* gene took part in the TAG synthesis from the conversion of phospholipid and diacylglycerol (Dahlqvist et al. 2000; Oelkers et al. 2000).
3. TAGs could be formed using free fatty acids as substrates and diacylglycerol as the acceptor directly (Wagner and Paltauf 1994; Sorger and Daum 2002).

Among these three pathways, *dga1* and *lro1* played a predominant role in the synthesis of total cellular TAGs, and the *dga1*-regulated pathway was the only one that directly depended on the fatty acyl-CoAs as substrates.

While TAGs are valuable, they cannot be used as fuels or for other applications directly. Therefore, the conversion of TAGs to fatty acid-derived chemicals suffers from a practical standpoint. In yeast cells, fatty acyl-CoAs serve as a source for more activated starting materials. The fatty acyl-CoAs released from the FAS complex are either used for the synthesis of steryl ester (Yang et al. 1996; Yu et al. 1996) or destined for the TAG storage pathway. In this study, the fatty acyl-CoA-dependent pathway of TAG synthesis was modified to be significantly blocked. The metabolic engineering of the conversion process from fatty acyl-CoAs to lipids has been demonstrated as one of the more efficient means to accumulate precursors for fatty acid derivative synthesis. The deletion of *dga1* resulted in a distinct decrease of the total fatty acids included in the lipid content, and the turnover of fatty acyl-CoAs to lipid was therefore concluded to be reduced. It was verified by the increased production levels of fatty alcohols in the engineered strains with disrupted *dga1* and overexpressed FAR. Compared with that in the wt-FAR strain, the fatty alcohol yield in the Δ*dga1*-FAR strain increased more than 1.5 times, reflected in both the intracellular and extracellular contents.

The FAR activity herein exhibited as the alcohol-forming type, and no fatty aldehydes were detected. We compared its amino acid sequence with that of the previously reported MaFAR from *Marinobacter aquaeolei* VT8 (YP_959769), which reduces the fatty acyl-CoAs to the corresponding alcohols by a single four-electron reduction (Willis et al. 2011). Two specific domains were found with high similarity. The first domain, responsible for reducing fatty acyl-CoA to the fatty aldehyde, shared about 60% similarity between residues 375 and 660 of the MaFAR, and the second domain, responsible for the fatty aldehyde reductase, shared 39% similarity between residues 2 and 260 of the MaFAR. These similarities were consistent with the MaFAR activity and made the FAR in this study more similar in reactivity to convert the fatty acyl-CoAs to fatty alcohols. The increased production levels in the engineered strain were evident in the production of long-chain fatty alcohols (C14–C18), and the saturated ones predominated in the total fatty alcohols.

Various converting enzymes have been reported to have the ability to convert fatty acyl-CoAs to corresponding fuels in *S. cerevisiae* cells. The FAEEs could be produced by the acyl-CoA alcohol transferase, which uses fatty acyl-CoAs and ethanol as substrates (Stöveken et al. 2005; Yu et al. 2013). The fatty aldehydes could be generated by the aldehyde-forming FAR (Wahlen et al. 2009), and it could be further transformed to alkanes or alkenes by aldehyde decarbonylase (Schirmer et al. 2010). These kinds of tool enzymes make the production of valuable chemicals from fatty acids highly efficient. The metabolic engineering of the TAG synthesis pathway studied here is expected to play a crucial role in providing precursors for all these bioreactions and to realize the idea of enhanced production of fatty acid–derived chemicals.

In conclusion, this study not only successfully improved the production of fatty alcohols in *S. cerevisiae* cells, which are used in a wide range of products and increasingly demanded by a large market, but also provided a more direct strategy to transform fatty acids included in lipids into useful fuels and chemicals, to solve the bottleneck of utilizing TAGs as precursors.

10.3.5 Conclusion

This case study emphasizes the genetic engineering of *S. cerevisiae* to lead it to the production of fatty alcohol, which is the desired target product, rather than the detailed industrial biofuel production process. The modified yeast cell can be applied in industry for microbial biofuel production under proper conditions, scale-up to 500 L pilot-scale production or further expanded industrial-scale production, which is the ultimate goal of the research. Different from lab-scale research, more factors need to be taken into consideration when the cells are cultured at the industrial scale. Section 10.1.4.3 exhibited the industrial biofuel production process by yeast in general.

At the lab scale, commercial glucose is utilized as the carbon source, but it can be expensive for industrial-scale production. Wastewater, waste food, residual biomass, and many other by-products can be pretreated and utilized as the carbon source, which reduces the competition for resources with land and food production, and also is beneficial from economic and environmental points of view.

S. cerevisiae, like most yeast, is a facultative anaerobe. It converts sugars into CO_2 and water in an aerobic environment, but it produces CO_2 and ethanol when cultured in an anaerobic environment. In the industrial bioethanol production process, it is crucial to exclude significant oxygen during cultivation. An optimized culture condition needs to be obtained for the scale-up fermentation of the engineered *S. cerevisiae*, such as the nutrients, temperature, pH value, and oxygen content, to achieve the best capacity and productivity. The engineered yeast strain will be cultured in a propagation tank, which mixes nutrients, water, enzymes, and yeast cells with good control of the temperature, pH value, and oxygen concentration, to activate yeast and shorten the fermentation lag time.

After fermentation, the downstream separation and purification procedures are also important in the production of bioethanol at the industrial scale. Efficient distillation and dehydration processes are needed to separate and purify the bioethanol from the industrial fermentation tanks.

Different from a lab-scale or medium-term study, the biggest concern in industrial bioethanol production will be the cost. The productive efficiency is important, and the balance between cost and effectiveness is even more crucial. The environmental aspect also needs to be taken into consideration, such as the chemicals and wastes generated during the process. Much more effort is required to develop renewable biofuels for economic sustainability and environmental concerns.

References

Alper H, Moxley J, Nevoigt E, Fink GR, Stephanopoulos G. 2006. Engineering yeast transcription machinery for improved ethanol tolerance and production. *Science* 314: 1565–1568.

Alvarez H, Steinbüchel A. 2002. Triacylglycerols in prokaryotic microorganisms. *Appl Microbiol Biotechnol* 60:367–376.

Azócar L, Ciudad G, Heipieper HJ, Navia R. 2010. Biotechnological processes for biodiesel production using alternative oils. *Appl Microbiol Biotechnol* 88:621–636.

BP. 2015. Statistical energy review of world energy. London, BP. www.bp.com/statisticalreview.

Carlsson AS, van Beilen JB, Möller R, Clayton D. 2007. In *Micro- and Macro-Algae: Utility for Industrial Applications, Outputs from the EPOBIO Project.* Newbury (UK), ed. D Bowles. Realising the Economic Potential of Sustainable Resources—Bioproducts from Non-food Crops. University of York, CPL Press, London, p. 86. Available at http://www.biofuelstp.eu/downloads/epobio_aquatic_report.pdf.

Chisti Y. 2007. Biodiesel from microalgae. *Biotechnol Adv* 25:294–306.

Cho C, Choi SY, Luo ZW, Lee SY. 2015. Recent advances in microbial production of fuels and chemicals using tools and strategies of systems metabolic engineering. *Biotechnol Adv* 33:1455–1466.

Choi YJ, Lee SY. 2013. Microbial production of short-chain alkanes. *Nature* 502:571–574.

Choi YJ, Park JH, Kim TY, Lee SY. 2012. Metabolic engineering of *Escherichia coli* for the production of 1-propanol. *Metab Eng* 14:477–486.

Dahlqvist A, Ståhl U, Lenman M, Banas A, Lee M, Sandager L, Ronne H, Stymne S. 2000. Phospholipid: Diacylglycerol acyltransferase: An enzyme that catalyzes the acyl-CoA-independent formation of triacylglycerol in yeast and plants. *Proc Natl Acad Sci USA* 97(12):6487–6492.

Dai ZJ, Nielsen J. 2015. Advancing metabolic engineering through systems biology of industrial microorganisms. *Curr Opin Biotechnol* 36:8–15.

Demirbas A. 2008. Comparison of transesterification methods for production of biodiesel from vegetable oils and fats. *Energy Convers Manage* 49:125–130.

Demirbas A. 2009. Biodiesel from waste cooking oil via base-catalytic and supercritical methanol transesterification. *Energy Convers Manag* 50:923–927.

Elshahed MS. 2010. Microbiological aspects of biofuel production: Current status and future directions. *J Adv Res* 1:103–111.

Fukuda H, Kondo A, Noda H. 2001. Biodiesel fuel production by transesterification of oils. *J Biosci Bioeng* 92(5):405–416.

Guo D, Zhu J, Deng Z, Liu T. 2014. Metabolic engineering of *Escherichia coli* for production of fatty acid short-chain esters through combination of the fatty acid and 2-keto acid pathways. *Metab Eng* 22:69–75.

Ha SJ, Galazka JM, Kim SR, Choi JH, Yang X, Seo JH. 2011. Engineered *Saccharomyces cerevisiae* capable of simultaneous cellobiose and xylose fermentation. *Proc Natl Acad Sci USA* 108:504–509.

Hassan M, Blanc PJ, Granger L-M, Pareilleux A, Goma G. 1996. Influence of nitrogen and iron limitations on lipid production by *Cryptococcus curvatus* grown in batch and fed-batch culture. *Process Biochem* 31:355–361.

Hegemann JH, Güldener U, Köhler GJ. 2006. Gene disruption in the budding yeast *Saccharomyces cerevisiae*. *Methods Mol Biol* 313:129–144.

Hellenbrand J, Biester EM, Gruber J, Hamberg M, Frentzen M. 2011. Fatty acyl-CoA reductases of birds. *BMC Biochem* 12:64.

Hu Q, Sommerfeld M, Jarvis E, Ghirardi M, Posewitz M et al. 2008. Microalgal triacylglycerols as feedstocks for biofuel production: Perspectives and advances. *Plant J* 54:621–639.

Huang G, Chen F, Wei D, Zhang X, Chen G. 2010. Biodiesel production by microalgal biotechnology. *Appl Energy* 87:38–46.

IEA (International Energy Agency). 2015. CO_2 emissions from fuel combustion highlights 2015. Paris: IEA.

Jang YS, Lee JY, Lee J, Park JH, Im JA et al. 2012. Enhanced butanol production obtained by reinforcing the direct butanol-forming route in *Clostridium acetobutylicum*. *MBio* 3:e00314-12.

Joachimsthal EL, Rogers PL. 2000. Characterization of a high-productivity recombinant strain of *Zymomonas mobilis* for ethanol production from glucose/xylose mixtures. *Appl Biochem Biotechnol* 84–86:343–356.

Jullesson D, David F, Pfleger B, Nielsen J. 2015. Impact of synthetic biology and metabolic engineering on industrial production of fine chemicals. *Biotechnol Adv* 33:1395–1402.

Kalscheuer R, Stölting T, Steinbüchel A. 2006. Microdiesel: Escherichia coli engineered for fuel production. *Microbiology* 152:2529–2536.

Kamisaka Y, Kimura K, Uemura H, Yamaoka M. 2013. Overexpression of the active diacylglycerol acyltransferase variant transforms *Saccharomyces cerevisiae* into an oleaginous yeast. *Appl Microbiol Biotechnol* 97(16):7345–7355.

Klasson KT, Ackerson MD, Clausen EC, Gaddy JL. 1992. Bioconversion of synthesis gas into liquid or gaseous fuels. *Enzyme Microb Technol* 14(8):602–608.

Largeau C, Casadevall E, Berkaloff C, Dhamelincourt P. 1980. Sites of accumulation and composition of hydrocarbons in *Botryococcus braunii*. *Phytochemistry* 19:1043–1051.

Leman J. 2009. Lipids, production. In *Encyclopedia of Microbiology*, ed. M Schaechter. 3rd ed. Oxford: Academic Press, pp. 393–406.

Li Q, Du W, Liu D. 2008. Perspectives of microbial oils for biodiesel production. *Appl Microbiol Biot* 80:749–756.

Li YH, Liu B, Zhao ZB, Bai FW. 2006. Optimized culture medium and fermentation conditions for lipid production by *Rhodosporidium toruloides*. *China J Biotechnol* 22:650–656.

Liu SJ, Yang WB, Shi AH. 2000. Screening of the high lipid production strains and studies on its flask culture conditions. *Microbiology* 27:93–97.

Liu ZL, Saha BC, Slininger PJ. 2008. Lignocellulosic biomass conversion to ethanol by *Saccharomyces*. In *Bioenergy*, ed. CS Harwood, AL Demain, JD Wall. Washington, DC: ASM Press, pp. 17–36.

Meher LC, Vidya Sagar D, Naik SN. 2006. Technical aspects of biodiesel production by transesterification—A review. *Renew Sustain Energy Rev* 10:248–268.

Meng X, Yang J, Xu X, Zhang L, Nie Q, Xian M. 2009. Biodiesel production from oleaginous microorganisms. *Renew Energy* 34:1–5.

Metz JG, Pollard MR, Anderson L, Hayes TR, Lassner MW. 2000. Purification of a jojoba embryo fatty acyl-coenzyme A reductase and expression of its cDNA in high erucic acid rapeseed. *Plant Physiol* 122(3):635–644.

Metzger P, Rager M, Fosse C. 2008. Braunicetals: Acetals from condensation of macrocyclic aldehydes and terpene diols in *Botryococcus braunii*. *Phytochemistry* 69:2380–2386.

Mora G, Scharnewski M, Fulda M. 2012. Neutral lipid metabolism influences phospholipid synthesis and deacylation in *Saccharomyces cerevisiae*. *PloS One* 7(11):e49269.

Naik SN, Goud VV, Rout PK, Dalai AK. 2010. Production of first and second generation biofuels: A comprehensive review. *Renew Sustain Energy Rev* 14:578–597.

Nielsen J. 2009. Systems biology of lipid metabolism: From yeast to human. *FEBS Lett* 583(24):3905–3913.

Nigam PS, Singh A. 2011. Production of liquid biofuels from renewable resources. *Prog Energy Combust Sci* 37:52–68.

Oelkers P, Cromley D, Padamsee M, Billheimer JT, Sturley SL. 2002. The *DGA1* gene determines a second triglyceride synthetic pathway in yeast. *J Biol Chem* 277(11):8877–8881.

Oelkers P, Tinkelenberg A, Erdeniz N, Cromley D, Billheimer JT, Sturley SL. 2000. A lecithin cholesterol acyltransferase-like gene mediates diacylglycerol esterification in yeast. *J Biol Chem* 275(21):15609–15612.

Ostergaard S, Olsson L, Nielsen J. 2000. Metabolic engineering of *Saccharomyces cerevisiae*. *Microbiol Mol Biol Rev* 34–50.

Panwar AS, Jugran J, Joshi GK. 2013. Microbial production of biofuels. In *Biofuels Production*, ed. B Vikash, T Ashish, K Girijesh. New York: John Wiley & Sons, pp. 147–166.

Papanikolaou S, Aggelis G. 2009. Biotechnological valorization of biodiesel derived glycerol waste through production of single cell oil and citric acid by *Yarrowia lipolytica*. *Lipid Technol* 21:83–87.

Papanikolaou S, Aggelis G. 2011. Lipids of oleaginous yeasts. Part I: Biochemistry of single cell oil production. *Eur J Lipid Sci Technol* 113:1031–1051.

Papanikolaou S, Komaitis M, Aggelis G. 2004. Single cell oil (SCO) production by *Mortierella isabellina* grown on high-sugar content media. *Bioresour Technol* 95:287–291.

Patrik L, Erlandsson P, Taherzadeh M. 2014. Integration of the first and second generation bioethanol processes and the importance of by-products. *Bioresour Technol* 165:3–8.

Peralta-Yahya PP, Zhang FZ, del Cardayre SB, Keasling JD. 2012. Microbial engineering for the production of advanced biofuels. *Nature* 488:320–328.

Posten C, Schaub G. 2009. Microalgae and terrestrial biomass as source for fuels—A process view. *J Biotechnol* 142:64–69.

Prasad S, Singh A, Jain N, Joshi HC. 2007. Ethanol production from sweet sorghum syrup for utilization as automotive fuel in India. *Energy Fuel* 4:2415–2420.

REN21. 2014. Renewables 2014 global status report. Paris: REN21 Secretariat, p. 43.

Schenk MP, Thomas-Hall SR, Stephens E, Marx UC, Mussgnug JH, Posten C. 2008. Second generation biofuels: High-efficiency microalgae for biodiesel production. *Bioenergy Res* 1:20–43.

Schirmer A, Rude MA, Li X, Popova E, Cardayre SB. 2010. Microbial biosynthesis of alkanes. *Science* 329(5991):559–562.

Schnepf R, Yacobucci BD. 2013. Renewable fuel standard (RFS): Overview and issues. Washington, DC: Congressional Research Service, p. 1.

Shi JH, Feng HX, Lee J, Chen WN. 2013. Comparative proteomics profile of lipid-cumulating oleaginous yeast: An iTRAQ-coupled 2-D LC-MS/MS analysis. *PLoS One* 8(12):e85532.

Shi S, Rodriguez H, Siewers V, Nielsen J. 2011. Prospects for microbial biodiesel production. *J Biotechnol* 6(3):277–285.

Singh A, Nigam PS. 2014. Microbial biofuels production. In *Microbial Biotechnogy Progress and Trends*, ed. FD Harzevili, HZ Chen. Boca Raton, FL: CRC Press, pp. 155–168.

Solovchenko A, Khozin-Goldberg I, Didi-Cohen S, Cohen Z, Merzlyak M. 2008. Effects of light and nitrogen starvation on the content and composition of carotenoids of the green microalga *Parietochloris incisa*. *Russ J Plant Physiol* 55:455–462.

Sorger D, Daum G. 2002. Synthesis of triacylglycerol by the acyl-coenzyme A: Diacyl-glycerol acyltransferase Dga1p in lipid particles of the yeast *Saccharomyces cerevisiae*. *J Bacteriol* 184(2):519–524.

Stephanopoulos G, Vallino JJ. 1991. Network rigidity and metabolic engineering in metabolite overproduction. *Science* 252:1675–1681.

Stöveken T, Kalscheuer R, Malkus U, Reichelt R, Steinbüchel A. 2005. The wax ester synthase/acyl coenzyme A: Diacylglycerol acyltransferase from *Acinetobacter sp.* strain AD P1: Characterization of a novel type of acyltransferase. *J Bacteriol* 187(4):1360–1376.

Taherzadeh M, Lennartsson P, Teichert O, Nordholm H. 2013. Bioethanol production processes. In *Biofuels Production*, ed. V Babu, A Thapliyal, G Patel. Beverly, MA: Scrivener Publishing, pp. 211–253.

Tang X, Chen WN. 2014. Enhanced production of fatty alcohols by engineering the TAGs synthesis pathway in *Saccharomyces cerevisiae*. *Biotechnol Bioeng* 112(2):386–392.

Tang X, Feng H, Chen WN. 2013. Metabolic engineering for enhanced fatty acids synthesis in *Saccharomyces cerevisiae*. *Metab Eng* 16:95–102.

Tehlivets O, Scheuringer K, Kohlwein SD. 2007. Fatty acid synthesis and elongation in yeast. *Biochim Biophys Acta* 1771(3):255–270.

Thompson RA, Trinh CT. 2014. Enhancing fatty acid ethyl ester production in *Saccharomyces cerevisiae* through metabolic engineering and medium optimization. *Biotechnol Bioeng* 111:2200–2208.

Thornley P, Rogers J, Huang Y. 2008. Quantification of employment from biomass power plants. *Renew Energy* 33:1922–1927.

Tolmac D, Prulovic S, Lambic M, Radovanovic L, Tolmac J. 2014. Global trends on production and utilization of biodiesel. *Energy Sources Part B* 9:130–139.

Trinh CT, Unrean P, Srienc F. 2008. Minimal *Escherichia coli* cell for the most efficient production of ethanol from hexoses and pentoses. *Appl Environ Microbiol* 74:3634–3643.

Ugarte DG, Walsh ME, Shapouri H, Slinsky P. 2003. The economic impacts of bioenergy crop production in US agriculture. USDA Agricultural Economic Report No. 816. Washington, DC: U.S. Department of Agriculture, p. 41.

Vicente G, Bautista L, Rodriguez R, Gutiérrez F, Sádaba I, RuizVázquez R, Torres-Martínez S, Garre V. 2009. Biodiesel production from biomass of an oleaginous fungus. *Biochem Eng J* 48:22–27.

Wagner S, Paltauf F. 1994. Generation of glycerophospholipid molecular species in the yeast *Saccharomyces cerevisiae*. Fatty acid pattern of phospholipid classes and selective acyl turnover at sn-1 and sn-2 positions. *Yeast* 10(11):1429–1437.

Wahlen BD, Oswald WS, Seefeldt LC, Barney BM. 2009. Purification, characterization, and potential bacteria wax production role of an NADPH-dependent fatty aldehyde reductase from *Marinobacter aquaeolei* VT8. *Appl Environ Microbiol* 75(9):2758–2764.

Willis RM, Wahlen BD, Seefeldt LC, Barney BM. 2011. Characterization of a fatty acyl-CoA reductase from *Marinobacter aquaeolei* V T8: A bacterial enzyme catalyzing the reduction of fatty acyl-CoA to fatty alcohol. *Biochemistry* 50(48):10550–10558.

Yang H, Bard M, Bruner DA, Gleeson A, Deckelbaum RJ, Aljinovic G, Pohl TM, Rothstein R, Sturley SL. 1996. Sterol esterification in yeast: A two gene process. *Science* 272(5266):1353–1356.

Yu C, Kennedy NJ, Chang CC, Rothblatt JA. 1996. Molecular cloning and characterization of two isoforms of *Saccharomyces cerevisiae* acyl-CoA: Sterol acyltransferase. *J Biol Chem* 271(39):24157–24163.

Yu KO, Jung J, Kim SW, Park CH, Han SO. 2012. Synthesis of FAEEs from glycerol in engineered *Saccharomyces cerevisiae* using endogenously produced ethanol by heterologous expression of an unspecific bacterial acyltransferase. *Biotechnol Bioeng* 109(1):110–115.

Yu KO, Jung J, Ramzi AB, Choe SH, Kim SW, Park C, Han SO. 2013. Development of a *Saccharomyces cerevisiae* strain for increasing the accumulation of triacylglycerol as a microbial oil feedstock for biodiesel production using glycerol as a substrate. *Biotechnol Bioeng* 110(1):343–347.

Zhang FZ, Rodriguez S, Keasling J. 2011. Metabolic engineering of microbial pathways for advanced biofuels production. *Curr Opin Biotechnol* 22:775–783.

11

Designing a Plant for Bioethanol Production from Different Raw Materials:
The Biowanze Ethanol Plant as a Case Study

Olivier Janssens

Contents

11.1 Introduction

The first generation of bioethanol production processes is based on sugar or starch. These substrates are recovered from crops that currently represent a significant proportion of the total world crop harvest. Each ethanol producer has chosen its feedstock depending on market prices and local availability. However, the available volumes and prices of sugar, cereals, and protein vary in time, and it can be a competitive advantage to produce ethanol from different sources (Table 11.1). In this chapter, the Biowanze fuel ethanol plant in Belgium has been chosen to illustrate the design procedure of an ethanol plant able to process different raw materials, such as starch from wheat grain and sucrose from sugar beet. Since valuable fractions other than the ethanol itself are produced from these agricultural bioresources, this process can be considered as a kind of biorefinery. Indeed, a large part of the business benefits are based on the gluten, that is, the proteins are extracted from the wheat grain to be sold for bakery use. Moreover, both the wheat bran and stillage (distilled liquor) produced in the process are also valorized. The bran (i.e., 15%–17% of wheat grain) is burned in a 21 MW biomass furnace for heat production reaching 85% of the whole process energy needs. The stillage at 5% dry weight (DW) is partially dehydrated up to 30% DW for animal feed. Therefore, the

TABLE 11.1 Agricultural Yield and Potential Bioethanol Yield for Different Crops

	Agricultural Yield	Plant Sugar Content	Ethanol Yield	
	(t/ha)	(% Mass)	(L/t Crop Plant)	(m³/ha)
Sugar beet	70–75	15	92	6.4–6.9
Sugarcane	80–90	14–15	85	6.8–7.7
Sugar sorgho	50–60	16	80	4.0–4.8
Wheat	7.2–8.3	62–65	370	2.6–3.1
Maize	7.2–8.5	72	400	2.9–3.4
Barley	5.0–7.0	56–59	320	1.6–2.2

Source: Ballerini, D., and Alazard-Toux, N., *Les biocarburants. Etats des lieux, perspectives et enjeux du développement*, 1ère éd., Editions TECHNIP, Paris, 2006.

fermentation facilities of such a biorefinery represent less than 25% of the total process implantation area, whereas the total area reaches about 50% in other bioethanol processes from maize starch or sugarcane sucrose. The global investment for the Biowanze facilities was also significantly higher than in other cases. The overheads were justified by considerations other than only environmental aspects regarding the renewable biofuel production or employment aspects regarding direct and indirect manpower. Indeed, many agricultural advantages were also associated with the Biowanze process, for example, sugar beet culture continuation in the sugar open market, diversification of local crop cultures, and improvement of agriculture revenues.

11.2 Basic Design of the Bioethanol Production Plant

The basic design of a process is typically established backwards to forwards: from ethanol to the incoming raw materials (Figure 11.1). The starting point is the targeted production output, that is, the ethanol volume expected per day or per year. This parameter will determine the characteristics of the distillation and dehydration installations.

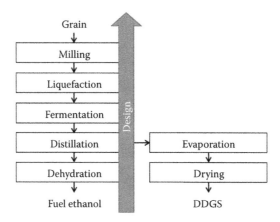

Figure 11.1 Design sequence for a typical grain ethanol plant. DDGS, distillers dried grains with solubles.

11.2.1 Design of the Distillation and Dehydration Installations

In order to cope with the standards of the fuel industry, a certain number of distillation steps are added to the classical water–ethanol separation columns (Figure 11.2). When operating, these additional columns will lead to by-products, which contain a certain proportion of impurities, separated from the main ethanol stream. Typically, head spirits contain light impurities, and fusel oil heavier products than ethanol. The final purification step is dehydration on molecular sieves, where the last traces of water are separated from azeotropic ethanol (at 95% v/v alcohol content). Distillation and dehydration can be considered the heart of an ethanol plant, and probably the part where the know-how of an expert engineering office is the most valuable. In the case of process modification, for example, substitution of raw materials, ethanol producers will try to keep these last purification steps unchanged (distillation and dehydration), because of the intensive investment required.

The mash column will be dimensioned according to the specifications of the incoming fermented mash (i.e., the culture medium containing ethanol produced by yeast, similar to the beer recovered from a brewery process). Despite high-gravity yeast strains able to produce ethanol up to 18% v/v in a culture medium, most process suppliers base their design on an ethanol percentage close to 12% by volume in fermented mash. Indeed, the ethanol content must be determined as a compromise between optimal fermentation characteristics and optimal distillation characteristics. The latest suggests reaching a high ethanol percentage, whereas these conditions would induce slower fermentation and more stress on yeast, leading to a lower

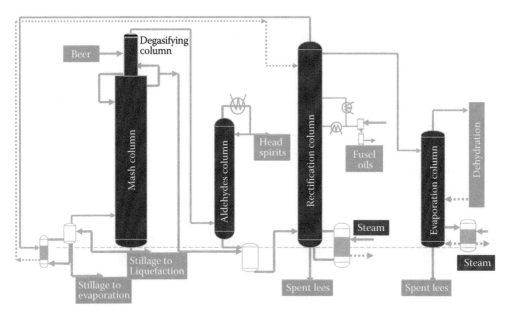

Figure 11.2 Schema of the distillation process to recover a 95% v/v ethanol mixture from the fermented "beer-like" mash recovered from the fermentation process.

productivity. By contrast, a low ethanol percentage in the distillation process means more energy needs per ethanol unit volume and more capital investment.

Another important parameter for the mash column design is the maximum temperature allowable on the column surfaces, taking into account the composition of the fermented mash. Typically, high protein contents will lead to fouling, and high mineral contents will lead to scaling problems. Despite curative solutions (acid and basic cleaning in place [CIP]), a preventive approach, whether based on experience values or pilot trials, should reduce downtime and increase plant reliability. For instance, wheat is known to be a difficult product compared with other cereals because of the presence of gluten (about 8%–12% wheat DW). When a high proportion of wheat is used in fermentation, it is recommended to keep the surface temperatures below 80°C. This can be achieved in distillation and evaporation by adjusting the vacuum to the desired value.

According to the McCabe–Thiele theory, these design values (flow, alcohol percentage, and temperature) determine where the fermented mash will enter in the distillation plant and fix the other parameters (steam pressure, flow rate, and cooling power). At this point, the basic design must integrate some utilities: vacuum, steam production, and cooling water.

11.2.2 Design of the Fermentation Installations

The next step in the design sequence (Figure 11.1) is the conversion of sugar (mono-, di-, or polysaccharides) into ethanol according to the following chemical equation:

$$C_6H_{12}O_6 \rightarrow 2\,C_2H_5OH + 2\,CO_2$$

The theoretical values obtained from the stoichiometric mass ratio are given in Table 11.2.

Process suppliers can generally guarantee an efficiency of 90% based on these values for the entire process. Therefore, the basic calculation for the determination of raw material input can be summarized as dividing the projected ethanol output by different factors:

- Theoretical sugar/ethanol conversion ratio
- Process guarantee ratio
- Sugar content of raw material
- Eventual separation yields (milling and gluten separation)

TABLE 11.2   Carbohydrate Theoretical Conversion Ratio to Ethanol

100 kg of hexose	Generate theoretically	64.8 L of pure ethanol
100 kg of disaccharide	Generate theoretically	68.0 L of pure ethanol
100 kg of polysaccharides	Generate theoretically	71.3 L of pure ethanol

Source: Heiss, R., *Lebensmitteltechnologie: Biotechnologische, chemische, mechanische und thermische Verfahren der Lebensmittelverarbeitung*, Springer, Berlin, 2003.

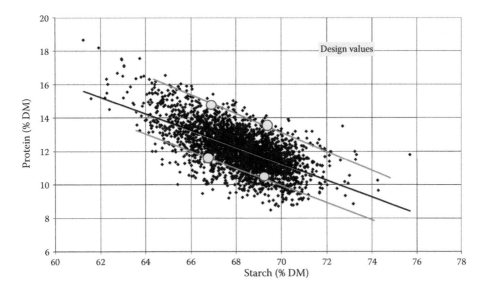

Figure 11.3 Wheat quality classified by protein and starch content in percent dry matter (DM). (From Walloon Agronomical Research Center, Gembloux, Belgium, 2005.)

Consequently, determining the sugar (or starch) content of the raw material must be done carefully. However, whereas mono- and disaccharides can easily and accurately be measured by polarimetry, determining polysaccharides is much more complicated and subject to discussions since, for example, acid hydrolysis at temperatures higher than 100°C may lead to some carbohydrate denaturation before chemical analysis. In addition, seasonal and geographical variation in the supply of raw material should be considered. In order to estimate the variability, and the impact on the process, different scenarios have to be calculated. In the case of the Biowanze plant, statistical data were provided by the Walloon Agronomical Research Center (CRA-W) to assess the quality of available raw material. As shown in Figure 11.3, starch content and protein content are roughly inversely proportional. In order to fix the capacities of the different units, and to identify eventual bottlenecks, four scenarios were chosen as design values for the preliminary mass balance.

11.3 Design of the Other Elements of the Biowanze Biorefinery

Once the figures for ethanol to raw material have been determined, the other side streams can be calculated by using similar ratios. In the case of the Biowanze plant, the block flow diagram (Figure 11.4) shows that bran is separated from flour in a mill. The milling process, similar to the bakery flour production, can be described as a succession of milling and sieving steps (Figure 11.5). The bran is used as fuel in a biomass

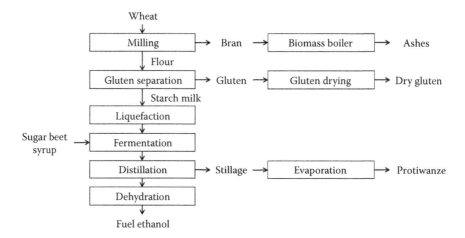

Figure 11.4 Block flow diagram of the Biowanze plant.

(a)

Figure 11.5 Roller mills (a). (From Biowanze, *Wanze, Belgium*.)   (*Continued*)

(b)

(c)

Figure 11.5 (Continued)　Plansichters (b). (From Biowanze, *Wanze, Belgium.*) (c) Roller assembly and roller surface. (From Buhler Group, Uzwil, Switzerland, www.buhlergroup.com.)

boiler providing steam to a cogeneration power plant. The flour is mixed with water, and the dough is then centrifuged, enabling the sedimentation of different fractions by density (Figure 11.6). The A-starch has the highest density and will be separated from gluten and the B-starch and from a lighter fraction containing pentosanes and water. The B-starch will be washed out from gluten and, after concentration, mixed with the A-starch and pentosanes; this stream is called starch milk and will be lique-fied and fermented. As a consequence, the removal of gluten from wheat also leads to a lower protein content in distillation, reducing the fouling phenomenon. After separation, gluten is dried in a ring drier, and then milled or pelletized before being sold as a high-value by-product.

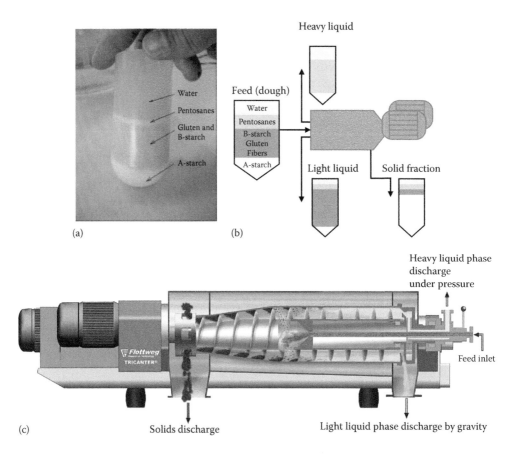

Figure 11.6   Sedimented fractions of wheat flour at the lab scale (a) and through industrial-scale separation (b). (c) Three-phase centrifuge decanter. (From Flottweg, Vilsbiburg, Germany, www.flottweg.com.)

11.4 Operation Conditions of the Simultaneous Saccharification and Fermentation

The starch milk obtained after gluten separation is liquefied at 90°C for 2 hours in the presence of alpha-amylase (0.1 kg/ton grain). The aim of this enzyme is the production of shorter chains of sugars, the so-called "dextrins." After cooling, the liquefied slurry is mixed in a fermentation tank with sugar beet syrup, a fraction of distillation stillage, and other nutrients, such as ammonia. The complex carbohydrates of the mixture are hydrolyzed enzymatically, and consequently, the simple sugars are fermented anaerobically to ethanol by yeast cells. Glucoamylase is added simultaneously to the different fermentation reagents at a rate of 0.5 kg/ton grain. During hydrolysis, the glucoamylase activity removes glucose units in a step-by-step manner from the nonreducing end of the substrate molecule. Enzyme activity is controlled in order to achieve simultaneous saccharification and fermentation (SSF), preventing undesirable biodegradation of sugars by contaminant microorganisms, such as lactic acid bacteria, since the fermentation medium is not sterilized. This process is carried out for about 56 hours in six fermenters containing 3000 m^3 of liquid volume (Figure 11.7). Every 60 hours, the operation of two fermenters is started with the addition of the yeast grown in a 600 m^3 aerobic prefermenter. At the end of the process, the stillage leaving the distillation is evaporated from 5% to about 35% DW and sold as a liquid foodstuff, called Protiwanze, with a general composition as shown in Table 11.3.

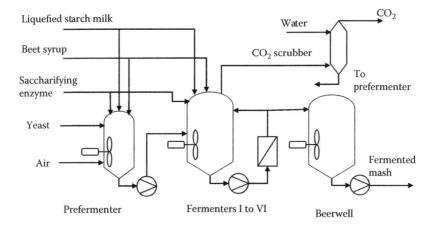

Figure 11.7 Schema of bioethanol fermentation plant with three prefermenters for yeast growth, six fermenters for ethanol production and a "beerwell" intermediary tank before mash distillation for ethanol recovery. ADY, alcohol dehydrated yeast; CWS, cooling water stream; CWR, cooling water recovery.

TABLE 11.3   General Composition of Concentrated Stillage from Biowanze Industrial Bioethanol Process (Protiwanze)

Parameter	Unit	Range
Dry matter (DM)	%	30–35
pH		4–4.5
Proteins	% DM	30–35
Starch	% DM	5–8
Sugar	% DM	1–3
Lipids	% DM	4–5
Ashes	% DM	6–8
Sodium	% DM	≤1
Potassium	% DM	≤1.5
Feed energy (dairy cattle)	kJ/kg DM	6,500–7,200
Feed energy (pork)	kJ/kg DM	9,600–10,100

11.5 Mass and Energy Balance and Other Crucial Parameters

When several raw materials for bioethanol production are combined in different proportions, a separate balance has to be established for each design case. Regarding the Biowanze plant, three cases were studied: 100% sugar syrup, 100% wheat, and 85% wheat/15% syrup. The advantage with the sugar beet syrup was its availability (from a nearby sugar refinery) and its opportunity to be added to fermentation in any proportion. By contrast, wheat is limited by the capacity of the milling and gluten separation lines, and thus represents a bottleneck for the whole plant.

Once the mass flow of the main streams is determined, a tabulated mass and energy balance (MEB) has to be established. This is the starting point for the detailed engineering, where equipment is dimensioned according to indications of the balance. The common form of this documentation is a table, where each column represents a product stream and each line a parameter (Table 11.4). In order to identify the streams, a detailed process flow diagram (PFD) is added to the document, where the

TABLE 11.4   Example of Tabulated Mass and Energy Balance for an Industrial Stream

State	Liquid/Gas/Solid
Temperature	°C
Pressure	Bar/Pa
Mass flow	Kg/h
Density/specific mass	Kg/m^3
Specific heat	kJ/kg/K
Enthalpy	kJ/kg

different streams of the table are labeled. The tabulated MEB varies in the different sections of the plant; for example, in the distillation section, the chemical composition is of great importance, while in the liquefaction or evaporation section, the solids content and viscosity are indicated. Usually, these parameters are present in different sections.

Indeed, viscosity is a crucial parameter, because it will determine the heat transfer coefficients and pressure drops. Since starch solution, fermented mash, and stillage are non-Newtonian fluids, viscosity must be measured with a rheometer at different shear rates. If experience values are not available, pilot trials have to be conducted in order to produce sufficient product quantity. As a first approach, simple formulas correlating viscosity with the heat transfer coefficient can be used. In the sugar industry, the following formula is of common use:

$$k = \frac{230}{\eta^{0.305}}$$

where k is the product-side heat transfer coefficient (W/m²/K) and η is the dynamic viscosity (Pa·s).

Specific heat can be measured on a physical sample, if available, or estimated as a weighted average of its constituents. Note that water is the most predominant constituent in the calculation, with the highest specific heat.

11.6 Detailed Engineering

Based on the MEB, detailed engineering can start. Each piece of equipment is dimensioned, detailed in PFDs, and finally, positioned on arrangement drawings. Start-up and shutdown procedures must also be studied, especially for plants where the concept of heat integration has been applied extensively. What happens if one stream must be heated or cooled by another not yet available? In that case, redundancy, spare boilers and coolers, have to be installed.

Because several sections of an ethanol plant contain flammable or explosive products, ATEX (European legislation for EXplosive ATmospheres) and HAZOP (HAZard and Operability) studies are mandatory before production can start. These safety studies must, of course, be revised if modifications occur, especially in classified sections.

Finally, the commissioning of the ethanol plant has to be performed with the different suppliers. Due to the size of such a factory (Figure 11.8 and Table 11.5), the logic is to ensure an output to the different streams transformed in the plant. The worst scenario would be the accumulation and disposal of tons of badly liquefied or fermented streams. Therefore, the commissioning tests occur generally backwards to forwards; that is, the distillation must be ready and tested before the fermentation section is operated.

Figure 11.8 Aerial view of the Biowanze plant (during construction).

TABLE 11.5　Key Figures of the Biowanze Plant

Raw material	Wheat: 750,000 t/year
	Beets: 400,000 t equivalent (transferred from sugar refinery as a thick juice)
Finished goods	Ethanol: 300,000 m³/year
	Gluten: 55,000 t/year
	Protiwanze: 200,000 t/year
Investment	More than €250 million
Employment	Direct employees: 123 persons

11.7 Upgrading an Existing Ethanol Plant to a Lignocellulosic Process

Producing ethanol from lignocellulosic sources has been intensively investigated over the last decades. These processes are called second generation, by comparison with the first-generation process using sucrose or starch as raw material. The second-generation processes are more complex since in lignocellulosic compounds, the carbohydrates suitable for ethanol fermentation are closely entrapped in lignin, which is relatively recalcitrant and not ethanolgenic. Another issue to address compared with the hexose fermentation that occurs in the first-generation process is related to both hexose and pentose carbohydrates involved in lignocellulosic biomass. Indeed, *Saccharomyces* yeast is not naturally able to ferment pentose, and pentose-fermenting microorganisms are not as efficient as *Saccharomyces cerevisiae*. As a consequence, after several processing steps (mechanical pretreatment, hydrolysis, and biocatalyzation) to recover the carbohydrates from the initial bioresource, different fermentation strategies had to be developed (Abbas 2003):

- Single or pure culture: Hexose- and pentose-fermenting ethanolgenic yeast (e.g., *Pachysolen*) or recombinant bacteria or yeast
- Coculture: Ethanolgenic pentose-fermenting yeast in combination with a brewing yeast
- Pentose isomerization, followed by fermentation using brewing yeast
- SSF by combining the use of free cellulases, hemicellulases, and ligninases with an ethanolgenic organism

Besides the equipment required for the long and complex treatment of lignocellulosic feedstock, the common point of all these processes is the production of ethanol by fermentation. Ethanol can be obtained from hexose or pentose from different microorganisms, after a cofermentation step (Lu and Mosier 2008). If the ethanol flow and content in the mash differ significantly from the initial design, the distillation section will require major adaptations. In that case, would an upgrade be economically viable? Many investigations have been carried out worldwide at the pilot or a larger scale, but the question is still pending.

11.8 Conclusion

The strategy for the production of biofuels within the concept of biorefinery is quite promising. Using the value of grain to produce renewable energy for transport, while using the remaining protein content of the grain as a valuable protein source for livestock and fish, can be seen as complementary and optimal use of all the grain constituents (Scholey et al. 2016). This production scheme, implemented in the Biowanze facility since 2009, could be transposed to other crops where high-value by-products could be separated and marketed separately. Beside proteins, cellulosic by-products could be processed in the sense of a second-generation ethanol plant or, alternately, valorized by combustion in a power plant. The Biowanze research team is currently working on the use of miscanthus as a nonfood feedstock because of the rapid growth, low mineral content, and high biomass yield of this plant. Developing feed and food alternatives for wheat bran also seems to be a promising development for the factory.

Acknowledgment

Our special thanks to Biowanze for supporting this publication.

References

Abbas C.A. (2003). Lignocellulosics to ethanol: Meeting ethanol demand in the future. In *The Alcohol Textbook*, eds. K.A. Jacques, T.P. Lyons, D.R. Kelsall. 4th ed. Nottingham, UK: Nottingham University Press, pp. 41–57.

Ballerini D., Alazard-Toux N. (2006). *Les biocarburants. Etats des lieux, perspectives et enjeux du développement.* 1ère éd. Paris: Editions TECHNIP.

Heiss R. (2003). *Lebensmitteltechnologie: Biotechnologische, chemische, mechanische und thermische Verfahren der Lebensmittelverarbeitung.* Berlin: Springer.

Lu Y., Mosier N.S. (2008). Current technologies for fuel ethanol production from lignocellulosic plant biomass. In *Genetic Improvement of Bioenergy Crops*, ed. W. Vermerris. Berlin: Springer, pp. 161–184.

Scholey D.V., Burton E.J., Williams P.E.V. (2016). The bio refinery; producing feed and fuel from grain. *Food Chemistry* 197: 937–942.

12

Microbial Lipids as Diesel Replacement:
Current Challenges and Recent Advances

Hatim Machrafi, Christophe Minetti, and Carlo Saverio Iorio

Contents

12.1 Introduction

The continuously increasing energy demand, especially for the last decades, has caused an important need to find solutions regarding the exhaustion of the extensive but limited fossil fuel reserves. This current problem also implies a dramatic increase of energy costs, as well as environmental, social, and economic problems. This has motivated the search for new, cleaner energy sources that are able to meet the energy demand while being renewable. As a consequence, many new research topics have emerged. Being of recent great interest, biofuels are considered a promising renewable energy source for the replacement of fossil-based fuels. Biofuel is the general name for a variegated class of fuels, including biodiesel, bio-oil, biosyngas, biobutanol, bio-ethanol, biomethanol, and biohydrogen. Biodiesel as a vegetable oil is a mixture of esters with long-chain fatty acids with properties such as renewability, biodegradability, accessibility, sustainability, and nontoxicity. Compared with conventional diesel fuel, it has a very similar energy content, and unlike other fuels, it burns cleanly. Therefore, biodiesel is a suitable aspirant to substitute for oil. Moreover, it is derived from sources that allow for fixing carbon dioxide. It helps to decrease greenhouse gas emissions, the hydrocarbon percentage, and particulate emissions. Additionally, biodiesels are supposed to relieve the environmental problem of acid rain, because of the near-zero sulfur emissions. Many nonfossil sources have been identified as primary ones for biofuel production. Examples are wheat, corn, sugar beets, and more generally, sugar-rich biomasses (Tilman et al. 2009; Chen and Zhang 2015). By contrast, they are in great conflict with food supply and agricultural health.

Basically, the production of biofuels is based on the storage of renewable energy sources, such as sun energy. Typically, this is related to biologically produced materials. As a principle, the majority of routes start with photosynthesis, which can be followed by diverging methods. In general, we can divide these routes into three categories (Rude and Schirmer 2009; Cheon et al. 2016):

1. Direct biofuel production by photosynthesis organisms, such as plants and algae
2. Fermentative or nonfermentative production of plant or algae biomass by heterotrophic microorganisms, such as bacteria, yeast, or fungi
3. Chemical conversion of biomass to fuels

Oleaginous microorganisms, that is, yeast, filamentous fungi, and microalgal species, are very interesting organisms that are able to accumulate lipids at levels greater than 20% of their cellular dry weight (Meng et al. 2009; Capus et al. 2016). Hence, these organisms are useful sources of lipids, which can subsequently be used for producing "green" diesel, or biodiesel (Li et al. 2008). Presently, oil from rapeseed, sunflower, or palm, or even animal, fats is used worldwide as the main feedstock for biodiesel production. Nonetheless, the second generation of biodiesel, being based on the fermentation of oleaginous microorganisms, has the advantage of not being linked to a particular season and does not compete with food production for the use of agricultural land. Therefore, biodiesel from oleaginous microorganisms is generally considered to be an ideal green fuel candidate for the replacement of fossil diesel due to its high oil accumulation and rapid growth rate of microorganisms (Chisti 2007). These neutral lipid stocks are composed of hydrophobic lipid molecules lacking charged groups that mostly consist of triacylglycerol (TAG) in the form of intracellular lipid droplets (LDs) surrounded by a monolayer of phospholipids.

12.2 Oleaginous Microorganisms

12.2.1 Different Classes of Oleaginous Microorganisms

The oleaginous microorganisms are classified into four main types: bacteria, yeast, fungi, and microalgae. Microorganisms have long been used in industrial processes, starting with alcohol fermentation using *Saccharomyces cerevisiae*. Basically, micro-organisms require organic feedstocks for growth and metabolism, producing fuels that are nevertheless close to carbon neutrality. The reason lies in the fact that, contrary to petroleum, the feedstocks are produced using the principle of plant-based carbon dioxide consumption. However, since the feedstock production itself is a high-cost driver, it is convenient to prevent any organic carbon waste and, consequently, to operate the processes of fermentation and fuel production as efficiently as possible (Stephanopoulos 2007; Fischer et al. 2008). In addition, whereas it has been highlighted that some of the new microbial fuels truly have the potential to fulfill the criteria of efficiency (Rude and Schirmer 2009), most of them are currently produced from simple carbohydrate feedstocks, such as glucose. These should be extracted from biomass feedstocks, leading to some wastes and a limited conversion yield, to consider in the efficiency of the whole process, from the raw biomass to biofuel. Much research has focused on the efficient production of plant cellulosic biomass (Somerville 2007; Zheng et al. 2007) and the breakdown thereof, and specifically on the production of lignocellulosic material toward simple carbohydrates that are readily usable (Kumar et al. 2008; Lynd et al. 2008). Such developments have great potential in the future to develop further procedures that are able to produce high-value microbial fuels from only low-cost biomass.

Yeasts are also an interesting lipid source. Many yeast species, such as *Cryptococcus albidus*, *Lipomyces lipofera*, and *Yarrowia lipolytica*, are able to produce lipids, if certain cultivation conditions are respected, but depend also on the type of yeast species (Liu et al. 2000; Liang et al. 2006). These studies have revealed that the main fatty acids in yeast oils, interesting for their production of biodiesel, are myristic acid, palmitic acid, stearic acid, oleic acid, linoleic acid, and linolenic acid (Meng et al. 2009). Such oils can be feedstocks for biodiesel production, following the catalysis principle, by using either lipase or a chemical catalyst (W. Li et al. 2007). Cultivation conditions (the carbon/nitrogen ratio in the yeast oils, temperature, pH, and oxygen concentration) appear to have different influences on the oil production (Meng et al. 2009). Different nitrogen sources also have a varied influence on oil production. Both inorganic and organic nitrogen sources are used for yeast cultivation, each having different outcomes, which are reviewed in Liu et al. (2007).

Apart from bacteria and yeasts, fungi (such as *Aspergillus oryzae*, *Humicola lanuginose*, and *Mortierella vinacea*) can also produce oil under special cultivation conditions (Ma 2006; Du et al. 2007; Meng et al. 2009). However, they are mostly investigated for the production of some special lipids. There are only a few reports on using oils stemming from fungi for the application of biodiesel production (Ma 2006; Du et al. 2007).

Finally, the fourth main oleaginous type of microorganism is microalgae. The term *algae* describes an extensive and diverse group of aquatic plants whose size spans from a few micrometers (unicellular microalgae) to several meters (multicellular algae, also known as seaweeds). The latter can create extremely large and showy colonies, often referred to as algae forests, because they can quickly reach a height of about 60 m. At the other end, microalgae's size ranges from 1–2 μm to a few hundred micrometers. They also present fairly different shapes, such as spheres, filaments, asymmetric prolates, and/or oblate ellipsoids (Greenwell et al. 2010). Autotrophic microalgae have the ability to use carbon dioxide as the carbon source for oil production and sunlight as the energy, when some conditions are met; examples of such microalgae are *Chlorella vulgaris*, *Botryococcus braunii*, and *Tetraselmis sueica* (Li et al. 2008; Meng et al. 2009). Apart from algae species influencing oil production, cultivation parameters also have an influence on oil production, which is generally the case for microorganisms. One of the big differences between autotrophic and heterotrophic microalgae is that the scaling up for autotrophic microalgae is more complicated, for the simple reason that light is needed during the cultivation. In order to minimize the cost, the oil production from autotrophic microorganisms must rely on available free sunlight, taking into consideration the daily or seasonal light variations (Li et al. 2008). By changing the conditions of cultivation or by using genetic modification, some autotrophic microalgae can be converted to heterotrophic ones. These have the ability to produce oil by using organic carbon, instead of sunlight, as the source. For instance, *Chlorella potothecoides*, usually known to be autotrophic microalgae, is reported to be converted to heterotrophic microalgae, using only organic carbon sources for oil production (Miao and Wu 2004). The oil content is about four times that for the corresponding autotrophic cells. Some of the heterotrophic microalgae are even able to use cheap organic carbon sources for the production of oil (Han et al. 2006). Heterotrophic microalgae are easily cultivated and controlled in normal fermenters. However, despite the easiness of the fermentation process, it still requires nonfree organic carbon sources for oil production, which might limit, to some extent, the application of heterotrophic microalgae for biodiesel production.

12.2.2 Microalgae

Many crops have been identified as possible sources for the production of biofuel: first-generation biofuels (made from sugar, food crops, starch, or vegetable oils), second-generation biofuels (nonfood crops, corn, or wood), third-generation biofuels (algal oil), and fourth-generation biofuels (vegetable oil and biodiesel to make biogasoline) (Demirbas and Demirbas 2009). One of the main reasons to give more attention to the cultivation of microalgal biomass, as a source of third-generation biofuels, is that the first and second generations of biofuels take agricultural land and use products that were intended for food. This can strongly affect the food supply chain in some countries.

The principal advantages of microalgae, with respective to other crop plants, are the efficient conversion of sunlight, carbon dioxide, and nutrients into energy-rich compounds; their massive production on nonarable land; and their short growth time. In addition, microalgae are productive almost all year—their cultivation should be stopped only for short periods in order to clean and prepare the substrates for a new production cycle. Because of this, microalgae, which are also a source of proteins, carbohydrates, lipids, and vitamins, are a highly promising source for biodiesel compared with other oil crops. Microalgae are rich in some products, such as pigments and chlorophyll, which make them particularly attractive for other applications, like cosmetics, pharmaceuticals, and food.

Algae oil is not the only component with intriguing properties. The biomass residues are a protein-rich compound that can be used, for example, in the manufacture of animal feed. On the other hand, when biomass is subjected to anaerobic digestion, it produces biogas, a high-energy product.

Microalgae can be classified into four main classes: diatoms, green algae, blue-green algae, and golden algae. They can also be divided into eukaryotic or prokaryotic. The first class is characterized by the presence of a well-defined nucleus and includes both diatoms and green algae. Diatoms are the most abundant group of microalgae, and also the one having the highest production of biomass. Prokaryotes, as etymology suggests, do not have a distinct nucleus. For that reason, they are often not considered algae. A substantial class belonging to this latter group is the cyanobacteria, known for their potential to release oxygen as a by-product of photosynthesis.

Although the number of collected algal species is tremendous and possibly not exhaustive, not all of them are suitable for biodiesel production, and only a few have been thoroughly investigated. The following list shows some microalgae considered to be appealing for oil extraction: *Botryococcus braunii, Chaetoceros muelleri, Nitzschia communis, Scenedesmus dimorphus, Euglena gracilis, Prymnesium parvum, Phaeodactylum tricornutum, Pleurochrysis carterae, Tetraselmis chui, Tetraselmis suecica, Neochloris oleoabundans, Isocharysis galbana, Nannochloropsis salina,* and *Dunaliella tertriolecta* (Domingo 2014).

Although many species of microalgae are remarkably robust and easily acclimatized in a great variety of environments, their culture represents a decisive step in the context of industrial exploitation. There are two types of culture modes, depending on the light requirement. When light is required, the mode is called photoautotrophic, and when algae are able to live in darkness, the cultivation mode employed is heterotrophic, as was mentioned earlier in the chapter. Besides not requiring light, heterotrophic cultivation offers elevated lipid production.

12.3 Fermentation Strategies for Single-Cell Oil Production

Single-cell oils are lipids that are extracted from unicellular microorganisms. Some microorganisms have a better ability to convert carbohydrates and other biological sources into intracellular lipids. As was said before, microorganisms, ranging from bacteria, yeasts, and molds to algae, have the capability to produce within them lipids at a quantity that is more than 20% of their own biomass, this especially being the case in oleaginous yeasts and molds (Papanikolaou and Aggelis 2002; Zhao et al. 2008). There are, however, a few yeast species that can produce lipids as high as 50% of their own dry weight. A majority of them are of the TAG type, which contain long-chain fatty acids of 14–20 carbon atoms (Meher et al. 2006). The advantage is that such microbial lipids can be produced in manufacturing systems that are closed without requiring the use of any extensive cultivable land. This is indeed important for a sustainable biodiesel industry (Meher et al. 2006) and provides for great potential concerning the biotechnology sector (Ratledge 2004). Indeed, modern molecular biology techniques have the ability to improve the capacity of lipid storage, thereby generating lipids with rare fatty acid profiles (Mličková et al. 2004). Also, huge amounts of neutral lipids under various conditions, governed by new technology, were reported to be accumulated (Suzuki and Hasegawa 1974; Yamauchi et al. 1983). Further research showed that the biochemical basis of lipid accumulation appears to stem from an imbalanced metabolism with conditions that consist of a carbon-rich and nitrogen-deficient environment, while de novo lipid biosynthesis occurs (Zhao et al. 2008). Lipid production and intracellular accumulation become primarily an anabolic activity, accompanied by biomass formation, when the substrates are made of hydrophobic material. Since lipid contents of oleaginous microorganisms are generally influenced by the environment in which their formation takes place, it thus makes sense to search for good fermentation strategies. The design of the fermentation media is therefore critical for biotransformation, product yield, and volume productivity. The fermentation strategies are discussed in more detail in the next sections.

12.4 Bioreactors

The complexity of the biological medium has a great influence on the operation mode of bioreactors and needs to be discussed further (Al-Mashhadani et al. 2015; Tao et al. 2017). Generally, this medium consists of cells, nutrients, solvents, and gases, albeit for feeding purposes. This means that there is a multiphase system where solids, liquids, and gases interact. Therefore, the flow mixing patterns need to be well understood, as well as solid–liquid interactions and gas–liquid exchanges. This will help to provide information on optimal conditions for the growth and product formation in a bioreactor, assisted by control systems and the monitoring of some characteristic parameters, such as pH and temperature. Many shapes are possible for a bioreactor configuration, depending on several parameters, including costs and the efficiency of mixing. Cylindrical bioreactors are the standard, egg-shaped configurations and have been widely used; the rectangular ones are more limited to poor mixing efficiency (Metcalf 2003). In order to better understand the mixing process in a bioreactor, one could visualize the mixing by using high-speed cameras. However, due to the many solid and gas components in the bioreactors, this is quite difficult (Al-Mashhadani et al. 2015). Therefore, computer simulations can offer useful information. They provide for mapping of the internal flow patterns and evaluation of the overall bioreactor performance in terms of mixing, energy cost, and yield. Therefore, computational fluid dynamics (CFD) has been used more and more to simulate, design, and troubleshoot bioreactors. Several works on this can be consulted (Meroney and Colorado 2009; Wu 2010; Moraveji et al. 2011; Šimčík et al. 2011; Pfaffinger et al. 2016).

When light is involved, the bioreactors are often called photobioreactors (PBRs). Figure 12.1 shows an example of a PBR in our laboratory.

A typical PBR (Posten 2009) is also a system composed of three phases, namely, the liquid phase containing the culture; the solid phase, which is the cells themselves; and the gas phase. In the case where the type of microorganism requires light, the latter can be defined as a fourth phase. The design of a PBR also demands a good

Figure 12.1 Photobioreactor.

understanding of its relation with the surrounding environmental parameters and the response of the biological components (Cornet et al. 1995; Tichy et al. 1995; Grobbelar et al. 1996; Posten 2009; Shariff and Chakraborty 2017). For instance, the CO_2 supply and aeration are to be justly adapted to the generation of unwanted gases or inhibition of side products. This can be calculated on a stoichiometric basis by taking into account the carbon content of the biomass. However, in a PBR, next to the mixing effects, light exposure will also play an important role. Then, a possible gradient of dissolved gases from the bioreactor wall to its center could emerge. The energy required for the mixing also has to be taken into account. Even though the volumetric mass transfer in photobioprocesses is more than two orders of magnitude lower than in heterotrophic stirred reactors, mixing is still an important issue for both. Next to the light gradient (if applicable), it is known to be a key parameter for efficient PBR operation (Pruvost et al. 2006). Equally important is the mass transfer along the axis of the reactor. In the case of vertical tubes, the problems that can arise are the settling of the cells on the bottom and fouling. The use of mechanical energy or bubbling of gas as a means for better mixing was mentioned earlier. Although this holds, it should be recalled that cells may get damaged, or at the least stressed, by high local mechanical energies. Energy cost remains a problem, if one gets the cost sheet into the picture. A proper study of hydrodynamics can prevent the aforementioned problems, or at least reduce them. Fluid dynamics calculation is a useful tool here. An example of the latter is a study on a tubular reactor with particle trajectories (Perner-Nochta and Posten 2007), which shows from a frequency analysis that mixing in the radial direction increases the quality of light exposure. It appears that to achieve good mixing with a minimum use of auxiliary energy, while at the same time avoiding turbulence, one specific frequency of several tours per second is sufficient.

Until now, (photo-)bioreactors that are of the (cylindrical) closed type have been discussed. In continuous culture systems, sometimes referred to as chemostats, the cells are grown in a tightly controlled environment, where the chemically constant culture density is controlled by nutrients (Miller et al. 2013). Since the data from chemostats are very reproducible, as they provide constant growth rates and steady-state operating conditions, chemostats have become useful tools for characterizing physiology systems (Miller et al. 2013), which allow monitoring of the evolution of microbial populations.

The diversity of bioreactors suggests that there is still much to be discovered, and that for proper and optimal culture production, the design of the bioreactor is important.

12.4.1 Design and Techniques for Fermentation

Already in the 1950s, the procedure of microalgae cultivation in shallow open ponds or in closed PBRs was well established (Preisig and Andersen 2005; Greenwell et al. 2010). In the decades that followed, these principles were refined with advanced multidisciplinary research and technological development in biology, physics, mathematics, and process engineering. The methods that require light for algae growth are performed in open or covered ponds or in closed PBRs. Open-pond systems are superficial ponds with some limitations, because if they are in open air, the weather conditions (temperature and illumination intensity) influence the culture parameters. Additionally, algae subjected to this kind of systems are vulnerable to contamination or soiling (Halim et al. 2012). For these reasons, open ponds will be acceptable only when the strain survives in hard conditions, like *Spirulina*, which are able to live in high-pH media (from a pH of 3.5 up to at least 10) (Huang et al. 2010).

Closed bioreactors are generally more expensive. There is a large design variety, such as flat, plate, or tubular bioreactor. The last one needs long tubes to increase the light absorption. These complex designs and their cost only make their use possible in small scales (Huang et al. 2010) when they are producing fine chemical products. PBRs offer some advantages over open ponds, such as less light requirement, no need for agricultural land, and a more controlled environment. However, nutrients, carbon dioxide, and water must be appropriately provided (Demirbas and Demirbas 2009).

Much has already been studied concerning the operating procedures and designs for the cultivation procedures in unmixed ponds and stirred raceways (artificial channels used to culture the microorganisms) (Borowitzka 2005). Typically, shallow water depths of 0.2–0.3 m are used. The biomass may be harvested by methods of flocculation or centrifugation (Del Campo et al. 2007). Now, in terms of avoiding contamination, as well as giving a higher culture yield and better controlling the operating conditions, closed bioreactors are usually considered. Although quite a number of PBR designs have been enumerated (Carvalho et al. 2006; Eriksen 2008), only a few have been commercialized. These generally concern photoautotrophic production that uses artificial lighting, although some heterotrophic species can be cultivated without light (Greenwell et al. 2010). Complete PBR (Figure 12.1) systems are generally composed of the following entities (Greenwell et al. 2010):

- The vessel that contains the culture, which is usually transparent and designed to allow a short optical path under illumination
- The system that provides for the light (in the absence of natural light), which generally consists of fluorescent or metal halide lamps with photosynthetically active radiation ($\lambda = 400$–700 nm)
- The system that controls the gas exchange (typically delivering carbon dioxide and removing generated oxygen)
- The system for concentrating the microorganisms for downstream processing and recovery of the product

Of course, the entities that include illumination are only essential to the microorganisms that need light. Closed PBRs can be operated completely manually or, which is more and more the case, include automatization and feedback systems for more stable culture conditions. As is the case for other types of bioreactors, PBRs can operate in batch, semicontinuous, or continuous (chemostat) modes. This will be discussed in the next section. The most commonly used PBR is a semienclosed transparent column, with a CO_2-containing gas inlet at the bottom and external illumination (Greenwell et al. 2010; Perrin et al. 2016). More advanced closed PBRs have the ability to offer shorter optical pathways when illuminated externally, due to new transparent materials (Carvalho et al. 2006; Eriksen 2008), which are designed to minimize light attenuation between the center of the culture and the wall itself.

Another important point is mixing, as the physical mixing of the liquid in which the culture is present is necessary to ensure that the cells are moved appropriately through the bioreactor. This will ensure that the distribution of nutrients, carbon sources, metabolites, heat, gas transfer, and illumination, if applicable to the type of microorganism, is as efficient as possible (Carvalho et al. 2006). Mixing can also be assisted by mounting devices that alter the fluid flow, such as baffles, discs, or inert particles. Attention should also be paid to the surface characteristics of the walls, thereby preventing possible cell destruction due to shear stress during mixing. Surfactants have been reported to be used to lessen such cell damage (Eriksen 2008; Greenwell et al. 2010). If the microorganisms need proper gas transfer, this becomes a critical aspect of the design. Also, the removal of any generated gases that can inhibit the product yield, such as oxygen (Carvalho et al. 2006), can then be of concern. The introduction of CO_2 to the PBR can be done either passively, via permeable or microporous membranes, or actively, via bubbling the gas from the bottom or again introducing the gas via a gas exchanger (Carvalho et al. 2006). pH-based monitoring is typically included to control the content of CO_2 and regulate its delivery by either simple on–off dosing or a model-based controller (Carvalho et al. 2006). It is clear that the design of the PBR plays an important role.

12.4.2 Fermentation Process and Its Operating Conditions

A good bioreactor design should address improved productivity and validation of desired parameters toward obtaining consistent and higher-quality products in a cost-effective manner (Singh et al. 2014). The way a bioreactor is designed and operates depends not only on the kind of medium, as was said earlier, but also on the production itself, its added value, and the scale of production. An efficient bioreactor should positively influence the biological reaction and prevent contamination. It should also take into account the investment and operating costs that are involved. During fermentation, a clean environment should be kept, while at the same time providing for optimal mixing, but with low shear rates and good aeration (Eibl et al. 2008). Hereby, adequate mass transfer, heat transfer, and appropriate feeding are to be considered. These features are discussed briefly as follows (Singh et al. 2014). The working volume of a bioreactor, being typically about 70%–80% of the total volume, comprises three elements: the liquid medium, the microbes, and the gas bubbles (Singh et al. 2014). The remaining volume is called the headspace. The temperature and pH are kept at certain levels by classical control systems. The cleaning system should keep the environment sterilized so that contamination is avoided. Thermal sterilization by steam is the preferred option due to its low cost and high applicability for high-scale operations. Sterilization can also be performed by chemical substances or ultraviolet (UV) radiation. The agitator system is driven by an external power source, impeller, and baffles for intense mixing and increased heat and mass transfer between the liquid and gas boundary layers. The air delivery system is operated by a compressor and consists of an inlet, a sterilization system, and an exit sterilization system for preventing contamination.

Generally, the types of reactions that take place in the fermentation process are divided into three: batch, continuous, and fed batch (Singh et al. 2014). This division is based on the feeding strategy of the culture and the medium to the bioreactor (McNeil and Harvey 2008). Batch and continuous stirred tank reactors (STRs and CSTRs) have existed for a long time and are still widely employed in the chemical and bioengineering industries (McNeil and Harvey 2008). The fermentation process runs for a predetermined amount of time, and the end product is then eventually harvested (Singh et al. 2014). Batch bioreactors are generally operated in such a way that a single tank is capable of carrying out a sequence of reactions and is easy to operate. They are nowadays the most usual type of aerobic bioreactors, providing high flexibility with varying product systems. Also, the risk of contamination is rather low due to a relatively low growth period with lower capital investment when compared with their continuous counterpart.

In a CSTR, the fresh medium is continuously added with the nutrients and the products are continuously extracted, together with the culture, at the same rate. Hereby, a theoretically constant concentration of nutrients and cells is maintained (Abbott et al. 2013; Singh et al. 2014). Continuous processes are often used for high-volume productions and soluble solid substrates. The advantage of a CSTR over a STR is the obvious ability to maintain a continuous feeding process, which is of great interest to large-quantity production. The result is continuous productivity and output (McNeil and Harvey 2008). However, a CSTR requires large volumes in order to obtain acceptable conversion rates. Using automated control, the time requirements normally involved in such large volumes can be significantly reduced. A fed-batch reactor can provide advantages for both a batch and a continuous reactor, thereby reducing their limitations to some extent. In a fed-batch process, additional nutrients are progressively added to the reactor, while the bioreactions are taking place. This results in a higher yield and better selectivity. The products are collected at the end of the process, as in a batch process. Fed-batch processes are more stable and safer than batch processes and require less operating costs than continuous ones. Algae use photosynthesis as a biochemical process to convert solar energy, water, and carbon dioxide into chemical energy. In a broader sense, photosynthesis helps to balance the amount of carbon dioxide present in the environment and mitigate anthropogenic carbon emissions.

Algae store all their energy in the form of lipids, carbohydrates, and proteins. Depending on the strain, the lipid content usually differs from 2% to 40% of the dry weight of algal biomass. However, algae may be found with higher lipid contents, reaching 80% of their dry weight (Demirbas and Demirbas 2009).

Microalgae are used for the environmentally friendly production of biofuel because they minimize greenhouse gas emissions and need nonarable land use for their cultivation. There are three requirements that help in the selection of strains for biofuel production based on the following attributes (Greenwell et al. 2010). A strain should

- Present a high lipid content
- Present no difficulty for extraction using standard techniques
- Grow rapidly

A high lipid content is the starting point to achieve a viable exploitation of micro-algae for biofuel production. Strains that are not difficult to extract using simple extraction techniques are also preferable. Lastly, a rapid growth of algae means more efficient biomass production. Regarding the type of algae used for the production of biofuel, green algae are among the most suitable because they have rapid growth. Diatoms are also used for their abundance.

From a chemical point of view, algae contain proteins, carbohydrates, lipids, and nucleic acids in a proportion strongly dependent on the type of algae (Greenwell et al. 2010). The membrane consists of algae lipids and fatty acids. The cell membrane toughness is a feature that indicates how easy the lipid extraction will be.

Lipids are a group of organic molecules composed mainly of carbon and hydrogen, and oxygen to a lesser degree, although they may also contain phosphorus, sulfur, and nitrogen. As they are not soluble in water, lipids are typically extracted through the use of organic solvents (mixtures of chloroform, methanol, and water) (Domingo 2014). Indeed, like many biochemical compounds, they are quite soluble in such solvents.

Microalgae present an appreciable chemical variety of lipids depending on the strains. So a detailed chemical characterization is needed, depending on the targeted application. In considering their chemical composition, lipids are divided into saponifiable and unsaponifiable. Those containing fatty acids are called saponifiable lipids, and in turn, they can be simple (constituting only C, H, and O; fatty acids and acylglycerides are included) or complex (containing C, H, and O and other elements, such as P, N, or S). Unsaponifiable lipids are not hydrolyzed in the presence of a base because they do not contain fatty acids.

Fatty acids are the basic units of the saponifiable lipids and consist of molecules formed by a long hydrocarbon chain with an even number of carbon atoms, usually ranging from 12 to 24, and a terminal carboxyl group. They are divided into saturated or unsaturated lipids, depending on the presence or not of double bonds, strongly influencing the melting point. Regarding the polarity, lipids can be divided into neutral and polar. The former have no electric charge, and they are fatty acids; tri-, di-, and monoglycerides; waxes; carotenoids; sterols; and tocopherols. The latter are phospholipids and glycolipids (Greenwell et al. 2010).

Lipids are found in the form of triglycerides (TAGs), formed by the esterification of three molecules of fatty acids with a glycerine molecule. They accumulate in oleaginous microorganisms and are the principal source for biodiesel production. Consequently, only the nonpolar TAGs show a commercial interest for the production of fuel. Among the neutral lipid algae, *Chaetoceros muelleri*, *C. vulgaris*, and Chlorophyceae are considered the best candidates for lipid extraction (Domingo 2014).

Microalgae, due to several advantages, such as the use of photosynthesis for energy production, their high biomass product with elevated oil content, and a fast growth rate, are an ideal biofuel candidate.

12.5 Downstream Processing

12.5.1 Cell Recovery

The oil extraction and purification of single cell oils still represent major costs in the production process. It is therefore important to have efficient downstream processing. We discuss this further since downstream processing in bioprocess engineering is often the most costly step (Probst et al. 2015). These high costs are sometimes a drawback in the whole process being commercialized. Microalgae biomass has a large percentage of water (close to 99% water). Therefore, before lipid extraction, it is mandatory to reduce the water content in the algal yield. Harvesting costs are about 20%–30% of the total cost (Pragya et al. 2013), requiring exhaustive water elimination. The most used harvesting and dewatering operations are centrifugation, sedimentation, flocculation, flotation, filtration, and electrophoresis, and any combination of these. These operations are briefly summarized:

- Centrifugation is employed as a rapid and efficient, although relatively expensive, method, with regard to equipment costs and maintenance. This process should not be applied for a long time because, besides the relatively high energy consumption, it could damage the cells, masking the efficiency of subsequent disruption methods.
- Sedimentation is an energetically efficient process to separate the algae from the water based on gravity. Densely packed algae, such as *Spirulina*, settle easily (Pragya et al. 2013). Also, the addition of flocculants (such as aluminum sulfate, ferric chloride, and ferric sulfate) accelerates the process, because they create aggregates with an increase in the characteristic agglomerate size.
- In flotation, the cultures are displaced in the opposite direction with respect to gravity. Bubbles drag algal biomass to the surface in order to facilitate its removal. This process can also be performed in conjunction with flocculation.
- There are different types of filtration, and they show good results. Since this separation is based on size differences between the particles, larger microalgae will be more efficiently separated than smaller ones.
- Finally, electrophoresis is a nontoxic method that, on the laboratory scale, is cost-effective and efficient. It uses an electric field to displace the algae to the surface (Pragya et al. 2013).

Once the cells are recovered, they can be treated in order to extract the lipids. A suitable method will have economic and environmental influences on biofuel production. To correctly choose a technique, thorough knowledge of all the algae features (density and size) is required, knowing that there are a large variety of microalgae, as mentioned in Section 12.2.2. This is less the case for yeasts and bacteria. The challenge is to find a method that is independent of the chosen strain, nontoxic, and as cheap as possible. Therefore, if a disruption cell method can be successfully applied on wet microalgae, this harvesting step will suppose a requirement for less energy. It should be noted that most extraction methods have only been successfully performed at the laboratory scale, with only little effort made for developing standardized methods on the industry scale. Some microorganisms represent difficulties toward lipid extraction due to their robust wall, which often requires a lysis step for extracting the lipids that have accumulated inside. A summary of some current extraction methods is given in the next section.

12.5.2 Cell Disruption

Since algae are rich in chemical and biological molecules with attractive properties, they should be properly released by a rapid and effective cell disruption process, in order to maximize the value of extracted products without harming or, in the case of proteins, denaturing them. More information can be found on downstream processing in Probst et al. (2015), Lesage and Bussey (2006), and Angles et al. (2017). Lysis is a method that conditions the cell in such a way that the tough, outer wall gets disrupted. There are a huge variety of cell disruption methods, and they can be divided into mechanical and nonmechanical.

Mechanical methods are widely used to completely break the cell and release all the compounds in the medium using huge amounts of energy. This means that all products will be mixed, hampering separation between them. Acids or alkalis can be added to decrease the wall strength and facilitate mechanical methods, but this could promote a contamination of the medium (Lee et al. 2012). The main advantage of these methods is their nontoxicity.

Examples of such methods are homogenization, cavitation, or cell disruption by bead mills.

Homogenizers can be divided into two types, depending on the control parameters: high-pressure and high-speed homogenizers. In the first type, an algal culture is forced to pass with the help of some pressure through a hole, to make the algae explode due to a fast pressure drop of about 1500 bar. In the latter type, high-speed homogenizers comprise a utensil rotating fast in a static tube. Different results can be achieved, depending on the chosen speed, the viscosity, and the concentration of the medium (Lee et al. 2012). Homogenizers have the highest efficiency: 77%–96% (Domingo 2014). Pressure disruption through mechanical stress, another mechanical method, of bacteria and yeast is used at the industrial scale at very high pressures (~1500 bar).

Other methods are based on cavitation phenomena. Sonication (with frequencies from 20 to 100 kHz) has the lowest efficiency (Lee et al. 2012), and it is based on sonic waves electrically produced by a metallic tip. These waves disintegrate the cell (Halim et al. 2012). In addition, it is a high-cost method, which excludes its use at an industrial scale.

Equipment for cell disruption also involves ball mills; in this case, cells are in suspension with small abrasive particles that generate shear forces, breaking these cells. Glass can also be used (bead mills) to release biomolecules by cell disruption. The breakage depends on the speed, size, residence time, and composition, which make the collision more efficient (Lee et al. 2012). Other physical, nonmechanical, methods comprise freeze–thawing and chemical and biological methods, such as organic solvent, hydrolysis, and enzymatic treatments. More innovative disruption methods are also being explored, like using metal nanoparticles (Costas et al. 2011) or the so-called milking procedure (Kleinegris et al. 2010; Vinayak et al. 2015). The concept of using metal nanoparticles consists of propelling the nanoparticles at high velocities by means of an electromagnetic field, allowing them to penetrate the cell wall. Since these nanoparticles are charged, they can be easily separated and reused by a magnetic field. Milking is a process that uses one of the classic extraction methods, such as an organic solvent, but it operates in such a manner that the cells do not die and can be regrown and reused. It should be noted that the cell toughness is believed to be caused by its high elasticity (Lesage and Bussey 2006), while bud scar formation and cell aging also contribute to the strength of the cell wall (Probst et al. 2015). Along with the optimum extraction method, attention should also be given to the cell wall strength and its dependence on external factors.

The extraction and purification processes are typically described by multiple steps that use organic solvents so that the oil gets separated from other cellular components. Even though in some extraction processes (Probst et al. 2015) the cell biomass has to be dried before solvent extraction, which is an energy-intensive step, alternative methods, including liquid–liquid extraction, have become more and more preferred. Hexane is an industry-adopted and -accepted solvent that is used for extraction, generally recognized as safe, and relatively inexpensive, readily available, and reusable (which is also useful for the milking principle [Kleinegris et al. 2010]). Other solvents of the same category, such as heptane, decane, and dodecane, are also used. A better alternative, still under study but gaining more and more interest, is the use of supercritical CO_2 for extraction (Adam et al. 2012). This extraction method has gained much attention due to its renewable character, and being an inert gas, it is recognized as safe. However, a major drawback is the high energy requirements it brings, which makes this method difficult to apply for large-scale processes. Others have proposed "softer," renewable terpene-based solvents, such as p-cymene and d-limonene (Tanzi et al. 2013). Also, ionic liquids seem to be viable green solvents when it comes to lipid extraction processes. They are nonaqueous solutions that have low vapor pressures, reducing the amount of evaporation, which can stay in liquid form from 0°C to 140°C (Probst et al. 2015). Overall, it is clear that in order to have a low-cost, solvent-based extraction method, the key considerations are renewable solvents, improvement of solvent recovery and/or recycling, minimization of the generation of waste, and reduction of the energy requirements.

However, disruption of cells is affected by many variables: strength, elasticity and size of cell walls (Greenwell et al. 2010), contamination, cost, and energy requirements. Processes requiring large amounts of energy for cell disruption represent a major cost, but on the other hand, processes that take advantage of the addition of a solvent to break the cells can contaminate the extracted element, due to the difficulty in separating the solvent. Cell disruption must be performed very precisely, because a suitable process selection will prevent the release of uncontrolled products and increase the efficiency of lipid recovery. Many of these methods, such as ultrasonication or osmosis (explained below), do not require total water extraction in algae, so they could be performed before the drying process and are, in that sense, cost-effective (Halim et al. 2012).

Nonmechanical methods are based on permeabilization. They can be divided into three main classes: physical, chemical, and enzymatic permeabilization.

One of the most known physical permeabilization method is osmotic shock or osmotic stress. With this method, the cells are mechanically damaged by rapid changes in the external concentrations. The dramatic change in the salinity of the culture media creates an abrupt change in the osmotic pressure at the level of the cell membrane. When the induced stress is higher than the mechanical resistance of the cell membrane, disruption occurs.

As said in Section 12.5.1, the applicability of this method to wet microalgae is the main advantage; the drying process is the higher energetic step. This feature makes the osmosis method economically feasible for biofuel production, increasing lipid recovery by a factor of 2 (Yoo et al. 2012). The fact that osmotic shock does not need extreme conditions, such as high temperature, pressure, or toxic solvents, makes it an attractive method.

Nevertheless, because of the difference in morphology and composition, it should be expected that this method is not suitable for all the potentially commercial strains. In this respect, the study by Yoo et al. (2012) shows good results for osmotic shock on *Chlamydomonas reinhardtii*, *Botryococcus* sp., *C. vulgaris*, and *Scenedesmus* sp. Moreover, it was proved in that study that lipid conversion into fatty acids was successful, which is the best process option for biofuel production. As a conclusion, the osmotic methods require the control of several variables, such as cell morphology or the agents used (NaCl or sorbitol).

Other physical methods involve the use of microwaves, autoclaving, or freeze drying. Concerning chemical permeabilization, many chemical products can be employed to release intracellular components. These can be, among others, antibiotics, chelates, or organic solvents. However, the addition of these chemicals, such as detergent or EDTA, can injure the lipids by contamination (Lee et al. 2012).

Lastly, in enzymatic permeabilization, enzymes (such as cellulose and papain) are used to permeabilize and efficiently disrupt cells; the main problem for this class is the difficult recovery of the enzymatic chains from the mixture of protoplasmic material, as well as its high cost.

12.6 Technicoeconomic Evaluation of Single-Cell Oil Production

Present technology uses the process of transesterification of TAGs with methanol to produce fatty acid methyl esters (FAMEs) (Koutinas et al. 2014). One of the useful end uses of microbial oil (MO) is to substitute vegetable oils for biodiesel and oleochemical production. The latter finds application in, for instance, pharmaceuticals, paints, and polymer additives. The focus here is on biodiesel as the end use. MOs that are produced through the use of different oleaginous microorganisms will, under different cultivation conditions, give a diversified panel of fatty acid compositions. These could be used for feedstocks that have either multiple or a specific purpose, depending on the complexity of the fatty acid compositions (Wynn and Ratledge 2005). MOs that consist of common, more basic fatty acids can be more easily transformed in several end products. However, MOs that are made of unusual fatty acids, with higher molecular complexity, can only be deployed for case-specific feedstocks with special properties. Unusual and case-specific processes naturally bring the cost question into the picture. In fact, whether the use of MOs is beneficial depends, among other factors, on the price range of petroleum oil, as the price of individual fatty acids (derived from petroleum oil) can vary from as little as €0.25 to more than €80/kg (Wynn and Ratledge 2005). Therefore, the identification of microorganisms that have the ability to produce high-value oils can lead to a large-scale substitution of petroleum-derived lipids (Wynn and Ratledge 2005). Nonetheless, the development of biodiesel from oleaginous microorganisms demands an efficient control of the fermentation process, as discussed earlier, as well as well-thought-out cost-saving downstream processing. The economic aspect of the present technologies, which focus on the production of biodiesel, is important in order to assess the possibility of commercialization. For the next example, it is assumed that the selling price of oil is €0.6/kg. Assuming a continuous fermentation of 20 m³/h of whole whey for 250 days/year of plant operation, this results in a production of 18,000 tons of yeast or whey protein dry mix (Davies 1988). In such a case, the plant operation would cost around €2.5 million, which would provide an internal rate of return of 140% (after taxes). It should be noted, however, that there is no detailed study to be found in the open literature on the exact estimation of what the production of biodiesel from MO would cost. Although the studies on microalgae exploitation are numerous, their conclusions do not always tackle the economic and feasibility aspects.

Also, despite the many varieties of algae that are cited as possible strains, only a few have been extensively investigated or tested. The main drawback at present is related to production costs, since the production of biodiesel from microalgae has a high operational cost, which makes it, for the time being, economically unviable. Indeed, the estimated costs for the production of a barrel of algae-based fuel are around €250–€2500 using current technology, compared with €20–€50 (2016) for petroleum (Dismukes et al. 2008). It is a fact that biodiesel production can be done in two ways, by the so-called wet and dry processes, but in both cases, there are operational steps that require intensive energy consumption, such as a drying or extraction process when the amount of water present remains high.

Despite the high oil prices, these energy costs still make the price of biofuel too expensive (Dismukes et al. 2008). The challenge of all these studies is to reduce energy costs in order to close the price gap between biofuel from algae and conventional fuel. Therefore, there are certain obstacles that must be overcome for biofuel production: economic, supply, and environmental issues, as well as competition with other sectors, such as food or cosmetic.

12.7 Case Study: Lipid Extraction from Microalgae

The case study in this section present several investigations on lipid extraction from microalgae that have been performed in our laboratory, where, among other things, the feasibility of this process was demonstrated (Domingo 2014) on a laboratory scale.

12.7.1 Introduction

The main objective of this case study was to analyze the effectiveness of osmotic shock in lipid extraction from microalgae. In the frame of this work, five different strains were cultivated. Among them, only two have been extensively tested for their positive and robust growth. This allows a better understanding of their behavior and better statistics (due to their robust growth). The species cultivated were *C. reinhardtii*, *C. vulgaris*, and *Scenedesmus dimporphus* (freshwater), and *P. tricornutum* and *Spirulina* (salty water). The species tested were *C. vulgaris* and *P. tricornutum*.

The alga culture is for sure an important stage since it determines many characteristics of the yield. In addition, this process is limited by the type of equipment needed for the cultivation of algae since eventually the equipment employed will have complex and expensive designs that limit the production to a small scale. The choice of the most suitable strain for biofuel production is difficult. However, there are some criteria that can help when making a decision. A high lipid amount is the first and foremost requirement, as it is the material needed to produce the biofuel (see Section 12.4.2). The lipids must be extracted and separated from algae, so strains that allow easy lipid extraction with the typical cell disruption methods are appropriate candidates (see Section 12.5.2). These processes have a cost and are energy-consuming, especially when it is necessary to dewater the samples (see Section 12.6). Lastly, fast growing rates are essential to biomass production. The lipids have to be extracted by rapid, economic, and effective cell disruption processes. In that way, osmotic shock does not require total water extraction, which is economically favorable. The cells treated by osmotic shock are damaged by rapid changes in the external concentration (dramatic change of salinity by using salt). Knowledge of the type and amount of the extracted lipids is very important for characterizing the treatment efficiency. Regarding the analytical techniques employed in this work, fluorescence seemed to be the most suitable for its ability to detect and quantify the actual compound at very low concentrations. However, molecules of interest are not necessarily fluorescent. Therefore, dyes have to be added in the process to mark such molecules. Throughout this work, two dyes were employed: Nile Red (NR) to stain lipids and Sytox Green to distinguish among living or dead cells by labeling damaged cells. Another technique that could be useful in this context is spectrophotometry; it is an analytical technique that is often used for determining the concentration of a compound in solution using an absorbance spectrum. The sample should show a colored product to be detected (such as chlorophyll and/or suitable dyes). The combination of the above-mentioned techniques allows for estimating the osmotic shock effectiveness to disrupt algae cells and evaluate the lipid content.

12.7.2 Materials and Methods

When studying algae, it is important to take into account the specific conditions that allow improving the yield of algae and facilitating their growth. All algae in our case study (*C. reinhardtii*, *C. vulgaris*, *P. tricornutum*, *Spirulina* sp., and *S. dimorphus*) were cultivated in a 400 mL tissue culture flask. Figure 12.2 shows microscope photos for the algae considered in this work.

C. reinhardtii and *Spirulina* sp. can grow at room temperature, while *C. vulgaris*, *P. tricornutum*, and *S. dimorphus* need a more controlled environment since they should live at 16°C during the day and 14°C during the night. Given these constraints, they were conserved in an aquarium to maintain the desired temperature with a

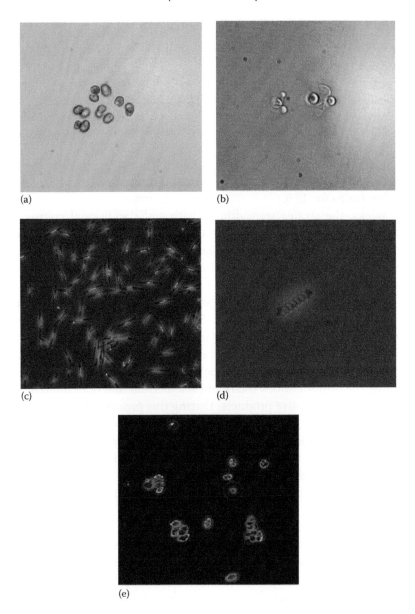

Figure 12.2 Pictures taken with a contrast phase microscope for (a) *C. reinhardtii* (20× magnification), (b) *C. vulgaris* (20× magnification), (c) *P. tricornutum* (40× magnification), (d) *Spirulina* sp. (100× magnification), and (e) *Scenedesmus dimorphous* (40× magnification).

TABLE 12.1   Culture Conditions for Each Algal Strain

Strain	Medium	Growth Condition
C. reinhardtii	Bg 11	Room temperature
C. vulgaris	Bg 11	16°C (day) and 14°C (night)
P. tricornutum	MDV	16°C (day) and 14°C (night)
Spirulina sp.	T+	Room temperature
S. dimorphus	Bg 11 + V8	16°C (day) and 14°C (night)

temperature controller. Concerning the optimal illumination, an artificial light was shined on the strains for 16 h.

Every 2 weeks, the strains were diluted with some additional culture media to provide additional nutrients and avoid abrupt colony degradation. The growth media include all the necessary elements, such as nutrients and proteins. The culture medium for each algal strain is reported in Table 12.1, which also summarizes their optimal growth conditions.

When these required conditions were suitably met, it was possible to maximize the algal growth and quantity of lipids. So, a correct cultivation environment is essential to magnify the cost-effectiveness of biofuel from microalgae (Demirbas and Demirbas 2009). The majority of these methods require the use of chemicals. We present in our case study the use of osmotic pressure, which means that no chemicals were involved. The osmotic shock treatment is a method that does not require a large amount of energy or expensive products. During this process, the cells were exposed to dramatic salinity changes, which caused the cell membranes to break. Moreover, as pointed out, osmotic shock can be applied directly to wet algae. This is one of the biggest advantages compared with the other techniques. The following process was carried out. A sample of 60 mL of each strain was taken. The necessary amount of NaCl was weighed in order to create the conditions for osmotic shock. The weighed salt was then introduced, with a magnetic stirrer, into a 40 mL disposable container for cell culture. The container was closed and the culture left in agitation. For each successive time step (1, 2, 4, 6, and 8 h), a set of three measurements was performed.

One milliliter of alga was mixed with 2 μL of NR, another one was mixed with 0.4 mL of Sytox, and the last one was divided into three wells of a fluorometer microplate to quantify the chlorophyll content. Each measurement was repeated three times. Furthermore, the content of each well was automatically measured repetitively 10 times so that a total of 30 values were obtained for each sample point.

The next step was to check if the NR was able to stain the lipids without breaking the algae. In order to make this happen, the algae were mixed with 0.5 μg/mL of NR and then kept in the dark for 10 min. Six strains with and without NR were studied; for each well, the signal was measured 10 times successively. The results show that there is always an increase in the signal when the algae are mixed with NR, which means that the dye is able to stain lipids even when they are inside the algae. As expected, the signal was saturated in the wells that contain only NR.

The effect of the solute ions should be discussed. In considering that the osmotic process implies a dramatic change on salinity, the effects of introducing additional ions in the culture media were thoroughly studied. The effect of mixing the strains with NaCl should also be taken into account because this compound can also affect the staining efficiency of the employed dyes. NR is able to stain lipids when the cell is alive or dead. Sytox staining enabled us to distinguish among living and dead ones by penetrating cells with plasma membranes in distress. As a fluorometer needs to use different filters for each dye, two experiments were executed to study the NaCl effect with these dyes. In this test, the NR concentration needed to stain the cells was lowered (0.33 µg/mL) to achieve a greater gain, and the salt concentration was 0.045 g NaCl/mL algae solution. The obtained results indicate that the salt did not affect all the stained strains in the same way. The salt affected the cells in order to liberate the lipids, but it also reacted with the dye, provoking amplification or diminution of the NR signal. This was increasing clearly observed in salty species, *P. tricornutum* (Figure 12.2c) and *Spirulina* sp. (Figure 12.2d), and in *C. vulgaris* (Figure 12.2b), but on the contrary, the freshwater strains, *Scenedesmus* sp. (Figure 12.2e) and *C. reinhardtii* (Figure 12.2a), showed a decrease of the signal with NaCl and NR. Regarding the osmotic experiments, two algae with the same behavior with respect to the salt were tested (one saltwater and one freshwater species): *C. vulgaris* (Figure 12.2b) and *P. tricornutum* (Figure 12.2c). The effect of salt mixed with Sytox was also studied. This dye stains dead cells, so the amount of salt employed in this case was superior (0.1 g/mL) in order to kill more algae, and the Sytox concentration was 0.03 µM Sytox/mL wet algae. The addition of salt to freshwater strains should increase the Sytox penetration and a higher signal was expected. However, the opposite happened, indicating that Sytox itself is inactivated by the change in salty content. For the salty species, the effect was less pronounced.

To counteract this effect, and in order to have a detectable signal of Sytox, higher concentrations of the dyes were tested. Also, a calibration test was performed to better understand the way the Sytox signal could be interpreted. It seemed logical to refer the calibration to a microalgae suspension where the external membranes of algae were definitely disrupted. In this condition, Sytox (staining only dead cells) should give the maximum signal. Microwaves were used to achieve that goal. In this light, two samples of each strain were taken. The four samples (10 mL each) were inserted into the centrifuge to further concentrate the algae for 5 min at 2000 rpm. Later, half of the total volume, mostly containing microalgal cells, was retrieved. Subsequently, for each strain, one of the concentrated samples was microwaved for 1 min at 600 W, while the other was used as a reference. Also, different Sytox concentrations were analyzed. It appeared that the best concentration for the next experiments was 0.4 µM Sytox/mL concentrated algae.

All the culture media were also tested independently to check for their own specific fluorescence signals. The culture media were a mix of different compounds, and their excitation and emission spectra were nonnegligible in the wavelength range of the filters. The goal of these series of experiments was to quantify the influence of the culture media. A small signal from the culture media evidenced that all the fluorescence was mainly coming from the lipids or the algae. All the culture media were mixed separately with 0.5 µg NR/mL and 0.4 µM Sytox. Five samples per culture media were analyzed, each with a different dye. In addition, 10 measures were performed for each well. For all the experiments, small signals were obtained that confirmed that the fluorescence was coming quasi-exclusively from the algae or lipid.

Spectrophotometry was mainly used to pinpoint the excitation range of the different compounds. It is interesting to point out that the near-zero signal for *Spirulina* may be caused by the low concentration of this strain, as it is very difficult to take a concentrated sample thereof. Chlorophylls typically have two absorption peaks in the visible spectrum, one in the vicinity of the blue light (400–500 nm) and another in the red region of the spectrum (600–700 nm), but reflect the middle part of the spectrum, corresponding with green (500–600 nm) and explaining the green color of most photosynthetic plants. Also, chlorophyll is the common name for a class of pigments, whose most known representatives are chlorophyll α and β. The spectrophotometer allowed us to check the efficiency of the fluorometer filter chosen for Sytox. An excitation spectrum corresponding to this dye was generated. The dye had a bandwidth of 20 nm and centered at 485 nm. Even though this wavelength does not correspond to the maximum, the absorbance is relatively important.

Another test that seemed meaningful to be performed (considering that in the osmotic treatment salt is continuously added) was the possible shifting in the excitation spectra due to the presence of additional ions in the culture media. For that purpose, the strains were mixed with salt (100 g salt/L) and their spectra measured. Generally, the addition of salt diminished the absorbance signal. This effect should then be taken into account via a calibration test. For *Spirulina*, the lines were very similar due to the low concentration of this strain in the sample.

In principle, the addition of salt should not increase or decrease the signal emitted by the NR. However, when the amount of salt was increased, we marked that the signal was increased too, which makes a calibration curve mandatory to correct this effect in the results. To accomplish this, the untreated algae signal, that is, without salt, was taken as a reference, and thus the increase in the signal when the salt was added was normalized. The obtained calibration curve is presented in Figure 12.3.

By using the curves in Figure 12.3, it is possible to know the correction value that must be applied to reduce the increase caused by the salt addition. Each new point obtained should be decreased by multiplying it by the corresponding correction

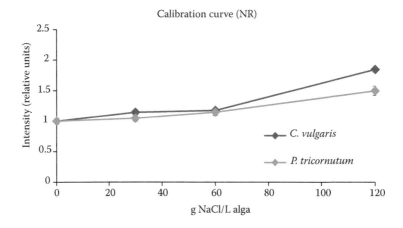

Figure 12.3 NR calibration curve.

value. After calibration, the values for each duration time of the experiments and each concentration could be obtained.

12.7.3 Results and Discussion

The aim of this work was to try to analyze the effectiveness of osmotic shock in the cell disruption step needed to extract lipids from microalgae strains. To this purpose, lipids and disrupted cells were characterized after being stained with two different specific stains: NR helped to quantify the lipid amount, while Sytox Green allowed for evaluating the proportion of disrupted cells, that is, how effective the osmotic treatment was.

For the reasons mentioned in Section 12.7.1, two strains were employed: *C. vulgaris* (freshwater) and *P. tricornutum* (salty water). They were preconcentrated to ensure that both had the same volume concentration (0.002 mL alga/mL solution). As each sample was 15 mL, the water content in both cases was higher than 99%.

12.7.3.1 Lipid Amount
The NR was able to quantify the neutral lipid amount in a sample, and as those types of lipids are the ones needed for biofuel production, it is very important to know which strain is richer in neutral lipids. To calculate this, four untreated samples of the same concentration were mixed with NR and introduced into the fluorometer. Since it was important to see how much lipids were being stained by the dye (in order to see which strain had the highest lipid content), the increased ratio for each signal, when the NR was added, was calculated. This was done by calculating $(I - I_0)/I_0$. Here, I is the intensity signal given by the sample with NR, and I_0 is the intensity signal given by the same sample without NR; thus, the base autofluorescence signal was deleted. Figure 12.4 shows the lipid content (expressed in fluorescence intensity) for two strains. The standard deviations, based on the three measurements performed

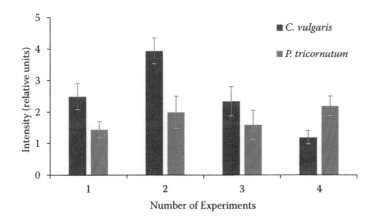

Figure 12.4 Lipid content for both strains.

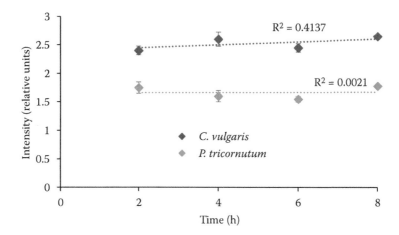

Figure 12.5 Time evolution of the lipid content. Linear regression lines are added.

for each strain, are indicated by error bars. The average standard deviation of the intensities (in relative units) is 0.36.

Since the initial volume concentration of the alga solution was the same, it was possible to spot which strain contained more lipids per milliliter of alga. However, the growth stage of the culture could not be precisely known, which explains the variability of the relative units between the different experiments. Nonetheless, the graph shows that in the case of *C. vulgaris*, the dye stained, on average, larger lipid amounts. The NR was able to quantify the amount of lipids independently of the status of the algae, that is, whether they are disrupted or not. As previously mentioned, the influence of the addition of salt on the measured lipid content should be assessed. For the purpose of this work, this influence should be weak. Figure 12.5 presents the lipid evolution in time for the untreated algae, when salt was not added.

Figure 12.5 shows that between 2 and 8 h, no significant change was observed in the measured lipid content. To show that the NR is independent of the salt content, Figure 12.5 should be compared with the case where salt was added. Figure 12.6 shows the signal obtained for all the salt concentrations after 2 h of lipid release (after using the calibration curves).

Figure 12.6 shows that different salt concentrations hardly affect the measured fluorescence signal, indicating that, indeed, the NR results are independent of the salt content. It should be noted that the R^2 values are much smaller than 1. This simply indicates that the trend is far from linear. However, the purpose was not to find the tendency the intensity has as a function of the salt content, but rather to confirm that there is no significant dependence, which is the case.

12.7.3.2 Dead Cell Amount
Sytox can stain those cells that have been damaged by salt, allowing the lipids, as well as other protoplasmic biomolecules, to be released. For that reason, it could be used to characterize the effectiveness of the osmotic shock treatment. Since salt is responsible

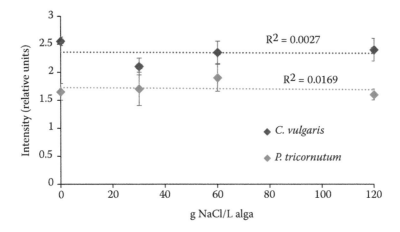

Figure 12.6 Lipid content for different concentrations of salt at 2 h of treatment. Linear regression lines are added.

for algae degradation or disruption, the higher the amount of salt, the greater the number of dead cells and the greater the signal that should be emitted by Sytox. For the results to be comparable, and due to the influence that additional ions have on the emission spectra of Sytox, a calibration curve has been created to correct for the effects of the increased salt content in osmotic treatment. To have a meaningful calibration curve, it is necessary to be sure that the cells in our sample are definitely disrupted, to ensure that the Sytox signal measured compares with a fixed disrupted cell concentration. Although energy demanding, microwaves are one of the most efficient methods for cells disruption; for generating our calibration curve, two samples (one per strain) with the same concentration employed during the experiments were microwaved at 600 W for 1 min. As the strains undergo a change in color when disrupted due to the release of protoplasmic material, these criteria have been used to validate the assumption that all cells could be considered disrupted. This treatment involves a volume loss in the sample because the heat causes water evaporation, so this volume loss should be filled with media to maintain the same concentration. Then the samples were divided into four shares of 3 mL each (a total of eight samples, four per strain). Since they are homogeneous, the dead cell amount per sample should be the same, and the same applies for the Sytox signal emitted. Therefore, the signal decrease obtained by the salt addition is used to correct the final results. In this case too, the reference is the untreated alga. It appeared that *C. vulgaris* is affected to a greater extent by the salt. Taking this into account, the results must be corrected by multiplying by a correction value that was obtained by the same principle as in Section 12.7.3.1. After calibration with respect to the salt content, the intensity of the Sytox signal can be directly correlated to the disrupted cell concentration, allowing spotting of the treatment efficiency. The intensity is expressed as $(I - I_0)/I_{alldead}$, where I is the intensity signal given by the treated sample with Sytox, I_0 is the intensity signal given by the same sample without Sytox, and $I_{alldead}$ is the intensity signal given by the sample where all the cells are dead. Generally, it seemed that the rising tendency of the Sytox signal proves that the salt is efficient in disrupting the algae. It is interesting to note that at the beginning of the experiments, the dead cell percentage was higher for *Chlorella*; however, at the end of the experiments, when 8 h had elapsed, the dead cell percentage was higher for *Phaeodactylum*. This means that the latter strain needs more time to be broken, but at the end, it shows better results for the osmotic treatment. Unlike *Chlorella*, it has a nonspherical shape, which possibly facilitates its breakage. Figure 12.7a shows the dead cell count as a function of time for different salt concentrations for the strain *C. vulgaris*. Figure 12.7b shows the same, but for the strain *P. tricornutum*.

These two plots with an increased trend, more marked in the case of *P. tricornutum*, corroborate the fact that this strain needs more time to be broken, and that the treatment for *Chlorella* can be reduced to a few hours because the values obtained for 4 h are similar to the last ones (8 h) when 60 g/L salt was used, which is the concentration with the best results for this strain. In the best case, the osmotic treatment kills

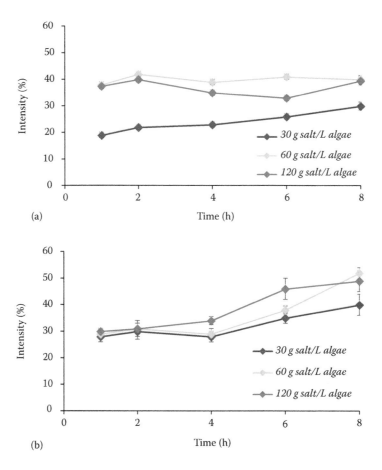

Figure 12.7 Time evolution of the fluorescence intensity by employing Sytox for (a) *C. vulgaris* and (b) *P. tricornutum*.

TABLE 12.2 Dead Cell Concentration for *C. Vulgaris* and *P. Tricornutum*

Strain Type	Mass Concentration (g/L)	% of Dead Cells after Treatment	Dead Cell Concentration (g/L)	40% of Lipids (g/L)
C. vulgaris	0.3224	40.4%	0.1302	0.0521
P. tricornutum	0.5807	51.6%	0.2996	0.1199

around 40.4% of the cells in *Chlorella* with 60 g/L salt. In *Phaeodactylum*, the osmotic treatment is able to disrupt 51.6% of the cells with the same salt concentration.

12.7.3.3 Scale-Up

Depending on the strain employed, the lipid content typically differs from 2% to 40% of the dry weight of algal biomass (Demirbas and Demirbas 2009). However, a few species may be found with a higher oil content, reaching 80%. In this case study, we take 40% as the typical value. As it is the usual unit, we present the results of the biomass in grams per liter. The amount of dead cells obtained from the biomass is shown in percentage with the corresponding concentration. Taking the typical value of 40%, we can give estimates of the typical lipid contents that can be obtained using the method from the case study. The results are shown in Table 12.2.

Table 12.2 represents the lipid amount that could be extracted by osmotic shock when a salt concentration of 60 g/L is employed. If this experiment were done at the industrial scale, the volume of all the samples would be much larger; for example, if 1 m^3 of these strains (with the same concentrations employed during the experiments) were analyzed, the lipid amounts would be 52 and 120 g, respectively. Although these amounts seem small, it is important to remember that the analyzed samples contain more than 99% water, and that if some of this water is removed, which means an increase in the alga concentration, the results could be greatly enhanced.

12.8 Conclusions

In this chapter, four main types of microorganisms that are used for the production of biodiesel have been discussed: bacteria, yeast, fungi, and microalgae. We have set out the different culture environments, ranging from open ponds to complicated (photo-)bioreactors. The discussion was continued by focusing on fermentation techniques and processes, including primarily batch, continuous, and fed-batch processes. Downstream processes have been analyzed and directly linked to the cost picture. Indeed, this latter step of the whole process seems to be one of the most expensive and technologically difficult. Although more effort is needed for investigating the mixing characteristics in bioreactors, the fermentation strategies are already reasonably well understood. However, it appears that more analysis is needed for improving and rendering more efficient the extraction methods of lipids. In our case study, we discussed investigations from our laboratory concerning lipid extraction from microalgae by osmotic shock. Two techniques have been used in order to analyze the efficiency of this extraction method. On the one hand, spectroscopy is able to determine the concentration of the samples. On the other hand, fluorometry allows analysis of the lipid presence using NR and the dead cell amount using Sytox Green. The osmotic shock is produced by creating a dramatic change in the sample's salinity by adding salt. The efficiency of the osmotic shock treatment is strongly dependent on the employed strain.

In general, the choice of strain to be the appropriate candidate for biofuel production is based on several criteria:

- High lipid content: The highest in this study appeared to be by *C. vulgaris*.
- Good performance to standard extraction techniques: Osmotic shock is performing relatively well for the strains tested in this work. Both samples perform well with the osmotic shock. Although *Phaeodactylum* needs more time to be disrupted, it has shown better results with this treatment, with 51.6% of the cells being dead, instead of 40.4% in the case of *Chlorella*.
- The chosen alga must have rapid growth. *C. vulgaris*, as a green alga, presents this characteristic, which makes it very favorable for biofuel production. On the other hand, *P. Tricornutum*, as a diatom, is very abundant.

The lipid amount that could be obtained if 1 m^3 of these strains is treated has also been calculated, also showing a better, although small, result for *Phaeodactylum*. This value can be enhanced by removing some water or by using a more powerful technique, but always keeping in mind that it is difficult to improve the economic aspects related to osmotic shock. It is an efficient treatment that can be applied on wet algae with more than 99% water, and also which does not need a huge amount of energy, which is the case for ultrasonic or microwave treatments.

Since salt reacts with the dyes, making some calibration curves necessary, it would be interesting to investigate this effect in more detail. Future studies could be based on the use of other agents for osmotic shock, such as sorbitol, which does not release as many ions as NaCl. Salt has been employed for salty and fresh species treatment. The salty species are used to culture media with a high salt content; therefore, another experiment could be based on the opposite effect: carrying out osmotic shock by using distilled water instead of salt. In all the studies, it has been proven that the dyes do not perform well for all the strains, since they are influenced by the media characteristics in which they are introduced. Due to this, other dyes could be used to check if they perform better with the studied strains or if they are more stable against media changes. Finally, it can be concluded that the results show that osmotic shock is an efficient treatment for lipid extraction.

References

Abbott MSR, Harvey AP, Perez GV, Theodorou MK. Biological processing in oscillatory baffled reactors: Operation, advantages and potential. Newcastle upon Tyne, UK: Bioprocessing Biopharmaceutical Technology Centre, Newcastle University, 2013.

Adam F, Abert-Vian M, Peltier G, Chemat F. "Solvent-free" ultrasound-assisted extraction of lipids from fresh microalgae cells: A green, clean and scalable process. *Bioresour Technol* 2012, 114:457–465.

Al-Mashhadani MKH, Wilkinson SJ, Zimmerman WB. Airlift bioreactor for biological applications with microbubble mediated transport processes. *Chem Eng Sci* 2015, 137:243–253.

Angles E, Jaouen P, Pruvost J, Marchal L. Wet lipid extraction from the microalga Nannochloropsis sp.: Disruption, physiological effects and solvent screening. *Algal Res* 2017, 21:27–34.

Borowitzka MA. 2005. Culturing microalgae in outdoor ponds. In *Algal Culturing Techniques*, ed. RA Andersen. London: Academic Press, pp. 205–218.

Carvalho AP, Meireles LA, Malcata FX. Microalgal reactors: A review of enclosed system designs and performances. *Biotechnol Prog* 2006, 22(6):1490–1506.

Capus A, Monnerat M, Ribeiro LC, De Souza W, Martins JL, Sant'Anna C. Application of high-content image analysis for quantitatively estimating lipid accumulation in oleaginous yeasts with potential for use in biodiesel production. *Bioresour Technol* 2016, 203:309–317.

Chen HG, Zhang YHP. New biorefineries and sustainable agriculture: Increased food, biofuels, and ecosystem security. *Renew Sustain Energy Rev* 2015, 47:117–132.

Cheon S, Kim HM, Gustavsson M, Lee SY. Recent trends in metabolic engineering of microorganisms for the production of advanced biofuels. *Curr Opin Microbiol* 2016, 35:10–21.

Chisti Y. Biodiesel from microalgae. *Biotechnol Adv* 2007, 25:294–306.

Cornet JF, Dussap CF, Gros JB, Binois C, Lasseur C. A simplified monodimensional approach for modelling coupling between radiant light transfer and growth-kinetics in photobioreactors. *Chem Eng Sci* 1995, 50(9):1489–1500.

Costas CR, Eck CR, Raytheon Co. Algal cell lysis and lipid extraction using electromagnetic radiation-excitable metallic nanoparticles. U.S. Patent 20110217748. 2011.

Davies RJ. Yeast oil from cheese whey-process development. In *Single Cell Oil*, ed. RS Moreton. London: Longman, 1988, pp. 99–145.

Del Campo JA, García-González M, Guerrero MG. Outdoor cultivation of microalgae for carotenoid production: Current state and perspectives. *Appl Microbiol Biotechnol* 2007, 74(6):1163–1174.

Demirbas A, Demirbas MF. *Algae Energy*. Turkey: Springer. 2009.

Dismukes GC, Carrieri D, Bennette N, Ananyev GM, Posewitz MC. Aquatic phototrophs: Efficient alternatives to land-based crops for biofuels. *Current Opin Biotech* 2008, 19:235–240.

Domingo PM. Concept and validation of lipid extraction from microalgae by osmotic shock technique. Brussels: Université Libre de Bruxelles, 2014.

Du J, Wang HX, Jin HL, Yang KL, Zhang XY. Fatty acids production by fungi growing in sweet potato starch processing waste water. *Chin J Bioprocess Eng* 2007, 5(1):33–36.

Eibl R, Eibl D, Portner R, Catapano G, Czermak P. *Cell and Tissue Reaction Engineering*. New York: Springer, 2008.

Eriksen NT. The technology of microalgal culturing. *Biotechnol Lett* 2008, 30(9):1525–1536.

Fischer CR, Klein-Marcuschamer D, Stephanopoulos G. Selection and optimization of microbial hosts for biofuels production. *Metab Eng* 2008, 10:295–304.

Greenwell HC, Laurens LML, Shields RJ, Lovitt RW, Flynn KJ. Placing microalgae on the biofuels priority list: A review of the technological challenges. *J R Soc Interface* 2010, 7:703–726.

Grobbelar J, Neddal L, Tichy V. Influence of high frequency light/dark fluctuations on photosynthetic characteristics of microalgae photo acclimated to different light intensities and implications for mass algal cultivations. *J Appl Phycol* 1996, 8:335–343.

Halim R, Danquah M, Webley P. Extraction of oil from microalgae for biodiesel production: A review. *Biotechnol Adv* 2012, 30:709–732.

Han X, Miao XL, Wu QY. High quality biodiesel production from heterotrophic growth of *Chlorella protothecoides* in fermenters by using starch hydrolysate as organic carbon. *J Biotechnol* 2006, 126(4):499–507.

Huang G, Chen F, Wei D, Zhang X, Chen G. Biodiesel production by microalgal biotechnology. *Appl Energy* 2010, 87:38–46.

Kleinegris DMM, Janssen M, Brandenburg WA, Wijffels RH. The selectivity of milking of *Dunaliella salina*. *Mar Biotechnol* 2010, 12:14–23.

Koutinas AA, Chatzifragkou A, Kopsahelis N, Papanikolaou S, Kookos IK. Design and techno-economic evaluation of microbial oil production as a renewable resource for biodiesel and oleochemical production. *Fuel* 2014, 116:566–577.

Kumar R, Singh S, Singh OV. Bioconversion of lignocellulosic biomass: Biochemical and molecular perspectives. *J Ind Microbiol Biotechnol* 2008, 35:377–391.

Lee A, Lewis D, Ashman P. Disruption of microalgal cells for the extraction of lipids for biofuels: Processes and specific energy requirements. *Biomass and Bioenergy* 2012, 46:89–101.

Lesage G, Bussey H. Cell wall assembly in *Saccharomyces cerevisiae*. *Microbiol Mol Biol Res* 2006, 70:317–343.

Li Q, Du W, Liu D. Perspectives of microbial oils for biodiesel production. *Appl Microbiol Biotechnol* 2008, 80:749–756.

Li W, Du W, Li YH, Liu DH, Zhao ZB. Enzymatic transesterification of yeast oil for biodiesel fuel production. *Chin J Process Eng* 2007, 7(1):137–140.

Liang XA, Dong WB, Miao XJ, Dai CJ. Production technology and influencing factors of microorganism grease. *Food Res Dev* 2006, 27(3):46–47.

Liu SJ, Yang WB, Shi AH. Screening of the high lipid production strains and studies on its flask culture conditions. *Microbiology* 2000, 27(2):93–97.

Lynd LR, Laser MS, Bransby D, Dale BE, Davison B, Hamilton R, Himmel M et al. How biotech can transform biofuels. *Nat Biotechnol* 2008, 26:169–172.

Ma YL. Microbial oils and its research advance. *Chin J Bioprocess Eng* 2006, 4(4):7–11.

McNeil B, Harvey LM. *Practical Fermentation Technology*. Hoboken, NJ: John Wiley & Sons, 2008.

Meher LC, Sagar DV, Naik SN. Technical aspects of biodiesel production by transesterification—A review. *Renew Sustain Energy Rev* 2006, 10:248–268.

Meng X, Yang J, Xu X, Zhang L, Nie Q, Xian M. Biodiesel production from oleaginous microorganisms. *Renew Energy* 2009, 34:1–5.

Meroney RN, Colorado PE. CFD simulation of mechanical draft tube mixing in anaerobic digester tanks. *Water Res* 2009, 43(4):1040–1050.

Metcalf E. *Wastewater Engineering Treatment and Reuse*. New York: McGraw Hill, 2003.

Miao XL, Wu QY. Bio-oil fuel production from microalgae after heterotrophic growth. *Renew Energy Resour* 2004, 4(116):41–44.

Miller AW, Befort C, Kerr EO, Dunham MJ. Design and use of multiplexed chemostat arrays. *J Vis Exp* 2013, 72:e50262.

Mličková K, Luo Y, Andrea S, Peč P, Chardot T, Nicaud JM. Acyl-CoA oxidase, a key step for lipid accumulation in the yeast *Yarrowia lipolytica*. *J Mol Catal B Enzym* 2004, 28:81–85.

Moraveji MK, Sajjadi B, Jafarkhani M, Davarnejad R. Experimental investigation and CFD simulation of turbulence effect on hydrodynamic and mass transfer in a packed bed airlift internal loop reactor. *Int Commun Heat Mass* 2011, 38:518–524.

Papanikolaou S, Aggelis G. Lipid production by *Yarrowia lipolytica* growing on industrial glycerol in a single-stage continuous culture. *Bioresour Technol* 2002, 82:43–49.

Perner-Nochta I, Posten C. Simulations of light intensity variation in photobioreactors. *J Biotechnol* 2007, 131(3):276–285.

Perrin L, Probert I, Langer G, Aloisi G. Growth of the coccolithophore Emiliania huxleyi in light- and nutrient-limited batch reactors: Relevance for the BIOSOPE deep ecological niche of coccolithophores. *Biogeosciences* 2016, 13:5983–6001.

Pfaffinger CE, Schöne D, Trunz S, Löwe H, Weuster-Botz D. Model-based optimization of microalgae areal productivity in flat-plate gas-lift photobioreactors. *Algal Res* 2016. 20:153–163.

Posten C. Design principles of photo-bioreactors for cultivation of microalgae. *Eng Life Sci* 2009, 9(3):165–177.

Pragya N, Pandey KK, Sahoo PK. A review on harvesting, oil extraction and biofuels production technologies from microalgae. *Renew Sustain Energy Rev* 2013, 24:159–171.

Preisig HR, Andersen RA. Historical review of algal culturing techniques. In *Algal Culturing Techniques*, ed. RA Andersen. London: Academic Press, 2005, pp. 1–12.

Probst KV, Schulte LR, Durrett TP, Rezac ME, Vadlani PV. Oleaginous yeast: A value-added platform for renewable oils. *Critical Rev Biotechnol* 2015, 36:942–955.

Pruvost J, Pottier L, Legrand J. Numerical investigation of hydrodynamic and mixing conditions in a torus photobioreactor. *Chem Eng Sci* 2006, 61(14):4476–4489.

Ratledge C. Fatty acid biosynthesis in microorganisms being used for single cell oil production. *Biochimie* 2004, 86:807–815.

Rude MA, Schirmer A. New microbial fuels: A biotech perspective. *Curr Opin Microbiol* 2009, 12:274–281.

Shariff S, Chakraborty S. Two-scale model for quantifying the effects of laminar and turbulent mixing on algal growth in loop photobioreactors. *Applied Energy* 2017, 185:973–984.

Šimčík M, Mota A, Ruzicka MC, Vicente A, Teixeira J. CFD simulation and experimental measurement of gas holdup and liquid interstitial velocity in internal loop airlift reactor. *Chem Eng Sci* 2011, 66:3268–3279.

Singh J, Kaushik N, Biswas S. Bioreactors—Technology & design analysis. *Scitech J* 2014, 1:28–36.

Somerville C. Biofuels. *Curr Biol* 2007, 17:R115–119.

Stephanopoulos G. Challenges in engineering microbes for biofuels production. *Science* 2007, 315:801–804.

Suzuki T, Hasegawa K. Lipid molecular species of *Lipomyces starkeyi*. *Agric Biol Chem* 1974, 38:1371–1376.

Tanzi CD, Vian MA, Chemat F. New procedure for extraction of algal lipids from wet biomass: A green clean and scalable process. *Bioresour Technol* 2013, 134:271–275.

Tao Q, Gao F, Qian CY, Guo XZ, Zheng Z, Yang ZH. Enhanced biomass/biofuel production and nutrient removal in an algal biofilm airlift photobioreactor. *Algal Res* 2017, 21:9–15.

Tichy V, Poulson M, Grobbelaar JU, Xiong F, Nedbal L. Photosynthesis, growth and photoinhibition of microalgae exposed to intermittent light. In *Photosynthesis: From Light to Biosphere*, ed. P Mathis. Vol. 5. Amsterdam: Kluwer Academic, 1995, pp. 1029–1032.

Tilman D, Socolow R, Foley JA, Hill J, Larson E, Lynd L, Pacala S, Reilly J, Searchinger T, Somerville C, Williams R. Beneficial biofuels—The food, energy, and environment trilemma. *Science* 2009, 325:270–271.

Vinayak V, Manoylov KM, Gateau H, Blanckaert V, Hérault J, Pencréac'h G, Marchand J, Gordon R, Schoefs B. Diatom milking: A review and new approaches. 2015, 13:2629–2665.

Wu B. CFD simulation of mixing in egg-shaped anaerobic digesters. *Water Res* 2010, 44(5):1507–1519.

Wynn JP, Ratledge C. Oils from microorganisms. In *Bailey's Industrial Oil and Fat Products*, vol. 3, *Edible Oil and Fat Products: Specialty Oils and Oil Products*, ed. F Shahidi. 6th ed. Hoboken, NJ: John Wiley & Sons, 2005, pp. 121–153.

Yamauchi H, Mori H, Kobayashi T, Shimizu S. Mass production of lipids by *Lipomyces starkeyi* in microcomputer-aided fed-batch culture. *J Ferment Technol* 1983, 61:275–280.

Yoo G, Park W, Kim C, Choi Y, Yang J. Direct lipid extraction from wet *Chlamydomonas reinhardtii* biomass using osmotic shock. *Bioresour Technol* 2012, 123:717–722.

Zhao X, Kong X, Hua Y, Feng B, Zhao Z. Medium optimization for lipid production through co-fermentation of glucose and xylose by the oleaginous yeast *Lipomyces starkeyi*. *Eur J Lipid Sci Technol* 2008, 110:405–412.

Zheng Y, Pan Z, Zhang R, Labavitch JM, Wang D, Teter SA, Jenkins BM. Evaluation of different biomass materials as feedstock for fermentable sugar production. *Appl Biochem Biotechnol* 2007, 137–140:423–435.

Section III

Prospects for Future Development of Biofuels

13

New Tools for Bioprocess Analysis and Optimization of Microbial Fuel Production

Isabelle France George, Philippe Bogaerts, Dimitri Gilis,
Marianne Rooman, and Jean-François Flot

Contents

13.1 State-of-the-Art in Biofuel Production by Microbial Cell Factories

Biofuels, that is, nonconventional liquid and gaseous fuels derived from renewable sources, such as crop plants, forest products, algae, or waste materials, are widely promoted as a sustainable alternative to fossil fuels and a means to secure our energy supply (Tilman et al. 2009). Other potential benefits of biofuel production are the creation of new, local employment (e.g., in rural areas) and the reduction of emissions of greenhouse gases (Fargione et al. 2008; Duke et al. 2013). A broader definition of biofuels (that we will use in this chapter) includes fuels produced from other renew-able sources, such as carbon dioxide (CO_2), exploited by biodiesel- or hydrogen (H_2)-producing phototrophic organisms.

Presently, thanks to their great metabolic diversity, bacteria, fungi, and microalgae are largely involved in the production of several biofuels, and they are expected to play an even greater role in the future. Most biofuel production processes rely on pure strains or their enzymes.

The most obvious example is first-generation bioethanol produced by fermenta-tion of sugar beet, sugarcane, and corn starch by the yeasts *Saccharomyces cerevisiae* or *Zymomonas mobilis* (Goldemberg 2007; Sánchez and Cardona 2008). Second-generation bioethanol from nonfood feedstocks (wood, sawdust, perennial plants like switchgrass or *Miscanthus*, agricultural residues, or municipal waste [Sánchez and Cardona 2008]) is more challenging to produce because of their lignocellulosic nature (except the last one). Therefore, their conversion requires a combination of physico-chemical and biological treatments to deconstruct lignocellulosic biomass into sugars that can be transformed into end products (Lynd et al. 2002). The standard proce-dure consists of a physical, chemical, or combined pretreatment to make cellulose and hemicellulose more accessible to hydrolytic enzymes, followed by the biological breakdown of (hemi-)cellulose by multienzyme cocktails, in which cellulase enzymes are the major components (Medie et al. 2012). The fungus *Trichoderma reesei* is the current gold standard for production of commercialized cellulases, but competitive fungal cellulase producers like *Myceliophthora thermophila* have appeared on the market (Gusakov 2013). Finally, monomeric hexose (and pentose) sugars are trans-formed by the yeast *S. cerevisiae*, the bacterium *Z. mobilis*, or genetically modified *Escherichia coli* into desirable products: mostly ethanol, but also longer-chain alco-hols, alkanes, and fatty acid esters (reviewed in Liao et al. 2016). The aforementioned organisms are utilized primarily because (1) their physiology and metabolism have been extensively studied and (2) dedicated genetic engineering toolkits have been developed. As a possible exception, *S. cerevisiae* does have several traits that are advantageous, such as its high fermentative capacity, natural ethanol tolerance, and ability to grow at acidic pH; however, it remains incapable of fermenting C5 sugars and of surviving at tem-peratures optimal for exogenous hydrolytic enzymes (Akinosho et al. 2014). Last but not least, lignocellulose degradation and sugar fermentation can be performed by pure bacterial cultures, among which the most exploited strain is thermophilic *Clostridium thermocellum*, which produces an extracellular enzymatic complex—the cellulosome—with remarkable enzymatic hydrolysis efficiency compared with that of free cellulases (Akinosho et al. 2014; Scully and Orlygsson 2014). Unfortunately, it can only tolerate a low ethanol concentration (maximum 5 g/L) and is unable to ferment C5 sugars.

Production of biofuels needs to be very efficient because of their low unit price. As metabolic pathways in natural microorganisms turn out to be insufficient for conver-sion of cellulosic biomass to a level that is compatible with industrial applications (i.e., titer > 40 g/L, yield > 90% of the theoretical maximum, and rate > 1 g/L/h [Dien et al. 2003]), they have been optimized or redesigned to improve production effi-ciency, either by genetic engineering (i.e., the manipulation of an organism's genome to alter its traits) or by metabolic engineering (i.e., the modification of an organism's metabolic network to produce a specific compound) via the optimization of their genes and regulatory processes (Stephanopoulos et al. 1998; Nielsen 2003; Kim et al. 2013). Recent breakthroughs in the high-throughput sequencing of genomes and genome-editing tools have accelerated our capability to modify genetic information (Esvelt and Wang 2013). Recently, the potential of metabolic engineering has been boosted by coupling it with microfluidics, which allows high-throughput selection of the most promising individuals in genetically modified microbial populations (Wang et al. 2014; Huang et al. 2015). So far, microbial engineering for the production of biofuels has largely consisted of overexpressing the desired pathway genes and knocking out competing pathways. Biofuel production pathways have even been introduced in microorganisms that do not natively ferment sugars or hydrolyze cellulose, but whose genetic and metabolic pathways are better known and easier to manipulate, such as *E. coli*, *Lactobacillus* sp., *Pseudomonas putida*, and *Bacillus subtilis* (Liu and Khosla 2010; Zhang et al. 2016). These approaches and more advanced ones based on the artificial regulation of pathways and creation of new pathways have been reviewed elsewhere (Liao et al. 2016).

Microorganisms have also been engineered to perform both cellulose deconstruction and fermentation of the resultant released sugars, a process called consolidated bioprocessing (CBP) (Lynd et al. 2005; Schuster and Chinn 2012; Hasunuma et al. 2013; Mbaneme and Chinn 2015; Salehi Jouzani and Taherzadeh 2015). This way, inhibition of cellulose breakdown by accumulation of its intermediary and final products is avoided. CBP has been implemented either by making efficient cellulose-degrading bacteria like *C. thermocellum* able to produce ethanol (Akinosho et al. 2014; Chung et al. 2014; Huang et al. 2014) or by transferring cellulase-encoding genes into standard strains used for ethanol production, like *S. cerevisiae* (Tsai et al. 2009; Hasunuma and Kondo 2012). Although highly promising, these engineered microbes still harbor insuffient cellulose-to-ethanol conversion yields, and therefore they have not been used for industrial ethanol production yet. *C. thermocellum* is, however, a good CBP candidate, provided that its ethanol yield can be improved (Akinosho et al. 2014).

A lot of effort has also been put into the biological conversion of sugars into various biofuels beyond ethanol. In this respect, bacterial and fungal strains (mostly *E. coli* [Clomburg and Gonzalez 2010], but also *B. subtilis*, *C. thermocellum*, *S. cerevisiae*, *P. putida*, and *Yarrowia lipolytica*) have been engineered to convert sugars into isobutanol (an alternative to ethanol with higher energy content for blending with gasoline or for transformation into kerosene), fatty acids, or the isoprenoid hydrocarbon farnesene (reviewed in Hasunuma and Kondo 2012; Liao et al. 2016).

Four decades of research has strived to produce cheap enzymes and develop robust engineered microorganisms able to operate at high rate and/or at high temperature, to cope with complex hydrolysate streams, to use all sugars produced from (hemi-) cellulose degradation (and not only hexoses), to maintain high viability, and to be resistant to multiple stresses (such as inhibition by high product [ethanol] concentrations, high acid concentrations or high temperature, or contamination by bacteria or phages). However, despite these advances, the process of lignocellulosic biomass conversion to biofuels—primarily bioethanol—remains commercially unviable (Kricka et al. 2015). In the remainder of this chapter, we discuss some of the ways to circumvent the current obstacles standing in the way of converting efficiently complex biomass into biofuels.

Last but not least, the integration of chemical catalysis into biological catalysis has opened up the possibility to transform products of sugar fermentations other than ethanolic fermentation. For example, acetone and 1-butanol, resulting from solventogenic acetone-butanol-ethanol (ABE) fermentation (Ezeji et al. 2007), or acetate from acetogenic fermentation, can be converted into long-chain alkanes for use as gasoline, jet fuel, and diesel (Anbarasan et al. 2012; Sreekumar et al. 2015), or into alcohols, acids, esters, or olefins, respectively. Such flexible approaches allowing the synthesis of a variety of bio-based products, in addition to low-value biofuels, are necessary to make biorefineries economically viable (Bozell and Petersen 2010).

Another high-profile example of microbial conversion of a nonfood substrate into biofuel is the production of oil by eukaryotic microalgae or cyanobacteria. Using sunlight, CO_2, and water, they produce biomass, including triglycerides and lipids (up to 70% lipid per dry weight), that can be converted into biodiesel, or sugars that can be fermented into other products. Genetic or metabolic engineering has allowed us to increase the lipid storage capacity in a handful of algal strains, or to expand the range of molecules produced by cyanobacteria (liquid fuels like ethanol, isobutanol, and isopropanol, but also hydrogen, bioplastics, or commodity chemicals). In addition, algal biomass can be converted into biogas through anaerobic fermentation. Algal cultures in ponds or photobioreactors offer great advantages over agricultural oleaginous crops, primarily a higher lipid productivity per ground area, as well as lack of competition for arable land, therefore avoiding the "food or fuel" dilemma (Wijffels and Barbosa 2010). In particular, photobioreactors, although more expensive to set up and operate, produce a much more concentrated algal broth than do open ponds, which decreases the cost of dewatering substantially (Chisti and Yan 2011). Much effort has been devoted to developing pure strains with characteristics of an "ideal" photosynthetic cell factory: high photosynthesis efficiency, ability to grow and produce oil at the same time, excretion of oil outside the cells, and insensitivity to oxygen concentration (Wijffels and Barbosa 2010). A good ability to tolerate heat and salts is desired as well, as many regions of nonarable land where algal growth could extend are arid and have no surface freshwaters, but alkaline or saline water reservoirs beneath them (Hannon et al. 2010). In that respect, cyanobacteria, whose genome is smaller and more easily edited than eukaryotic microalgae, hold great promise for the production of various fuels and chemicals (Angermayr et al. 2015; Liao et al. 2016). However, production of microalgae for biofuels is presently low scale and expensive compared with that of conventional energy crops, and it is therefore not viable at an industrial scale yet (Lardon et al. 2009; Chisti and Yan 2011). Life cycle assessment studies suggest that algal fuels may be better than fossil fuels, but there are no clear answers in this emerging field (Chisti and Yan 2011). Despite large research programs focused on developing microalgal energy production systems since the late 1970s, substantial improvements in productivity are still needed. They are expected to arise in the next 10–15 years from technological improvements (bioreactor design, nutrient and light regimen control, nutrient recycling, biomass harvesting, extraction, and exploitation of all these ingredients into a biorefinery infrastructure) and strain improvements (detailed physiological and genomic characterization, dedicated genetic engineering toolkits, and/or exploitation of new strains) (Brennan and Owende 2010; Hannon et al. 2010; Wijffels and Barbosa 2010). With regard to strain improvement, a great step forward is expected with the development of algal species engineered to use atmospheric nitrogen instead of petroleum-derived nitrogen fertilizers (Chisti and Yan 2011). A low-tech, recently revisited alternative is to grow algae in wastewater, which is rich in CO_2 and nutrients, thereby coupling algal biomass production with wastewater treatment (Brennan and Owende 2010).

The two aforementioned examples illustrate the longtime effort devoted to develop liquid biofuels, which offer the advantages of high energy density and compatibility with current vehicle motors. In parallel, gaseous biofuels are of interest as well, either because they already have a market (as in the case of biogas, a mixture of methane [CH_4] and CO_2, with applications similar to natural gas) or because they are expected to find one in the middle-term future (as for molecular hydrogen).

Biogas is produced by complex microbial communities that digest agricultural and municipal organic wastes. In the first stage, organic matter is hydrolyzed and fermented by anaerobic bacteria into acetate, H_2, CO_2, alcohols, and volatile fatty acids, such as propionate or butyrate. The higher volatile fatty acids are further converted into acetate, H_2, and CO_2, by anaerobic acetogenic bacteria. In the second stage, the three latter molecules are transformed into methane and CO_2 by acetotrophic or hydrogenotrophic methane-producing archaea. Both stages need to be balanced in terms of degradation rate: the temperature, pH, ammonia concentration, and retention time must be carefully controlled (Weiland 2010). The by-product of this treatment, the digester residue, can be used as fertilizer. Globally speaking, biogas production has been evaluated as one of the most energy-efficient and environmentally friendly technologies for bioenergy production (Weiland 2010). However, despite a remarkable development of farm-scale or domestic biogas plants and large-scale codigestion plants (where biogas is produced from a mixture of substrates expected to provide a more balanced nutrient pool to microorganisms), more research is still needed to achieve the full potential of biogas as an alternative to fossil fuels (Sárvári Horváth et al. 2016). Indeed, it is mainly used for generating heat and electricity, but its use as a vehicle fuel or as an alternative to natural gas in the grid requires its conversion into a liquid fuel like biodiesel or its upgrade to biomethane, respectively. Conversion of CH_4 to biodiesel is possible—although far from commercially viable—using a group of bacteria called methanotrophs. They use methane as their sole carbon source and convert methane into cellular compounds, among which are biolipids (Strong et al. 2015). Upgrade to near-pure biomethane is hampered by the CH_4 content of biogas, which is highly dependent on the characteristics of feedstocks and the anaerobic microbial community, and can be as low as 50% for carbohydrate substrates. Therefore, the worldwide expansion of biogas production will require technological and biological improvements. The former include new technologies for the pretreatment of recalcitrant substrates, mixing, retention of slow-growing methanogens, process monitoring, and process control. The latter rely on a better understanding of the microbial communities responsible for biogas production. The basic metabolisms in anaerobic digestion processes have been known for decades, but it is only recently (during the late 1990s) that the great diversity of microbes at play has been progressively unraveled. Such studies have shed light on the highly dynamic nature of communities involved in biogas production, as discussed in Section 13.4.1.2.

Another promising fuel is molecular hydrogen, a nearly ideal energy carrier because of its high energy content, high efficiency of conversion to usable power, and nonpolluting oxidation product (water). Moreover, it has wider industrial applications than CH_4 (such as the synthesis of ammonia and hydrogenation of fossil fuels). Biohydrogen has the potential to replace current energy-intensive hydrogen production technologies based on fossil fuels (Elsharnouby et al. 2013). Four biological pathways can lead to hydrogen production: (1) biophotolysis by green microalgae (where solar energy is directly converted to H_2 via photosynthetic reactions), (2) photofermentation by anoxygenic phototrophs (where organic substrates like fatty acids or alcohols are converted to H_2 and CO_2 thanks to the energy of sunlight), (3) dark fermentation by heterotrophs (where organic substrates are converted to simpler substrates and CO_2 in anaerobiosis), and (4) microbial electrolysis cells (a very recent technology discussed in Section 13.5). The use of biophotolysis is challenged by the low transfer efficiency of light, complexity in reactor design, and low hydrogen production rates. Harnessing photofermentation to produce H_2 from organic acids is hindered by its need for light, by its low solar conversion efficiency, by the sensitivity of the nitrogenase enzyme (responsible for H_2 production) to ammonia and oxygen, and by its high ATP demand. As far as we know, photofermentation has been mostly tested in lab reactors inoculated with pure strains; therefore, its efficiency when upscaled to industrial bioreactors remains uncertain. In contrast, hydrogen production using dark fermentation is flexible in regard to substrate range, cheaper, and easier to implement, but several parameters need to be considered to ensure a high hydrogen yield and production rate: (1) the inoculum composition (as end products will depend on the bacterial metabolisms involved in the process); (2) the substrate complexity and biodegradability; and (3) operational bioreactor parameters, such as pH (pH < 4.5 negatively affects hydrogen production), nutrient concentration, partial H_2 pressure, and temperature (typically, [hyper-]thermophilic cultures exhibit superior performance to mesophilic ones because the hydrolysis rate is favored by high temperatures). Despite its potential, the development of dark fermentation for hydrogen production at the industrial scale is presently limited by its low hydrogen yield. One of the reasons is that this process relies on undefined inocula (e.g., from anaerobic digesters, compost piles, or soils) with interconnected degradation pathways, some of them producing little or no hydrogen. Diverting the process exclusively toward high-yield hydrogen-producing pathways remains challenging (Ghimire et al. 2015). A solution could be the use of monocultures or cocultures of pure strains, which are less robust than complex communities, but offer the possibility to detect more easily metabolic shifts to find conditions that are optimal for biohydrogen production (Elsharnouby et al. 2013) (see Section 13.4.2). Another hurdle in dark fermentation is the production of residues besides H_2, primarily volatile fatty acids, that need to be utilized to achieve complete conversion of the organic biomass. For this purpose, dark fermentation can be coupled with anaerobic digestion or photofermentation to convert the by-products to CH_4 or H_2, respectively. In particular, photofermentation harbors high substrate-to-H_2 conversion efficiency; therefore, the combined hydrogen yield is usually greatly improved in comparison with the sole dark fermentation (Ghimire et al. 2015).

In conclusion, despite a tremendous effort and public money investment over the last decades to improve biological processes of liquid and gaseous biofuels, most of them are still not commercially viable. There is fierce debate regarding their ability to compete economically with fossil fuels and, if so, when (e.g., Biello 2011; Ramos et al. 2016); their environmental impact (Zah et al. 2007); and how to measure the latter properly (McKone et al. 2011). In an era of exploding biofuel-oriented research and development (R&D) teams, it is important to remember the basic criteria that should drive research in this area: cost in the short term and energy efficieny in the long term (Zhang 2011). Nevertheless, there is a general consensus that we need to develop more efficient or new conversion technologies for biofuel production. As mentioned briefly in this chapter, a vast literature deals with engineering pure strains for fuel production (see Chapter 10 for a more detailed discussion). In the present chapter, we focus on four specific approaches that could contribute to further improving microbial biofuel production and that reflect the authors' scientific interests: (1) the rational design of robust enzymes and tolerant strains, (2) the use of natural or engineered microbial consortia for fuel production, (3) the production of biofuels by microbial electrosynthesis (MES), and (4) the development of more efficient dynamic macroscopic models of conversion bioprocesses.

13.2 Targeted Approach 1: The Rational Design of Optimized Enzymes for Fuel Production

The term *biofuel* encompasses a wide variety of products: bioethanol, biodiesel, biogas, biohydrogen, biomethanol, biomethyl-ether, biobutanol, and so on. Among them, bioethanol—and biodiesel to a lesser extent—represents the largest industrial production. Their bioproduction relies on microbial or enzymatic reactors, and uses several types of sources.

The enzymes involved in biofuels are various and numerous, depending on the type of biofuel required and the starting material. For instance, biodiesel can be obtained by transesterification of vegetal, animal, or waste oils (triglycerides and fatty acids), using lipases as catalyst. In the case of first-generation bioethanol production from arable crops, enzymes such as α-amylases or glucoamylases are used (Harris et al. 2014), whereas the second-generation production exploits enzymes such as endoglucanases, β-glucosidases, or xylanases (Harris et al. 2014; Uday et al. 2016) to degrade lignocellulosic sources into fermentable sugars. Note that laccases or lytic polysaccharide monoxygenases can help the work of the latter enzymes (Harris et al. 2014; Walton and Davies 2016). This is not an exhaustive list of enzymes used in this field, but it illustrates the large number of pathways to obtain biofuels with biomolecules as catalysts.

In general, the physicochemical conditions of these industrial processes are remote from the physiological conditions: higher temperatures, acidic or basic conditions, nonaqueous solvent, and so forth. The enzymes must therefore be carefully selected or modified to be efficient in these conditions. In particular, the optimization of the thermal or thermodynamic stability, the solvent stability, and the substrate specificity is often required. Concerning the latter, the adaptation of the enzyme to a substrate slightly different from the original one, or to a larger range of substrates, can be aimed. The optimization of these enzymes can be achieved by directed evolution (Jemli et al. 2016; Kaushik et al. 2016), by rational design, or by combination of rational and random mutation approaches. In directed evolution (Dalby 2011), the natural processes that operate during evolution (i.e., random mutations followed by selection) are mimicked in the laboratory. Several molecular biology techniques are used for that purpose: mutations are randomly introduced in the gene coding for the enzyme using, for instance, error-prone polymerase chain reactions (PCRs); the most optimal enzyme variants according to the desired property are screened and selected; and this is iteratively repeated until the desired degree of optimization is reached. This method relies on random mutations and is quite powerful, but it is also quite expensive and requires high-throughput screening and selection. In the case of rational design (Kaushik et al. 2016), mutations likely to endow the enzyme with interesting properties are selected on the basis of a sequence–structure–function relationship analysis using several bioinformatics tools. These computational tools range from structural analysis and sequence alignments to more dedicated software that predict the effect of mutations on the thermodynamic stability (Guerois et al. 2002; Zhou and Zhou 2002; Capriotti et al. 2005; Cheng et al. 2006; Yin et al. 2007; Masso and Vaisman 2008; Dehouck et al. 2009, 2011; Worth et al. 2011; Pires et al. 2014; Laimer et al. 2015), on the thermal stability (Masso and Vaisman 2008, 2014; Pucci and Rooman 2014; Pucci et al. 2016), or on the solubility (Tian et al. 2010; Agostini et al. 2012) (Figure 13.1).

As a first example of rational design of enzymes with improved thermostability for bioethanol production, let us consider the works of Bayram Akcapinar et al. (2015) and Larsen et al. (2015). Lignocellulosic biomass is a possible source of fermentable sugars, but it is difficult to degrade. Working at higher temperatures is generally needed to decrease the viscosity of the medium, to increase the solubility of the substrate, and to reduce the risk of microbial contamination. Enzymes with improved thermostability are therefore desirable. In a study aimed at thermostabilizing endoglucanase I from *T. reesei*, which degrades cellulose (Bayram Akcapinar et al. 2015), a first step was to select candidate mutations using three bioinformatics tools that compute the effect of single-point mutations on the thermodynamic stability of the protein: I-mutant (Capriotti et al. 2005), PoPMuSiC (Dehouck et al. 2009, 2011) (Figure 13.1), and SDM (Worth et al. 2011). A subset of possible mutations was then subjected to molecular dynamic simulations at 310 K and 550 K to confirm, using this different approach, the effect of the variants on the thermostability of the protein. Finally, three mutants were characterized experimentally and showed a better thermostability than the wild type, but unfortunately a lower specific activity (Figure 13.1). Another study (Larsen et al. 2015) focused on endo-1,4-β-galactanase from *Talaromyces stipitatus*. Their approach combined sequence alignments with thermostable enzymes, the use of PoPMuSiC (Dehouck et al. 2009, 2011), and an analysis of crystallographic

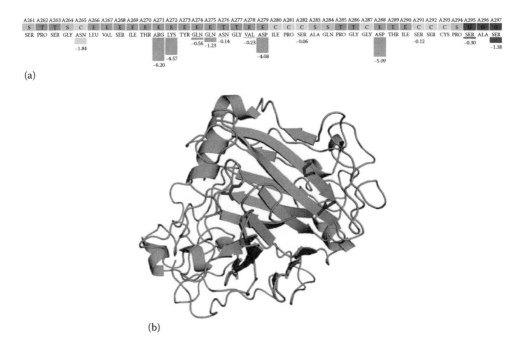

(a)

(b)

Figure 13.1 (a) Histogram obtained with the PoPMuSiC program (Dehouck et al. 2009, 2011). This histogram shows the sum of the $\Delta\Delta G$ of the stabilizing mutations at each sequence position. Colors represent the secondary structure (blue, β-strand; red, α-helix; other colors, turn and coil). The letter in each colored rectangle is the DSSP code of the secondary structure (Kabsch and Sander 1983). (b) 3D structure of the endoglucanase I from *Trichoderma reesei* (PDB code 1EG1). Residues at position 126, 272, and 274 have been mutated to thermostabilize the enzyme and are represented in red sticks.

B-factors to identify possible variants. The sequence alignments aimed to identify residues that were conserved in the thermostable enzymes and different in the mesostable ones. Then PoPMuSiC was used to compute the stability changes triggered by the introduction of these mutations in the endo-1,4-β-galactanase, and the B-factor analysis revealed the most flexible residues. The authors finally obtained mutations that increase the thermostability of the enzyme.

The stability of an enzyme in acidic conditions is another property that could be tuned. Beliën et al. (2009) worked on the endo-β-1,4-xylanase from *B. subtilis*, an enzyme that hydrolyzes xylans. This hydrolysis does not produce fermentable sugars but helps other enzymes access and degrade other parts of the lignocellulosic material, thereby generating sugars that can be turned into ethanol. They combined molecular modeling techniques with software that predicts the mutations that will change the pKa value of a given residue (Tynan-Connolly and Nielsen 2006). The different variants that have been designed showed experimentally an increase in their functional stability at lower pH.

Rational design is also helpful to improve the catalytic activity and the enantioselectivity, or to change the substrate specificity of an enzyme. For instance, secondary alcohols are poor substrates of the lipase from *Burkholderia cepacia*, but Ema et al. (2012) engineered a double mutation that overcomes that. For that purpose, they modeled the transition state of the enzyme–substrate reaction and performed docking simulations of the substrate on this transition state model. Other authors (Santarossa et al. 2005) modified the substrate specificity of a *Pseudomonas fragi* lipase. In this study, they modeled the structure of the lipase and combined visual inspection and structural superimpositions to design the variant. This mutated lipase showed an increased relative activity on C8 substrates and a higher thermostability.

Note finally that enzyme production is expensive, and having a process that permits the reuse of the enzymes can make it more economically viable. Working with immobilized enzymes on a solid support is an alternative to reactors with free enzymes that will facilitate the separation from the products and thus their recovery. We will not describe in this chapter the wide variety of immobilization supports and techniques, which are detailed in different reviews (Franssen et al. 2013; Sheldon and van Pelt 2013; Eş et al. 2015). We will just mention that the immobilization process can improve some of the physicochemical properties of the enzyme, such as its thermostability or its stability in nonaqueous solvents. Together with the bioinformatics technique described above, immobilization has been reported to yield very efficient biocatalysts for use in biofuel production (Zhang et al. 2012; Franssen et al. 2013; Singh et al. 2013; Poppe et al. 2015).

13.3 Targeted Approach 2: The Rational Design of Tolerant Strains for Fuel Production

Among the various features that are expected to be improvable by genetic engineering, the increase of biofuel tolerance in engineered microbes appears as an important challenge for the success of biofuel production at the industrial scale (Nicolaou et al. 2010; Dunlop 2011; Fu et al. 2016). Indeed, the accumulation of biofuel molecules in the cell envelope of producing microorganisms is in general toxic to them. These molecules are highly hydrophobic; hence, they modify the membrane properties and influence the physiological processes that are essential to the cell. This leads to inhibition of cell growth and even to cell death. Efflux pumps increase the tolerance, but burden the cells when overexpressed. Therefore, a trade-off must be found between the toxicity of the biofuel and the cost of expressing the corresponding efflux pumps.

Efflux pumps are membrane proteins that export toxic compounds such as bile salts, solvents, and antimicrobial drugs, and thus play an important role in cell survival (Paulsen et al. 1996; Putman et al. 2000; Nikaido and Takatsuka 2009) (Figure 13.2a). Some are quite specific to certain molecules, whereas others are able to transport a large range of them. The expression of efflux pumps has been shown to be a promising engineering strategy for the production of long-chain alcohols (Dunlop et al. 2011); in contrast, there is evidence that efflux pumps are not effective for exporting short-chain alcohols (Ankarloo et al. 2010; Minty et al. 2011). Clearly, the tolerance mechanisms through efflux pumps are specific to particular classes of biofuels

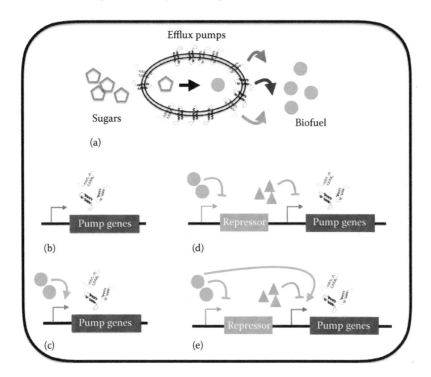

Figure 13.2   (a) Sugars enter a microorganism, are fermented into biofuel, and are then exported by different types of efflux pumps. (b–e) Different ways of mathematically modeling the efflux pump expression (Dunlop et al. 2010). (b) The pump expression is driven by a constitutive promoter. (c) The promoter is regulated by a transcription factor (an activator) that detects the amount of biofuel present in the cell. (d) Biofuel molecules repress the expression of a transcription factor that in turn represses the expression of the efflux pumps. (e) A feedforward loop controls pump expression, depending on the amount of both the biofuel and the repressor molecules.

(Nicolaou et al. 2010; Dunlop 2011). Another aspect that complicates the increase of tolerance is that the overexpression of efflux pumps—of membrane proteins in general—changes the membrane content and can therefore be toxic for the cell, or at least inhibit cell growth (Wagner et al. 2007). Moreover, all types of biofuels are not equally toxic (Sikkema et al. 1994; Isken and de Bont 1998; Ramos et al. 2002; Nicolaou et al. 2010); in particular, longer-chain alcohols are usually more toxic than short-chain alcohols. Finally, toxicity also depends on the microorganism. For example, *P. putida* can tolerate higher butanol concentrations (Rühl et al. 2009).

In general, microorganisms that exhibit high tolerance levels use multiple mechanisms in synergy: several types of efflux pumps (Segura et al. 2003), as well as other mechanisms, such as altered membrane properties and energy metabolism (Weber and de Bont 1996; Ramos et al. 2002; Nicolaou et al. 2010), and changes in cell size and shape to decrease the cell surface-to-volume ratio (Neumann et al. 2005). Finally, note that while tolerance improvement is a necessary condition to improve production yields, it is far from being a sufficient condition (Dunlop 2011).

A first approach to improve biofuel tolerance consists in using genomic DNA from microorganisms that are not (or less) affected by hydrocarbon-rich environments, such as *Marinobacter aquaeolei* (Singer et al. 2011), and transplant them in, for example, *E. coli* (Tomko and Dunlop 2015).

Mathematical models that take into account the detrimental effects of both biofuel toxicity and pump overexpression have been designed with the objective of improving the understanding of microbial biofuel production and robustness against intrinsic noise (i.e., random variability of the number of molecules) and parameter variations (Dunlop et al. 2010). These include feedback loops similar to those that are commonly used by organisms as versatile control mechanisms, for example, in response to changing conditions and to maintain homeostasis (Alon 2006). An example of a negative feedback loop is given by cells producing a biofuel that inhibits their growth; positive feedback would instead increase their growth. In synthetic biology, such loops, whether positive or negative, are used as key building blocks for synthetic rewiring of gene circuits (Nandagopal and Elowitz 2011).

For example, Dunlop's model (Dunlop et al. 2010) consists of a set of differential equations. The first one models the growth of the cellular population, with a degradation term that depends on the quantity and type of biofuel. The second equation quantifies biofuel production, which is proportional to the number of cells in the population. The introduction of efflux pumps in the cellular membrane adds an additional degradation term to the first equation, accounting for the toxicity of the pumps, as well as a degradation term in the second equation, which represents the export of biofuel to the extracellular medium. A third equation models the expression and degradation of the pump itself. This simple model includes a negative feedback mechanism due to the toxicity of the biofuel for the cell, which limits biofuel production. It describes the case where the pump's expression is controlled by a constitutive promoter and is thus constant (Figure 13.2b). The number of cells and the intracellular biofuel level reach a constant value in the stationary state, whereas the extracellular biofuel levels increase.

In the second model (Figure 13.2c), the constitutive promoter of the pump is replaced by a promoter that depends on the intracellular biofuel level: the higher the level, the higher the number of expressed pumps, with saturation happening above some threshold value of intracellular biofuel concentration. A slightly more complex model (Figure 13.2d) introduces an additional gene, the expression of which depends on the biofuel concentration, and acts as a repressor on the efflux pump expression. The fourth model (Figure 13.2e) is a mixture of the latter two models: a feedforward loop is introduced so that both the biofuel and the transcription repressor protein control pump expression.

The four controllers were simulated for different parameter values and were shown to be capable of producing similar levels of biofuel, although the first model (with constant pump expression) is less effective. Moreover, this simplest model was shown to be more sensitive to the parameters, and to produce biofuel only for a limited range of parameter values. Moreover, the last, most complex model, with a feedforward loop, appeared more robust to noise. These models were improved using a slightly more complex feedback mechanism from the biofuel concentration by introducing a biosensor, which delays the production of pumps until necessary (Harrison and Dunlop 2012). More complete models could include diffusion, several types of efflux pumps, or other tolerance mechanisms.

Among the other possible tolerance mechanisms that are controlled by genes and could be added to the model, changes in the cellular wall or membrane are interesting candidates (Sikkema et al. 1995). For example, the overexpression of targeted genes such as *mur* that are involved in cell wall biosynthesis increases ethanol tolerance by up to more than 100% in *E. coli* (Goodarzi et al. 2010; Nicolaou et al. 2012). Heat shock proteins are also key targets, as they are involved in the response to many types of environmental stress, among which is stress caused by high biofuel concentrations. Indeed, proteins of this family appear to be upregulated in response to biofuel stress (Rutherford et al. 2010), and their overexpression has been shown to increase biofuel tolerance in *E. coli* (Zingaro and Terry Papoutsakis 2013) and *Clostridium acetobutylicum* (Mann et al. 2012).

13.4 Targeted Approach 3: Analysis and Engineering of Multispecies Microbial Communities Involved in Fuel Production

For decades, synthetic biologists have selected and tweaked molecular pathways to make cells produce useful products like biofuels in sufficient amounts. Genetic and metabolic engineering tools have been extensively applied to a handful of workhorse model strains (such as *E. coli* or *S. cerevisiae*), but they are not yet easily transferrable to other potentially interesting microorganisms (Liao et al. 2016). In addition, engineered strains are often fragile, might suffer from self-intoxication by the molecules they produce (an issue discussed in Section 13.3), and are vulnerable to environmental changes. Last but not least, the use of genetically modified microorganisms (GMMs) is strictly regulated in most countries. In Europe, they are solely allowed in confined environments (like fermenters), and prior risk assessment for humans and the open environment is required (European Council Directive 2009/41/EC). The U.S. legislation (the Toxic Substances Control Act of the Environmental Protection Agency) is less strict, and GMMs are allowed for environmental processes (like cleanup of contaminated soils).

For all these reasons, exploiting the properties of multispecies communities might be an interesting alternative. In nature, microbes do not live as pure strains, but in complex, dynamic communities in which they interact with other members via the exchange of genes (Frost et al. 2005; Thomas and Nielsen 2005; Choi and Kim 2007; Boto 2010) and communication signals and metabolites (Seth and Taga 2014). Living in the vicinity of other microorganisms can have detrimental effects (increased competition for substrates, and antagonistic interactions through production of antimicrobials) (Hibbing et al. 2010), but a growing number of examples of species synthesizing molecules that are beneficial to other community members (e.g., amino acids, cofactors, and scavengers of reactive oxygen species) suggest that positive interactions are ubiquitous (Seth and Taga 2014). Very frequently, the benefit for the donor (if any) has not been clearly identified. Identification of the exact nature of interactions can be tricky, because there is a whole range of interactions between competition (where both species are negatively affected) and cooperation (where both species are positively affected) (Faust and Raes 2012), and because interactions can involve more than two species. In addition, detecting metabolite cross-feedings is difficult due to their intrinsically dynamic nature and the complexity of communities (Ponomarova and Patil 2015). To unravel the balance between competition and cooperation, simple models based on metabolic pathways have been proposed for pairs of species (Freilich et al. 2011). In complex communities, the best documented example of positive interactions within complex communities is the sequential transformation of organic matter by metabolic cross-feeding in anaerobic digestion. It involves a cascade of metabolic reactions (hydrolysis, acidogenic and acetogenic fermentations, and methane production) performed synergistically by different functional guilds of microbes. Their metabolic cooperation is thermodynamically constrained, and one of the steps, acetogenesis (performed by so-called synthrophs), would be impossible without methanogens consuming its products (H_2, formate [Stams 1994; Schink 1997], or—as recently discovered—electrons [Kouzuma et al. 2015]). Examples of cross-feeding have been reported in other microbial systems as well (Morris et al. 2013), such as the human gut (Louis et al. 2014; Rakoff-Nahoum et al. 2016) or consortia-degrading xenobiotics (Dejonghe et al. 2003; Lykidis et al. 2011; Men et al. 2011).

In this chapter, we discuss how a better understanding of the functioning of the multispecies communities involved in biofuel production could contribute to better process performance, and how genetic and metabolic engineering toolkits initially developed for pure strains are progressively being applied to low-complexity consortia to enhance process yield and stability.

13.4.1 Study of Community Structure and Dynamics, and Exploitation of Community Properties in the Context of Biofuel Production

13.4.1.1 Why Analyze Community Structure and Dynamics?

There is a vast literature on the beneficial properties of mixed communities compared with pure strains, which range from their greater functionality (i.e., their potential to be more performative or to realize more diverse functions) (Bell et al. 2005) to their greater biomass productivity and their superior robustness (i.e., their ability to cope with perturbations) (Stenuit and Agathos 2015).

Robustness and its variants, resistance (the degree to which a community remains unchanged after a perturbation) or resilience (the degree to which a community changes after a perturbation but eventually returns to its predisturbance state), are key properties of community dynamics because they guarantee the maintenance of the community's structural organization and/or functional performance in the face of (un-)predictable disturbances (Shade et al. 2012; Stenuit and Agathos 2015). These terms apply to the phylogenetic composition of a community and/or to its functional traits (encoded in a set of metabolic genes). Interestingly, both can be uncoupled, and there are many examples of fluctuations or shifts in taxonomic composition while metabolic traits are preserved (Fernández et al. 1999; Bull 2010; Kraft et al. 2014). Such shifts are not necessarily the result of disturbances, as they are frequently observed in undisturbed conditions. They reflect a certain degree of functional redundancy among community members for a broad function, for example, denitrification (Kraft et al. 2014), heterotrophic activity (Ofiţeru et al. 2010) in wastewater treatment plants, or fermentation in biogas plants (Werner et al. 2011). As a result, such communities are more prone to stochasticity with regard to their taxonomic composition.

On the other hand, in systems with cross-feedings between community members, the performance of the community is tightly linked to its structure, that is, the number of taxa (richness) and their relative abundance (evenness). In such cases, community dynamics is either quite predictable (as in the case of anaerobic digestion [Vanwonterghem et al. 2014]) or chaotic (as in the case of nitrification, which relies on a fragile interaction between two partners sensitive to environmental conditions [Graham et al. 2007]). In a remarkable large-scale comparative study of anaerobic reactors, Werner et al. (2011) demonstrated that bioreactor performance (methane production) was more related to phylogenetic community structure than to operating conditions. They showed that syntrophic populations behaved—as expected—as functional specialists, and that they were stable, resilient after changes in operating conditions (an important characteristic for functional maintenance), and very site (i.e., reactor) specific. In contrast, the communities of fermentative Clostridia and Bacteroidetes were highly dynamic, relying on functional redundancy to maintain the overall community function. Better reactor performances (methane production and organic solid removal) were observed when total communities were more even and more dynamic over time, respectively. The latter suggests that community dynamics helped maintain efficient performance.

Finally, initial community assembly is often a critical step in community dynamics. In closed systems like reactors, the same initial assemblage of microbes can lead to different community structures, a phenomenon called "ecological drift" (e.g., wastewater microbes colonizing electrodes in microbial fuel cells [MFCs] [Zhou et al. 2013]). Drift corresponds to stochastic changes in the relative abundance of different taxa over time and is favored in communities with low diversity and when selection is weak. In open systems, community assembly is also largely dependent on immigration or dispersal processes, that is, movements across space (Ofiţeru et al. 2010). Beyond niche differentiation (a deterministic process), drift (a neutral process), and dispersal, diversification (i.e., the evolution of community members' genome via mobile genetic elements or mutations) also influences community structure (Nemergut et al. 2013). Unfortunately, diversification remains largely understudied. As pointed out before, it is often more appropriate to study community assembly at the functional than taxonomical level, as the dynamics of both can be nonsynchronized (Burke et al. 2011). Because of these four factors in action (niche differentiation, drift, dispersal, and diversification), assessing "benchmark" values of community stability (i.e., in undisturbed conditions) is difficult—but necessary—before analyzing community robustness. Generally speaking, a reasonable range of variation around a reference value is tolerated. This reference value is easier to define for microbial functions than for phylotypes, as the identification of a "core" microbiome responsible for the function of interest can be difficult in communities with high phylotype turnover and functional redundancy.

When facing a stress, the response of a community is a function of (1) the community properties (species' plasticity, species' physiological state, community taxonomic richness and evenness, interspecies connectivity, and functional redundancy) and (2) the nature of disturbances applied, such as frequency, intensity, and duration ("pulse" vs. "press" disturbance) (Bender et al. 1984; Shade et al. 2012; Stenuit and Agathos 2015). For example, a greater richness (Cook et al. 2006) or evenness (Wittebolle et al. 2009) was shown to favor functional robustness after a perturbation. Beyond robustness, the response of a community can be evolution to one or multiple alternative stable states (with a different community structure and/or function) or to irremediable dysfunctionality and collapse. The existence of multiple stable states and possibly fluctuations in between has been reported in microbial communities (Shade et al. 2012). For example, in bioreactors, a new community configuration was shown to positively (Luo et al. 2015) or negatively (Bürgmann et al. 2011) impact the targeted function (biogas production and nitrogen removal, respectively). The magnitude of robustness, a key parameter for microbial communities in engineered ecosystems, can be assessed using various metrics (Stenuit and Agathos 2015).

The recent development of high-throughput molecular and computational tools to analyze the metagenome (collective set of genes), transcriptome (collective set of expressed genes), proteome (collective set of expressed proteins), and metabolome (collective set of secreted molecules) of microbial communities has undoubtedly boosted our understanding of complex microbial ecosystems. These techniques and their recent development are discussed in Section 13.4.1.2. We also provide examples of useful input from such techniques for biofuel production, from the discovery of new enzymes to the manipulation of environmental parameters for increased microbial fuel production based on advanced knowledge of the communities in action. Finally, we also discuss how community-integrated "omics" and high resolution time and space analyses of microbial ecosystems, which hold promise to unravel community functioning and dynamics, must be integrated into a more holistic approach that also includes mathematical predictive modeling. Such a unifying perspective, obviously useful to understand complex, natural ecosystems, could be translated into practical applications in renewable energy supply, but also in environmental safety and health, management of environmental biogeochemical cycles, and provision of new materials, a concept globally named "microbial resource management" (Verstraete et al. 2007).

13.4.1.2 Tools to Analyze Community Structure and Dynamics and How They Contribute to a Better Understanding of Biofuel-Producing Communities

One of the greatest challenges in environmental microbiology today is undoubtedly unraveling and modeling microbial interactions in complex and often dynamic communities. This task is hampered by our inability to cultivate 90%–99% of environmental microbes in laboratory conditions (Rappe and Giovannoni 2003) and to recreate complex environmental conditions in the lab. Although novel cultivation techniques led to remarkable successes in the isolation of recalcitrant microbes (Pham and Kim 2012; Stewart 2012), it is above all the spectacular development over the last 15 years of high-throughput sequencing, "meta-omics" techniques (Figure 13.3), and

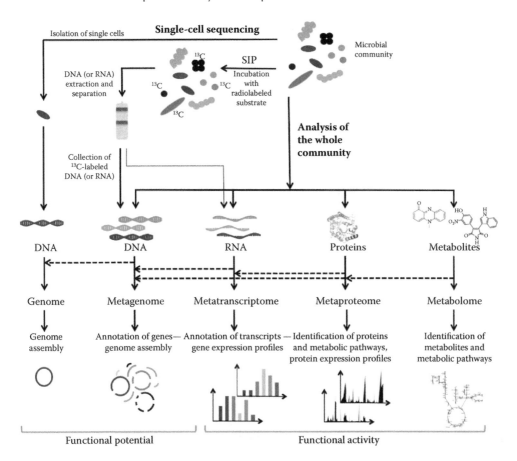

Figure 13.3 Different techniques used to characterize the phylogenetic and functional diversity of a microbial community. Metagenomics provides information about the taxa present in the samples and their functional potential (alternatively, only 16S rRNA marker genes are sequenced when the goal is to obtain deep phylogenetic information—not shown). The functional activity of the community is measured by metatranscriptomics, metaproteomics, and/or metabolomics. Metatranscriptomics measures gene expression, which can be compared across samples, and helps identify key metabolic pathways. Metabolic pathways may also be inferred from metaproteomics, which reflects functional protein expression, and metabolomics, which focuses on metabolites. In single-cell sequencing, individual cells are selected by optical tweezers or fluorescence-activated cell sorting (FACS) before DNA sequencing, allowing genome analysis of single community members. Finally, stable isotope probing (SIP) is based on the incorporation of a ^{13}C-radiolabeled substrate into the active fraction of the community, followed by recovery of labeled DNA (or less often RNA) by isopycnic centrifugation and sequencing. When integrated, omic techniques can be mapped against each other (dotted arrows, each arrow pointing to the reference dataset), thereby providing more powerful analyses. (Adapted from Vanwonterghem, I. et al., *Curr. Opin. Biotechnol.*, 27, 55–64, 2014.)

integrated bioinformatic pipelines that has nurtured recent breakthroughs in quantifying microbial community composition, function, and spatiotemporal dynamics (Zengler and Palsson 2012; Segata et al. 2013).

The central focus of all studies on microbial relationships is to understand the link between community diversity, community function, and community robustness. For this purpose, emerging technologies like metagenomics, metatranscriptomics, and metaproteomics (Figure 13.3) are precious allies to analyze the composition of complex natural communities in pre- and postdisturbance situations (the latter being mostly changes in physicochemical parameters: temperature, oxygen concentration, and nutrient concentration). Over two decades (since the early 1990s), these technologies were preceded by the application of low-throughput, culture-independent approaches to study the diversity of phylogenetic marker genes (e.g., 16S rRNA genes for bacteria and archaea and 18S rRNA genes for eukaryotes) (Schmidt et al. 1991; Hugenholtz et al. 1998) or, to a lesser extent, specific functional genes. They included clone libaries, quantitative PCR (qPCR), and fingerprinting techniques such as DGGE, FAME, ARISA, or (T-)RFLP (reviewed in Nocker et al. 2007). Their wide application to microbial ecosystems has shed light on the tremendous and ubiquitous diversity of microbes (Hugenholtz et al. 1998; Huisman and Weissing 1999; Cases and de Lorenzo 2002) and provided a partial glimpse of microbial community dynamics. In the mid-2000s, the popularity and effectiveness of 16S-based surveys grew dramatically with the increased throughput of next-generation sequencing (NGS) methods, like 454 and then Illumina sequencing, and the concomitant development of bioinformatic pipelines to analyze the 16S data (Schloss 2009; Caporaso et al. 2010, 2012; Logares et al. 2012; Bálint et al. 2016). Today, "routine" analyses consist of sequencing simultaneously tens of thousands of 16S rRNA marker genes per sample using NGS. A remarkable discovery resulting from deep sequencing of 16S rRNA genes was the existence, in most environments, of a "rare" biosphere made of numerous taxa in low relative abundance (Lynch and Neufeld 2015). In the specific case of anaerobic digesters, the application of NGS has allowed a better characterization of microbial communities in action than with older methods (reviewed in Sárvári Horváth et al. 2016). It has also provided a better understanding of how reactor performance (substrate removal and methane production) is correlated with community structure (Werner et al. 2011). Anaerobic digesters were shown to be typically dominated by previously uncharacterized microorganisms (Sundberg et al. 2013), which may influence the degradation process. In addition, phylotypes could be classified into core taxa common to most of the digesters, shared taxa among a few digesters, and specific taxa to each digester (Rivière et al. 2009; Stolze et al. 2016).

Likewise, high-throughput sequencing of targeted functions has revealed a much greater diversity of genes potentially important in biogeochemical cycles and depollution than previously realized (Iwai et al. 2010; Bowen et al. 2013). However, a better understanding of the metabolic capabilities of these uncultured microorganisms, the level of functional redundancy within communities, and the fundamental mechanisms of interspecies interactions was still needed. In the past 10 years, continuous advances in sequencing have made it possible to (almost) achieve this goal using omic technologies to analyze entire collections of genomes and transcriptomes from a sample, and thereby infer community composition and function. Coupled to technological advances in protein and metabolite analysis, imaging, and labeling techniques, such as stable isotope probing (Chen and Murrell 2010), these provide altogether a solid technological basis for a better understanding of complex microbiomes. Recent examples include the soil microbiome, the wastewater microbiome (Muller et al. 2014), the microbiome of anaerobic digesters (Vanwonterghem et al. 2014), and the highly publicized human gut microbiome (Franzosa et al. 2015), which might be manipulated to treat gastrointestinal disorders (reviewed in Grogan et al. 2015).

13.4.1.2.1 Metagenomics.   Metagenomics is the analysis of the metagenome, that is, the collection of genomes, of a given microbial community at a given point in time. DNA is extracted from the environment, and either directly sequenced using NGS or cloned into a vector transferred to a host to build large metagenomic libraries. Clone libraries are then screened for target genes (genetic screening) or for metabolic activities (functional screening) resulting from heterologous expression of cloned DNA by the host. Library-based metagenomics is tedious (especially the cloning step of large DNA fragments) and time-consuming, and suffers from barriers to heterologous gene expression (that can be partly overcome by the use of phylogenetically diverse hosts) and slow advances in screening technology. Despite such limitations, it has allowed the discovery of novel gene products that lacked sequence similarity to genes of known function (e.g., Craig et al. 2009; George et al. 2010; Jacquiod et al. 2014). However, this approach now lags behind direct sequencing of environmental DNA, also referred to as "shotgun metagenomics," which has been boosted by the advent of high-throughput sequencing (NGS) and is now widely adopted in environmental microbiology laboratories worldwide. Metagenomics goes beyond 16S rRNA gene surveys, as it provides concrete linkages between microbial community composition ("who is there") and potential function ("who is able to do what"). In addition, because it is not based on PCR amplification of specific genes, common PCR biases, like those observed in 16S rRNA gene-based surveys, are avoided and no prior knowledge of the community is required. This latter point explains why metagenomics has enabled unprecedented discovery of new taxa and genes. The ultimate goal of metagenomics is to reconstruct large DNA fragments and complete genomes (Sharon and Banfield 2013). Historically, this was not possible because of insufficient sequencing depth (except for low-diversity samples) and the challenge of assembling sequence data originating from complex genomic mixtures; hence, gene-centric approaches based on unassembled data were favored. Thanks to continuing advances in sequencing depth at decreasing cost, read binning (classification), and read assembly (reviewed in Thomas et al. 2012), genome reconstruction is now within reach, even in natural and engineered environments harboring high taxonomical diversity (Albertsen et al. 2013). Another technique, meta3C (see Section 13.4.1.2.2), is emerging as a powerful alternative to sequence microbial genomes from complex environments. Last but not least, the issue of assembling (near-)complete genomes from complex communities can now be tackled using a third technique called single-cell genomics. This technique has been reviewed elsewhere (Blainey 2013; Lasken and McLean 2014).

Briefly, the workflow consists of (1) cell isolation by flow cytometry and fluorescence-activated cell sorting or by use of optical tweezers, (2) whole-genome amplification of single cells by multiple displacement amplification (random DNA amplification using phi29 polymerase and random hexamers) (Binga et al. 2008), and (3) NGS of amplicons and the read assembly (Rinke et al. 2014). Single-cell genomics is still facing technical challenges, such as physical separation of cells from the environmental matrix, uneven amplification of genomic fragments, and subsequent assembly of sequencing data with variable read depth. This being said, it has already proved its potential to access new genomes of uncultivated strains across the tree of life (Rinke et al. 2013) and to reveal intraspecies genetic variations (Lasken and McLean 2014) or unknown interactions between organisms (Yoon et al. 2011). In addition, single-cell genomics and metagenomics can be used on the same samples, thereby allowing cross-validation of the results.

By analyzing all the genomes in a community, it is possible to build up a catalogue of reference genomes for other meta-omic approaches (e.g., metatranscriptomics and metaproteomics), and find out whether the different microbial guilds harbor specific metabolism or if there are vital interactions between them or some functional redundancy.

For example, metagenomics has been extensively applied to characterized natural communities degrading cellulosic plant material (Hess et al. 2011; Lelie et al. 2012). The main drivers of such studies have been (1) the fundamental understanding of species- and community-level mechanisms responsible for efficient cellulose degradation, in order to help advance biofuel production, and (2) the discovery of as yet uncharacterized lignocellulosic biomass-degrading enzymes. Indeed, the paucity of known enzymes that efficiently deconstruct plant polysaccharides represents a major bottleneck for industrial-scale conversion of lignocellulosic biomass into biofuels. Analyzing the enzyme-encoding genes in the metagenome of an anaerobic poplar wood-degrading community allowed us to identify numerous (hemi-)cellulases, debranching enzymes, and enzymes homologous to fungal lignolytic enzymes that could be assigned to Bacteroidetes and Clostridiales (Lelie et al. 2012). Unexpectedly, enzymes assigned to *Magnetospirillum* were also identified, which were hypothesized to play a role in the degradation of lignin-derived aromatic compounds, thereby preventing these toxic compounds from accumulating in the environment. More recently, an analysis of the metagenome of the sugarcane bagasse microbiome brought to light the existence of a large repertoire of lignocellulolytic enzymes and newly defined auxiliary activity proteins (Mhuantong et al. 2015). In addition, comparison with other lignocellulolytic metagenomes revealed the existence of a core of conserved genes despite very different species composition. Besides looking at communities that are found on the surface of lignocellulosic material, several metagenomic studies were focused on host-associated microbes living in the gut of cellulose-degrading organisms. For example, the metagenomic analysis of the cow rumen microbiome provided a substantially expanded catalogue of genes coding for carbohydrate-active enzymes, and allowed us to generate draft genomes of 15 uncultured strains involved in cellulose deconstruction (Hess et al. 2011). Similar studies were performed on the panda microbiome (Zhu et al. 2011), leaf-cutter ant microbiome (Suen et al. 2010), and termite microbiome (Scharf 2015).

13.4.1.2.2 Meta3C, a New Technique to Sequence Microbial Genomes in Complex Communities.   Obtaining genomic information on the microorganisms involved in the production of microbial fuels is nowadays the fastest way to characterize their metabolic activities, thereby paving the way for improvements in yield, titer, and productivity (Orphan 2009; Warner et al. 2009; Gowen and Fong 2011). However, most naturally occurring members of microbial communities are very difficult, if not impossible, to cultivate in isolation. Although sequencing the genomes of bacterial species from pure cultures is a relatively simple task that can be performed efficiently and at a relatively low cost using the latest sequencing techniques (Koren and Phillippy 2015), sequencing the genomes of mixed bacterial species is much more complicated: one ends up with hundreds of thousands of genome fragments that are mixed with one another in the assembly software output, with little to no indication of which microorganism they come from.

For this reason, people have long contented themselves with "gene-centric" metagenomic analyses in which the main aim is to produce a catalogue of the genes present in a given environment (Tringe et al. 2005). Complete or near-complete genome assemblies of noncultivable members of microbial communities have been sometimes obtained using iterative assembly procedures (Pelletier et al. 2008); however, this only works for members that are largely overrepresented in the community. Recently, the availability of new methods to "bin" metagenomes into sets of groups of contiguous sequences ("contigs" [Staden 1980]) hypothetically coming from different species has triggered a shift from gene-centric to "genome-centric" approaches, where the aim is now to assemble and characterize the genome of each member of the community in order to understand its metabolic activities (Waldor et al. 2015). The most popular binning approaches are based on the guanine + cytosine (GC) content and/or on coverage (i.e., how often a given stretch of DNA is represented among the sequence reads), assuming that the abundance of each species in the mix should endow it with a characteristic coverage signature (Albertsen et al. 2013; Alneberg et al. 2014). However, microbial genomes have varying and overlapping GC contents, and the coverage signal can be obscured by the presence of repeated genome regions or by natural and/or artefactual variations in coverage along genomes, notably as a function of the distance to its origins of replication (Semova et al. 2012; Hawkins et al. 2013).

To overcome these problems, a promising new approach has been to turn to "contact genomics" (Flot et al. 2015) (Figure 13.4). This term encompasses several approaches that rely on the use of a cross-linking agent (generally formaldehyde) to "record" the physical contacts between chromosomes in cells. The first studies using such a technique, under the name "chromosome conformation capture" (3C), aimed to investigate the tridimensional structure of yeast chromosomes (Dekker et al. 2002): first, a genomic library was generated from a cross-linked chromatin fragment, and then the frequency of cross-linking events between chosen genome regions was investigated using semiquantitative PCRs. Nowadays, an overall picture of the

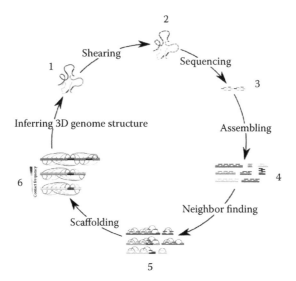

Figure 13.4   In this much simplified circular representation of the (meta-)3C procedure, one starts from (1) a genome or metagenome where the 3D contacts between chromatin regions have been "frozen" by addition of a cross-linking agent (typically, formaldehyde). This DNA is then sheared to obtain fragments (2) that are sequenced, resulting in a collection of reads (3). The reads are subsequently assembled into contigs, yielding an assembly that is still fragmented (4) as identical or nearly identical regions found in several places in the genome or metagenome are assembled together (in black). To improve this preliminary assembly, one can first use the local information contained in 3C data to find the missing neighbors of each contig in the assembly (5), thereby replacing each repeated region in its proper location and binning the dataset into putative species (if dealing with a metagenome). As a second step, one can use the long-distance information contained in the 3C data to scaffold the extended contigs into supercontigs covering entire chromosomes (6). Finally, one may even use the 3D information contained in the 3C data to reconstruct the 3D structure of the chromosomes, thereby closing the loop.

physical interactions in a genome can be gleaned in a single step by applying NGS to the genomic library mentioned above. These genome-wide approaches are called "Hi-C" (Lieberman-Aiden et al. 2009) or "3C-seq" (chromosome conformation capture sequencing [Marbouty et al. 2014]), depending on whether an additional enrichment step is performed to ensure that most read pairs contain information on the three-dimensional (3D) interactions of the chromosomes.

Contact genomic libraries contain information regarding the physical linkage of DNA regions on chromosomes across very long distances (typically at least 150–250 kb); as a result, they are very useful to scaffold contigs and improve assemblies, with the potential to yield chromosome-scale scaffolds if the data are in sufficient amount and quality (Flot et al. 2015). For example, Marie-Nelly et al. (2014) used 3C-seq to finish the genome of *T. reesei*, a fungus used extensively in the biomass-to-biofuel industry (Gusakov 2011); starting from a published draft assembly fragmented into 77 scaffolds (Martinez et al. 2008), they obtained seven large scaffolds containing 99.8% of the initial assembly, in perfect agreement with karyotypic data, suggesting that this species possesses seven chromosomes. Several programs have been devised to scaffold contigs using contact genomic information: dnaTri (Kaplan and Dekker 2013), Lachesis (Burton et al. 2013), GRAAL (Marie-Nelly et al. 2014), and HiRISE (Putnam et al. 2016). Among these, the latter two are the most appealing, as they are based on explicit probabilistic models. At present, there is no program available to assemble directly the reads obtained by sequencing contact genomic libraries; instead, a preliminary assembly must first be obtained, and then 3C data are used to improve it. This two-step process is a major limitation, as generating a decent preliminary assembly can consequently be a daunting task when dealing with highly repetitive and/or heterozygous genomes. These problems could potentially be solved by taking the contact genomic information into account already during contig formation (Flot et al. 2015).

Contact genomics also empowers metagenomic analyses: as cell membranes prevent the chromosomes of different organisms from interacting spatially (even if they live in close symbiotic relationships), quantifying 3D interactions makes it possible to divide metagenomic datasets into bins reflecting faithfully the different components of the microbial community under investigation (Beitel et al. 2014; Marbouty et al. 2014; Marbouty and Koszul 2015). When dealing with microbial communities (either natural or evolved in the laboratory), a possible workflow is therefore to first generate a metagenomic assembly using a suitable program such as IDBA-UD (Peng et al. 2012), and then bin the resulting assembly by performing a network analysis of 3C data (Marbouty et al. 2014), and finally reassemble each genome separately using GRAAL (Marie-Nelly et al. 2014). In the case of diploid microorganisms, this may be followed by a final step during which 3C data are used to phase the scaffolds, that is, to determine which genetic variants are parts of the same haplotype (Selvaraj et al. 2013; Flot et al. 2015). Although such meta3C approaches have not been applied yet to microbial communities relevant to biofuel production, this represents a promising research direction.

13.4.1.2.3 Metatranscriptomics.   Metatranscriptomics is the analysis of all mRNAs (expressed genes) in a microbial community at a given point in time. It involves mRNA extraction, followed by reverse transcription and high-throughput sequencing of cDNAs. This technique focuses on active members in a community, and therefore the level of complexity of analyses is reduced compared with metagenomics. To infer differences in levels of gene expression, metatranscriptomic reads are mapped against reference genomes or a metagenome of the same environment. This method has advantageously replaced microarrays to measure *in situ* gene expression, as it is faster and more sensitive, and allows the identification of expressed genes without prior knowledge of their sequence. Technical challenges include the recovery of sufficient amounts of high-quality mRNA (a task made difficult by the overwhelming contribution of rRNA to the total RNA pool), the inherent liability of mRNA, and the biases linked to cDNA synthesis and amplification. The first application of metatranscriptomics to a production-scale biogas plant dates back from 2012 (Zakrzewski et al. 2012). In that study, the comparison of metatranscriptome data and 16S rRNA gene data indicated a high transcriptional activity of archaeal species. More recently, several studies of the metagenome and metatranscriptome of biogas reactors have shown that they contain a tremendous number of as yet uncharacterized taxa, and that the highly abundant taxa are not necessarily the most active ones (Campanaro et al. 2016; Maus et al. 2016; Stolze et al. 2016). In some digesters, high functional redundancy was detected—especially in the early steps of anaerobic digestion (Cai et al. 2016)—raising the question of whether this phylogenetic diversity is needed to ensure efficient anaerobic degradation of organic matter. In all, there are still few metatranscriptomic studies of communities involved in biofuel production, but this technique has great potential to identify the most active pathways in different operating conditions (and therefore drive the community toward pathways of interest), and to measure the immediate regulatory response of the entire community to a perturbation. On the other hand, metatranscriptomics is the method of choice for the genetic analysis of eukaryotic communities, as illustrated by recent metatranscriptomic studies on organic matter degradation by fungi in forest soils (Damon et al. 2012; Kuske et al. 2015).

13.4.1.2.4 Metaproteomics. Metaproteomics is the large-scale characterization of the entire protein complement of environmental microbiomes at a given point in time (Wilmes et al. 2015). Compared with metatranscriptomics, it takes into account posttranslational events. Proteins are extracted from the environmental sample, fractionated, separated using liquid chromatography (LC), and detected with tandem mass spectrometry (MS/MS) (VerBerkmoes et al. 2009). This protocol has not changed much for the last decade, after spectacular progress in analytical capacities. In about a year, proteomics moved from separation of proteins by two-dimensional (2D) polyacrylamide gel electrophoresis, in-gel enzymatic digestion, identification of a few proteins by LC-MS/MS, and *de novo* peptide sequencing to the identification of thousands of proteins by the above-mentioned pipeline (Wilmes et al. 2015). As other omic techniques, metaproteomics is facing technical challenges, including (1) cell lysis and quantitative protein extraction from complex matrices, (2) standardized protein fractionation methods exploiting both molecular weight and charge, (3) the need for mass spectrometers with faster scan speeds and higher mass accuracies, and (4) the need for good reference databases (sample-derived metagenomic and metatranscriptomic datasets rather than public protein databases) (Wilmes et al. 2015). Solutions to these challenges already exist, but they are not routinely applied to metaproteomic studies yet (Wilmes et al. 2015). One exception might be coupling metaproteomics to metagenomics to analyze the same sample, an approach called "community proteogenomics" (VerBerkmoes et al. 2009). This procedure is now widely applied, since its usefulness to improve protein identification was proved over and over. It is based on the construction of a protein sequence database from the sequenced metagenome. *In silico* trypsin digest is performed on the predicted proteins, resulting in a peptide database. MS/MS spectra are then matched to the peptides in the database, and after filtering, a list of identified peptides is obtained. Peptides that are present in only one protein in the whole database can be unambiguously tracked back to their corresponding protein, and thus permit reliable protein identification, and hopefully the microbial species it is derived from. In addition, in case the protein mixture was fractionated (into extracellular, soluble, and membrane fractions) after the initial extraction, information about protein localization is available. There are other difficulties linked to the application of metaproteomics to complex communities, like detection of strain variability (amino acid substitutions) and follow-up of protein turnover in highly dynamic microbial assemblages (VerBerkmoes et al. 2009).

Metaproteomics has been successfully applied to many natural environments with a moderate level of complexity (the last problematic environment might be soils); in engineered ecosystems, activated sludge has been extensively studied with this technique from the perspective of polymeric substance production and biological phosphorus removal (Wilmes et al. 2015). The metaproteome expressed by thermophilic communities producing methane from cellulose (paper) has been investigated too (Lü et al. 2014); syntrophic acetate oxidation and hydrogenotrophic methanogenesis were revealed to be the dominant pathways for methane production. One unexpected result was, however, the high proteolytic activity in that community. The potential of metaproteomics to understand microbial ecology of biogas plants has been recently reviewed in Heyer et al. (2015).

13.4.1.2.5 Metabolomics.   Metabolomics is the qualitative and quantitative analysis of low-molecular-weight molecules produced by a microbial community at a given point in time. It relies on mass spectrometry methods. Noninduced metabolite production was historically studied by adding the cell-free culture filtrate of a donor strain to the recipient culture, assaying filtrate activity, and identifying the molecules in the spent medium. Recently, imaging mass spectrometry (Watrous et al. 2012) has been applied to explore the metabolite production of strains stimulated by the presence of other organisms in petri dishes. A recent analysis of the secreted metabolome of *Streptomyces coelicolor* growing in the vicinity of five other actinomycetes revealed that for each interaction, a specific chemical response was produced (Traxler et al. 2013). Other recent instrumentation includes microfluidics coupled to MS. The range of small molecules produced by microbes is vast, from amino acids to nucleosides, polyketides, alkaloids, signaling molecules, enzymes, or cofactors (Phelan et al. 2012). They are suspected to be produced by small subsets of the community, either passively by cell lysis or leaking or actively by efflux, and benefit a large number of neighbors. How these neighbors contribute to community stability or diversity is usually unknown. Microbial interactions not only influence the survival of community members, but also control their behavior and differentiation. Identifying metabolic exchanges is feasible for simple cocultures (Nakanishi et al. 2011), but it is an overwhelming task in natural communities with hundreds of species living in fluctuating conditions (Ponomarova and Patil 2015): first, because metabolites cannot be directly attributed to a particular species; second, because a large proportion of microbes have never been characterized in regard to their metabolic needs and biosynthetic capabilities; and third, because only a few percent of spectra can currently be annotated. Fortunately, new computational tools have been recently developed to annotate spectra from unknown molecules based on automated comparison with chemical structures found in public chemistry databases (Silva et al. 2015).

13.4.1.2.6 Perspectives in the Analysis of Complex Communities. When combined, metagenomics, metatranscriptomics, metaproteomics, and/or metabolomics provide a formidable suite of tools (meta-omics) to study the direct link between microbial community composition or genetic potential and final phenotype. Many authors predict that such integrated platforms will become the future standard for large-scale characterization of microbial consortia, including those performing important biological processes (Raes and Bork 2008; Narayanasamy et al. 2015). For example, meta-omics is now revealing the complexity of prokaryotic degradative activity in lignocellulose-rich environments (Cragg et al. 2015; Scharf 2015). Its application to reactor communities has just begun, from anaerobic digestors (Maus et al. 2016) to wastewater activated sludge (Roume et al. 2015). The goal of these two studies was to harness such communities in the future for optimal CH_4 production and lipid biofuel production (Muller et al. 2014), respectively. Transitioning from research that produces lists of genes, proteins, and genomes to research that unlocks their functional significance is not an easy task, though, and more emphasis on function is badly needed (Cragg et al. 2015; Scharf 2015).

Another limitation is that at present, such techniques are mostly used to provide a "snapshot" of a community, which does not say much about its adaptive response to disturbances, potential for self-stabilization, or phylogenetic and functional dynamics (all key parameters in biotechnological applications). Therefore, in a foreseeable future, they will hopefully be integrated in space- and time-series analyses in order to unravel structure–function relationships within microbial communities in stable and/or perturbed conditions. For example, meta-omic analysis of microbial communities over time in a larger number of biogas plants operating in different conditions would make a significant contribution to a better understanding of the relationship between microbial community structure, operating conditions (operational parameters and nature of substrates), process efficiency (e.g., CH_4 yield), and process stability. In addition, metatranscriptomics and metaproteomics could be used to indirectly determine which environmental stressors or growth-limiting factors affect each species' growth (Konopka and Wilkins 2012). Indeed, determining such factors in most environments is analytically difficult, as nutrient concentrations are extremely low and biomass increase is usually modest. Practically speaking, it would consist of sampling across environmental gradients and querying the physiological state of the different microbes, that is, searching for transcripts or proteins correlated with gradient intensity.

Despite the impressive accumulation of experimental data over the last decade using meta-omic approaches, several authors warn that an important gap remains: the identification of broadly applicable principles that can then be used to develop conceptual (and eventually predictive) models of microbial community dynamics (Konopka et al. 2014; Prosser 2015). Therefore, one could question the usefulness of describing in detail community structure ("who is there" and "who is doing what"), as broad principles to describe interactions between microbes ("who is doing what with whom") and community-level properties derived from such interactions are still largely missing. We think that this assertion must, however, be nuanced.

First, taking into account microbial community structure without knowledge of microbial interactions can significantly increase the accuracy of community *function* prediction. In a recent paper, Graham et al. (2016) performed a comprehensive analysis of 82 literature datasets on carbon and nitrogen cycling across different ecosystems and showed that incorporating microbial diversity metrics in statistical analyses could improve model accuracy of processes like denitrification, nitrogen mineralization, or carbon respiration rates, compared with analyses where only environmental parameters or gross physiological parameters (like biomass) were used to predict the function.

Second, this knowledge gap is presently being filled by the development of tailored mathematical tools to model community dynamics based on high-resolution meta-omics datasets coupled to extensive ecological time series. The development of such tools is still in its infancy but is going fast, because it is driven by (1) a fundamental need to understand the common mechanistic bases that explain how the collection of species properties drive high-order community properties and generate the endogenous community dynamics (Prosser et al. 2007; Little et al. 2008; Konopka et al. 2014; Widder et al. 2016), and (2) an applied need to predict the functioning of specific microbial systems (i.e., communities performing useful tasks for humankind: waste treatment, nutrient cycling, bioremediation, etc.), and how to design more robust communities. Mathematical tools to exploit large microbial datasets and identify patterns of (non-)association of certain ecotypes and forces that maintain community function have been reviewed elsewhere (Stenuit and Agathos 2015). In summary, they are based on (1) the comparison of independent datasets from different environments (via multivariate statistical tools, calculation of interaction coefficients based on correlations in phylotype abundance, buildup of cooccurrence networks [Faust et al. 2015] to identify keystone species, etc.) and (2) the analysis of time-series datasets from the same environment for dynamic modeling, detection of cause-and-effect relationships, and predictive purposes. Last but not least, such tools are progressively being integrated with engineering and biochemistry models to predict fluxes of molecules (like flux balance analysis, a computational method for analyzing the steady-state flow of metabolites through the metabolic network of an organism or a community and evaluating the production of targeted metabolites— see Section 13.6).

Third, a more recent approach is being implemented to unravel species interactions. It consists of manipulating the structure of the community to understand the function of each member, its connections with other members, and its relative importance in the response to disturbances. To answer such questions, the most common experimental setup is to work with simplified, artificial communities in controllable reproducible systems, like batches or chemostats (Bull 2010; Goers et al. 2014), via the initial addition or substraction of species in the mix. Such model systems are oversimplified, but they (1) are highly flexible and reproducible; (2) offer the possibility to apply single or multiple perturbations to the targeted community, from changes in physicochemical parameters to invasion by foreign, potentially competing microbes or by predators (protozoa, phages, etc.); and (3) allow us to easily study community heterogeneity, redundancy, modularity, internal control (via feedback loops), and species- and multispecies-level responses to perturbations (Stenuit and Agathos 2015), all features that could hopefully be extrapolated to more complex microbiomes. More recently, it has been suggested that alteration of the microbiomes in their natural context would be possible in the future by *in situ* microbiome engineering (Sheth et al. 2016). This approach is based on a suite of genetic tools to manipulate community composition and functions: probiotics (addition of foreign microbes), antibiotics (unspecific killing of microbes), plasmids or insertion sequences (for the stable propagation of exogenous DNA in the community [Springael et al. 2002]), or phages (for the elimination of specific strains). In particular, phage host range could be modularly engineered by swapping phage tail components, and used to deliver a lethal CRISP-Cas RNA-guided nuclease system in the target strain. However, such engineered vectors have not been successfully implemented to natural microbial communities yet and will not be in the near future.

13.4.2 Management and Genetic Engineering of Microbial Consortia

Years ago, scientists advocated the benefits of using complex multispecies microbial communities for bioenergy production (Rittmann et al. 2008). The idea was to benefit from their low cost and high resilience toward fluctuations in environmental conditions or intrusions by microbial invaders (Table 13.1). In regard to this last point, working with mixed communities represents a considerable advantage: it eliminates the need to work under stringent sterile conditions. In addition, the authors suggested ways to manage mixed communities for maximal production of desired metabolites: (1) CH_4 production by funneling the electron flow through H_2 and acetate in the anaerobic digestion process, (2) H_2 production by biomass fermentation in the absence of methanogens coupled to rapid harvest of H_2, and (3) algal lipid production by coupling algae with heterotrophic bacteria that recycle the organic soluble compounds released by algae into inorganic nutrients and CO_2 made available again for the algae. Other benefits from algae–bacteria interactions in the context of biofuel production include a better protection of algae against harmful bacteria and fungi (therefore avoiding algal population collapse), and initiation of algal flocculation by bacteria (therefore improving biomass harvesting) (Ramanan et al. 2016).

Less than a decade later, it must be recognized that the scientific community did not pursue much in that direction. One exception is the study of Kohn and Kim (2015), where the product spectrum in a mixed culture fermentation process was determined based on thermodynamic considerations. The authors succeeded in favoring the growth of microorganisms from the cow's rumen microbiome that were able to produce high concentrations of fuel alcohols and alkanes, or H_2 and CO_2, by manipulating the ratio of H_2 to CO_2 and the total gas pressure. But apart from biomass conversion into CH_4 production, natural complex communities have been largely underutilized in biofuel production applications, because they are often too tricky to manage when the production of one specific molecule is targeted. A more successful alternative is the development of artifical consortia of a few wild-type or (most often) engineered strains. Indeed, the lower level of complexity of artificial

TABLE 13.1   Comparative SWOT Analysis of Monocultures versus Mixed Consortia for Biofuel Production

Monocultures		Mixed Consortia	
Strengths	**Weaknesses**	**Strengths**	**Weaknesses**
• The most widely used strains are (almost) perfect cell factories: metabolic processes are understood, modeled, and controlled	• Sensitivity to stresses (change in substrate type, temperature, viral invasion etc.); therefore, maintaining monocultures at large scale can be impractical and cost-prohibitive • Available genetic tools developed to engineer specific strains are not easily transferrable to other strains	• Usually greater functionality (faster growth, greater performance, or greater flexibility in terms of substrates used) than pure strains • Increased robustness, resilience, and self-stability when facing perturbations • Enables cultivation and exploitation of the metabolism of microbes that are unable to grow as pure strains • Can accomodate complex waste streams a substrates • Mixed communities may harbor community-level functional traits that are not found in individual members	• Current techniques like metabolomics are not powerful enough to study mixed communities in their whole complexity • No general methods are available to engineer consortia of different organisms • Final products are rarely pure
Opportunities	**Threats**	**Opportunities**	**Threats**
• Omics revolution could rapidly lead to more diverse genetic toolkits	• None	• Fast development of omics techniques, decreasing price of analysis • Possibility to tweak environmental parameters to direct community activity toward a desired function	• Omics techniques do not provide a basic mechanistic understanding of community functioning and its emergent properties • Techniques to manipulate consortia are still in their infancy; their future is uncertain • Some techniques, like metabolomics, might remain for a long time insufficient to disentangle interspecies connections in complex communities

Note: SWOT, strengths, weaknesses, opportunities, and threats.

consortia is easier to handle, and they can be analyzed with current technologies like metabolomics (Ponomarova and Patil 2015) with enough resolution and coverage. In contrast, tools developed for genome-scale editing (Esvelt and Wang 2013) are not transposable yet to metagenome engineering of complex communities in order to add, remove, or modify their functional profile and alter their community-level properties (Sheth et al. 2016).

Compared with monocultures, metabolic exchange in multispecies consortia can be viewed as a strategy for group success: they favor the emergence of community properties, such as enhanced process performance and faster growth (Morris et al. 2013). In biotechnological applications, compartmentalization of parts of metabolic pathways in different community members ("division of labor") can enhance the production of a valuable metabolite, as recently demonstrated for the synthesis of a precursor of the antitumor drug paclitaxel (Zhou et al. 2015). Division of labor was also reported in the conversion of pollutants with inhibitory intermediates (Dejonghe et al. 2003). However, predicting which of the two strategies (engineering several strains to catalyze specific steps of a pathway vs. engineering one strain to catalyze the complete pathway) maximizes the performance of a process is difficult (Lindemann et al. 2016). In that respect, understanding the biochemical causes of metabolic specialization (i.e., the fact that species often specialize at metabolizing only subsets of the available substrates rather than all of them) could provide a foundation for rational design of consortia in synthetic ecology (Johnson et al. 2012).

Emergence and maintenance of metabolic exchanges depend on particular circumstances, such as the spatial structure of the microbial community, nutrient availability, diffusion constraints, and cost-effectiveness of the biosynthetic processes in question (Morris et al. 2013). In particular, the existence of a spatial structure was shown to stabilize consortia. Using microfluidics, Kim et al. (2008) showed indeed that imposing a miscroscale spatial structure was both necessary and sufficient for the coexistence of three soil bacterial species involved in reciprocal syntrophy and forming a consortium that was otherwise unstable. The authors foresaw that controlling spatial structure, rather than matching metabolic and growth rates, would expand the range of systems amendable to synthetic biology approaches. In addition, aggregation in flocs or biofilms maximizes the efficiency of metabolite transfer and stimulates otherwise thermodynamically unfavorable metabolic processes (Agapakis et al. 2012). For example, syntrophic hydrogen-producing fermenters and methanogens produce large aggregative flocs in anoxic lake sediments or sewage sludge. On the other hand, cell populations that communicate through diffusible molecules may be separated to a larger degree without disrupting communication; in this case, coculture systems can be treated as comprising "monoculture modules" that can be connected with each other (Goers et al. 2014). Nutrient-poor environments favor interspecies relationships because reciprocal exchange is more likely to be necessary. Remarkably, it can arise between unrelated strains providing that complementation of each other's biosynthetic capabilities becomes vital for the survival of the consortium (Hom and Murray 2014).

There are many examples of consortia tailored for specific functions. In practice, cocultures rarely exceed two or three species, because larger numbers of interacting populations, although potentially more valuable for industrial applications, often show unstable and unpredictable behavior (Goers et al. 2014).

In the context of cellulosic biofuels, several studies have explored the issue of using mixed cultures for ethanol conversion compared with engineered monocultures for CBP. The majority of published papers reported that fermentation consortia produced a higher yield of ethanol than did a single microorganism (Alper and Stephanopoulos 2009). For example, dilute acid softwood hydrolysate (which contains glucose and xylose as the dominant sugars) was fermented to ethanol with exceptional yield by a coculture of *S. cerevisiae*, *Pachysolen tannophilis*, and a genetically modified *E. coli* strain that carried both *pdc* and *adhB* genes derived from *Z. mobilis* (Qian et al. 2006). At the enzymatic level, an enzyme cocktail expressed by the fungus *Hypocrea jecorina* and a cellulosome preparation secreted from *C. thermocellum* showed great activity on pretreated biomass and purified cellulose, respectively (Resch et al. 2013). When combined, these systems displayed dramatic synergistic activity of cellulose deconstruction. Xia et al. (2012) set up a consortium of three *E. coli* strains able to remove acetate (a growth inhibitor contained in lignocellulosic hydrolysates) and to degrade simultaneously glucose, xylose, and arabinose. Kato et al. (2004) reported improved cellulose degradation by a coculture of an anaerobic cellulolytic *Clostridium* and aerobic noncellulolytic bacteria. The latter probably consumed all available oxygen and degraded inhibitory substances toward *Clostridium*. In another study, the same authors observed that a cellulose-degrading community of five strains showed functional and structural stability (Kato et al. 2005). Interactions between community members were unraveled by removing each community member separately and recording community dynamics and degradation efficiency. Finally, Minty et al. (2013) designed and tested a synthetic two-species consortium for the production of isobutanol from cellulosic biomass. In this coculture, the fungus *T. reesei* secreted cellulase enzymes to hydrolyze lignocellulosic biomass into soluble saccharides, which were metabolized by a genetically engineered bacterium *E. coli* into desired products. Isobutanol titer and yield reached 1.88 g/L and 62% of theoretical maximum, respectively, which are low in absolute value but greater than those reported to date for CBP production of advanced biofuels.

Regarding algal biofuels, several recent papers reported that algal growth and lipid accumulation were enhanced when algae were cocultured with specific strains of bacteria compared with axenic cultures (Bashan et al. 2002; Higgins and VanderGheynst 2014). Cocultivation of microalgae with filamentous fungi was shown to enhance total biomass production and lipid yield as well, but also biomass flocculation (Wrede et al. 2014; Muradov et al. 2015). The authors suggested that oil composition could be optimized via such interactions, as shifts in the fatty acid profile were observed. Finally, a modeling study of cocultivation of an algal strain with an oleaginous yeast predicted that the algae–yeast coculture could produce biodiesel at competitive prices (Gomez et al. 2016). The model was built on genome-based metabolic networking. In this system, the algae benefited from a local source of CO_2 (the respiration of lignocellulosic sugars by yeast metabolism). At the same time, the yeast benefited from the O_2 produced by algal photosynthesis, and therefore increased lipid production.

Regarding the production of H_2 by dark fermentation, cocultures were reported to achieve systematically better results than monocultures (reviewed in Bader et al. 2010; Elsharnouby et al. 2013). In the first type of coculture, a strict anaerobe is mixed with a facultative one, which consumes traces of oxygen and therefore allows stable H_2 production without any added reducing agents. In the second type of coculture, a cellulolytic strain (e.g., *C. thermocellum* or *C. acetobutylicum*) is grown with a noncellulolytic high-hydrogen-producing strain, allowing efficient hydrogen production from cellulosic waste without expensive hydrolysis treatment. In the third type of coculture, a hydrogen-producing aciduric strain is grown with a high-hydrogen-producing strain, which allows us to extend the range of pH values at which biohydrogen can be produced, and therefore eliminates the need for buffering the growth medium.

The question that remains, however, is how to design productive microbial consortia from building blocks (microbial populations) in a rational way. Mechanistic understanding of interactive behaviors in cocultures remains very limited, and to date, the design of consortia more productive than monocultures was either due to luck or based on intuitive ecological principles like division of labor (Lindemann et al. 2016). In the future, advancements in the field will increasingly rely on mathematical modeling, either population-based modeling for the prediction of the interspecies dynamics without detailed information of intracellular metabolism, or metabolic network modeling based on energy and material fluxes within and between cells (Lindemann et al. 2016).

13.5 Targeted Approach 4: Microbial Electrosynthesis of Biofuels

MES is the production of chemical commodities by microbes using electrons derived from the cathode of a bioelectrochemical system. This technology is increasingly focusing attention on a context of transition from an oil economy to an electric economy. Electrobiocommodities include acetate, but also biofuels like H_2, CH_4, ethanol, butanol, and 1,3-propanediol.

In their natural habitats, microorganisms use various forms of electron acceptors (O_2, nitrate, sulfate, ferric iron oxide, and heavy metals), as well as of electron donors (organic compounds, ferrous iron oxide, H_2S, and H_2), to sustain their energy requirements. In particular, some of them have the capacity to use natural insoluble electron donors or acceptors. The best-characterized ones are *Geobacter sulfurreducens* and *Shewanella oneidensis*, reducing ferric iron oxides, and *Acidithiobacillus ferroxidans*, oxidizing ferous iron minerals. In addition, there is growing evidence that a variety of anaerobic microbes can oxidize metallic iron (Fe(0)), a form of iron uncommon in nature that has become widespread due to human activity during the last millennium. The best documented Fe(0)-oxidizing microbes are sulfate-reducing bacteria corroding metallic pipes (Dinh et al. 2004; Venzlaff et al. 2013), but some methanogenic strains have been shown recently to use metallic iron as the sole electron donor as well (Dinh et al. 2004).

Microorganisms exploiting insoluble compounds for their oxidoreduction reactions have been harnessed in microbial electrochemical technologies, because they can give and receive electrons to and from electrodes. Initially, in the early 2000s, such technologies were developed and operated to *produce* electrical current (reviewed in Logan 2009). The basic configuration of these so-called MFCs consists of two chambers containing an electrode, filled in with an electrolyte (an aqueous solution), and separated by an ion exchange membrane. Both electrodes are connected to an external electric circuit. Current is generated at the anode from the oxidation of organic compounds from wastewater or, less frequently, the conversion of toxic H_2S into elemental sulfur or sulfate. At the cathode, oxygen is reduced chemically or biologically. It is also possible to expose the cathode directly to air and therefore eliminate the need of a cathode chamber. MFCs basically work as a voltage source with internal resistance; the difference in potential between the oxidation reaction at the cathode and the reduction reaction at the anode leads to a spontaneous electrical current from the anode to the cathode. The actual potential is lower than this difference because of various losses related to the electrochemical properties of the electrodes and the biofilm, the accumulation of reaction products, and the depletion of reactants near the two electrodes, and ohmic losses in the electrical circuit itself (reviewed in Rinaldi et al. 2008). Over a decade, the performance level of MFCs has been raised by several orders of magnitude, reaching 2400 mW/m^2 (Logan 2009) and 200–250 W/m^3 (Rinaldi et al. 2008). This was made possible thanks to substantial research effort to optimize aspects related to materials, cell architecture, and buffer solution (Rinaldi et al. 2008), but also thanks to a better understanding of the biological aspects of MFCs. In this context, electron transfer mechanisms between bacteria and the electrodes have been extensively studied. Historically, they were mainly examined in two strains: *G. sulfurreducens* and *S. oenidensis* (reviewed in Lovley 2012). First, direct contact between electrogenic microbes and the electrode is possible via c-type cytochromes or "nanowires." Nanowires are either type IV electroconductive pili (Reguera et al. 2005) made of the aromatic amino acids phenylalanine and tyrosine, or outer membrane and periplasmic extensions containing cytochromes used as "hopping" sites by electrons. Microbes able to make direct contact with an electrode usually develop thick biofilms at its surface. Second, transfer can be mediated by electron shuttles (also called mediators) secreted by the microbes (phenazine derivatives or flavins). In mixed communities, shuttles produced by one species can help other species in their interactions with the electrode (Rabaey et al. 2004). Alternatively, artificial shuttles (like neutral red, methylene blue, methyl viologen, anthraquinone-2,6-disulfonate [AQDS], and resazurin) can be added in the medium. A new generation of putative electron transfer molecules, called conjugated oligoelectrolytes, has been recently introduced into this research field (Philips et al. 2015). These molecules integrate into the microbial membrane, making it leakier, which favors the release of electroactive components. In a great number of studies, artificial shuttles have been successfully supplied to nonelectroactive microbes in order to allow them to exchange electrons with the electrode without the need for attachment to it. However, long-term use of artificial shuttles is questionable; they represent an extra cost, are nonselective (they can be used by other microorganisms), may be unstable (e.g., methyl viologen) or toxic, and can be lost in the medium on their way to the electrode surface (Philips et al. 2015). In some cases, MFCs have been extended to two biological compartments; the electrons microbiologically generated at the anode can be used by electrotrophic microbes at the cathode (for the elimination of NO_3^- and NO_2^- from wastewater, for the precipitation of soluble U(VI) into insoluble U(IV), etc.) (reviewed in Logan 2009).

Today, MFCs remain noncompetitive compared with chemical fuel cells in terms of efficiency and power output. However, they represent a more sustainable way to produce energy, which explains the tremendous effort that has been devoted so far to developing upscaled MFC systems to combine electricity production and wastewater treatment. Interestingly, another potentially more successful application of MFC is emerging: their integration into robotic or electronic systems as sole power source (i.e., in remote sensors, autonomous robots, or electronic gadgets). For this purpose, different strategies are being tested to miniaturize MFCs (while keeping enough power output) and improve their autonomy (Ieropoulos et al. 2012, 2016; Liu et al. 2016; Tommasi et al. 2016). For example, Ren et al. (2015) recently developed a miniaturized MFC equipped with an anode made of carbon nanotubes that reached a power density of 3320 W/m³.

MES cells are a variant of microbial electrochemical cells, where focus is on the reductive microbial reactions taking place at the cathode to produce specific chemical compounds (Rabaey and Rozendal 2010; Lovley and Nevin 2013; Tremblay and Zhang 2015). Various electron acceptors can be reduced at the cathode, ranging from O_2 to protons and CO_2. In particular, MES cells based on electroautotrophic microbes developing at the cathode have received increasing attention over the last 5 years due to their ability to transform CO_2, a greenhouse gas, into various organic compounds: CH_4, acetate, or other multicarbon compounds. For CH_4 production, another advantage of MES cells is that the gas they produce is enriched in methane (about 85%) compared with biogas resulting from anaerobic digestion, because of the consumption of carbon dioxide in MES cells. The utility of autotrophic MES cells for biological production of H_2 has been reported as well. Their initial functioning is a bit different, though, as the first prototypes (in the early 2000s) to generate H_2 relied on coating the cathode with hydrogenase enzymes. Such enzymatic cells typically possess power densities that are generally orders of magnitude greater than MES cells using whole microbes, but they are not self-regenerating and therefore losing activity. Therefore, as far as we know, they have been replaced by MES cells based on living microbes in all applications.

In MES cells, electrotrophic microbes are inoculated either as mixed communities (usually undefined cultures from wastewater, sludge, or sediment) or pure cultures. Mixed-community reactors were reported to have a remarkable production rate, but most of the time they produce a mix of compounds (e.g., alcohols–H_2–organic acids or CH_4–acetate–H_2), illustrating the difficulty to generate a single specific product (Marshall et al. 2012; Tremblay and Zhang 2015). Therefore, most studies have focused on pure strains.

Compared with electron transfer to the anode in MFCs, electron transfer from the cathode to microbes is very poorly characterized. It is currently the object of much study, as understanding the underlying mechanisms of electron transfer could lead to major breakthroughs in the development of more efficient MES cells. Similarly

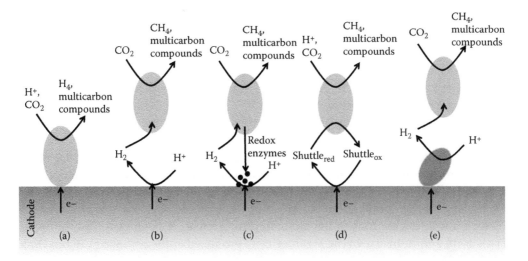

Figure 13.5 Possible mechanisms for electron transfer from the cathode to microorganisms producing methane or multicarbon compounds (e.g., acetate, ethanol, butanol, or 2,3-butanediol): (a) direct electron transfer, (b) abiotic production of H_2 for subsequent conversion, (c) enzyme (hydrogenase)-mediated biotic production of H_2 for subsequent conversion, (d) shuttle-mediated electron transfer, and (e) in mixed communities, biotic production of H_2 by some community members for subsequent conversion by other ones. In cases b, c, and e, formate can be produced (from H^+ and CO_2) as well, depending on the cathode potential (involving formate dehydrogenase in case c).

to anodes, evidence for both electron shuttle-mediated and direct electron transfer mechanisms has been found (Figure 13.5), although discrimination between both is often challenging.

The first MES cells for methane, acetate, or H_2 production relied on the addition of artificial shuttles (reviewed in Tremblay and Zhang 2015). Later, studies showed that the same processes could take place without artificial shuttles (Rozendal et al. 2008; Cheng et al. 2009; Villano et al. 2010). In the study of Rozendal et al. (2008), H_2 was produced after reverting the polarity of an anode, on which a mixed biofilm of acetate and hydrogen-oxidizing microbes had developed, to make it a biocathode. Their strategy was based on the well-known reversibility of hydrogenases. This experimental setup raised questions on the autotrophy of microbes producing H_2 (as they could have been exploiting organic compounds derived from the biofilm), but a recent study showed that it is possible to grow a self-regenerating electroactive biofilm at the cathode that uses CO_2 as the sole carbon source to produce H_2 as the sole compound (Jourdin et al. 2015). Indirect transfer via other exogenous shuttles (formate, Fe(II), NH_4^+, and H_2) or via shuttles secreted by bacteria (phenazine, riboflavine, B12 vitamin, or DNA) has been reported as well (Philips et al. 2015). Among these, molecular hydrogen is a unique shuttle in MES cells. First, it can be used as an electron donor by all mixed communities and pure strains producing CH_4 or the multicarbon compounds that have been studied so far (Philips et al. 2015). Second, it is abiotically generated in the cathode vicinity from the protons migrating from the anodic chamber, as long as cathode potential is sufficiently negative (i.e., lower than the formal potential of the $2H^+/H_2$ couple). This potential determines, in theory, which mechanism is used by bacteria and archaea to acquire the necessary electrons for their metabolism, that is, H_2-mediated transfer versus direct electron transfer (Tremblay et al. 2016). Direct electron transfer via c-type cytochromes has been demonstrated for metal-oxidizing bacteria, but the latter do not produce biofuels (unless they can be engineered to do so). It is suspected for electromethanogenic archaea and electroautotrophic bacteria (including electroacetogenic ones) can get electrons directly from a cathode, but the underlying mechanism is unknown. In practice, discrimination between H_2-mediated and direct electron transfer is challenging. For example, depending on the potential set at the cathode, CH_4 can be produced by direct transfer of electrons to the methanogenic archaea (Cheng et al. 2009; Villano et al. 2010) and/or by use of the H_2 abiotically produced at the cathode (Villano et al. 2010). The hypothesis of direct electron transfer is supported by observations in nature that some methanogens can corrode Fe(0) much faster than related hydrogenotrophic methanogens (Dinh et al. 2004; Uchiyama et al. 2010), suggesting the existence of a direct mechanism to oxidize metallic iron. In addition, a methanogenic strain (*Methanococcus maripaludis*) mutated to lack hydrogenases could still produce CH_4 (although at a low rate) (Lohner et al. 2014). However, some researchers (Deutzmann et al. 2015) suspect that Fe(0)-oxidizing methanogens actually exploit H_2 and formate in MES cells by releasing extracellular redox enzymes, such as hydrogenases and formate dehydrogenases, which associate with the electrode surface and catalyze the generation of H_2 and formate. Therefore, it is likely that in MES cells, most of the electrons used for CH_4 production are indirectly derived from H_2 produced at the cathode, either abiotically or biotically via extracellular enzymes or other microbes (in mixed biofilms). For example, in their study, Villano et al. (2010) used an uncharacterized methanogenic culture in which interspecies H_2 transfer between electroactive H_2-producing microorganisms and H_2-utilizing methanogenic archaea could have sustained CH_4 production. Future research is needed to investigate the involvement of mechanisms of direct electron transfer, production of redox enzymes, and interspecies interactions in biomethane production.

The same holds true for the production of acetate and other multicarbon compounds in MES cells. The latter include ethanol, butanol, or 2-3 butanediol, which can be produced by natural or genetically engineered Clostridia under appropriate conditions (Schiel-Bengelsdorf and Dürre 2012; Claassens et al. 2016). Several pure acetogenic cultures (*Sporomusa* species, *Clostridium* species, or *Morella thermoacetica*), as well as mixed acetogenic communities (dominated by *Acetobacterium* species), were shown to convert CO_2 into acetate with electrons drawn from an electrode, but the mechanism of electron transfer was unclear (Nevin et al. 2011; Marshall et al. 2012). In pure cultures, direct electron uptake was hypothesized based on the absence of detectable concentrations of molecular H_2 and the observation of attachment of bacteria on the cathode surface (Nevin et al. 2011). However, all the aforementioned bacteria are capable of using H_2 as an electron donor for the reduction of CO_2, and therefore they could use an indirect electron uptake mechanism mediated by H_2 abiotically generated by the cathode (and not detected). In addition, Clostridia, Acetobacteria, and *M. thermoacetica* are gram-positive bacteria, and it is unclear how electrons could be transferred through their thick cell wall; in addition, some have no cytochromes encoded in their genome. However, the idea of direct electron transfer has recently returned to the fore by the recent description of an acetogenic strain capable of using metallic iron as a sole electron donor (Kato et al. 2015). This strain was highly related to *Sporomusa sphaeroides* and was capable of significantly enhancing Fe(0) corrosion in comparison with abiotic corrosion, while hydrogenotrophic acetogens could not. This suggested a direct, as yet unraveled electron uptake mechanism. As far as acetogenic communities enriched on cathodes are concerned, they most likely rely on (1) hydrogen abiotically generated at the cathode when low electrode potentials are used, and (2) hydrogen produced by other, nonacetogenic bacteria drawing electrons from the cathode and then used by hydrogenotrophic acetogens (Patil et al. 2015). Interspecies H_2 transfer within a community characterized by a high rate or efficiency of acetate synthesis was recently reported as well (Jourdin et al. 2016). The authors suggested that such a high rate of efficiency was because consumption of all H_2 and its transformation into acetate took place within the biofilm, that is, between closely located microbes, and that H_2-producing bacteria improved the electrode surface by producing copper nanoparticles.

Besides their capacity to use CO_2 directly as feedstock, a major strength of MES cells is their versatility with regard to the source of electrons. Indeed, the necessary electricity can be generated from multiple renewable sources, like solar energy, wind, or wastewater. Critics of renewable energy often cite the fact that such electricity sources are intermittent. In this context, MES becomes a complementary technology that allows direct storage of electricity into value-added chemical commodities. Another strength is their efficiency. Tremblay and Zhang (2015) compared the efficiency of plant photosynthesis versus MES coupled with solar cells. In nature, plants convert sunlight to biomass with an efficiency of 1%–2% (3% for microalgae). The conversion efficiency of electricity to chemical commodities by MES processes is about 80%–90%. As photovoltaic solar panels are at least six times more efficient at capturing the sun energy than plants, they concluded that powering MES with electricity from solar cells was a more potent strategy for storing the sun energy into multicarbon compounds than photosynthesis. A similar conclusion was drawn by Claassens et al. (2016), who compared the solar-to-product energy conversion efficiency of plant biomass fermentation, for example, into ethanol (in total <1%) and of solar panels coupled to electrocatalysis (to produce H_2) and to chemoautotrophic MES cells fed with H_2 and CO_2 (in total ≈7.6%).

Despite all its potential, MES is not an economically viable technology yet, because of the high electron demand for the reduction of CO_2, and therefore low production rates. Recent advances could, however, help their transition into commercial applications (Rosenbaum and Franks 2014). First, to minimize internal power losses, cathodes could be engineered to improve their specific surface area, topography, electric conductivity, and biocompatibility. To date, stainless steel is the most efficient cathode material, but cathodes are generally made of cheaper carbonaceous material (plain graphite, graphite granules, or carbon felt, cloth or granules, and vitreous carbon). There are different possibilities to increase electron transfer at moderate cost (reviewed in Rinaldi et al. 2008; Tremblay and Zhang 2015). One of them is to increase the electrode surface available for bacterial colonization using nanostructured materials like carbon nanotubes or graphene. Another is coating the electrode with posivitely charged compounds (e.g., chitosan) to favor attachment of bacteria (which are negatively charged). Optimization of other parts of the electrochemical hardware is also ongoing, such as the removal of the expensive ion exchange membrane or the improvement of the electrical circuit. Such technical improvements have contributed to the recent 400- to 500-fold increase in organic compound production rates (in mM/day/m^2) and electron transfer rates (in mA/m^2) (Tremblay and Zhang 2015). Nevertheless, this trend will only continue if technical improvements are coupled with careful selection of strains and a better understanding of all the interactions between microorganisms and electrodes. In that respect, the wide variety of autotrophs in nature provide a largely untapped pool of potential hosts that could be converted into autotrophic cell factories, as illustrated by the recent contribution of metallic iron-oxidizing strains to MES. Their performance could be improved by genetic engineering, providing that the required genetic tools are available (Rosenbaum and Henrich 2014; Claassens et al. 2016). Alternatively, the ability to interact with electrodes could be introduced into strains that are both metabolically versatile and easy to engineer, as attempted in *E. coli* (Jensen et al. 2010). In addition, as mentioned earlier, a lot of research is currently dedicated to characterizing electron transfer between microbes and the cathode, as this represents a cornerstone for more efficient MES production of chemical commodities. New transfer mechanisms could be discovered, but there is plenty of space to improve known ones as well. For example, Tan et al. (2016) recently succeeded in modifying a strain of *G. sulfurreducens* so that it could produce nanowires containing the amino acid tryptophane, which can promote fast electron transport, resulting in a 2000-fold increase in nanowire conductivity.

Another way of diversifying MES applications and obtaining sufficient productivity rates is electrofermentation (reviewed in Rabaey and Rozendal 2010; Schievano et al. 2016), a hybrid metabolism in which electrons from the cathode are utilized as coreducing equivalents in addition to carbon source-derived reducing equivalents (e.g., NADH) during heterotrophic fermentation. This way, fermentation is redirected toward more electron-dense desired products (e.g., propionate, glutamic acid, butyric acid, ethanol, or butanol). In early electrofermentation based on pure strains, the addition of shuttles was necessary due to the lack of an electroactive heterotroph. For example, Steinbusch et al. (2010) tested different shuttles to stimulate acetate reduction into ethanol by a mixed inoculum growing at the cathode. Mediators are not necessary anymore: Choi et al. (2014) showed that the heterotroph *Clostridium pasteurianum* could simultaneously use a soluble organic substrate (glucose or glycerol) *and* a cathode as electron donors. Usually, *C. pasteurianum* produces acetate and butyrate (and little butanol) from glucose, and butanol and 1,3-propanediol from glycerol. When grown in the presence of glucose or glycerol and a small electrode supply by the cathode, its fermentation profile was steered to produce a greater amount of electron-dense products (butanol and 1,3-propanediol, respectively). The mechanism of direct electron transfer in this gram-positive bacterium was not unraveled. The authors speculated that such results were opening up the possibility of using electricity to efficiently produce electron-dense chemicals and fuels.

13.6 Targeted Approach 5: Dynamic Macroscopic Models of Bioprocesses and Their Application to Microbial Fuel Production

13.6.1 Introduction

Many dynamic models have been used for modeling key component concentration time profiles in bioprocesses. Most of these bioprocess models can be found in the framework of biopharmaceutical and agro-food applications where the products of interest are obtained through cultures of microorganisms (bacteria and yeasts) or mammalian cells in bioreactors. Dynamical models at the macroscopic scale allow reproducing the concentration time profiles of the cells, main substrates, metabolites, and products of interest (Bastin and Dochain 1990). Models that consider the cells as a global entity (without paying attention to intracellular metabolism) and their population as homogeneous are generally classified as unstructured and unsegregated. This kind of macroscopic model typically consists of a set of differential equations that represent the mass balances of the above-mentioned main components. Such a macroscopic description allows the building of a model-based control and monitoring tools. For instance, bioreactor feeding time profiles can be determined for optimizing the bioprocess (e.g., maximum yield of substrate to cells or maximum cell productivity) (Banga et al. 2005; Amribt et al. 2014; Richelle and Bogaerts 2014). Closed-loop controllers can be tuned for maintaining some variables at their optimal set point despite the presence of process disturbances and/or modeling errors (e.g., maintaining microorganism cultures at the limit of overflow metabolism) (Van Impe and Bastin 1998; Smets et al. 2004; Dewasme et al. 2010). Software sensors (also called state observers) can be used for online estimation of key variable concentrations that are not directly measured through sensor probes (e.g., a cell density software sensor based on output gas analysis and base feeding flow rate) (Bogaerts and Vande Wouwer 2003; Ali et al. 2015). All these engineering tools have the common feature of being model-based solutions, and they all require a mathematical description of the process that is sufficiently detailed for accurately reproducing the basic biological phenomena linking the essential process variables, but sufficiently simple so that they can be efficiently used for reaching the engineering goal.

With the increasing number of microbial fuel production processes, the use of these dynamic macroscopic models (and the associated model-based engineering tools) has naturally been extended to this new application field. Although these processes have, of course, their own specificities, the general framework described above remains applicable with some adaptations. Section 13.6.2 provides to the reader some basic concepts about dynamic macroscopic models of bioprocesses and shows how general kinetic laws (which are the key issues in this kind of model) can provide very flexible and generic model structures. Sections 13.6.3 and 13.6.4 present two case studies (bioethanol production from potato peel wastes and microalgae cultures) that illustrate how the general framework of Section 13.6.2 can be adapted to specific applications.

13.6.2 Dynamic Macroscopic Models of Bioprocesses and General Kinetic Model Structures

The first step in macroscopic modeling of bioprocesses is to summarize the main biological phenomena (e.g., cell growth, maintenance, and death) that link the cells and the main extracellular components (substrates, metabolites, and products of interest) (Bastin and Dochain 1990). This leads to the definition of a macroscopic reaction scheme, which contains M reactions involving N products:

$$\sum_{i \in R_k} K_{ik}\xi_i \xrightarrow{\varphi_k} \sum_{j \in P_k} K_{jk}\xi_j \quad k \in [1, M], \quad i, j \in [1, N] \tag{13.1}$$

where φ_k is the rate of the k-th reaction, K_{ik} and K_{jk} are pseudostoichiometric (or yield) coefficients, ξ_i and ξ_j are the extracellular components (substrates, metabolites, products, or cells), R_k is the set of component indices that are reactants (or catalysts) in the k-th reaction, and P_k is the set of component indices that are products (or [auto-]catalysts) in the k-th reaction.

Note that there exist methods for determining the minimal number of reactions M and even for determining the global reaction scheme on the basis of experimental measurements of the different component concentrations (Bernard and Bastin 2005a, 2005b; Hulhoven et al. 2005). The general dynamic macroscopic model (Bastin and Dochain 1990) is made of the ordinary differential equations (ODEs) describing the mass balances of each of the N components involved in the reaction scheme (Equation 13.1):

$$\frac{d\xi(t)}{dt} = K\varphi(\xi(t)) - D(t)\xi(t) + F(t) - Q(t) \tag{13.2}$$

where $\xi(t) \in \Re^N$ is the vector of the N component concentrations, $\varphi(t) \in \Re^M$ is the vector of the M reaction rates, $K \in \Re^{N \times M}$ is the (pseudo-)stoichiometric or yield matrix (K_{ij} is the yield coefficient of the i-th component in the j-th reaction), $D(t) \in \Re$ is the dilution rate (feeding flow rate divided by the volume), $F(t) \in \Re^N$ is the vector of external inflow rates, and $Q(t) \in \Re^N$ is the vector of gaseous outflow rates.

Given some assumptions, the yield matrix K (or at least part of it) can be identified on the basis of experimental measurements of $\xi(t)$ through simple linear regressions and without any knowledge about the reaction rates $\varphi(t)$ (Chen and Bastin 1996; Bogaerts et al. 2003). The key issue in Equation 13.2 is the mathematical structure of the reaction rates $\varphi(t)$, which are typically nonlinear functions (of the different component concentrations $\xi(t)$) representing different biological phenomena. There exist plenty of kinetic models representing particular phenomena (activation, saturation, and inhibition) for particular components (substrates, metabolites, etc.) with particular mathematical functions. In order to circumvent the different choices that have to be made for a given application, it is possible to use general kinetic model structures that allow representing in a systematic way the different biological phenomena due to any component in any reaction (Savageau 1969a, 1969b; Haag et al. 2005; Grosfils et al. 2007; Richelle and Bogaerts 2015). A first example is the extended Monod law framework, where each reaction rate is modeled as

$$\varphi_k(\xi) = \mu_{\max,k} \prod_{l \in A_k} \frac{\xi_l}{\xi_l + K_{S,lk}} \prod_{m \in I_k} \frac{K_{I,mk}}{\xi_m + K_{I,mk}} X \tag{13.3}$$

where $\mu_{\max,k}$ is the maximum specific rate of the k-th reaction, ξ_l is the concentration of the l-th component in vector ξ, X is the cell density, A_k is the set of indices of the components that activate (as reactants or catalysts) the k-th reaction, I_k is the set of indices of the components that inhibit (as metabolites or reactants) the k-th reaction, $K_{S,lk}$ is the saturation constant of the l-th component in the k-th reaction, and $K_{I,mk}$ is the inhibition constant of the m-th component in the k-th reaction.

A second example of a general kinetic model is given by (Grosfils et al. 2007; Richelle and Bogaerts 2015)

$$\varphi_k(\xi) = \mu_{\max,k} \prod_{l \in A_k} \xi_l^{\gamma_{lk}} \prod_{m \in I_k} \xi_m^{\beta_{mk}} X \tag{13.4}$$

where γ_{lk} is the activation coefficient of the l-th component (reactant or catalyst) in the k-th reaction, and β_{mk} is the inhibition coefficient of the m-th component (metabolite or reactant) in the k-th reaction.

Note that this latter general kinetic model structure has the advantage that it can be rigorously linearized, up to a logarithmic transformation, hence allowing the identification of the kinetic parameters $\mu_{\max,k}$, γ_{lk}, and β_{mk} based on simple linear regressions. It is also worth noting that it is possible to go back to the extended Monod formalism (Equation 13.3) based on a model identified with Equation 13.4, as shown by Richelle and Bogaerts (2015).

The concept of the macroscopic reaction scheme (Equation 13.1), the mathematical model consisting of the corresponding mass balances (Equation 13.2), and general kinetic model structures like Equations 13.3 and 13.4, can be used for modeling bioprocesses in many different fields of applications. The next two sections focus on two particular case studies in the framework of microbial fuels.

13.6.3 Case Study 1: Bioethanol Production from Potato Peel Wastes

In this case study, potato peel residues are used as a source of fermentable sugars that are subsequently used as carbon source in S. cerevisiae batch cultures for producing bioethanol (Richelle et al. 2015). The additional use of ammonium as an inorganic nitrogen source provides a possibility to boost the fermentation metabolism. However, too high ammonium concentrations may lead to an inhibition (or at least saturation) effect. A macroscopic model of the process has been proposed and validated and allowed, determining the optimal initial ammonium concentration to be provided to a batch reactor for maximizing the bioethanol production, and which will be fully depleted at the end of the batch run.

The initial macroscopic reaction scheme, corresponding to the general form given in Equation 13.1, that was proposed by Richelle et al. (2015), contained three reactions involving four components (fermentable sugars, ammonium, yeast cells, and ethanol), and the corresponding dynamic model (Equation 13.2) (using extended Monod kinetics [Equation 13.3]) presented 14 parameters to be identified. Based on experimental data coming from six batch cultures (differing by their initial ammonium concentrations) and a multistep parameter identification procedure, it was shown that the model could be strongly reduced (based on the parameter uncertainty analysis), leading to the final very simple macroscopic model:

$$S \xrightarrow{\varphi_1} \alpha_1 X + \beta_1 E$$
$$N \xrightarrow{\varphi_2} \beta_2 E \tag{13.5}$$

where S, N, X, and E represent, respectively, the fermentable sugars, ammonium, yeast cells, and ethanol; α_1, β_1, and β_2 are yield coefficients; and φ_1 and φ_2 are the reaction rates.

Note that the first reaction represents the classical fermentation of the sugars, leading to yeast growth and ethanol production (not coupled with a respiration reaction after model reduction), and that the second reaction represents the boost provided by nitrogen consumption on the ethanol production (not coupled with biomass growth after model reduction).

The dynamic model corresponding to the mass balances of the components involved in (Equation 13.5) is given by

$$\frac{dS}{dt} = -\varphi_1$$
$$\frac{dN}{dt} = -\varphi_2$$
$$\frac{dX}{dt} = \alpha_1 \varphi_1 \tag{13.6}$$
$$\frac{dE}{dt} = \beta_1 \varphi_1 + \beta_2 \varphi_2$$

The reaction rates (still after model reduction) use finally simple Monod laws:

$$\varphi_1 = \mu_{max,S}\, \frac{S}{S+K_S}\, X$$

$$\varphi_2 = \mu_{max,N}\, \frac{N}{N+K_N}\, X \qquad (13.7)$$

The values of the six parameters, α_1, β_1, β_2, $\mu_{max,S}$, $\mu_{max,N}$, and K_S, were identified based on different subsets of the six batch cultures. Note that an arbitrarily low value has been fixed for K_N, as it was associated to high variation coefficients but could not be canceled, so that the model stops the nitrogen consumption when nitrogen is fully depleted. The identified values can be found in the paper by Richelle et al. (2015).

Figure 13.6 presents cross-validation results, that is, comparisons between concentrations estimated with Equations 13.6 and 13.7, compared with experimental data that were not used for the parameter identification. For instance, the first two rows corresponding to Cross-Validation 1 present a comparison of model estimates with experimental data of, respectively, Experiments 1 and 4, the model parameters having been identified on the basis of the measurements from Experiments 2, 3, 5, and 6. The model leads to satisfactory predictions of the process behavior.

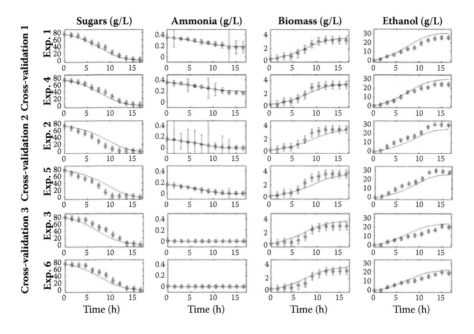

Figure 13.6 Comparison between model simulation (blue curves) and measurements of all subsets of two different batch experiments (Exp.) used for cross-validation (read circles with 95% confidence intervals). (Reproduced from Richelle, A. et al., *Bioprocess Biosyst. Eng.*, 38(9), 1819–33, 2015.)

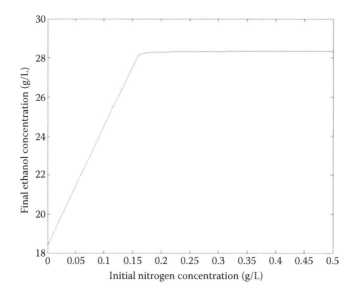

Figure 13.7   Final ethanol concentration obtained at the end of a batch culture in function of initial ammonium concentration. (Reproduced from Richelle, A. et al., *Bioprocess Biosyst. Eng.*, 38(9), 1819–33, 2015.)

Using the model identified with the whole set of experiments, the final ethanol concentration was computed for different initial ammonium concentrations, all the other experimental variables being fixed (batch duration of 17 hours and initial concentrations of fermentable sugars, yeast cells, and ethanol corresponding to the mean values of the six experiments). The results are plotted in Figure 13.7 and show that there is an optimal initial ammonium concentration (170 mg/L) that maximizes the final bioethanol production and above which no fermentation boost can be obtained anymore under these experimental conditions.

This case study illustrates how a very simple dynamic macroscopic model (two reactions, four components, and six parameters to be identified) can exhibit very satisfactory prediction abilities, and how it can be used to determine optimal experimental conditions.

13.6.4 Case Study 2: Microalgae Cultures

This second case study concerns microalgae cultures that are used, besides other applications, to produce large quantities of lipids (even above 50% dry weight, at least at the lab scale), which can then be transformed into biodiesel through a transesterification reaction. Dynamic macroscopic models of microalgae cultures are generally based on the well-known Droop model (Droop 1968, 1983). The key concept of this model is to assume that the uptake of the limiting substrate (e.g., a nitrogen source) leads to the accumulation of an internal quota, and the microalgae growth depends on (and starts only above a minimal value of) that internal quota. Note that by considering this intracellular quota, the model should not be classified as "unstructured" anymore and should be considered a "structured and unsegregated" model. Still making the parallel with the general framework described in Section 13.2, the reaction scheme (Equation 13.1) becomes

$$S \xrightarrow{\varphi_1} Q$$
$$Q \xrightarrow{\varphi_2} Q + X$$

$$(13.8)$$

where S, Q, and X represent, respectively, the limiting substrate, the internal quota, and the microalgae, and φ_1 and φ_2 are the reaction rates.

The first reaction represents the substrate uptake leading to internal quota accumulation. The second one represents the microalgae growth catalyzed by the internal quota.

The dynamic model corresponding to the mass balances of the components involved in Equation 13.8 is given by

$$\frac{dS}{dt} = -\varphi_1 - DS + DS_{in}$$

$$\frac{d(QX)}{dt} = \varphi_1 - D(QX)$$

$$(13.9)$$

$$\frac{dX}{dt} = \varphi_2 - DX$$

where S and X are the volumetric concentrations of the limiting substrate (mass of substrate per unit of volume) and of the biomass (mass of microalgae cells per unit of volume), Q is the intracellular quota concentration (mass of intracellular substrate per unit of microalgae cell mass), D is the dilution rate, and S_{in} is the substrate concentration in the feeding.

The reaction rates are given by

$$\varphi_1 = \rho(S)X = \rho_{max} \frac{S}{S + K_S} X$$

$$\varphi_2 = \mu(Q)X = \mu_{max} \left(1 - \frac{Q_{min}}{Q}\right) X$$

$$(13.10)$$

where ρ_{max} and μ_{max} are the maximum specific rates of substrate uptake and biomass growth, K_S is a saturation constant, and Q_{min} stands for the minimum quota for microalgae growth. Bernard and Gouzé (1995) showed that the internal quota is bounded as

$$Q_{min} \leq Q \leq Q_{max}$$

$$(13.11)$$

where the maximum internal quota is given by

$$Q_{max} = Q_{min} + \frac{\rho_{max}}{\mu_{max}}$$ (13.12)

Injecting Equation 13.10 into Equation 13.9 and developing the time derivative of the product $Q X$ leads to the usual Droop model:

$$\frac{dS}{dt} = -\rho(S)X - DS + DS_{in}$$

$$\frac{dQ}{dt} = \rho(S) - \mu(Q)Q$$ (13.13)

$$\frac{dX}{dt} = \mu(Q)X - DX$$

Benavides et al. (2015) proposed a dedicated methodology for the parameter identification of the model (Equations 13.10 and 13.13) (with four kinetic parameters, ρ_{max}, K_S, μ_{max}, and Q_{min}, and three initial conditions, $S(0)$, $Q(0)$, and $X(0)$), based on two batch experiments. A cross-validation test (validation with two other batch experiments, not used for parameter estimation) is reproduced in Figure 13.8.

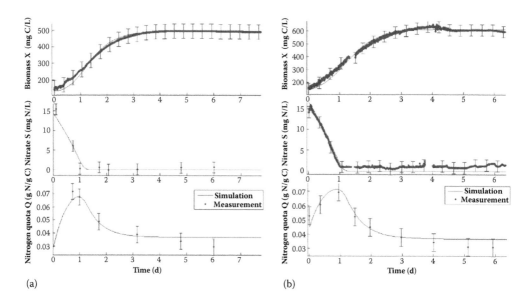

Figure 13.8 (a and b) Are two independent batch cultures used to test the predictive capability of the model by cross-validation. Comparison between model simulation (blue curves) and measurements of two different batch experiments used for cross-validation (red curves for online measurements and red dots with 95% confidence intervals for off-line measurements). (Reproduced from Benavides, M. et al., *Bioprocess Biosyst. Eng.*, 38(9), 1783–93, 2015.)

The goodness of fit is quite acceptable, apart from the internal quota Q, which decays experimentally and not in the model simulation. The authors propose a solution to that problem by considering a consumption of the internal quota with the biomass growth (leading to an additional negative term in the mass balance of Q).

The Droop model can be extended in many ways for taking into account light limitation effects in photobioreactors, temperature limitation effects, and so forth (Bernard et al. 2016). It can also be extended for reproducing the production of lipids, like triacylglycerols, which can be transformed into biodiesel (Bernard et al. 2016).

The Droop model has been used for building software sensors, for example, interval observers for estimating substrate and internal quota concentrations based on microalgae concentration measurements (Goffaux et al. 2009). An extended version was also used to test in simulation adaptive control laws aimed at maximizing the microalgae production (Mairet et al. 2015).

13.7 Conclusion

In this chapter, we have provided an overview of different ways to improve biofuel production that seemed relevant to us. Genetic engineering and bioinformatic modeling tools are presently driving substantial improvements in the design of efficient enzymatic catalysts and whole-cell factories. In parallel, the application of molecular biology tools (omic techniques) to unravel the diversity and dynamics of complex communities historically used for biofuel production (i.e., fermentative and methanogenic microbes used in anaerobic digestion) has led to a much better understanding of their functioning that can help optimize their performance. These tools have also shed light on an untapped diversity of enzymes and microorganisms in natural environments that could be exploited for biofuel production. In the forthcoming years, we expect biofuels to be increasingly produced by low-complexity consortia rather than pure strains, as the latter have repeatedly proved to be more robust and more versatile. Their internal functioning is, however, neither fully characterized nor mastered. In this context, omic techniques could provide a solid analytical plaform to understand the interactions between partners and guide their metabolism toward targeted products. Finally, the versatility of microbial metabolism is currently exploited in bioelectrochemical applications for the production of a variety of chemical commodities, including biofuels. This technology is young and is still being investigated at the laboratory scale, but it could be scaled up and integrated into biorefineries, providing that the current spectacular improvements in production rate and electron transfer rates observed over the last couple of years persist. Such improvements imply a better characterization of electron transfer between microbes and electrodes, improved electrode materials, and better management of microbial metabolisms involved in such processes. Time will tell, among these different ways of exploiting microbial metabolism, which ones will lead to commercial production of microbial biofuels. At the macroscopic level of the biofuel production process, several techniques allow the building of dynamic models that are able to predict the concentration time profiles of the main culture species (substrates, cells, metabolites, and products of interest). They can be succesfully used for model-based process optimization, control, and soft sensing. Future research will help to build efficient bridges between this macroscopic level of description and the available knowledge at the metabolic level (e.g., metabolic networks). Although many research efforts have already been devoted to some microorganisms and mammalian cell lines, many open problems remain, for instance, in the field of microalgae.

References

Agapakis, Christina M., Patrick M. Boyle, and Pamela A. Silver. 2012. Natural Strategies for the Spatial Optimization of Metabolism in Synthetic Biology. *Nature Chemical Biology* 8 (6): 527–35.

Agostini, Federico, Michele Vendruscolo, and Gian Gaetano Tartaglia. 2012. Sequence-Based Prediction of Protein Solubility. *Journal of Molecular Biology* 421 (2–3): 237–41.

Akinosho, Hannah, Kelsey Yee, Dan Close, and Arthur Ragauskas. 2014. The Emergence of *Clostridium thermocellum* as a High Utility Candidate for Consolidated Bioprocessing Applications. *Frontiers in Chemistry* 2.

Albertsen, Mads, Philip Hugenholtz, Adam Skarshewski, Kåre L. Nielsen, Gene W. Tyson, and Per H. Nielsen. 2013. Genome Sequences of Rare, Uncultured Bacteria Obtained by Differential Coverage Binning of Multiple Metagenomes. *Nature Biotechnology* 31 (6): 533–38.

Ali, Jarinah Mohd, Ngoc Ha Hoang, Mohd Azlan Hussain, and Denis Dochain. 2015. Review and Classification of Recent Observers Applied in Chemical Process Systems. *Computers & Chemical Engineering* 76: 27–41.

Alneberg, Johannes, Brynjar Smári Bjarnason, Ino de Bruijn, Melanie Schirmer, Joshua Quick, Umer Z. Ijaz, Leo Lahti, Nicholas J. Loman, Anders F. Andersson, and Christopher Quince. 2014. Binning Metagenomic Contigs by Coverage and Composition. *Nature Methods* 11 (11): 1144–46.

Alon, Uri. 2006. *An Introduction to Systems Biology: Design Principles of Biological Circuits.* Boca Raton, FL: CRC Press.

Alper, Hal, and Gregory Stephanopoulos. 2009. Engineering for Biofuels: Exploiting Innate Microbial Capacity or Importing Biosynthetic Potential? *Nature Reviews Microbiology* 7 (10): 715–23.

Amribt, Z., L. Dewasme, A. Vande Wouwer, and Ph. Bogaerts. 2014. Optimization and Robustness Analysis of Hybridoma Cell Fed-Batch Cultures Using the Overflow Metabolism Model. *Bioprocess and Biosystems Engineering* 37 (8): 1637–52.

Anbarasan, Pazhamalai, Zachary C. Baer, Sanil Sreekumar, Elad Gross, Joseph B. Binder, Harvey W. Blanch, Douglas S. Clark, and F. Dean Toste. 2012. Integration of Chemical Catalysis with Extractive Fermentation to Produce Fuels. *Nature* 491 (7423): 235–39.

Angermayr, S. Andreas, Aleix Gorchs Rovira, and Klaas J. Hellingwerf. 2015. Metabolic Engineering of Cyanobacteria for the Synthesis of Commodity Products. *Trends in Biotechnology* 33 (6): 352–61.

Ankarloo, Jonas, Susanne Wikman, and Ian A. Nicholls. 2010. *Escherichia coli* Mar and acrAB Mutants Display No Tolerance to Simple Alcohols. *International Journal of Molecular Sciences* 11 (4): 1403–12.

Bader, Johannes, Edeltraud Mast-Gerlach, Milan K. Popović, Rakesh Bajpai, and Ulf Stahl. 2010. Relevance of Microbial Coculture Fermentations in Biotechnology. *Journal of Applied Microbiology* 109 (2): 371–87.

Bálint, Miklós, Mohammad Bahram, A. Murat Eren, Karoline Faust, Jed A. Fuhrman, Björn Lindahl, Robert B. O'Hara et al. 2016. Millions of Reads, Thousands of Taxa: Microbial Community Structure and Associations Analyzed via Marker Genes. *FEMS Microbiology Reviews* 40 (5): 686–700.

Banga, Julio R., Eva Balsa-Canto, Carmen G. Moles, and Antonio A. Alonso. 2005. Dynamic Optimization of Bioprocesses: Efficient and Robust Numerical Strategies. *Journal of Biotechnology* 117 (4): 407–19.

Bashan, Luz E. de, Yoav Bashan, Manuel Moreno, Vladimir K. Lebsky, and Jose J. Bustillos. 2002. Increased Pigment and Lipid Content, Lipid Variety, and Cell and Population Size of the Microalgae *Chlorella* Spp. When Co-Immobilized in Alginate Beads with the Microalgae-Growth-Promoting Bacterium *Azospirillum brasilense*. *Canadian Journal of Microbiology* 48 (6): 514–21.

Bastin, Georges and Denis Dochain. 1990. *On-Line Estimation and Adaptive Control of Bioreactors*. 1st ed., Process Measurement and Control. Amsterdam: Elsevier.

Bayram Akcapinar, Gunseli, Alessandro Venturini, Pier Luigi Martelli, Rita Casadio, and Ugur O. Sezerman. 2015. Modulating the Thermostability of Endoglucanase I from Trichoderma Reesei Using Computational Approaches. *Protein Engineering, Design & Selection: PEDS* 28 (5): 127–35.

Beitel, Christopher W., Lutz Froenicke, Jenna M. Lang, Ian F. Korf, Richard W. Michelmore, Jonathan A. Eisen, and Aaron E. Darling. 2014. Strain- and Plasmid-Level Deconvolution of a Synthetic Metagenome by Sequencing Proximity Ligation Products. *PeerJ* 2: e415.

Beliën, Tim, Iris J. Joye, Jan A. Delcour, and Christophe M. Courtin. 2009. Computational Design-Based Molecular Engineering of the Glycosyl Hydrolase Family 11 *B. subtilis* XynA Endoxylanase Improves Its Acid Stability. *Protein Engineering, Design & Selection: PEDS* 22 (10): 587–96.

Bell, Thomas, Jonathan A. Newman, Bernard W. Silverman, Sarah L. Turner, and Andrew K. Lilley. 2005. The Contribution of Species Richness and Composition to Bacterial Services. *Nature* 436 (7054): 1157–60.

Benavides, Micaela, Anne-Lise Hantson, Jan Van Impe, and Alain Vande Wouwer. 2015. Parameter Identification of Droop Model: An Experimental Case Study. *Bioprocess and Biosystems Engineering* 38 (9): 1783–93.

Bender, Edward A., Ted J. Case, and Michael E. Gilpin. 1984. Perturbation Experiments in Community Ecology: Theory and Practice. *Ecology* 65 (1): 1–13.

Bernard, Olivier, and Georges Bastin. 2005a. Identification of Reaction Networks for Bioprocesses: Determination of a Partially Unknown Pseudo-Stoichiometric Matrix. *Bioprocess and Biosystems Engineering* 27 (5): 293–301.

Bernard, Olivier, and Georges Bastin. 2005b. On the Estimation of the Pseudo-Stoichiometric Matrix for Macroscopic Mass Balance Modelling of Biotechnological Processes. *Mathematical Biosciences* 193 (1): 51–77.

Bernard, Olivier, Francis Mairet, and Benoît Chachuat. 2016. Modelling of Microalgae Culture Systems with Applications to Control and Optimization. *Advances in Biochemical Engineering/Biotechnology* 153: 59–87.

Biello, David. 2011. The False Promise of Biofuels. *Scientific American* 305 (2): 58–65.

Binga, Erik K., Roger S. Lasken, and Josh D. Neufeld. 2008. Something from (Almost) Nothing: The Impact of Multiple Displacement Amplification on Microbial Ecology. *ISME Journal* 2 (3): 233–41.

Blainey, Paul C. 2013. The Future Is Now: Single-Cell Genomics of Bacteria and Archaea. *FEMS Microbiology Reviews* 37 (3): 407–27.

Bogaerts, Philippe, Jean-Luc Delcoux, and Raymond Hanus. 2003. Maximum Likelihood Estimation of Pseudo-Stoichiometry in Macroscopic Biological Reaction Schemes. *Chemical Engineering Science* 58 (8): 1545–63.

Bogaerts, Ph., and A. Vande Wouwer. 2003. Software Sensors for Bioprocesses. *ISA Transactions* 42 (4): 547–58.

Boto, Luis. 2010. Horizontal Gene Transfer in Evolution: Facts and Challenges. *Proceedings of the Royal Society B: Biological Sciences* 277 (1683): 819–27.

Bowen, Jennifer L., Jarrett E. K. Byrnes, David Weisman, and Cory Colaneri. 2013. Functional Gene Pyrosequencing and Network Analysis: An Approach to Examine the Response of Denitrifying Bacteria to Increased Nitrogen Supply in Salt Marsh Sediments. *Frontiers in Microbiology* 4.

Bozell, Joseph J., and Gene R. Petersen. 2010. Technology Development for the Production of Biobased Products from Biorefinery Carbohydrates—The US Department of Energy's "Top 10" Revisited. *Green Chemistry* 12 (4): 539–54.

Brennan, Liam, and Philip Owende. 2010. Biofuels from microalgae—A Review of Technologies for Production, Processing, and Extractions of Biofuels and Co-Products. *Renewable and Sustainable Energy Reviews* 14 (2): 557–77.

Bull, Alan T. 2010. The Renaissance of Continuous Culture in the Post-Genomics Age. *Journal of Industrial Microbiology & Biotechnology* 37 (10): 993–1021.

Bürgmann, Helmut, Sarina Jenni, Francisco Vazquez, and Kai M. Udert. 2011. Regime Shift and Microbial Dynamics in a Sequencing Batch Reactor for Nitrification and Anammox Treatment of Urine. *Applied and Environmental Microbiology* 77 (17): 5897–907.

Burke, Catherine, Peter Steinberg, Doug Rusch, Staffan Kjelleberg, and Torsten Thomas. 2011. Bacterial Community Assembly Based on Functional Genes Rather Than Species. *Proceedings of the National Academy of Sciences of the United States of America*.

Burton, Joshua N., Andrew Adey, Rupali P. Patwardhan, Ruolan Qiu, Jacob O. Kitzman, and Jay Shendure. 2013. Chromosome-Scale Scaffolding of De Novo Genome Assemblies Based on Chromatin Interactions. *Nature Biotechnology* 31 (12): 1119–25.

Cai, Mingwei, David Wilkins, Jiapeng Chen, Siu-Kin Ng, Hongyuan Lu, Yangyang Jia, and Patrick K. H. Lee. 2016. Metagenomic Reconstruction of Key Anaerobic Digestion Pathways in Municipal Sludge and Industrial Wastewater Biogas-Producing Systems. *Frontiers in Microbiology* 7: 778.

Campanaro, Stefano, Laura Treu, Panagiotis G. Kougias, Davide De Francisci, Giorgio Valle, and Irini Angelidaki. 2016. Metagenomic Analysis and Functional Characterization of the Biogas Microbiome Using High Throughput Shotgun Sequencing and a Novel Binning Strategy. *Biotechnology for Biofuels* 9: 26.

Caporaso, Gregory, J., Justin Kuczynski, Jesse Stombaugh, Kyle Bittinger, Frederic D. Bushman, Elizabeth K. Costello, Noah Fierer et al. 2010. QIIME Allows Analysis of High-Throughput Community Sequencing Data. *Nature Methods* 7 (5): 335–36.

Caporaso, J. Gregory, Christian L. Lauber, William A. Walters, Donna Berg-Lyons, James Huntley, Noah Fierer, Sarah M. Owens et al. 2012. Ultra-High-Throughput Microbial Community Analysis on the Illumina HiSeq and MiSeq Platforms. *ISME Journal* 6 (8): 1621–24.

Capriotti, Emidio, Piero Fariselli, and Rita Casadio. 2005. I-Mutant2.0: Predicting Stability Changes upon Mutation from the Protein Sequence or Structure. *Nucleic Acids Research* 33: W306–10.

Cases, Ildefonso, and Victor de Lorenzo. 2002. The Grammar of (Micro)biological Diversity. *Environmental Microbiology* 4 (11): 623–27.

Chen, Libei, and Georges Bastin. 1996. Structural Identifiability of the Yield Coefficients in Bioprocess Models When the Reaction Rates Are Unknown. *Mathematical Biosciences* 132 (1): 35–67.

Chen, Yin, and J. Colin Murrell. 2010. When Metagenomics Meets Stable-Isotope Probing: Progress and Perspectives. *Trends in Microbiology* 18 (4): 157–63.

Cheng, Jianlin, Arlo Randall, and Pierre Baldi. 2006. Prediction of Protein Stability Changes for Single-Site Mutations Using Support Vector Machines. *Proteins* 62 (4): 1125–32.

Cheng, Shaoan, Defeng Xing, Douglas F. Call, and Bruce E. Logan. 2009. Direct Biological Conversion of Electrical Current into Methane by Electromethanogenesis. *Environmental Science & Technology* 43 (10): 3953–58.

Chisti, Yusuf, and Jinyue Yan. 2011. Energy from Algae: Current Status and Future Trends: Algal Biofuels—A Status Report. *Applied Energy* 88 (10): 3277–79.

Choi, In-Geol, and Sung-Hou Kim. 2007. Global Extent of Horizontal Gene Transfer. *Proceedings of the National Academy of Sciences of the United States of America* 104 (11): 4489–94.

Choi, Okkyoung, Taeyeon Kim, Han Min Woo, and Youngsoon Um. 2014. Electricity-Driven Metabolic Shift through Direct Electron Uptake by Electroactive Heterotroph *Clostridium pasteurianum*. *Scientific Reports* 4: 6961.

Chung, Daehwan, Minseok Cha, Adam M. Guss, and Janet Westpheling. 2014. Direct Conversion of Plant Biomass to Ethanol by Engineered *Caldicellulosiruptor bescii*. *Proceedings of the National Academy of Sciences of the United States of America* 111 (24): 8931–36.

Claassens, Nico J., Diana Z. Sousa, Vitor A. P. Martins dos Santos, Willem M. de Vos, and John van der Oost. 2016. Harnessing the Power of Microbial Autotrophy. *Nature Reviews Microbiology* 14 (11): 692–706.

Clomburg, James M., and Ramon Gonzalez. 2010. Biofuel Production in *Escherichia coli*: The Role of Metabolic Engineering and Synthetic Biology. *Applied Microbiology and Biotechnology* 86 (2): 419–34.

Cook, K. L., J. L. Garland, A. C. Layton, H. M. Dionisi, L. H. Levine, and G. S. Sayler. 2006. Effect of Microbial Species Richness on Community Stability and Community Function in a Model Plant-Based Wastewater Processing System. *Microbial Ecology* 52 (4): 725–37.

Cragg, Simon M., Gregg T. Beckham, Neil C. Bruce, Timothy D. H. Bugg, Daniel L. Distel, Paul Dupree, Amaia Green Etxabe et al. 2015. Lignocellulose Degradation Mechanisms across the Tree of Life. *Current Opinion in Chemical Biology*, 29: 108–19.

Craig, Jeffrey W., Fang-Yuan Chang, and Sean F. Brady. 2009. Natural Products from Environmental DNA Hosted in *Ralstonia metallidurans*. *ACS Chemical Biology* 4 (1): 23–28.

Dalby, Paul A. 2011. Strategy and Success for the Directed Evolution of Enzymes. *Current Opinion in Structural Biology* 21 (4): 473–80.

Damon, Coralie, Frédéric Lehembre, Christine Oger-Desfeux, Patricia Luis, Jacques Ranger, Laurence Fraissinet-Tachet, and Roland Marmeisse. 2012. Metatranscriptomics Reveals the Diversity of Genes Expressed by Eukaryotes in Forest Soils. *PLoS One* 7 (1): e28967.

Dehouck, Yves, Aline Grosfils, Benjamin Folch, Dimitri Gilis, Philippe Bogaerts, and Marianne Rooman. 2009. Fast and Accurate Predictions of Protein Stability Changes upon Mutations Using Statistical Potentials and Neural Networks: PoPMuSiC-2.0. *Bioinformatics (Oxford, England)* 25 (19): 2537–43.

Dehouck, Yves, Jean Marc Kwasigroch, Dimitri Gilis, and Marianne Rooman. 2011. PoPMuSiC 2.1: A Web Server for the Estimation of Protein Stability Changes upon Mutation and Sequence Optimality. *BMC Bioinformatics* 12: 151.

Dejonghe, Winnie, Ellen Berteloot, Johan Goris, Nico Boon, Katrien Crul, Siska Maertens, Monica Hofte, Paul De Vos, Willy Verstraete, and Eva M. Top. 2003. Synergistic Degradation of Linuron by a Bacterial Consortium and Isolation of a Single Linuron-Degrading Variovorax Strain. *Applied and Environmental Microbiology* 69 (3): 1532–41.

Dekker, Job, Karsten Rippe, Martijn Dekker, and Nancy Kleckner. 2002. Capturing Chromosome Conformation. *Science* 295 (5558): 1306–11.

Deutzmann, Jörg S., Merve Sahin, and Alfred M. Spormann. 2015. Extracellular Enzymes Facilitate Electron Uptake in Biocorrosion and Bioelectrosynthesis. *mBio* 6 (2): e00496-15.

Dewasme, Laurent, Anne Richelle, Philippe Dehottay, Patrice Georges, Marcel Remy, Philippe Bogaerts, and Alain Vande Wouwer. 2010. Linear Robust Control of *S. cerevisiae* Fed-Batch Cultures at Different Scales. *Biochemical Engineering Journal* 53 (1): 26–37.

Dien, B. S., M. A. Cotta, and T. W. Jeffries. 2003. Bacteria Engineered for Fuel Ethanol Production: Current Status. *Applied Microbiology and Biotechnology* 63 (3): 258–66.

Dinh, Hang T., Jan Kuever, Marc Mußmann, Achim W. Hassel, Martin Stratmann, and Friedrich Widdel. 2004. Iron Corrosion by Novel Anaerobic Microorganisms. *Nature* 427 (6977): 829–32.

Droop, Michaël R. 1968. Vitamin B12 and Marine Ecology. IV. The Kinetics of Uptake, Growth and Inhibition in *Monochrysis lutheri*. *Journal of the Marine Biological Association of the United Kingdom* 48 (3): 689–733.

Droop, Michaël R. 1983. 25 Years of Algal Growth Kinetics: A Personal View. *Botanica Marina* 26 (3): 99–112.

Duke, Clifford S., Richard V. Pouyat, G. Philip Robertson, and William J. Parton. 2013. Ecological Dimensions of Biofuels. *Issues in Ecology* 17: 1–16.

Dunlop, Mary J. 2011. Engineering Microbes for Tolerance to Next-Generation Biofuels. *Biotechnology for Biofuels* 4: 32.

Dunlop, Mary J., Zain Y. Dossani, Heather L. Szmidt, Hou Cheng Chu, Taek Soon Lee, Jay D. Keasling, Masood Z. Hadi, and Aindrila Mukhopadhyay. 2011. Engineering Microbial Biofuel Tolerance and Export Using Efflux Pumps. *Molecular Systems Biology* 7: 487.

Dunlop, Mary J., Jay D. Keasling, and Aindrila Mukhopadhyay. 2010. A Model for Improving Microbial Biofuel Production Using a Synthetic Feedback Loop. *Systems and Synthetic Biology* 4 (2): 95–104.

Elsharnouby, Omneya, Hisham Hafez, George Nakhla, and M. Hesham El Naggar. 2013. A Critical Literature Review on Biohydrogen Production by Pure Cultures. *International Journal of Hydrogen Energy* 38 (12): 4945–66.

Ema, Tadashi, Yasuko Nakano, Daiki Yoshida, Shusuke Kamata, and Takashi Sakai. 2012. Redesign of Enzyme for Improving Catalytic Activity and Enantioselectivity toward Poor Substrates: Manipulation of the Transition State. *Organic & Biomolecular Chemistry* 10 (31): 6299–6308.

Eş, Ismail, José Daniel Gonçalves Vieira, and André Corrêa Amaral. 2015. Principles, Techniques, and Applications of Biocatalyst Immobilization for Industrial Application. *Applied Microbiology and Biotechnology* 99 (5): 2065–82.

Esvelt, Kevin M., and Harris H. Wang. 2013. Genome-Scale Engineering for Systems and Synthetic Biology. *Molecular Systems Biology* 9: 641.

Ezeji, Thaddeus Chukwuemeka, Nasib Qureshi, and Hans Peter Blaschek. 2007. Bioproduction of Butanol from Biomass: From Genes to Bioreactors. *Current Opinion in Biotechnology* 18 (3): 220–27.

Fargione, Joseph, Jason Hill, David Tilman, Stephen Polasky, and Peter Hawthorne. 2008. Land Clearing and the Biofuel Carbon Debt. *Science* 319 (5867): 1235–38.

Faust, Karoline, Gipsi Lima-Mendez, Jean-Sébastien Lerat, Jarupon F. Sathirapongsasuti, Rob Knight, Curtis Huttenhower, Tom Lenaerts, and Jeroen Raes. 2015. Cross-Biome Comparison of Microbial Association Networks. *Frontiers in Microbiology* 6: 1200.

Faust, Karoline, and Jeroen Raes. 2012. Microbial Interactions: From Networks to Models. *Nature Reviews Microbiology* 10 (8): 538–50.

Fernández, Ana, Suiying Huang, Sherry Seston, Jian Xing, Robert Hickey, Craig Criddle, and James Tiedje. 1999. How Stable Is Stable? Function versus Community Composition. *Applied and Environmental Microbiology* 65 (8): 3697–3704.

Flot, Jean-François, Hervé Marie-Nelly, and Romain Koszul. 2015. Contact Genomics: Scaffolding and Phasing (Meta)genomes Using Chromosome 3D Physical Signatures. *FEBS Letters* 589 (20 Pt A): 2966–74.

Franssen, Maurice C. R., Peter Steunenberg, Elinor L. Scott, Han Zuilhof, and Johan P. M. Sanders. 2013. Immobilised Enzymes in Biorenewables Production. *Chemical Society Reviews* 42 (15): 6491–6533.

Franzosa, Eric A., Tiffany Hsu, Alexandra Sirota-Madi, Afrah Shafquat, Galeb Abu-Ali, Xochitl C. Morgan, and Curtis Huttenhower. 2015. Sequencing and Beyond: Integrating Molecular "Omics" for Microbial Community Profiling. *Nature Reviews Microbiology* 13 (6): 360–72.

Freilich, Shiri, Raphy Zarecki, Omer Eilam, Ella Shtifman Segal, Christopher S. Henry, Martin Kupiec, Uri Gophna, Roded Sharan, and Eytan Ruppin. 2011. Competitive and Cooperative Metabolic Interactions in Bacterial Communities. *Nature Communications* 2: 589.

Frost, Laura S., Raphael Leplae, Anne O. Summers, and Ariane Toussaint. 2005. Mobile Genetic Elements: The Agents of Open Source Evolution. *Nature Reviews Microbiology* 3 (9): 722–32.

Fu, Y., L. Chen, and W. Zhang. 2016. Regulatory Mechanisms Related to Biofuel Tolerance in Producing Microbes. *Journal of Applied Microbiology* 121 (2): 320–32.

George, I., B. Stenuit, and S. N. Agathos. 2010. Application of Metagenomics to Bioremediation. In *Metagenomics: Theory, Methods and Applications*, ed. D. Marco, 119–40. Norfolk, UK: Caister Academic Press.

Ghimire, Anish, Luigi Frunzo, Francesco Pirozzi, Eric Trably, Renaud Escudie, Piet N. L. Lens, and Giovanni Esposito. 2015. A Review on Dark Fermentative Biohydrogen Production from Organic Biomass: Process Parameters and Use of By-Products. *Applied Energy* 144: 73–95.

Goers, Lisa, Paul Freemont, and Karen M. Polizzi. 2014. Co-Culture Systems and Technologies: Taking Synthetic Biology to the Next Level. *Journal of the Royal Society Interface* 11 (96): 20140065.

Goffaux, Guillaume, Alain Vande Wouwer, and Olivier Bernard. 2009. Continuous-Discrete Interval Observers for Monitoring Microalgae Cultures. *Biotechnology Progress* 25 (3): 667–75.

Goldemberg, José. 2007. Ethanol for a Sustainable Energy Future. *Science* 315 (5813): 808–10.

Gomez, Jose A., Kai Höffner, and Paul I. Barton. 2016. From Sugars to Biodiesel Using Microalgae and Yeast. *Green Chemistry* 18 (2): 461–75.

Goodarzi, Hani, Bryson D. Bennett, Sasan Amini, Marshall L. Reaves, Alison K. Hottes, Joshua D. Rabinowitz, and Saeed Tavazoie. 2010. Regulatory and Metabolic Rewiring during Laboratory Evolution of Ethanol Tolerance in *E. coli*. *Molecular Systems Biology* 6: 378.

Gowen, Christopher M., and Stephen S. Fong. 2011. Applications of Systems Biology towards Microbial Fuel Production. *Trends in Microbiology* 19 (10): 516–24.

Graham, David W., Charles W. Knapp, Erik S. Van Vleck, Katie Bloor, Teresa B. Lane, and Christopher E. Graham. 2007. Experimental Demonstration of Chaotic Instability in Biological Nitrification. *ISME Journal* 1 (5): 385–93.

Graham, Emily B., Joseph E. Knelman, Andreas Schindlbacher, Steven Siciliano, Marc Breulmann, Anthony Yannarell, J. M. Beman et al. 2016. Microbes as Engines of Ecosystem Function: When Does Community Structure Enhance Predictions of Ecosystem Processes? *Frontiers in Microbiology* 7: 214.

Grogan, David, Moises Velasquez-Manoff, Rob Knight, Ruth E. Ley, Justin L. Sonnenburg, Charles Schmidt, and Stephanie L. Schnorr. 2015. Nature Supplement: Innovations in the Microbiome. *Nature* 518 (7540): S1–52.

Grosfils, Aline, Alain Vande Wouwer, and Philippe Bogaerts. 2007. On a General Model Structure for Macroscopic Biological Reaction Rates. *Journal of Biotechnology* 130 (3): 253–64.

Guerois, Raphael, Jens Erik Nielsen, and Luis Serrano. 2002. Predicting Changes in the Stability of Proteins and Protein Complexes: A Study of More Than 1000 Mutations. *Journal of Molecular Biology* 320 (2): 369–87.

Gusakov, Alexander V. 2011. Alternatives to *Trichoderma reesei* in Biofuel Production. *Trends in Biotechnology* 29 (9): 419–25.

Gusakov, Alexander V. 2013. Cellulases and Hemicellulases in the 21st Century Race for Cellulosic Ethanol. *Biofuels* 4 (6): 567–69.

Haag, Jens E., Alain Vande Wouwer, and Marcel Remy. 2005. A General Model of Reaction Kinetics in Biological Systems. *Bioprocess and Biosystems Engineering* 27 (5): 303–9.

Hannon, Michael, Javier Gimpel, Miller Tran, Beth Rasala, and Stephen Mayfield. 2010. Biofuels from Algae: Challenges and Potential. *Biofuels* 1 (5): 763–84.

Harris, Paul V., Feng Xu, Nathaniel E. Kreel, Connie Kang, and Shiro Fukuyama. 2014. New Enzyme Insights Drive Advances in Commercial Ethanol Production. *Current Opinion in Chemical Biology* 19: 162–70.

Harrison, Mary E., and Mary J. Dunlop. 2012. Synthetic Feedback Loop Model for Increasing Microbial Biofuel Production Using a Biosensor. *Frontiers in Microbiology* 3: 360.

Hasunuma, Tomohisa, and Akihiko Kondo. 2012. Development of Yeast Cell Factories for Consolidated Bioprocessing of Lignocellulose to Bioethanol through Cell Surface Engineering. *Biotechnology Advances* 30 (6): 1207–18.

Hasunuma, Tomohisa, Fumiyoshi Okazaki, Naoko Okai, Kiyotaka Y. Hara, Jun Ishii, and Akihiko Kondo. 2013. A Review of Enzymes and Microbes for Lignocellulosic Biorefinery and the Possibility of Their Application to Consolidated Bioprocessing Technology. *Bioresource Technology* 135: 513–22.

Hawkins, Michelle, Sunir Malla, Martin J. Blythe, Conrad A. Nieduszynski, and Thorsten Allers. 2013. Accelerated Growth in the Absence of DNA Replication Origins. *Nature* 503 (7477): 544–47.

Hess, Matthias, Alexander Sczyrba, Rob Egan, Tae-Wan Kim, Harshal Chokhawala, Gary Schroth, Shujun Luo et al. 2011. Metagenomic Discovery of Biomass-Degrading Genes and Genomes from Cow Rumen. *Science* 331 (6016): 463–67.

Heyer, Robert, Fabian Kohrs, Udo Reichl, and Dirk Benndorf. 2015. Metaproteomics of Complex Microbial Communities in Biogas Plants. *Microbial Biotechnology* 8 (5): 749–63.

Hibbing, Michael E., Clay Fuqua, Matthew R. Parsek, and S. Brook Peterson. 2010. Bacterial Competition: Surviving and Thriving in the Microbial Jungle. *Nature Reviews Microbiology* 8 (1): 15–25.

Higgins, Brendan T., and Jean S. VanderGheynst. 2014. Effects of *Escherichia coli* on Mixotrophic Growth of *Chlorella minutissima* and Production of Biofuel Precursors. *PLoS One* 9 (5): e96807.

Hom, Erik F. Y., and Andrew W. Murray. 2014. Niche Engineering Demonstrates a Latent Capacity for Fungal-Algal Mutualism. *Science (New York, N.Y.)* 345 (6192): 94–98.

Huang, Jun, Dong Chen, Yutuo Wei, Qingyan Wang, Zhenchong Li, Ying Chen, and Ribo Huang. 2014. Direct Ethanol Production from Lignocellulosic Sugars and Sugarcane Bagasse by a Recombinant *Trichoderma reesei* Strain HJ48. *ScientificWorldJournal* 2014: 798683.

Huang, Mingtao, Yunpeng Bai, Staffan L. Sjostrom, Björn M. Hallström, Zihe Liu, Dina Petranovic, Mathias Uhlén, Haakan N. Joensson, Helene Andersson-Svahn, and Jens Nielsen. 2015. Microfluidic Screening and Whole-Genome Sequencing Identifies Mutations Associated with Improved Protein Secretion by Yeast. *Proceedings of the National Academy of Sciences of the United States of America* 112 (34): E4689–96.

Hugenholtz, Philip, Brett M. Goebel, and Norman R. Pace. 1998. Impact of Culture-Independent Studies on the Emerging Phylogenetic View of Bacterial Diversity. *Journal of Bacteriology* 180 (18): 4765–74.

Huisman, Jef, and Franz J. Weissing. 1999. Biodiversity of Plankton by Species Oscillations and Chaos. *Nature* 402: 407–10.

Hulhoven, Xavier, Alain Vande Wouwer, and Philippe Bogaerts. 2005. On a Systematic Procedure for the Predetermination of Macroscopic Reaction Schemes. *Bioprocess and Biosystems Engineering* 27 (5): 283–91.

Ieropoulos, Ioannis, Iwona Gajda, Jiseon You, Chris Melhuish, and John Greenman. 2016. Microbial Fuel Cell Development for Practical Applications Robotics, Urinals and Novel Transistor Analogy. *ECS Meeting Abstracts* MA2016-01 (36): 1798.

Ieropoulos, Ioannis A., John Greenman, Chris Melhuish, and Ian Horsfield. 2012. Microbial Fuel Cells for Robotics: Energy Autonomy through Artificial Symbiosis. *ChemSusChem* 5 (6): 1020–26.

Isken, Sonja, and Jan A. de Bont. 1998. Bacteria Tolerant to Organic Solvents. *Extremophiles: Life under Extreme Conditions* 2 (3): 229–38.

Iwai, Shoko, Benli Chai, Woo Jun Sul, James R. Cole, Syed A. Hashsham, and James M. Tiedje. 2010. Gene-Targeted-Metagenomics Reveals Extensive Diversity of Aromatic Dioxygenase Genes in the Environment. *ISME Journal* 4 (2): 279–85.

Jacquiod, Samuel, Sandrine Demanèche, Laure Franqueville, Luka Ausec, Zhuofei Xu, Tom O. Delmont, Vincent Dunon et al. 2014. Characterization of New Bacterial Catabolic Genes and Mobile Genetic Elements by High Throughput Genetic Screening of a Soil Metagenomic Library. *Journal of Biotechnology* 190: 18–29.

Jemli, Sonia, Dorra Ayadi-Zouari, Hajer Ben Hlima, and Samir Bejar. 2016. Biocatalysts: Application and Engineering for Industrial Purposes. *Critical Reviews in Biotechnology* 36 (2): 246–58.

Jensen, Heather M., Aaron E. Albers, Konstantin R. Malley, Yuri Y. Londer, Bruce E. Cohen, Brett A. Helms, Peter Weigele, Jay T. Groves, and Caroline M. Ajo-Franklin. 2010. Engineering of a Synthetic Electron Conduit in Living Cells. *Proceedings of the National Academy of Sciences of the United States of America* 107 (45): 19213–18.

Johnson, David R., Felix Goldschmidt, Elin E. Lilja, and Martin Ackermann. 2012. Metabolic Specialization and the Assembly of Microbial Communities. *ISME Journal* 6 (11): 1985–91.

Jourdin, Ludovic, Stefano Freguia, Bogdan C. Donose, and Jurg Keller. 2015. Autotrophic Hydrogen-Producing Biofilm Growth Sustained by a Cathode as the Sole Electron and Energy Source. *Bioelectrochemistry* 102: 56–63.

Jourdin, Ludovic, Yang Lu, Victoria Flexer, Jurg Keller, and Stefano Freguia. 2016. Biologically Induced Hydrogen Production Drives High Rate/High Efficiency Microbial Electrosynthesis of Acetate from Carbon Dioxide. *ChemElectroChem* 3 (4): 581–91.

Kabsch, Wolfgang and Christian Sander. 1983. Dictionary of Protein Secondary Structure: Pattern Recognition of Hydrogen-Bonded and Geometrical Features. *Biopolymers* 22 (12): 2577–637.

Kaplan, Noam, and Job Dekker. 2013. High-Throughput Genome Scaffolding from In Vivo DNA Interaction Frequency. *Nature Biotechnology* 31 (12): 1143–47.

Kato, Souichiro, Shin Haruta, Zong Jun Cui, Masaharu Ishii, and Yasuo Igarashi. 2004. Effective Cellulose Degradation by a Mixed-Culture System Composed of a Cellulolytic *Clostridium* and Aerobic Non-Cellulolytic Bacteria. *FEMS Microbiology Ecology* 51 (1): 133–42.

Kato, Souichiro, Shin Haruta, Zong Jun Cui, Masaharu Ishii, and Yasuo Igarashi. 2005. Stable Coexistence of Five Bacterial Strains as a Cellulose-Degrading Community. *Applied and Environmental Microbiology* 71 (11): 7099–106.

Kato, Souichiro, Isao Yumoto, and Yoichi Kamagata. 2015. Isolation of Acetogenic Bacteria That Induce Biocorrosion by Utilizing Metallic Iron as the Sole Electron Donor. *Applied and Environmental Microbiology* 81 (1): 67–73.

Kaushik, Mahima, Prashant Sinha, Pragya Jaiswal, Swati Mahendru, Kapil Roy, and Shrikant Kukreti. 2016. Protein Engineering and De Novo Designing of a Biocatalyst. *Journal of Molecular Recognition: JMR* 29 (10): 499–503.

Kim, Hyo Jin, Timothy Lee Turner, and Yong-Su Jin. 2013. Combinatorial Genetic Perturbation to Refine Metabolic Circuits for Producing Biofuels and Biochemicals. *Biotechnology Advances* 31 (6): 976–85.

Kim, Hyun Jung, James Q. Boedicker, Jang Wook Choi, and Rustem F. Ismagilov. 2008. Defined Spatial Structure Stabilizes a Synthetic Multispecies Bacterial Community. *Proceedings of the National Academy of Sciences of the United States of America* 105 (47): 18188–93.

Kohn, Richard A., and Seon-Woo Kim. 2015. Using the Second Law of Thermodynamics for Enrichment and Isolation of Microorganisms to Produce Fuel Alcohols or Hydrocarbons. *Journal of Theoretical Biology* 382: 356–62.

Konopka, Allan, Stephen Lindemann, and Jim Fredrickson. 2014. Dynamics in Microbial Communities: Unraveling Mechanisms to Identify Principles. *ISME Journal* 9 (7): 1488–95.

Konopka, Allan, and Michael J. Wilkins. 2012. Application of Meta-Transcriptomics and -Proteomics to Analysis of In Situ Physiological State. *Frontiers in Microbiology* 3: 184.

Koren, Sergey, and Adam M. Phillippy. 2015. One Chromosome, One Contig: Complete Microbial Genomes from Long-Read Sequencing and Assembly. *Current Opinion in Microbiology* 23: 110–20.

Kouzuma, Atsushi, Souichiro Kato, and Kazuya Watanabe. 2015. Microbial Interspecies Interactions: Recent Findings in Syntrophic Consortia. *Frontiers in Microbiology* 6: 477.

Kraft, Beate, Halina E. Tegetmeyer, Dimitri Meier, Jeanine S. Geelhoed, and Marc Strous. 2014. Rapid Succession of Uncultured Marine Bacterial and Archaeal Populations in a Denitrifying Continuous Culture. *Environmental Microbiology* 16 (10): 3275–86.

Kricka, William, James Fitzpatrick, and Ursula Bond. 2015. Challenges for the Production of Bioethanol from Biomass Using Recombinant Yeasts. In *Advances in Applied Microbiology*, ed. Sima Sariaslani and Geoffrey Michael Gadd, 89–125. Vol. 92. London: Academic Press.

Kuske, Cheryl R., Cedar N. Hesse, Jean F. Challacombe, Daniel Cullen, Joshua R. Herr, Rebecca C. Mueller, Adrian Tsang, and Rytas Vilgalys. 2015. Prospects and Challenges for Fungal Metatranscriptomics of Complex Communities. *Fungal Ecology* 14: 133–37.

Laimer, Josef, Heidi Hofer, Marko Fritz, Stefan Wegenkittl, and Peter Lackner. 2015. MAESTRO—Multi Agent Stability Prediction upon Point Mutations. *BMC Bioinformatics* 16: 116.

Lardon, Laurent, Arnaud Hélias, Bruno Sialve, Jean-Philippe Steyer, and Olivier Bernard. 2009. Life-Cycle Assessment of Biodiesel Production from Microalgae. *Environmental Science & Technology* 43 (17): 6475–81.

Larsen, Dorte M., Christian Nyffenegger, Maria M. Swiniarska, Anders Thygesen, Mikael L. Strube, Anne S. Meyer, and Jørn D. Mikkelsen. 2015. Thermostability Enhancement of an Endo-1,4-β-Galactanase from *Talaromyces stipitatus* by Site-Directed Mutagenesis. *Applied Microbiology and Biotechnology* 99 (10): 4245–53.

Lasken, Roger S., and Jeffrey S. McLean. 2014. Recent Advances in Genomic DNA Sequencing of Microbial Species from Single Cells. *Nature Reviews Genetics* 15 (9): 577–84.

Lelie, Daniel van der, Safiyh Taghavi, Sean M. McCorkle, Luen-Luen Li, Stephanie A. Malfatti, Denise Monteleone, Bryon S. Donohoe et al. 2012. The Metagenome of an Anaerobic Microbial Community Decomposing Poplar Wood Chips. *PLoS One* 7 (5): e36740.

Liao, James C., Luo Mi, Sammy Pontrelli, and Shanshan Luo. 2016. Fuelling the Future: Microbial Engineering for the Production of Sustainable Biofuels. *Nature Reviews Microbiology* 14 (5): 288–304.

Lieberman-Aiden, Erez, Nynke L. van Berkum, Louise Williams, Maxim Imakaev, Tobias Ragoczy, Agnes Telling, Ido Amit et al. 2009. Comprehensive Mapping of Long-Range Interactions Reveals Folding Principles of the Human Genome. *Science* 326 (5950): 289–93.

Lindemann, Stephen R., Hans C. Bernstein, Hyun-Seob Song, Jim K. Fredrickson, Matthew W. Fields, Wenying Shou, David R. Johnson, and Alexander S. Beliaev. 2016. Engineering Microbial Consortia for Controllable Outputs. *ISME Journal* 10 (9): 2077–84.

Little, Ainslie E. F., Courtney J. Robinson, S. Brook Peterson, Kenneth F. Raffa, and Jo Handelsman. 2008. Rules of Engagement: Interspecies Interactions That Regulate Microbial Communities. *Annual Review of Microbiology* 62: 375–401.

Liu, Bingchuan, Isaiah Williams, Yan Li, Lei Wang, Amvrossios Bagtzoglou, Jeffrey McCutcheon, and Baikun Li. 2016. Towards High Power Output of Scaled-up Benthic Microbial Fuel Cells (BMFCs) Using Multiple Electron Collectors. *Biosensors & Bioelectronics* 79: 435–41.

Liu, Tiangang, and Chaitan Khosla. 2010. Genetic Engineering of *Escherichia coli* for Biofuel Production. *Annual Review of Genetics* 44: 53–69.

Logan, Bruce E. 2009. Exoelectrogenic Bacteria That Power Microbial Fuel Cells. *Nature Reviews Microbiology* 7 (5): 375–81.

Logares, Ramiro, Thomas H. A. Haverkamp, Surendra Kumar, Anders Lanzén, Alexander J. Nederbragt, Christopher Quince, and Håvard Kauserud. 2012. Environmental Microbiology through the Lens of High-Throughput DNA Sequencing: Synopsis of Current Platforms and Bioinformatics Approaches. *Journal of Microbiological Methods* 91 (1): 106–13.

Lohner, Svenja T., Jörg S. Deutzmann, Bruce E. Logan, John Leigh, and Alfred M. Spormann. 2014. Hydrogenase-Independent Uptake and Metabolism of Electrons by the Archaeon *Methanococcus maripaludis*. *ISME Journal* 8 (8): 1673–81.

Louis, Petra, Georgina L. Hold, and Harry J. Flint. 2014. The Gut Microbiota, Bacterial Metabolites and Colorectal Cancer. *Nature Reviews Microbiology* 12 (10): 661–72.

Lovley, Derek R. 2012. Electromicrobiology. *Annual Review of Microbiology* 66 (1): 391–409.

Lovley, Derek R., and Kelly P. Nevin. 2013. Electrobiocommodities: Powering Microbial Production of Fuels and Commodity Chemicals from Carbon Dioxide with Electricity. *Current Opinion in Biotechnology* 24 (3): 385–90.

Lü, Fan, Ariane Bize, Alain Guillot, Véronique Monnet, Céline Madigou, Olivier Chapleur, Laurent Mazéas, Pinjing He, and Théodore Bouchez. 2014. Metaproteomics of Cellulose Methanisation under Thermophilic Conditions Reveals a Surprisingly High Proteolytic Activity. *ISME Journal* 8 (1): 88–102.

Luo, Gang, Davide De Francisci, Panagiotis G. Kougias, Treu Laura, Xinyu Zhu, and Irini Angelidaki. 2015. New Steady-State Microbial Community Compositions and Process Performances in Biogas Reactors Induced by Temperature Disturbances. *Biotechnology for Biofuels* 8: 3.

Lykidis, Athanasios, Chia-Lung Chen, Susannah G. Tringe, Alice C. McHardy, Alex Copeland, Nikos C. Kyrpides, Philip Hugenholtz et al. 2011. Multiple Syntrophic Interactions in a Terephthalate-Degrading Methanogenic Consortium. *ISME Journal* 5 (1): 122–30.

Lynch, Michael D. J., and Josh D. Neufeld. 2015. Ecology and Exploration of the Rare Biosphere. *Nature Reviews Microbiology* 13 (4): 217–29.

Lynd, Lee R., Willem H. van Zyl, John E. McBride, and Mark Laser. 2005. Consolidated Bioprocessing of Cellulosic Biomass: An Update. *Current Opinion in Biotechnology* 16 (5): 577–83.

Lynd, Lee R., Paul J. Weimer, Willem H. van Zyl, and Isak S. Pretorius. 2002. Microbial Cellulose Utilization: Fundamentals and Biotechnology. *Microbiology and Molecular Biology Reviews: MMBR* 66 (3): 506–577.

Mairet, Francis, Rafael Muñoz-Tamayo, and Olivier Bernard. 2015. Adaptive Control of Light Attenuation for Optimizing Microalgae Production. *Journal of Process Control* 30: 117–24.

Mann, Miriam S., Zdravko Dragovic, Georg Schirrmacher, and Tina Lütke-Eversloh. 2012. Over-Expression of Stress Protein-Encoding Genes Helps *Clostridium acetobutylicum* to Rapidly Adapt to Butanol Stress. *Biotechnology Letters* 34 (9): 1643–49.

Marbouty, Martial, Axel Cournac, Jean-François Flot, Hervé Marie-Nelly, Julien Mozziconacci, and Romain Koszul. 2014. Metagenomic Chromosome Conformation Capture (meta3C) Unveils the Diversity of Chromosome Organization in Microorganisms. *eLife* 3: e03318.

Marbouty, Martial, and Romain Koszul. 2015. Metagenome Analysis Exploiting High-Throughput Chromosome Conformation Capture (3C) Data. *Trends in Genetics: TIG* 31 (12): 673–82.

Marie-Nelly, Hervé, Martial Marbouty, Axel Cournac, Jean-François Flot, Gianni Liti, Dante Poggi Parodi, Sylvie Syan et al. 2014. High-Quality Genome (Re)assembly Using Chromosomal Contact Data. *Nature Communications* 5: 5695.

Marshall, Christopher W., Daniel E. Ross, Erin B. Fichot, R. Sean Norman, and Harold D. May. 2012. Electrosynthesis of Commodity Chemicals by an Autotrophic Microbial Community. *Applied and Environmental Microbiology* 78 (23): 8412–20.

Martinez, Diego, Randy M. Berka, Bernard Henrissat, Markku Saloheimo, Mikko Arvas, Scott E. Baker, Jarod Chapman et al. 2008. Genome Sequencing and Analysis of the Biomass-Degrading Fungus *Trichoderma reesei* (Syn. *Hypocrea jecorina*). *Nature Biotechnology* 26 (5): 553–60.

Masso, Majid, and Iosif I. Vaisman. 2008. Accurate Prediction of Stability Changes in Protein Mutants by Combining Machine Learning with Structure Based Computational Mutagenesis. *Bioinformatics (Oxford, England)* 24 (18): 2002–9.

Masso, Majid, and Iosif I. Vaisman. 2014. AUTO-MUTE 2.0: A Portable Framework with Enhanced Capabilities for Predicting Protein Functional Consequences upon Mutation. *Advances in Bioinformatics* 2014: 278385.

Maus, Irena, Daniela E. Koeck, Katharina G. Cibis, Sarah Hahnke, Yong S. Kim, Thomas Langer, Jana Kreubel et al. 2016. Unraveling the Microbiome of a Thermophilic Biogas Plant by Metagenome and Metatranscriptome Analysis Complemented by Characterization of Bacterial and Archaeal Isolates. *Biotechnology for Biofuels* 9: 171.

Mbaneme, Veronica, and Mari Chinn. 2015. Consolidated Bioprocessing for Biofuel Production: Recent Advances. *Energy and Emission Control Technologies* 3: 23–44.

McKone, Thomas, William Nazaroff, Peter Berck, Maximilian Auffhammer, Timothy Lipman, Margaret Torn, Eric Masanet et al. 2011. Grand Challenges for Life-Cycle Assessment of Biofuels. *Environmental Science & Technology* 45 (5): 1751–56.

Medie, Felix Mba, Gideon J. Davies, Michel Drancourt, and Bernard Henrissat. 2012. Genome Analyses Highlight the Different Biological Roles of Cellulases. *Nature Reviews Microbiology* 10 (3): 227–34.

Men, Yujie, Helene Feil, Nathan C. VerBerkmoes, Manesh B. Shah, David R. Johnson, Patrick K. H. Lee, Kimberlee A. West, Stephen H. Zinder, Gary L. Andersen, and Lisa Alvarez-Cohen. 2011. Sustainable Syntrophic Growth of *Dehalococcoides ethenogenes* Strain 195 with *Desulfovibrio vulgaris* Hildenborough and *Methanobacterium congolense*: Global Transcriptomic and Proteomic Analyses. *ISME Journal* 6 (2): 410–21.

Mhuantong, Wuttichai, Varodom Charoensawan, Pattanop Kanokratana, Sithichoke Tangphatsornruang, and Verawat Champreda. 2015. Comparative Analysis of Sugarcane Bagasse Metagenome Reveals Unique and Conserved Biomass-Degrading Enzymes among Lignocellulolytic Microbial Communities. *Biotechnology for Biofuels* 8: 16.

Minty, Jeremy J., Ann A. Lesnefsky, Fengming Lin, Yu Chen, Ted A. Zaroff, Artur B. Veloso, Bin Xie et al. 2011. Evolution Combined with Genomic Study Elucidates Genetic Bases of Isobutanol Tolerance in *Escherichia coli*. *Microbial Cell Factories* 10: 18.

Minty, Jeremy J., Marc E. Singer, Scott A. Scholz, Chang-Hoon Bae, Jung-Ho Ahn, Clifton E. Foster, James C. Liao, and Xiaoxia Nina Lin. 2013. Design and Characterization of Synthetic Fungal-Bacterial Consortia for Direct Production of Isobutanol from Cellulosic Biomass. *Proceedings of the National Academy of Sciences of the United States of America* 110 (36): 14592–97.

Morris, Brandon E. L., Ruth Henneberger, Harald Huber, and Christine Moissl-Eichinger. 2013. Microbial Syntrophy: Interaction for the Common Good. *FEMS Microbiology Reviews* 37 (3): 384–406.

Muller, Emilie E. L., Abdul R. Sheik, and Paul Wilmes. 2014. Lipid-Based Biofuel Production from Wastewater. *Current Opinion in Biotechnology* 30: 9–16.

Muradov, Nazim, Mohamed Taha, Ana F. Miranda, Digby Wrede, Krishna Kadali, Amit Gujar, Trevor Stevenson, Andrew S. Ball, and Aidyn Mouradov. 2015. Fungal-Assisted Algal Flocculation: Application in Wastewater Treatment and Biofuel Production. *Biotechnology for Biofuels* 8: 24.

Nakanishi, Yumiko, Shinji Fukuda, Eisuke Chikayama, Yayoi Kimura, Hiroshi Ohno, and Jun Kikuchi. 2011. Dynamic Omics Approach Identifies Nutrition-Mediated Microbial Interactions. *Journal of Proteome Research* 10 (2): 824–36.

Nandagopal, Nagarajan, and Michael B. Elowitz. 2011. Synthetic Biology: Integrated Gene Circuits. *Science (New York, N.Y.)* 333 (6047): 1244–48.

Narayanasamy, Shaman, Emilie E. L. Muller, Abdul R. Sheik, and Paul Wilmes. 2015. Integrated Omics for the Identification of Key Functionalities in Biological Wastewater Treatment Microbial Communities. *Microbial Biotechnology* 8 (3): 363–68.

Nemergut, Diana R., Steven K. Schmidt, Tadashi Fukami, Sean P. O'Neill, Teresa M. Bilinski, Lee F. Stanish, Joseph E. Knelman et al. 2013. Patterns and Processes of Microbial Community Assembly. *Microbiology and Molecular Biology Reviews* 77 (3): 342–56.

Neumann, Grit, Y. Veeranagouda, T. B. Karegoudar, Ozlem Sahin, Ines Mäusezahl, Nadja Kabelitz, Uwe Kappelmeyer, and Hermann J. Heipieper. 2005. Cells of *Pseudomonas putida* and *Enterobacter* Sp. Adapt to Toxic Organic Compounds by Increasing Their Size. *Extremophiles: Life under Extreme Conditions* 9 (2): 163–68.

Nevin, Kelly P., Sarah A. Hensley, Ashley E. Franks, Zarath M. Summers, Jianhong Ou, Trevor L. Woodard, Oona L. Snoeyenbos-West, and Derek R. Lovley. 2011. Electrosynthesis of Organic Compounds from Carbon Dioxide Is Catalyzed by a Diversity of Acetogenic Microorganisms. *Applied and Environmental Microbiology* 77 (9): 2882–86.

Nicolaou, Sergios A., Stefan M. Gaida, and Eleftherios T. Papoutsakis. 2010. A Comparative View of Metabolite and Substrate Stress and Tolerance in Microbial Bioprocessing: From Biofuels and Chemicals, to Biocatalysis and Bioremediation. *Metabolic Engineering* 12 (4): 307–31.

Nicolaou, Sergios A., Stefan M. Gaida, and Eleftherios T. Papoutsakis. 2012. Exploring the Combinatorial Genomic Space in *Escherichia coli* for Ethanol Tolerance. *Biotechnology Journal* 7 (11): 1337–45.

Nielsen, Jens. 2003. Metabolic Engineering. *Advances in Biochemical Engineering/Biotechnology* 73: 184.

Nikaido, Hiroshi, and Yumiko Takatsuka. 2009. Mechanisms of RND Multidrug Efflux Pumps. *Biochimica et Biophysica Acta* 1794 (5): 769–81.

Nocker, Andreas, Mark Burr, and Anne K. Camper. 2007. Genotypic Microbial Community Profiling: A Critical Technical Review. *Microbial Ecology* 54 (2): 276–89.

Ofiţeru, Irina Dana, Mary Lunn, Thomas P. Curtis, George F. Wells, Craig S. Criddle, Christopher A. Francis, and William T. Sloan. 2010. Combined Niche and Neutral Effects in a Microbial Wastewater Treatment Community. *Proceedings of the National Academy of Sciences of the United States of America* 107 (35): 15345–50.

Orphan, Victoria J. 2009. Methods for Unveiling Cryptic Microbial Partnerships in Nature. *Current Opinion in Microbiology* 12 (3): 231–37.

Patil, Sunil A., Jan B. A. Arends, Inka Vanwonterghem, Jarne van Meerbergen, Kun Guo, Gene W. Tyson, and Korneel Rabaey. 2015. Selective Enrichment Establishes a Stable Performing Community for Microbial Electrosynthesis of Acetate from CO2. *Environmental Science & Technology* 49 (14): 8833–43.

Paulsen, Ian T., Melissa H. Brown, and Ronald A. Skurray. 1996. Proton-Dependent Multi-drug Efflux Systems. *Microbiological Reviews* 60 (4): 575–608.

Pelletier, E., A. Kreimeyer, S. Bocs, Z. Rouy, G. Gyapay, R. Chouari, D. Riviere et al. 2008. "Candidatus Cloacamonas Acidaminovorans": Genome Sequence Reconstruction Provides a First Glimpse of a New Bacterial Division. *Journal of Bacteriology* 190 (7): 2572–79.

Peng, Yu, Henry C. M. Leung, S. M. Yiu, and Francis Y. L. Chin. 2012. IDBA-UD: A De Novo Assembler for Single-Cell and Metagenomic Sequencing Data with Highly Uneven Depth. *Bioinformatics (Oxford, England)* 28 (11): 1420–28.

Pham, Van H. T., and Jaisoo Kim. 2012. Cultivation of Unculturable Soil Bacteria. *Trends in Biotechnology* 30 (9): 475–84.

Phelan, Vanessa V., Wei-Ting Liu, Kit Pogliano, and Pieter C. Dorrestein. 2012. Microbial Metabolic Exchange—The Chemotype-to-Phenotype Link. *Nature Chemical Biology* 8 (1): 26–35.

Philips, Jo, Kristof Verbeeck, Korneel Rabaey, and Jan B. A. Arends. 2015. Electron Transfer Mechanisms in Biofilms. In *Microbial Electrochemical and Fuel Cells: Fundamentals and Applications*, ed. Keith Scott and Eileen Yu, 67–113. 1st ed. Amsterdam: Elsevier Woodhead Publishing.

Pires, Douglas E. V., David B. Ascher, and Tom L. Blundell. 2014. mCSM: Predicting the Effects of Mutations in Proteins Using Graph-Based Signatures. *Bioinformatics (Oxford, England)* 30 (3): 335–42.

Ponomarova, Olga, and Kiran Raosaheb Patil. 2015. Metabolic Interactions in Microbial Communities: Untangling the Gordian Knot. *Current Opinion in Microbiology* 27: 37–44.

Poppe, Jakeline Kathiele, Roberto Fernandez-Lafuente, Rafael C. Rodrigues, and Marco Antônio Záchia Ayub. 2015. Enzymatic Reactors for Biodiesel Synthesis: Present Status and Future Prospects. *Biotechnology Advances* 33 (5): 511–25.

Prosser, James I. 2015. Dispersing Misconceptions and Identifying Opportunities for the Use of "Omics" in Soil Microbial Ecology. *Nature Reviews Microbiology* 13 (7): 439–46.

Prosser, James I., Brendan J. M. Bohannan, Tom P. Curtis, Richard J. Ellis, Mary K. Firestone, Rob P. Freckleton, Jessica L. Green et al. 2007. The Role of Ecological Theory in Microbial Ecology. *Nature Reviews Microbiology* 5 (5): 384–92.

Pucci, Fabrizio, Raphaël Bourgeas, and Marianne Rooman. 2016. Predicting Protein Thermal Stability Changes upon Point Mutations Using Statistical Potentials: Introducing HoTMuSiC. *Scientific Reports* 6: 23257.

Pucci, Fabrizio, and Marianne Rooman. 2014. Stability Curve Prediction of Homologous Proteins Using Temperature-Dependent Statistical Potentials. *PLoS Computational Biology* 10 (7): e1003689.

Putman, Monique, Hendrik W. van Veen, and Wil N. Konings. 2000. Molecular Properties of Bacterial Multidrug Transporters. *Microbiology and Molecular Biology Reviews: MMBR* 64 (4): 672–93.

Putnam, Nicholas H., Brendan L. O'Connell, Jonathan C. Stites, Brandon J. Rice, Marco Blanchette, Robert Calef, Christopher J. Troll et al. 2016. Chromosome-Scale Shotgun Assembly Using an In Vitro Method for Long-Range Linkage. *Genome Research* 26 (3): 342–50.

Qian, Mingyu, Shen Tian, Xuefeng Li, Jing Zhang, Yaping Pan, and Xiushan Yang. 2006. Ethanol Production from Dilute-Acid Softwood Hydrolysate by Co-Culture. *Applied Biochemistry and Biotechnology* 134 (3): 273–84.

Rabaey, Korneel, and René A. Rozendal. 2010. Microbial Electrosynthesis—Revisiting the Electrical Route for Microbial Production. *Nature Reviews Microbiology* 8 (10): 706–16.

Raes, Jeroen, and Peer Bork. 2008. Molecular Eco-Systems Biology: Towards an Understanding of Community Function. *Nature Reviews Microbiology* 6 (9): 693–99.

Rakoff-Nahoum, Seth, Kevin R. Foster, and Laurie E. Comstock. 2016. The Evolution of Cooperation within the Gut Microbiota. *Nature* 533 (7602): 255–59.

Ramanan, Rishiram, Byung-Hyuk Kim, Dae-Hyun Cho, Hee-Mock Oh, and Hee-Sik Kim. 2016. Algae–Bacteria Interactions: Evolution, Ecology and Emerging Applications. *Biotechnology Advances* 34 (1): 14–29.

Ramos, Juan L., Estrella Duque, Maria-Trinidad Gallegos, Patricia Godoy, Maria Isabel Ramos-Gonzalez, Antonia Rojas, Wilson Teran, and Ana Segura. 2002. Mechanisms of Solvent Tolerance in Gram-Negative Bacteria. *Annual Review of Microbiology* 56: 743–68.

Ramos, Juan-Luis, Miguel Valdivia, Francisco García-Lorente, and Ana Segura. 2016. Benefits and Perspectives on the Use of Biofuels. *Microbial Biotechnology* 9 (4): 436–40.

Rappe, Michael S., and Stephen J. Giovannoni. 2003. The Uncultured Microbial Majority. *Annual Review of Microbiology* 57: 369–94.

Reguera, G., K. D. McCarthy, T. Mehta, J. S. Nicoll, M. T. Tuominen, and D. R. Lovley. 2005. Extracellular Electron Transfer via Microbial Nanowires. *Nature* 435 (7045): 1098–1101.

Ren, Hao, Soonjae Pyo, Jae Ik Lee, Tae Jin Park, Forrest S. Gittleson, Frederick C. C. Leung, Jongbaeg Kim, André D. Taylor, Hyung Sool Lee, and Junseok Chae. 2015. A High Power Density Miniaturized Microbial Fuel Cell Having Carbon Nanotube Anodes. *Journal of Power Sources* 273: 823–30.

Resch, Michael G., Bryon S. Donohoe, John O. Baker, Stephen R. Decker, Edward A. Bayer, Gregg T. Beckham, and Michael E. Himmel. 2013. Fungal Cellulases and Complexed Cellulosomal Enzymes Exhibit Synergistic Mechanisms in Cellulose Deconstruction. *Energy & Environmental Science* 6 (6): 1858–67.

Richelle, Anne, and Philippe Bogaerts. 2014. Off-Line Optimization of Baker's Yeast Production Process. *Chemical Engineering Science* 119: 40–52.

Richelle, Anne, and Philippe Bogaerts. 2015. Systematic Methodology for Bioprocess Model Identification Based on Generalized Kinetic Functions. *Biochemical Engineering Journal* 100: 41–49.

Richelle, Anne, Imen Ben Tahar, M. Hassouna, and Philippe Bogaerts. 2015. Macroscopic Modelling of Bioethanol Production from Potato Peel Wastes in Batch Cultures Supplemented with Inorganic Nitrogen. *Bioprocess and Biosystems Engineering* 38 (9): 1819–33.

Rinaldi, Antonio, Barbara Mecheri, Virgilio Garavaglia, Silvia Licoccia, Paolo Di Nardo, and Enrico Traversa. 2008. Engineering Materials and Biology to Boost Performance of Microbial Fuel Cells: A Critical Review. *Energy & Environmental Science* 1 (4): 417–29.

Rinke, Christian, Janey Lee, Nandita Nath, Danielle Goudeau, Brian Thompson, Nicole Poulton, Elizabeth Dmitrieff, Rex Malmstrom, Ramunas Stepanauskas, and Tanja Woyke. 2014. Obtaining Genomes from Uncultivated Environmental Microorganisms Using FACS-Based Single-Cell Genomics. *Nature Protocols* 9 (5): 1038–48.

Rinke, Christian, Patrick Schwientek, Alexander Sczyrba, Natalia N. Ivanova, Iain J. Anderson, Jan-Fang Cheng, Aaron Darling et al. 2013. Insights into the Phylogeny and Coding Potential of Microbial Dark Matter. *Nature* 499 (7459): 431–37.

Rittmann, Bruce E., Rosa Krajmalnik-Brown, and Rolf U. Halden. 2008. Pre-Genomic, Genomic and Post-Genomic Study of Microbial Communities Involved in Bioenergy. *Nature Reviews Microbiology* 6 (8): 604–12.

Rivière, Delphine, Virginie Desvignes, Eric Pelletier, Sébastien Chaussonnerie, Sonda Guermazi, Jean Weissenbach, Tianlun Li, Patricia Camacho, and Abdelghani Sghir. 2009. Towards the Definition of a Core of Microorganisms Involved in Anaerobic Digestion of Sludge. *ISME Journal* 3 (6): 700–14.

Rosenbaum, Miriam A., and Ashley E. Franks. 2014. Microbial Catalysis in Bioelectrochemical Technologies: Status Quo, Challenges and Perspectives. *Applied Microbiology and Biotechnology* 98 (2): 509–18.

Rosenbaum, Miriam A., and Alexander W. Henrich. 2014. Engineering Microbial Electrocatalysis for Chemical and Fuel Production. *Current Opinion in Biotechnology* 29: 93–98.

Roume, Hugo, Anna Heintz-Buschart, Emilie E. L. Muller, Patrick May, Venkata P. Satagopam, Cédric C. Laczny, Shaman Narayanasamy et al. 2015. Comparative Integrated Omics: Identification of Key Functionalities in Microbial Community-Wide Metabolic Networks. *NPJ Biofilms and Microbiomes* 1: 15007.

Rozendal, René A., Adriaan W. Jeremiasse, Hubertus V. M. Hamelers, and Cees J. N. Buisman. 2008. Hydrogen Production with a Microbial Biocathode. *Environmental Science & Technology* 42 (2): 629–34.

Rühl, Jana, Andreas Schmid, and Lars Mathias Blank. 2009. Selected *Pseudomonas putida* Strains Able to Grow in the Presence of High Butanol Concentrations. *Applied and Environmental Microbiology* 75 (13): 4653–56.

Rutherford, Becky J., Robert H. Dahl, Richard E. Price, Heather L. Szmidt, Peter I. Benke, Aindrila Mukhopadhyay, and Jay D. Keasling. 2010. Functional Genomic Study of Exogenous N-Butanol Stress in *Escherichia coli*. *Applied and Environmental Microbiology* 76 (6): 1935–45.

Salehi Jouzani, Gholamreza, and Mohammad J. Taherzadeh. 2015. Advances in Consolidated Bioprocessing Systems for Bioethanol and Butanol Production from Biomass: A Comprehensive Review. *Biofuel Research Journal* 2 (1): 152–95.

Sánchez, Óscar J., and Carlos A. Cardona. 2008. Trends in Biotechnological Production of Fuel Ethanol from Different Feedstocks. *Bioresource Technology* 99 (13): 5270–95.

Santarossa, Gianluca, Pietro Gatti Lafranconi, Claudia Alquati, Luca DeGioia, Lilia Alberghina, Piercarlo Fantucci, and Marina Lotti. 2005. Mutations in The "Lid" Region Affect Chain Length Specificity and Thermostability of a *Pseudomonas fragi* Lipase. *FEBS Letters* 579 (11): 2383–86.

Sárvári Horváth, Ilona, Meisam Tabatabaei, Keikhosro Karimi, and Rajeev Kumar. 2016. Recent Updates on Biogas Production—A Review. *Biofuel Research Journal* 3 (2): 394–402.

Savageau, Michael A. 1969a. Biochemical Systems Analysis I: Some Mathematical Properties of the Rate Law for the Component Enzymatic Reactions. *Journal of Theoretical Biology* 25 (3): 365–69.

Savageau, Michael A. 1969b. Biochemical Systems Analysis II: The Steady-State Solutions for an N-Pool System Using a Power-Law Approximation. *Journal of Theoretical Biology* 25 (3): 370–79.

Scharf, Michael E. 2015. Omic Research in Termites: An Overview and a Roadmap. *Frontiers in Genetics* 6.

Schiel-Bengelsdorf, Bettina, and Peter Dürre. 2012. Pathway Engineering and Synthetic Biology Using Acetogens. *FEBS Letters* 586 (15): 2191–98.

Schievano, Andrea, Tommy Pepé Sciarria, Karolien Vanbroekhoven, Heleen De Wever, Sebastià Puig, Stephen J. Andersen, Korneel Rabaey, and Deepak Pant. 2016. Electro-Fermentation—Merging Electrochemistry with Fermentation in Industrial Applications. *Trends in Biotechnology* 34 (11): 866–78.

Schink, Bernhard. 1997. Energetics of Syntrophic Cooperation in Methanogenic Degradation. *Microbiology and Molecular Biology Reviews* 61 (2): 262–80.

Schloss, Patrick D. 2009. A High-Throughput DNA Sequence Aligner for Microbial Ecology Studies. *PLoS One* 4 (12): e8230.

Schmidt, Thomas M., Edward F. DeLong, and Norman R. Pace. 1991. Analysis of a Marine Picoplankton Community by 16S rRNA Gene Cloning and Sequencing. *Journal of Bacteriology* 173 (14): 4371–78.

Schuster, Brian G., and Mari S. Chinn. 2012. Consolidated Bioprocessing of Lignocellulosic Feedstocks for Ethanol Fuel Production. *Bioenergy Research* 6 (2): 416–35.

Scully, Sean Michael, and Johann Orlygsson. 2014. Recent Advances in Second Generation Ethanol Production by Thermophilic Bacteria. *Energies* 8 (1): 1–30.

Segata, Nicola, Daniela Boernigen, Timothy L. Tickle, Xochitl C. Morgan, Wendy S. Garrett, and Curtis Huttenhower. 2013. Computational Meta'omics for Microbial Community Studies. *Molecular Systems Biology* 9 (1): 666.

Segura, Ana, Antonia Rojas, Ana Hurtado, María-José Huertas, and Juan L. Ramos. 2003. Comparative Genomic Analysis of Solvent Extrusion Pumps in *Pseudomonas* Strains Exhibiting Different Degrees of Solvent Tolerance. *Extremophiles: Life under Extreme Conditions* 7 (5): 371–76.

Selvaraj, Siddarth, Jesse R. Dixon, Vikas Bansal, and Bing Ren. 2013. Whole-Genome Haplotype Reconstruction Using Proximity-Ligation and Shotgun Sequencing. *Nature Biotechnology* 31 (12): 1111–18.

Semova, Ivana, Juliana D. Carten, Jesse Stombaugh, Lantz C. Mackey, Rob Knight, Steven A. Farber, and John F. Rawls. 2012. Microbiota Regulate Intestinal Absorption and Metabolism of Fatty Acids in the Zebrafish. *Cell Host & Microbe* 12 (3): 277–88.

Seth, Erica C., and Michiko E. Taga. 2014. Nutrient Cross-Feeding in the Microbial World. *Terrestrial Microbiology* 5: 350.

Shade, Ashley, Hannes Peter, Steven D. Allison, Didier L. Baho, Merce Berga, Helmut Burgmann, David H. Huber et al. 2012. Fundamentals of Microbial Community Resistance and Resilience. *Frontiers in Microbiology* 3: 417.

Sharon, Itai, and Jillian F. Banfield. 2013. Genomes from Metagenomics. *Science* 342 (6162): 1057–58.

Sheldon, Roger A., and Sander van Pelt. 2013. Enzyme Immobilisation in Biocatalysis: Why, What and How. *Chemical Society Reviews* 42 (15): 6223–35.

Sheth, Ravi U., Vitor Cabral, Sway P. Chen, and Harris H. Wang. 2016. Manipulating Bacterial Communities by In Situ Microbiome Engineering. *Trends in Genetics* 32 (4): 189–200.

Sikkema, Jan, Jan A. de Bont, and Bert Poolman. 1994. Interactions of Cyclic Hydrocarbons with Biological Membranes. *Journal of Biological Chemistry* 269 (11): 8022–28.

Sikkema, Jan, Jan A. de Bont, and Bert Poolman. 1995. Mechanisms of Membrane Toxicity of Hydrocarbons. *Microbiological Reviews* 59 (2): 201–22.

Silva, Ricardo R. da, Pieter C. Dorrestein, and Robert A. Quinn. 2015. Illuminating the Dark Matter in Metabolomics. *Proceedings of the National Academy of Sciences of the United States of America* 112 (41): 12549–50.

Singer, Esther, Eric A. Webb, William C. Nelson, John F. Heidelberg, Natalia Ivanova, Amrita Pati, and Katrina J. Edwards. 2011. Genomic Potential of Marinobacter Aquaeolei, a Biogeochemical "Opportunitroph." *Applied and Environmental Microbiology* 77 (8): 2763–71.

Singh, Raushan Kumar, Manish Kumar Tiwari, Ranjitha Singh, and Jung-Kul Lee. 2013. From Protein Engineering to Immobilization: Promising Strategies for the Upgrade of Industrial Enzymes. *International Journal of Molecular Sciences* 14 (1): 1232–77.

Smets, Ilse, Johan Claes, Eva J. November, Georges P. Bastin, and Jan F. Van Impe. 2004. Optimal Adaptive Control of (Bio)chemical Reactors: Past, Present and Future. *Journal of Process Control* 14 (7): 795–805.

Springael, D., K. Peys, A. Ryngaert, S. Van Roy, L. Hooyberghs, R. Ravatn, M. Heyndrickx et al. 2002. Community Shifts in a Seeded 3-Chlorobenzoate Degrading Membrane Biofilm Reactor: Indications for Involvement of In Situ Horizontal Transfer of the clc-Element from Inoculum to Contaminant Bacteria. *Environmental Microbiology* 4 (2): 70–80.

Sreekumar, Sanil, Zachary C. Baer, Anbarasan Pazhamalai, Gorkem Gunbas, Adam Grippo, Harvey W. Blanch, Douglas S. Clark, and F. Dean Toste. 2015. Production of an Acetone-Butanol-Ethanol Mixture from *Clostridium acetobutylicum* and Its Conversion to High-Value Biofuels. *Nature Protocols* 10 (3): 528–37.

Staden, R. 1980. A New Computer Method for the Storage and Manipulation of DNA Gel Reading Data. *Nucleic Acids Research* 8 (16): 3673–94.

Stams, Alfons J. M. 1994. Metabolic Interactions between Anaerobic Bacteria in Methanogenic Environments. *Antonie Van Leeuwenhoek* 66 (1–3): 271–94.

Steinbusch, Kirsten J. J., Hubertus V. M. Hamelers, Joris D. Schaap, Christel Kampman, and Cees J. N. Buisman. 2010. Bioelectrochemical Ethanol Production through Mediated Acetate Reduction by Mixed Cultures. *Environmental Science & Technology* 44 (1): 513–17.

Stenuit, Ben, and Spiros N. Agathos. 2015. Deciphering Microbial Community Robustness through Synthetic Ecology and Molecular Systems Synecology. *Current Opinion in Biotechnology* 33: 305–17.

Stephanopoulos, Gregory N., Aristos A. Aristidou, and Jens Nielsen. 1998. *Metabolic Engineering: Principles and Methodologies.* London: Academic Press.

Stewart, Eric J. 2012. Growing Unculturable Bacteria. *Journal of Bacteriology* 194 (16): 4151–60.

Stolze, Yvonne, Andreas Bremges, Madis Rumming, Christian Henke, Irena Maus, Alfred Pühler, Alexander Sczyrba, and Andreas Schlüter. 2016. Identification and Genome Reconstruction of Abundant Distinct Taxa in Microbiomes from One Thermophilic and Three Mesophilic Production-Scale Biogas Plants. *Biotechnology for Biofuels* 9: 156.

Strong, P. James, Sihuang Xie, and William P. Clarke. 2015. Methane as a Resource: Can the Methanotrophs Add Value? *Environmental Science & Technology* 49 (7): 4001–18

Suen, Garret, Jarrod J. Scott, Frank O. Aylward, Sandra M. Adams, Susannah G. Tringe, Adrián A. Pinto-Tomás, Clifton E. Foster et al. 2010. An Insect Herbivore Microbiome with High Plant Biomass-Degrading Capacity. *PLoS Genet* 6 (9): e1001129.

Sundberg, Carina, Waleed A. Al-Soud, Madeleine Larsson, Erik Alm, Sepehr S. Yekta, Bo H. Svensson, Søren J. Sørensen, and Anna Karlsson. 2013. 454 Pyrosequencing Analyses of Bacterial and Archaeal Richness in 21 Full-Scale Biogas Digesters. *FEMS Microbiology Ecology* 85 (3): 612–26.

Tan, Yang, Ramesh Y. Adhikari, Nikhil S. Malvankar, Shuang Pi, Joy E. Ward, Trevor L. Woodard, Kelly P. Nevin, Qiangfei Xia, Mark T. Tuominen, and Derek R. Lovley. 2016. Synthetic Biological Protein Nanowires with High Conductivity. *Small* 12 (33): 4481–85.

Thomas, Christopher M., and Kaare M. Nielsen. 2005. Mechanisms of, and Barriers to, Horizontal Gene Transfer between Bacteria. *Nature Reviews Microbiology* 3 (9): 711–21.

Thomas, Torsten, Jack Gilbert, and Folker Meyer. 2012. Metagenomics—A Guide from Sampling to Data Analysis. *Microbial Informatics and Experimentation* 2: 3.

Tian, Ye, Christopher Deutsch, and Bala Krishnamoorthy. 2010. Scoring Function to Predict Solubility Mutagenesis. *Algorithms for Molecular Biology: AMB* 5: 33.

Tilman, David, Robert Socolow, Jonathan A. Foley, Jason Hill, Eric Larson, Lee Lynd, Stephen Pacala et al. 2009. Beneficial Biofuels—The Food, Energy, and Environment Trilemma. *Science* 325 (5938): 270–71.

Tomko, Timothy A., and Mary J. Dunlop. 2015. Engineering Improved Bio-Jet Fuel Tolerance in *Escherichia coli* Using a Transgenic Library from the Hydrocarbon-Degrader Marinobacter Aquaeolei. *Biotechnology for Biofuels* 8: 165.

Tommasi, Tonia, Gian Paolo Salvador, and Marzia Quaglio. 2016. New Insights in Microbial Fuel Cells: Novel Solid Phase Anolyte. *Scientific Reports* 6: 29091.

Traxler, Matthew F., Jeramie D. Watrous, Theodore Alexandrov, Pieter C. Dorrestein, and Roberto Kolter. 2013. Interspecies Interactions Stimulate Diversification of the *Streptomyces coelicolor* Secreted Metabolome. *mBio* 4 (4): e00459-13.

Tremblay, Pier-Luc, Largus T. Angenent, and Tian Zhang. 2016. Extracellular Electron Uptake: Among Autotrophs and Mediated by Surfaces. *Trends in Biotechnology* 35 (4): 360–71.

Tremblay, Pier-Luc, and Tian Zhang. 2015. Electrifying Microbes for the Production of Chemicals. *Microbial Physiology and Metabolism* 6: 201.

Tringe, Susannah Green, Christian von Mering, Arthur Kobayashi, Asaf A. Salamov, Kevin Chen, Hwai W. Chang, Mircea Podar et al. 2005. Comparative Metagenomics of Microbial Communities. *Science* 308 (5721): 554–57.

Tsai, Shen-Long, Jeongseok Oh, Shailendra Singh, Ruizhen Chen, and Wilfred Chen. 2009. Functional Assembly of Minicellulosomes on the *Saccharomyces cerevisiae* Cell Surface for Cellulose Hydrolysis and Ethanol Production. *Applied and Environmental Microbiology* 75 (19): 6087–93.

Tynan-Connolly, Barbara M., and Jens Erik Nielsen. 2006. pKD: Re-Designing Protein pKa Values. *Nucleic Acids Research* 34: W48–51.

Uchiyama, Taku, Kimio Ito, Koji Mori, Hirohito Tsurumaru, and Shigeaki Harayama. 2010. Iron-Corroding Methanogen Isolated from a Crude-Oil Storage Tank. *Applied and Environmental Microbiology* 76 (6): 1783–88.

Uday, Uma Shankar Prasad, Payel Choudhury, Tarun Kanti Bandyopadhyay, and Biswanath Bhunia. 2016. Classification, Mode of Action and Production Strategy of Xylanase and Its Application for Biofuel Production from Water Hyacinth. *International Journal of Biological Macromolecules* 82: 1041–54.

van Impe, Jan F. and George Bastin. 1998. Optimal Adaptive Control of Fed-Batch Fermentation Processes. In *Advanced Instrumentation, Data Interpretation, and Control of Biotechnological Processes*, ed. Jan F. M. Van Impe, Peter A. Vanrolleghem, and Dirk M. Iserentant, 401–35. Amsterdam: Springer.

Vanwonterghem, Inka, Paul D. Jensen, Dang P. Ho, Damien J. Batstone, and Gene W. Tyson. 2014. Linking Microbial Community Structure, Interactions and Function in Anaerobic Digesters Using New Molecular Techniques. *Current Opinion in Biotechnology* 27: 55–64.

Venzlaff, Hendrik, Dennis Enning, Jayendran Srinivasan, Karl J. J. Mayrhofer, Achim Walter Hassel, Friedrich Widdel, and Martin Stratmann. 2013. Accelerated Cathodic Reaction in Microbial Corrosion of Iron due to Direct Electron Uptake by Sulfate-Reducing Bacteria. *Corrosion Science* 66: 88–96.

VerBerkmoes, Nathan C., Vincent J. Denef, Robert L. Hettich, and Jillian F. Banfield. 2009. Systems Biology: Functional Analysis of Natural Microbial Consortia Using Community Proteomics. *Nature Reviews Microbiology* 7 (3): 196–205.

Verstraete, Willy, Lieven Wittebolle, K. Heylen, B. Vanparys, Paul de Vos, Tom van de Wiele, and Nico Boon. 2007. Microbial Resource Management: The Road to Go for Environmental Biotechnology. *Engineering in Life Sciences* 7 (2): 117–126.

Villano, Marianna, Federico Aulenta, Costanza Ciucci, Tommaso Ferri, Antonio Giuliano, and Mauro Majone. 2010. Bioelectrochemical Reduction of CO2 to CH4 via Direct and Indirect Extracellular Electron Transfer by a Hydrogenophilic Methanogenic Culture. *Bioresource Technology* 101 (9): 3085–90.

Wagner, Samuel, Louise Baars, A. Jimmy Ytterberg, Anja Klussmeier, Claudia S. Wagner, Olof Nord, Per-Ake Nygren, Klaas J. van Wijk, and Jan-Willem de Gier. 2007. Consequences of Membrane Protein Overexpression in *Escherichia coli*. *Molecular & Cellular Proteomics: MCP* 6 (9): 1527–50.

Waldor, Matthew K., Gene Tyson, Elhanan Borenstein, Howard Ochman, Andrew Moeller, B. Brett Finlay, Heidi H. Kong et al. 2015. Where Next for Microbiome Research? *PLoS Biology* 13 (1).

Walton, Paul H., and Gideon J. Davies. 2016. On the Catalytic Mechanisms of Lytic Polysaccharide Monooxygenases. *Current Opinion in Chemical Biology* 31: 195–207.

Wang, Benjamin L., Adel Ghaderi, Hang Zhou, Jeremy Agresti, David A. Weitz, Gerald R. Fink, and Gregory Stephanopoulos. 2014. Microfluidic High-Throughput Culturing of Single Cells for Selection Based on Extracellular Metabolite Production or Consumption. *Nature Biotechnology* 32 (5): 473–78.

Warner, Joseph R., Ranjan Patnaik, and Ryan T. Gill. 2009. Genomics Enabled Approaches in Strain Engineering. *Current Opinion in Microbiology* 12 (3): 223–30.

Watrous, Jeramie, Patrick Roach, Theodore Alexandrov, Brandi S. Heath, Jane Y. Yang, Roland D. Kersten, Menno van der Voort et al. 2012. Mass Spectral Molecular Networking of Living Microbial Colonies. *Proceedings of the National Academy of Sciences of the United States of America* 109 (26): E1743–52.

Weber, Frans J., and Jan A. M. de Bont. 1996. Adaptation Mechanisms of Microorganisms to the Toxic Effects of Organic Solvents on Membranes. *Biochimica et Biophysica Acta* 1286 (3): 225–45.

Weiland, Peter. 2010. Biogas Production: Current State and Perspectives. *Applied Microbiology and Biotechnology* 85 (4): 849–60.

Werner, Jeffrey J., Dan Knights, Marcelo L. Garcia, Nicholas B. Scalfone, Samual Smith, Kevin Yarasheski, Theresa A. Cummings, Allen R. Beers, Rob Knight, and Largus T. Angenent. 2011. Bacterial Community Structures Are Unique and Resilient in Full-Scale Bioenergy Systems. *Proceedings of the National Academy of Sciences of the United States of America* 108 (10): 4158–63.

Widder, Stefanie, Rosalind J. Allen, Thomas Pfeiffer, Thomas P. Curtis, Carsten Wiuf, William T. Sloan, Otto X. Cordero et al. 2016. Challenges in Microbial Ecology: Building Predictive Understanding of Community Function and Dynamics. *ISME Journal*.

Wijffels, René H., and Maria J. Barbosa. 2010. An Outlook on Microalgal Biofuels. *Science* 329 (5993): 796–99.

Wilmes, Paul, Anna Heintz-Buschart, and Philip L. Bond. 2015. A Decade of Metaproteomics: Where We Stand and What the Future Holds. *Proteomics* 15 (20): 3409–17.

Wittebolle, Lieven, Massimo Marzorati, Lieven Clement, Annalisa Balloi, Daniele Daffonchio, Kim Heylen, Paul De Vos, Willy Verstraete, and Nico Boon. 2009. Initial Community Evenness Favours Functionality under Selective Stress. *Nature* 458 (7238): 623–26.

Worth, Catherine L., Robert Preissner, and Tom L. Blundell. 2011. SDM—A Server for Predicting Effects of Mutations on Protein Stability and Malfunction. *Nucleic Acids Research* 39: W215–22.

Wrede, Digby, Mohamed Taha, Ana F. Miranda, Krishna Kadali, Trevor Stevenson, Andrew S. Ball, and Aidyn Mouradov. 2014. Co-Cultivation of Fungal and Microalgal Cells as an Efficient System for Harvesting Microalgal Cells, Lipid Production and Wastewater Treatment. *PloS One* 9 (11): e113497.

Xia, Tian, Mark A. Eiteman, and Elliot Altman. 2012. Simultaneous Utilization of Glucose, Xylose and Arabinose in the Presence of Acetate by a Consortium of *Escherichia coli* Strains. *Microbial Cell Factories* 11: 77.

Yin, Shuangye, Feng Ding, and Nikolay V. Dokholyan. 2007. Eris: An Automated Estimator of Protein Stability. *Nature Methods* 4 (6): 466–67.

Yoon, Hwan Su, Dana C. Price, Ramunas Stepanauskas, Veeran D. Rajah, Michael E. Sieracki, William H. Wilson, Eun Chan Yang, Siobain Duffy, and Debashish Bhattacharya. 2011. Single-Cell Genomics Reveals Organismal Interactions in Uncultivated Marine Protists. *Science* 332 (6030): 714–17.

Zah, Rainer, Heinz Böni, Marcel Gauch, Roland Hischier, Martin Lehmann, and Patrick Wäger. 2007. Life Cycle Assessment of Energy Products: Environmental Impact Assessment of Biofuels. EMPA Report. Gallen, Switzerland, EMPA.

Zakrzewski, Martha, Alexander Goesmann, Sebastian Jaenicke, Sebastian Jünemann, Felix Eikmeyer, Rafael Szczepanowski, Waleed Abu Al-Soud, Søren Sørensen, Alfred Pühler, and Andreas Schlüter. 2012. Profiling of the Metabolically Active Community from a Production-Scale Biogas Plant by Means of High-Throughput Metatranscriptome Sequencing. *Journal of Biotechnology* 158 (4): 248–58.

Zengler, Karsten, and Bernhard O. Palsson. 2012. A Road Map for the Development of Community Systems (CoSy) Biology. *Nature Reviews Microbiology* 10 (5): 366–72.

Zhang, Baohua, Yanqing Weng, Hong Xu, and Zhiping Mao. 2012. Enzyme Immobilization for Biodiesel Production. *Applied Microbiology and Biotechnology* 93 (1): 61–70.

Zhang, Mingzi M., Yajie Wang, Ee Lui Ang, and Huimin Zhao. 2016. Engineering Microbial Hosts for Production of Bacterial Natural Products. *Natural Product Reports* 33 (8): 963–87.

Zhang, Y.-H. Percival. 2011. What Is Vital (and Not Vital) to Advance Economically-Competitive Biofuels Production. *Process Biochemistry* 46 (11): 2091–110.

Zhou, Hongyi, and Yaoqi Zhou. 2002. Distance-Scaled, Finite Ideal-Gas Reference State Improves Structure-Derived Potentials of Mean Force for Structure Selection and Stability Prediction. *Protein Science: A Publication of the Protein Society* 11 (11): 2714–26.

Zhou, Jizhong, Wenzong Liu, Ye Deng, Yi-Huei Jiang, Kai Xue, Zhili He, Joy D. Van Nostrand, Liyou Wu, Yunfeng Yang, and Aijie Wang. 2013. Stochastic Assembly Leads to Alternative Communities with Distinct Functions in a Bioreactor Microbial Community. *mBio* 4 (2): e00584-12.

Zhou, Kang, Kangjian Qiao, Steven Edgar, and Gregory Stephanopoulos. 2015. Distributing a Metabolic Pathway among a Microbial Consortium Enhances Production of Natural Products. *Nature Biotechnology* 33 (4): 377–83.

Zhu, Lifeng, Qi Wu, Jiayin Dai, Shanning Zhang, and Fuwen Wei. 2011. Evidence of Cellulose Metabolism by the Giant Panda Gut Microbiome. *Proceedings of the National Academy of Sciences of the United States of America* 108: 17714–19.

Zingaro, Kyle A., and Eleftherios Terry Papoutsakis. 2013. GroESL Overexpression Imparts *Escherichia coli* Tolerance to I-, N-, and 2-Butanol, 1,2,4-Butanetriol and Ethanol with Complex and Unpredictable Patterns. *Metabolic Engineering* 15: 196–205.

14

Industrial Integration of Biotechnological Processes from Raw Material to Energy Integration:
Study by Modeling Approach

Grégoire Léonard, Andreas Pfennig, Ayse Dilan Celebi,
Shivom Sharma, and François Maréchal

Contents

14.1 Introduction: Background and Driving Forces

The design and deployment of biotechnological processes at the industrial level present many common points with the chemical engineering industry. In particular, the tools and methods used in chemical process engineering may advantageously be applied to evaluate and optimize biotechnological processes, and conversely, biotechnologies can be integrated in conventional industries to extend their range of action. In this chapter, we describe how a cross-disciplinary approach can transpose (after adaptation) relevant solutions from one sector to solve similar problems identified in another sector and improve its performance.

First, thermodynamic modeling approaches developed to increase understanding and improve the evaluation of reaction pathways is presented. A "group contribution method" that allows estimation of the Gibbs free energy of formation of (bio-)chemical products and their free energy of reaction is discussed. This approach is extended by comparing different (bio-)chemical products based on their exergy characteristics, where exergy losses may be a helpful indication for selecting efficient products and associated processes. Disposing of a precise evaluation method for thermodynamic properties, such as reaction enthalpy or component exergy, is a first and critical step to ensure robustness and confidence in the results achieved thanks to the modeling approach.

In the second part of this chapter, the tools developed in process system engineering are applied to biological processes. These tools may be applied to study and optimize complete process flowsheets, from the upstream steps (biomass pretreatments) to the downstream ones (purification of products). For instance, the modeling approach can help to optimize upstream processing, which is increasing in importance, as biomass pretreatment steps like drying, grinding, hydrolysis, and extraction occur in practically all biomass conversion processes. In the case of microbial fuels, these steps may result in a more efficient conversion of biomass into biogas or bioethanol, for instance. Similarly, the modeling approach can be applied to the conversion and downstream processing steps. In this section, as the downstream processing of bioreactor effluents is of critical importance for the economic viability of most biotechnological processes, special attention is first paid to the description of a simple approach to optimize the selection of downstream processing steps, a methodology called "cascaded option trees." Then, a systematic approach for the optimization of the design of complete processes is described and applied to the case of biorefineries. Indeed, biorefineries are complex systems processing many different biomass feedstocks to yield a variety of different products, from fuels to heat and value-added chemicals. They are thus a perfect example to demonstrate the potential of the modeling approach for process design, integration, and optimization of complex biochemical systems.

14.2 Thermodynamic Modeling of Cell Metabolism

The use of thermodynamics provides two significant advantages that can be applied to biotechnological processes (Prausnitz, 2003). First, it has a very wide range of applications due to its universal nature based on a small number of fundamental principles. Second, it is an integrative science, and the thermodynamic properties of compounds can be largely derived from their structures thanks to molecular and statistical considerations, delivering many answers with few experimental data. These advantages are crucial in biotechnological applications due to the large variety of chemical reactions occurring at the cell level and for which the experimental collection of data is far from trivial. In order to favor the industrial integration of biotechnological processes, tools to characterize system properties, possible reactions, and their thermodynamic limitations can provide useful information.

In this section, we first present some developments applying thermodynamics to predict feasible ranges for reaction occurrence at the level of biotechnological reactors. This approach is useful, for instance, when estimating the feasibility of biodegradation pathways in order to describe an existing system, or when designing biological reactors. Then, this thermodynamic approach is extended at the process level by performing exergetic analysis of biological products. Starting from a given feedstock, the estimation of exergy losses allows the identification of the most relevant reaction pathways for designing a new biotechnological process. These two approaches can bring significant help for the understanding and design of optimal microbial reactions and corresponding equipment.

14.2.1 Group Contribution Method

A good example of applying thermodynamic models for predicting the properties of biotechnological systems is provided by the group contribution method developed by Jankowski et al. (2008). This method allows for the determination of the standard Gibbs free energy of formation, $\Delta_f G°$, and of reaction, $\Delta_r G°$, in biotechnological processes based only on the molecular structure of the components implied in the reactions. These components are decomposed according to predefined molecular substructures to which corresponds a contribution value. Adding the contributions of all structural groups results in the total standard Gibbs free energy of formation of the molecule. The subdivision of a molecule into its structural groups must be performed with care, as the same atom could be included in several predefined groups. Thus, rules implying search priority numbers have been defined by Jankowski et al. (2008). A simple example to better understand the principle of this group contribution method consists of calculating the free energy of glucose that is used for bioethanol production. This example, shown in Table 14.1, leads to a $\Delta_f G°$ value for glucose of 218.28 kcal/mol, which is very close to the experimentally observed value of 219 kcal/mol. This method has been successfully applied for determining the $\Delta_f G°$ and $\Delta_r G°$ values for the majority of compounds and reactions described in three biochemical databases: (1) the iJR904 and iAF1260 genome-scale metabolic models of *Escherichia coli*, (2) the Kyoto Encyclopedia of Genes and Genomes, and (3) the University of Minnesota Biocatalysis and Biodegradation Database. In the last example, the standard free energy values determined by the group contribution method have been compared with experimental values for 914 compounds and 902 reactions, and the thermodynamic feasibility of 89 biodegradation pathways could be evaluated by estimating the total free energy change of each pathway (Finley et al.,

TABLE 14.1   Estimation of the Free Energy of Formation of Glucose Based on a Group Contribution Method

Identified Group	Number of Groups in the Glucose Molecule	Individual Group Contribution to $\Delta_f G°$ (kcal/mol)	Total Contribution to $\Delta_f G°$ (kcal/mol)
–OH	5	–41.5	–207.5
–O– (participating in a ring)	1	–36.6	–36.6
>CH– (participating in a ring)	5	4.84	24.2
–CH$_2$–	1	1.62	1.62
		Total	–218.28

Source: Jankowski, M. D. et al., *Biophys. J.*, 95, 1487–1499, 2008.

2009). This application proved to be a useful tool to compare competing biodegradation pathways and to identify the most thermodynamically favorable ones. As kinetic and enzymatic limitations may also play a role, this analysis is not sufficient to predict the actually occurring biodegradation pathway, but it can eliminate unfavorable candidates and provide some help in comparing the feasibility of reactions.

Finally, such thermodynamic methods can be envisaged in the case of genome-scale metabolic models (Ataman and Hatzimanikatis, 2015). It appears that a thermodynamic approach can bring significant help in the evaluation of the directionalities and of the possible flux ranges. Several examples of successful studies are mentioned by Ataman and Hatzimanikatis (2015). For instance, this approach was applied by Birkenmeier et al. (2014) to confirm glycerol-3-phosphate dehydrogenase as the enzyme controlling the glycerol anabolism in *Saccharomyces cerevisiae*. However, the development of this method is still ongoing, and many challenges remain in order to be able to identify complex bioenergetic behaviors based on thermodynamics. Such a method for estimating the Gibbs free energies of components and reactions observed in plants like maize has been used, for example, in the framework of the PlantSEED biochemistry pathways database developed by Seaver et al. (2014).

Finally, it should be mentioned that alternatives to group contribution methods have been developed for estimating thermodynamic properties in biochemical systems. Let us just mention, for instance, that Hoffmann et al. (2013) used the electrolyte perturbed-chain statistical associating fluid theory (PC-SAFT) equation of state to predict activity coefficients of methyl ferulate and the equilibrium constants of its hydrolysis reaction catalyzed by feruloyl esterase. In conclusion, it can be stated that a lot of research is still ongoing in the development of reliable models to predict thermodynamic properties in biochemical systems, leading to a better understanding and prediction of possible reaction pathways and equilibria.

14.2.2 Exergetic Evaluation of Process Alternatives

The design of new processes, including biotechnological process steps, has to include economic aspects. Unfortunately, energy prices are volatile, which directly influences operating costs and may also affect investment. These prices are strongly influenced by the political decisions of oil-producing countries, as well as political measures to ensure the long-term sustainability of the energy supply and to minimize climate change. Thus, economic evaluation of process alternatives in a way that the results remain valid in the future is difficult. Alternatively, an evaluation based on a more general measure of energy utilized may be helpful, namely, the evaluation of exergy devaluation in the sense of an exergoeconomic analysis, which is independent of energy prices (Tsatsaronis, 1996; Frenzel et al., 2013, 2014a, 2014b). Exergy is that fraction of energy which can be converted freely into any other form of energy.

The exergy of any stream in a process can be described as

$$E = E_{\text{phys}} + E_{\text{chem}} + E_{\text{mix}},$$

where the physical contribution E_{phys} results from the pressure and temperature conditions of the stream; the chemical contribution E_{chem} takes into account, for example, the free energy of formation of all components contained in the stream; and the mixing E_{mix} contribution accounts for the entropy of mixing (Frenzel et al., 2013). For a given stream, exergy is thus a state variable. For most streams in a chemical process, the chemical contribution is dominant, so that for a first evaluation, it is sufficient to regard only its change along the process. The mixing contribution is so small compared with the other contributions that it can generally be neglected. Since in biotechnological processes the conditions are typically close to ambient, which corresponds to the reference state in the definition of exergy, usually the physical contribution can be neglected as well. As a result, the exergy change associated with a biotechnological reaction directly corresponds to the free energy of the reaction at these standard conditions, as determined in Section 14.2.1. The result of such a first estimate of exergy requirement for a conversion gives the minimum energy requirement based only on the changes in chemical composition. Especially for fuel components, it is important that the energy content of the fuel is higher than the energy requirement of the process. Thus, early in process development, exergy analysis can be used to compare proposed processes against this benchmark.

Exergy as a general measure has the advantage that upon further progress in process development, all information can be used to refine the exergetic evaluation (Frenzel et al., 2014a, 2014b). This includes changes in the physical conditions, like temperature or pressure required, for which exergy losses can be estimated based on typical assumptions. For example, any heat exchanger can be evaluated assuming a typical minimum temperature difference as driving potential, and for most separation processes, reasonable assumptions lead to estimates on the exergy devaluation induced. In the next step of refinement, the exergy losses associated with individual pieces of equipment can also be accounted for, including all equipment nonidealities. Thus, exergy can be used as a general measure for energy efficiency on very different levels of detail, starting from a first rough estimate based on minimal assumptions and going to the final realization, including equipment-specific losses.

To give an impression on how exergy as a fundamental measure for process evaluation can be utilized, Figure 14.1 shows the chemical exergy of a variety of components, including fossil raw materials, biomass, intermediates, and some products of the chemical industry (Frenzel et al., 2014a, 2014b). What is striking is the fact that one of the major processes in the chemical industry, namely, the conversion of crude oil via ethylene to polyethylene, runs essentially horizontal in this diagram. It has been shown that this is a general requirement for efficient processes, because otherwise excessive energy is required to shift the exergy from a lower to a higher level (Frenzel et al., 2013). It has also been shown that this refers to the net conversion, that is, taking into account the overall exergy change of all reactants and all products. For example, if CO_2, which is shown in Figure 14.1 and has an exergetic value of essentially zero, is released during a reaction, the reactants can be converted into products of higher exergetic value without additional exergy input.

A typical example for such a process is the fermentation of sugar to ethanol, where CO_2 is released without requiring additional energy input. This process is included

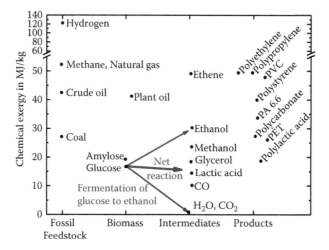

Figure 14.1 Chemical exergy of selected components from different component classes. (Reprinted with permission from Frenzel, P., and Pfennig, A., Methodik zur schnellen Bewertung von Syntheserouten auf Basis von Exergiebilanzen, presented at the ProcessNet-Jahrestagung und 32. DECHEMA-Jahrestagung der Biotechnologen, Aachen, Germany, September 12–15, 2016. http://orbi.ulg.ac.be/handle/2268/201884.)

in Figure 14.1. Of course, in this reaction one-third of the carbon originally contained in the sugar is devalued to CO_2, which corresponds roughly to the increase in exergetic value between glucose and ethanol. This example thus shows that for process evaluation, the carbon balance also has to be considered carefully, because the carbon-containing feedstock was originally produced with significant effort, which in itself can be evaluated exergetically, as well as with respect to land area used (Frenzel et al., 2014b).

For this example, the basic exergetic balances that can generally be applied for biofuels can be discussed. Pure ethanol has an exergy content of almost 30 MJ/kg. Producing the sugar as feedstock requires about 10 MJ/kg, the exergy losses of the reaction are around 3 MJ/kg as well, and the distillation requires around 4 MJ/kg, where all the values are specified with respect to the mass of the product ethanol. Without further measures, the effective exergy content of the final biofuel is thus less than half that of the ethanol, even if no further process losses are taken into account. Thus, it is apparent that the production of ethanol as biofuel can only be efficient if the energy for the process is supplied in a sustainable way, which in this example is typically realized by firing the distillation with the bagasse, which is the remainder of the original sugarcane as feedstock.

Closer inspection of Figure 14.1 also reveals that the components are essentially exergetically ordered according to their oxygen content. Components with a higher oxygen content have a lower specific exergy, which is to be expected, since the chemical exergy of a C=O bond in a molecule cannot be utilized for energy generation, since it is also contained in CO_2, which is the final product of any incineration (Frenzel et al., 2013, 2014b). Therefore, the general statements on exergy changes during reaction can be transferred into statements on oxygen content. If the oxygen content of a component is to be decreased, this will require either significant exergy input or the elimination of a component with high oxygen content, like water or CO_2. Both eliminations lead to a loss of hydrogen or carbon in the product, which is highly undesirable.

Thus, at first sight it may appear that such an oxygen balance could replace the full exergetic evaluation. Unfortunately, oxygen balancing does not allow further refinement, for example, with respect to the evaluation of losses in process steps or equipment. As a consequence, while such carbon, hydrogen, and oxygen balances are also important for the evaluation of processes, only exergetic evaluation leads to a systematic approach that can guide process design through all design stages, from first drafts to individual equipment design.

14.3 Process System Engineering Tools for the Design of Biotechnological Processes

Once the biochemical reaction pathways have been defined, the design of the industrial process has to be considered. Besides the biochemical reactor itself, classical processes include a reactant preparation section, as well as a downstream section. In the following, we have a closer look at the selection of downstream processing steps using a cascaded option trees method. Indeed, downstream steps are often critical for the economic viability of industrial biotechnological processes. Next, we apply process system engineering tools like process design, analysis, and optimization to the case of biorefineries, as they provide an interesting example of biotechnological process integration at the industrial scale.

14.3.1 Cascaded Option Trees

One of the challenges in developing new downstream processes is the large multitude of process options that may appear feasible at the beginning of the design process. In contrast, the reaction step is often already worked out to a certain extent at this point in the design process, because engineers start to ponder over the downstream process only when the reaction toward the desired product turns out to be feasible in principle. For the reaction, on the other hand, often only minor optimizations, for example, of microbial productivity, appear to be required. At the same time, especially for bio-based processes, the properties of the media are demanding, which results as a direct consequence of the higher oxygen content of the feedstock compared with crude oil and the specifics of biotechnological processes. If, for example, high titers of the products are desired, the fermentation broth will generally have a high viscosity, and components with an increased molecular oxygen content originating from the higher oxygen content in the biomass will lead to lower vapor pressures of the target components, compared with intermediates and products of similar molar mass from fossil feedstock. Thus, distillation may no longer be the preferred option for separation, which in turn will lead to a higher solids content, because solids have not been removed by passing through the vapor phase, and also, microbes act as solids in successive process steps. These increasingly demanding characteristics of the bio-based process streams also lead to challenges in the equipment design. Therefore, a design procedure is sought that allows systematic screening of the many process and equipment options, with the goal of finding the best option with the least design effort.

In a variety of applications, in cooperation with industry and academia, the method of cascaded option trees has been developed, which allows us to easily keep track of process options and their evaluation (Bednarz et al., 2014; the description of cascaded option trees is adopted with permission from Pfennig, 2016). The principle structure of characterizing an option in a cascaded option tree is shown in Figure 14.2. All options are listed on the regarded level of detail, and for each option, a variety of criteria can be evaluated. The evaluation can then be coded with either symbols or, for better visualization, green, yellow, and red. Application of the cascaded option trees method is shown in Figure 14.3 for an example of downstream process development, which resembles option trees developed in the mentioned cooperations. It is

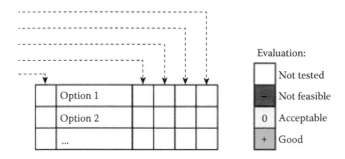

Figure 14.2 Principal representation of options in a cascaded option tree. (Reprinted with permission from Pfennig, A., Thermal unit operations, lecture manuscript. Lecture at University of Liège for Master Students in Chemical Engineering, pp. 838–848, 2016.)

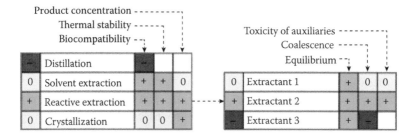

Figure 14.3 Application of cascaded option trees for an example process. (Reprinted with permission from Pfennig, A., Thermal unit operations, Lecture manuscript. Lecture at University of Liège for Master Students in Chemical Engineering, pp. 838–848, 2016.)

apparent that the method can be applied on very different cascading levels of process design, starting from overall process options and going down to specific equipment design. In the example shown, it is obvious that as soon as one of the criteria leads to the insight that an option is not feasible, this option does not need to be investigated further, so the effort for process design is minimized. The effort can be further reduced by choosing to evaluate the most critical criteria early in the evaluation process, because then the maximum number of options can be ruled out based on those criteria, avoiding effort in evaluating successive criteria.

One of the advantages of cascaded option trees is that any method can be used for the evaluation of option feasibility with respect to the design criteria. This can range from expert opinion, literature search, and simulations on different levels of detail to dedicated experiments. The source of the evaluation may be noted in the option trees as well, so that the evaluation can in principle be revised at any later time. Such flexibility also applies to the overall evaluation. Therefore, the best-suited option according to the regarded criteria is directly apparent in a well-structured way. Of course, only these optimal options will be investigated further as a first approach. However, if it then turns out in the evaluation of further criteria that these options are actually infeasible or pose significant problems, the second-best options are directly visible.

Working in project consortia, the method of cascaded option trees has also proven helpful to communicate the status of the design process, as well as the current results, in a very clearly organized way. Finally, it should be noted that upstream and reaction steps should preferably be included in the option trees from the start, since during the early stages of process development, the parameters of upstream and reaction steps, which actually may influence the downstream quite drastically, can still be adjusted relatively easily. If, for example, extractive steps are considered options in the downstream process, the nature of the fermentation medium may significantly influence coalescence behavior, which is a decisive parameter in extraction-equipment design. Thus, only if, for example, the different options for the buffer system and the fermentation medium for a biotechnological step are included while evaluating options for the downstream process, the overall optimum process can be found.

In biotechnological processes, product inhibition is frequently encountered, or the product is even toxic to the microorganisms. As a consequence, the product needs to be removed continually, ensuring a relatively low concentration in the fermenter. This can, for example, be realized by a pump-around, from which the product is removed in a separation step, preferably without the need to separate the microorganisms first. If a low product concentration in the fermenter is required, this will lead to a large flow rate of the pump-around and, correspondingly, large equipment for the separation step. Thus, while the separation may be feasible in principle, the equipment dimensions in such a case may not be. Thus, it cannot be overemphasized that in the evaluation of process options, the overall feasibility of process realization and equipment size on the desired technical scale needs to be taken into account for each option as well. Simple balances may help at this stage to gain significant insight, for example, on the flow rates required to keep product concentrations sufficiently low to avoid product inhibition at the production rate of the microorganisms.

14.3.2 Process System Engineering in Biorefineries

A biorefinery is an integrated processing facility where biomass is converted into a variety of products, ranging from value-added chemicals to fuels after many physical, chemical, and thermal conversion steps, with minimal waste and emissions. In a biorefinery, a variety of feedstocks, many products, and a large number of technologies exist. In addition, the energy needed for the conversion has to be supplied from the feedstocks or from the waste streams, so the process developers face great challenges while considering economic performance, energy requirements, and environmental impact simultaneously. A systematic approach for the design of a biorefinery integrates the use of advanced process synthesis, process analysis, and optimization methods (Antonis et al., 2012). Such comprehensive methods are crucial to developing, designing, and commercializing sustainable and cost-effective biorefineries (Yuan et al., 2013). Process synthesis methods collect the input data about different biorefinery processes and produce energetically and economically efficient biorefinery flowsheets with good operating conditions and design specifications. Process modeling and simulation are complementary approaches to analyze different designs obtained from process synthesis for the optimal configurations of biorefineries (Martín and Grossmann, 2012; Metzger et al., 2012). Over the last 45 years, researchers have studied process synthesis, and several have published extensive reviews on this subject (Hendry et al., 1973; Hlaváček, 1978; Westerberg, 1980; Nishida et al., 1981; Stephanopoulos, 1981). Recently, several researchers have been working in the field of process synthesis and the design of large-scale biorefinery systems. For example, the studies by Baliban et al. (2013) and Tay et al. (2011) addressed the determination of the single best technology for the production of single products, especially biofuels. Further, there are also some studies that take into account superstructures with multiple feedstocks, products, and conversion pathways (Santibañez-Aguilar et al., 2011; Kim et al., 2013; Murillo-Alvarado et al., 2013).

In the next sections, a brief review of advances in the area of process system engineering is provided, where available systematic techniques for process synthesis, design, integration, and optimization are introduced.

14.3.2.1 Process Flowsheeting or Synthesis

The process flowsheeting approaches are generally divided into three categories or some combination thereof: (1) methods that use heuristics, (2) methods that use thermodynamic targets and process integration, and (3) methods that use superstructures, mathematical programming, and optimization. All three approaches focus on flowsheet synthesis based on economics optimization. In the heuristic approach, one can decide whether the plant will be operated in batch or continuous, the type of reactor used, the recycle schemes for the material, the methods and sequence of separations, the energy integration applied, and so forth. In thermodynamic analysis, one has to decide the units and streams that are considered for the heat recovery, the thermodynamic targets to be used, and the level of utilities to be involved. In the optimization approach, the extent of the superstructure, the physical data included, the employed objective function, and the constraints and uncertainties to consider are the features to be decided by engineering (Seferlis and Georgiadis, 2004). Major contributions in the first two approaches (heuristics and thermodynamic targets) are hierarchical decomposition (Douglas, 1988) and pinch analysis, which is used to identify the possible heat recovery (Linnhoff, 1993). They have been successfully applied in many industrial applications (Martín and Grossmann, 2012). A more recent trend is to combine the mathematical programming approach with algorithmic methods (or optimization techniques) that can be used effectively in process synthesis (Grossmann et al., 1999). There are three steps in the combined approach: (1) the development of a representation for alternative superstructures, (2) the formulation of a mathematical program for the selection of the configuration and operating levels from the superstructure, and (3) the solution of the optimization model (Martín and Grossmann, 2012).

14.3.2.2 Process Modeling

In the context of process synthesis, process models are necessary for simulating process flowsheets. A mathematical model is used in process simulation that represents the behavior of the process. A mathematical process model consists of a set of variables that describe important properties of the process and a set of equations that build relationships among the variables to explain the behavior of the process (e.g., heat and mass balances). For instance, the thermodynamic models mentioned in Section 14.2 are included in the process model to describe the behavior of the (bio-)chemical systems, depending on the operating conditions. Process models also include mass and energy equations describing various equipment used in the process, such as reactors or separators. For complex systems found in the chemical industry, the process models represent mostly nonlinear behaviors. With the increased interest in process synthesis, the demand for process models with increased accuracy is also growing. One important concept in process system engineering is superstructure optimization. A superstructure contains most (or all) of the system alternatives. Superstructures are defined by the process modeler, and alternative superstructures can be derived for the same process. Then, a mathematical optimization approach seeks to find the optimal configuration among the proposed alternatives (Trespalacios and Grossmann, 2014).

14.3.2.3 Process Integration

Process integration is a design approach that deals with energy efficiency, waste minimization, and the efficient use of raw materials. The process integration stage is greatly facilitated by the use of process simulations. In biorefineries, process integration plays an important role in studying strong trade-offs (Antonis et al., 2012).

14.3.2.3.1 Heat Integration. To create a heat cascade in processes, pinch analysis is a very powerful thermodynamics-based technique with a structured approach to identifying minimum energy consumption targets for heating and cooling, and the maximum internal heat recovery within a process (Linnhoff, 1993; Smith, 1995; El-Halwagi, 2008). This concept is also applicable in other areas besides heat recovery. In fact, whenever an amount (e.g., heat or mass) has a quality (e.g., temperature or concentration), the concept of composite curves can provide a view of the problem related to the efficient recovery (or reuse) of resources. On the composite curves, the pinch point shows the location where there is an accumulated deficit of an amount above a certain quality (Gundersen, 2000). Maréchal and Kalitventzeff (1998) developed process integration techniques to study the energy supply and heat recovery in industrial processes. Duran and Grossmann (1986) proposed an algorithm for simultaneous heat integration and optimization of chemical processes. Pfeffer et al. (2007) applied process integration to a bioethanol production plant with minimization of heat demand as the optimization objective. More recently, Gassner and Maréchal (2009) developed a methodology that can be applied for the conceptual design of biofuel plants based on optimization with identification of promising flowsheets and process integration methods; they applied process synthesis on wood gasification processes.

14.3.2.3.2 Mass Integration.   Mass integration is a systematic methodology for generating, separating, and allocating streams and species along the process. Mass pinch was developed by El-Halwagi and Manousiouthakis (1989, 1990), and it is applicable to industrial processes where mass exchange occurs between process streams from a number of process units, such as extractors and absorbers. They considered the transfer of a single contaminant only from a rich stream to a lean stream, where a rich stream has a higher concentration of contaminant, while a lean stream represents one with a lower concentration of the contaminant. The target in their studies was to minimize the freshwater flow rate. Mass pinch can be applied in the area of wastewater minimization where the water and wastewater are optimally used by reuse, regeneration, and recycling. Wang and Smith (1994) aimed to minimize wastewater based on maximum water reuse, and they defined the water pinch concentration and targeted the minimum freshwater consumption. Combined methods considering the heat recovery and water savings at the same time are now emerging (Ahmetovic and Grossmann, 2011).

14.3.2.4 Costing and Life Cycle Analysis

With the growing interest in sustainability, biorefineries are likely to play significant roles in enhancing energy security and mitigating climate changes. Process modeling and simulation are essential for predicting the economic, environmental, and social performance of industrial processes (Gerber et al., 2011). In the recent past, researchers have mainly focused on the development of cost-effective biorefineries that can only be achieved by optimal mass and energy integrations. In order to make the biorefinery profitable, the production of value-added products (e.g., succinic acid and dimethyl ether), the valorization of waste mass and energy streams, and cogeneration are significantly important. If a biorefinery is only producing biofuels, environmental incentives may be required. The sustainable growth of a biorefinery requires not only cost-effective products, but also energy-efficient plants. The environmental assessment of a biorefinery includes land use changes, greenhouse gas emissions, the timing of emissions, waste production, and the environmental impact of products. Some studies in literature specifically focus on the life cycle assessment of biorefineries (Ahlgren et al., 2015).

Biorefineries can use much of traditional equipment from the petrochemical industry (e.g., distillation columns, pumps, heat exchanges, and compressors), and the cost functions for this equipment are well established (Turton et al., 2009). Conversely, some biorefinery processes are under development (e.g., gasifiers, fuel cells, and membrane separation), so their future costs are highly uncertain (Caliandro et al., 2014). Generally, the production cost goes down with an increase in plant size, and the optimal plant capacity depends on the economic value of the product. Since biomass is diluted and diversified in a vast area, a supply chain optimization and economic viability study for different sizes of biorefineries is critical for investment planning.

14.3.2.5 Process Optimization

The development of novel products from biorefineries usually focuses on the conversion of biomass, the selection of reaction pathway, and the maximum reaction yield. After that, different conversion and separation steps are included to form a biorefinery superstructure. The biorefinery superstructure can be optimized for different performance objectives, such as total cost and carbon dioxide emissions (Celebi et al., 2016). The optimization algorithm may generate different biorefinery configurations via product and technology selection, energy, and mass integration, depending on the objective of interest, raw material and utility costs, and product price. Gebreslassie et al. (2012) considered multiobjective and multiperiod optimization of biorefinery supply chains under supply and demand uncertainties. Geraili and Romagnoli (2015) used a multiobjective evolutionary algorithm to quantify the trade-offs between cost and financial risk for biorefineries. Stuart and El-Halwagi (2012) examined the integration of biorefineries into existing processes and infrastructure.

14.3.2.6 Case Study: Synthetic Natural Gas Production from Microalgae through Hydrothermal Gasification

Mian et al. (2015) have applied a systematic approach for the conceptual design of microalgae cultivation and hydrothermal gasification (Figure 14.4) by developing thermoeconomic and environmental models. Considering the energy integration principles, the multiobjective optimization (MOO) methodology has been applied to obtain a set of nondominated solutions (i.e., mathematically equally good solutions) by solving a mixed-integer nonlinear programming model. The set of nondominated solutions is referred to as a Pareto-optimal front, and each nondominated solution shows a different optimal configuration. There are three objective functions, namely, total annual cost, synthetic natural gas (SNG) production, and CO_2 emissions, in this

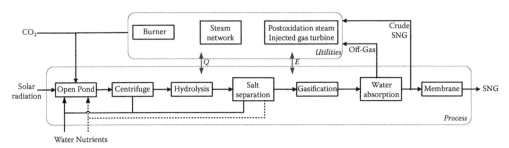

Figure 14.4 Simple flowsheet of microalgae to SNG production. (Adapted from Mian, A. et al., *Comput. Chem. Eng.*, 76, 170–171, 2015.)

optimization problem. Further, the formulated optimization problem has 11 decision variables, which are listed in Table 14.2. More details on the optimization procedure can be found in Mian et al. (2015). Figure 14.5 shows the Pareto-optimal front obtained and the selected optimal (nondominated) solutions for analyzing the effects of decision variables on the optimal designs. Table 14.2 presents the optimal values of decision variables for three selected optimal configurations. Solutions within the

TABLE 14.2   Details of Three Selected Optimal Configurations

Process		Selected Optimal Configurations		
Unit	Decision Variables	A	B	C
Salt separator	$T_{max,salts\ sep}$ [K]	744.21	787.18	851.42
	ΔT_{SSint} [K]	28.45	25.77	20.02
	$\Delta T_{SSbottom}$ [K]	20.00	29.83	30.72
	ΔT_{SStop} [K]	35.02	37.31	35.03
Hydrothermal gasification	HTG pressure	232.50	273.04	299.16
	Gasification reaction inlet temperature	623.29	623.15	623.15
Steam cycle utility	Steam pressure	66.96	61.84	60.29
	ΔT superheating	58.04	66.01	61.43
Pressure recovery expanders	Vapor high-pressure recovery	1	1	0
	Liquid high-pressure recovery	1	1	1
	Liquid low-pressure recovery	1	1	1

Source: Mian, A. et al., *Comput. Chem. Eng.*, 76, 170–183, 2015.
Note: HTG, hydrothermal gasification.

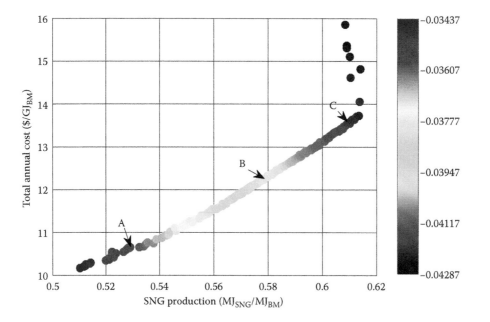

Figure 14.5   Pareto-optimal front for SNG production using microalgae; color map shows kg_{CO2}/MJ_{BM}. (Adapted from Mian, A. et al., *Comput. Chem. Eng.*, 76, 170–171, 2015.)

economic models gave a final SNG production cost of between \$20 and \$30/GJ SNG, which is in line with the price of fossil natural gas.

References

Ahlgren, S., Björklund, A., Ekman, A., Karlsson, H., Berlin, J., Börjesson, P., Ekvall, T., Finnveden, G., Janssen, M., and Strid I. (2015). Review of methodological choices in LCA of biorefinery systems—Key issues and recommendations. *Biofuels, Bioproducts & Biorefining*, 9(5), 606–619.

Ahmetovic, E., and Grossmann, I. E. (2011). Global superstructure optimization for the design of integrated process water networks. *AIChE Journal*, 57(2), 434–457.

Antonis, C. K., Aidong, Y., Marinella, T., and Ta-Chen, L. (2012). Systematic screening of multiple processing paths in biorefineries. In P. R. Stuart and M. El-Halwagi (eds.), *Integrated Biorefineries: Design, Analysis and Optimization*. Boca Raton, FL: CRC Press.

Ataman, M., and Hatzimanikatis, V. (2015). Heading in the right direction: Thermodynamics-based network analysis and pathway engineering. *Current Opinion in Biotechnology*, 36, 176–182.

Baliban, R. C., Elia, J. A., and Floudas, C. A. (2013). Biomass to liquid transportation fuels (BTL) systems: Process synthesis and global optimization framework. *Energy & Environmental Science*, 6, 267–287.

Bednarz, A., Rüngeler, B., and Pfennig A. (2014). Use of cascaded option trees in chemical-engineering process development. *Chemie Ingenieur Technik*, 86(5), 611–620.

Birkenmeier, M., Mack, M., and Roder, T. (2014). A coupled thermodynamic and metabolic control analysis methodology and its evaluation on glycerol biosynthesis in *Saccharomyces cerevisiae*. *Biotechnology Letters*, 37, 307–316.

Caliandro P., Tock L., Ensinas A. V., and Maréchal F. (2014). Thermo-economic optimization of a solid oxide fuel cell—Gas turbine system fueled with gasified lignocellulosic biomass. *Energy Conversion and Management*, 85, 764–773.

Celebi, A. D., Ensinas, A. V., Sharma, S., and Marechal F. (2016). Early-stage decision making approach for the selection of optimally integrated biorefinery processes. Presented at the 29th International Conference on Efficiency, Cost, Optimization, Simulation and Environmental Impacts of Energy System, Portoroz, Slovenia, June 19–23, 2016, p. 508.

Douglas, J. (1988). *Conceptual Design of Chemical Processes*. New York: McGraw-Hill.

Duran, M. A., and Grossmann, I. E. (1986). A mixed-integer nonlinear programming algorithm for process systems synthesis. *AIChE Journal*, 32(4), 592–606.

El-Halwagi, M. M. (2008). *Pinch Analysis and Process Integration: A User Guide on Process Integration for the Efficient Use of Energy* by Ian C. Kemp, 2nd ed. *AIChE Journal*, 54(11), 3031–3031.

El-Halwagi, M. M., and Manousiouthakis, V. (1989). Synthesis of mass exchange networks. *AIChE Journal*, 35(8), 1233–1244.

El-Halwagi, M. M., and Manousiouthakis, V. (1990). Automatic synthesis of mass-exchange networks with single-component targets. *Chemical Engineering Science*, 45(9), 2813–2831.

Finley, S. D., Broadbelt, L. J., and Hatzimanikatis V. (2009). Thermodynamic analysis of bio-degradation pathways. *Biotechnology and Bioengineering*, 103(3), 532–541.

Frenzel, P., Fayyaz, S., Hillerbrand, R., and Pfennig, A. (2013). Biomass as feedstock in the chemical industry—An examination from an exergetic point of view. *Chemical Engineering and Technology*, 36(2), 233–240.

Frenzel, P., Hillerbrand, R., and Pfennig, A. (2014a). Exergetical evaluation of biobased synthesis pathways. *Polymers*, 6, 327–345.

Frenzel, P., Hillerbrand, R., and Pfennig, A. (2014b). Increase in energy and land use by a bio-based chemical industry. *Chemical Engineering Research & Design*, 92, 2006–2015.

Frenzel, P., and Pfennig, A. (2016). Methodik zur schnellen Bewertung von Syntheserouten auf Basis von Exergiebilanzen. Presented at the ProcessNet-Jahrestagung und 32. DECHEMA-Jahrestagung der Biotechnologen, Aachen, Germany, September 12–15, 2016. http://orbi.ulg.ac.be/handle/2268/201884.

Gassner, M., and Maréchal, F. (2009). Thermo-economic process model for thermochemical production of synthetic natural gas (SNG) from lignocellulosic biomass. *Biomass and Bioenergy*, 33(11), 1587–1604.

Gebreslassie, B. H., Yao, Y., and You F. (2012). Multi-objective optimization of hydrocarbon biorefinery supply chain designs under uncertainty. In *51st IEEE Conference on Decision and Control*, 5560–5565. Maui, HI, December 10–13, 2012, p. 509.

Geraili, A., and Romagnoli J. A. (2015). A multi-objective optimization framework for design of integrated biorefineries under uncertainty. *AIChE Journal*, 61(10), 3208–3222.

Gerber, L., Gassner, M., and Maréchal, F. (2011). Systematic integration of LCA in process systems design: Application to combined fuel and electricity production from lignocellulosic biomass. *Computers & Chemical Engineering*, 35(7), 1265–1280.

Grossmann, I., Caballero, J., and Yeomans, H. (1999). Mathematical programming approaches to the synthesis of chemical process systems. *Korean Journal of Chemical Engineering*, 16(4), 407–426.

Gundersen, T. (2000). A process integration PRIMER. Trondheim, Norway: SINTEF Energy Research, International Energy Agency.

Hendry, J. E., Rudd, D. F., and Seader, J. D. (1973). Synthesis in the design of chemical processes. *AIChE Journal*, 19(1), 1–15.

Hlaváček, V. (1978). Synthesis in the design of chemical processes. *Computers & Chemical Engineering*, 2(1), 67–75.

Hoffmann, P., Voges, M., Held, C., and Sadowski, G. (2013). The role of activity coefficients in bioreaction equilibria: Thermodynamics of methyl ferulate hydrolysis. *Biophysical Chemistry*, 173–174, 21–30.

Jankowski, M. D., Henry, C. S., Broadbelt, L. J., and Hatzimanikatis, V. (2008). Group contribution method for thermodynamic analysis of complex metabolic networks. *Biophysical Journal*, 95, 1487–1499.

Kim, J., Sen, S. M., and Maravelias, C. T. (2013). An optimization-based assessment framework for biomass-to-fuel conversion strategies. *Energy & Environmental Science*, 6(4), 1093.

Linnhoff, B. (1993). Pinch analysis: A state-of-the-art overview: Techno-economic analysis. *Chemical Engineering Research & Design*, 71(5), 503–522.

Maréchal, F., and Kalitventzeff, B. (1998). Process integration: Selection of the optimal utility system. *Computers & Chemical Engineering*, 22(Suppl. 1), S149–S156.

Martín, M., and Grossmann, I. E. (2012). On the systematic synthesis of sustainable biorefineries. *Industrial & Engineering Chemistry Research*, 52(9), 3044–3064.

Metzger, M., Glasser, B., Patel, B., Hildebrandt, D., and Glasser, D. (2012). Teaching process design through integrated process synthesis. *Chemical Engineering Education*, 46(4), 260–270.

Mian, A., Ensinas, A. V., and Marechal, F. (2015). Multi-objective optimization of SNG production from microalgae through hydrothermal gasification. *Computers & Chemical Engineering*, 76, 170–183.

Murillo-Alvarado, P. E., Ponce-Ortega, J. M., Serna-González, M., Castro-Montoya, A. J., and El-Halwagi, M. M. (2013). Optimization of pathways for biorefineries involving the selection of feedstocks products, and processing steps. *Industrial & Engineering Chemistry Research*, 52(14), 5177–5190.

Nishida, N., Stephanopoulos, G., and Westerberg, A. W. (1981). A review of process synthesis. *AIChE Journal*, 27(3), 321–351.

Pfeffer, M., Wukovits, W., Beckmann, G., and Friedl, A. (2007). Analysis and decrease of the energy demand of bioethanol-production by process integration. *Applied Thermal Engineering*, 27(16), 2657–2664.

Pfennig, A. (2016). Thermal unit operations, lecture manuscript. Lecture at University of Liège for Master students in Chemical Engineering, pp. 838–848.

Prausnitz, J. M. (2003). Molecular thermodynamics for some applications in biotechnology. *Journal of Chemical Thermodynamics*, 35, 21–39.

Santibañez-Aguilar, J. E., González-Campos, J. B., Ponce-Ortega, J. M., Serna-González, M., and El-Halwagi, M. M. (2011). Optimal planning of a biomass conversion system considering economic and environmental aspects. *Industrial & Engineering Chemistry Research*, 50(14), 8558–8570.

Seaver, S., Gerdes, S., Frelin, O., Lerma-Ortiz, C., Bradbury, L., Zallot, R., Hasnain, G. et al. (2014). High-throughput comparison, functional annotation, and metabolic modeling of plant genomes using the PlantSEED resource. *Proceedings of the National Academy of Sciences of the United States of America*, 111, 26, 9645–9650.

Seferlis, P., and Georgiadis, M. (2004). *The Integration of Process Design and Control*. Amsterdam: Elsevier Science.

Smith, R. (1995). *Chemical Process Design*. New York: McGraw-Hill, Inc.

Stephanopoulos, G. (1981). Synthesis of process flowsheets: An adventure in heuristic design or a utopia of mathematical programming? In R. S. H. Mah and W. Seider (eds.), *Foundations of Computer-Aided Chemical Process Design*. Vol. 2. New York: Engineering Foundation, p. 439.

Stuart, P. R., and El-Halwagi, M. M. (eds.). (2012). *Integrated Biorefineries: Design, Analysis, and Optimization*. Boca Raton, FL: CRC Press.

Tay, D. H. S., Ng, D. K. S., Sammons, N. E., and Eden, M. R. (2011). Fuzzy optimization approach for the synthesis of a sustainable integrated biorefinery. *Industrial & Engineering Chemistry Research*, 50(3), 1652–1665.

Trespalacios, F., and Grossmann, I. E. (2014). Review of mixed-integer nonlinear and generalized disjunctive programming methods. *Chemie Ingenieur Technik*, 86(7), 991–1012.

Tsatsaronis G. (1996). Exergoeconomics: Is it only a new name? *Chemical Engineering & Technology*, 19, 163–169.

Turton, R., Bailie, R. C., Whiting, W. B., and Shaeiwitz, J. A. (2009). *Analysis, Synthesis and Design of Chemical Processes*. 3rd ed. Englewood Cliffs, NJ: Prentice Hall.

Wang, Y., and Smith, R. (1994). Wastewater minimization. *Chemical Engineering Science*, 49(7), 981–1006.

Westerberg, A. W. (1980). A review of process synthesis. In R. G. Squires and G. V. Reklaitis (eds.), *Computer Applications to Chemical Engineering*. http://pubs.acs.org/doi/10.1021/bk-1980-0124.ch003.

Yuan, Z., Chen, B., and Gani, R. (2013). Applications of process synthesis: Moving from conventional chemical processes towards biorefinery processes. *Computers & Chemical Engineering*, 49, 217–229.

15

Perspectives of Microbial Fuels for Low-Income and Emerging Countries:
Biogas Production

Puhulwella G. Rathnasiri

Contents

15.1 Introduction

Anaerobic digestion has recently gained wide acceptance among the public as a sustainable technology for environmental management, energy generation, and biofertilizer production. Although the major focus in developed countries is waste management, the priority for low-income countries is energy generation for heating and cooking, particularly for rural poor people. In addition, it improves health and produces nutrient-rich organic fertilizer. Instead of high-capital-intensive advanced technologies, low-income countries tend to develop low-capital-intensive anaerobic digestion reactors for domestic and institutional sectors. The promotion of biogas systems, based on raw materials such as animal waste and straw, has been largely targeted toward the rural household sector. Since the 1970s, the success rate of these ventures has been low for many reasons (De Alwis, 2012). The inefficiencies of some of these reactors are due to a lack of pretreatment of feedstock, a lack of process monitoring and control, and the reactor configuration itself. In developed countries, high-rate biogas systems are used, and most of these reactors are operated under controlled operating conditions, such as an optimal temperature higher than ambient. By contrast, reactor designs used in developing countries for the anaerobic digestion of livestock waste are classified as low-rate digesters, which lack heating and mixing capabilities. These digesters are usually sized to be fed human and animal waste generated in one household and to cater to the energy demand of the same household. In practice, this means that digester volumes are between 2 and 10 m^3, and that they produce around 0.5 m^3 of biogas per cubic meter of digester volume (Bond and Templeton, 2011).

Because of the abundance of a wide variety of renewable organic substrates, other biofuels, such as bioethanol and biohydrogen, can be identified as alternative options to biomethane. Bioethanol is produced from starch and sugars as sugar cane, corn, and sugar beet and is defined as a first-generation biofuel. When lignocellulosic residues and wastes, such as rice straw, bagasse, and forest residues, are used, bioethanol is defined as a second-generation biofuel. Both of these require extensive pretreatment of feedstock, such as steam explosion and acid hydrolysis (Levin et al., 2009). In addition, a separation technique such as distillation, with high energy input, is needed to recover bioethanol. As a clean biofuel, the production of biohydrogen from dark fermentation has gained significant advances and has been intensively investigated in Asian countries for the last decade. In dark fermentation, microorganisms anaerobically break down carbohydrate-rich substrates into organic acids and alcohols releasing H_2 and CO_2 (Brenter et al., 2010). One of the challenges of applying this technology to treat waste streams is the competing microorganisms present in the nonsterile conditions. The major drawbacks of dark fermentation are the low hydrogen yield and the formation of by-products, such as acids and alcohols, which again lead to waste disposal problems (Gupta et al., 2013). Moreover, this technology needs delicate control and monitoring of process parameters and culture conditions, and even removal of biogas is needed to maintain low pressures inside reactors.

As a consequence, although advanced large-scale technology is available for bioethanol production, biohydrogen production is advancing mostly at the laboratory scale. Therefore, due to economic constraints and poor technology management practices, implementation of sophisticated technologies for bioethanol and biohydrogen production, especially at the household and institutional levels in low-income countries, still poses a great challenge. By contrast, anaerobic digestion is more appropriate for low-income countries because it is less capital-intensive, has less reactor complexity, and has broad operability and applicability on a variety of feedstocks. In addition, stable digestate produced from the anaerobic digestion process improves sanitary conditions.

The objective of this review is to evaluate the anaerobic digestion technologies practiced in developing countries in the context of technological, socioeconomic, and environmental factors. As well as the above broad categories, different feedstocks, reactor configurations, performance and operational problems, technology dissemination, and the multiple benefits of anaerobic digestion are discussed.

15.2 Classification of Biogas Reactors

Biogas reactors are classified on the basis of operating temperatures as psychrophilic (0°C–20°C), mesophilic (20°C–45°C), and thermophilic (40°C–60°C) (Mungwe and Colombo, 2015), according to feedstock loading cycle (batch, semicontinuous, or continuous), total solids (TS) content in the digester (wet process at 5%–15% TS), or dry process (more than 15% TS in the digester). Digesters are also classified according to the place where they are installed, that is, domestic type, position (vertical tank or horizontal plug flow), digester size, construction material, and institutional and industrial scale. Biogas reactors are broadly classified according to their performance efficiency, that is, high-rate and low-rate reactors. When there is no mixing or heating is provided, the efficiency is low and the reactors are termed low rate. The main low-rate biogas reactors operating in developing countries are fixed-dome reactors, floating-drum reactors, and low-cost plug-flow reactors. Fixed-dome reactors were originally developed in China for manure treatment. By improving the original model, two alternative fixed-dome models were developed in India, namely, the *Janata* and *Deenbandhu* models (Rajendran et al., 2012). The floating-drum model was developed by Khadi and Village Industries Commission (KVIC) in 1962. To overcome the drawbacks of the above major reactor models, a low-cost portable plug-flow reactor was developed in Taiwan.

15.3 Dissemination of Biogas Reactor Technologies in Low-Income Countries

Anaerobic digestion technology at the household level has been largely practiced in India and other parts of Asia, Africa, and Latin America, and the driving force is government involvement in the subsidy, planning, design, construction, operation, and maintenance of biogas plants (Lohan et al., 2015). At the household level, animal manure is the major feedstock for small-scale biogas reactors, and cooking and lighting are the major applications. Including China, the status of the biogas digester in south Asian countries is given in Table 15.1.

In Bangladesh, fixed-dome digesters are the most popular, and about 90% of digesters are cow dung based, 6% utilize poultry waste, and 1%–2% utilize other substrates (Khan and Martin, 2016).

Even though numerous feedstock types are available, cow manure is the primary feedstock, as applied to sub-Saharan Africa (Rupf, 2016). In Ethiopia, a large portion of the population is located in rural villages, and their main source of income is agriculture. Ethiopia's national-level biogas program launched in 2008 and was able to establish more than 8000 family-level fixed-dome biogas digesters island-wide (Kamp and Forn, 2016). Kenya is a country confronting issues with climate, food, and energy. Under the International Fund for Agriculture Development (IFAD), 500 flexi

TABLE 15.1   Status of Biogas Digester in South Asian Countries in 2012

Country	Year of First Digester Installation	No. of Digesters Installed
China	1921	42×10^6
India	1900	50×10^5
Nepal	1955	27×10^4
Vietnam	1964	50×10^4
Bangladesh	1972	70×10^3
Cambodia	1986	20×10^3
Thailand	1960	2×10^2

Source: Adapted from Khan, E.U., and Martin, A.R., *Renew. Sustain. Energy Rev.*, 62, 247–259, 2016.

biogas systems (FBSs) were established at the household level (Sovacool et al., 2015). FBS is a low-cost tubular reactor fabricated from polyethylene or a polyvinyl chloride (PVC) bag, and it resembles plug-flow characteristics. Similar tubular digester technology has been practiced in Vietnam, and 20,000 units had been installed by 2006. Latin American countries, such as Mexico, Nicaragua, Colombia, Costa Rica, Peru, and Bolivia, have implemented and adopted this technology particularly under cold-climate conditions (Martí-Herrero et al., 2014). According to the survey conducted by Surendra et al. (2014), based on Asian and African countries, Nepal installed the largest number of domestic digesters. Among the African countries, Kenya has the highest installation. Number of biogas digesters installed in Africa in year 2012 increased by 44% compared in year 2011.

15.4 Low-Rate Biogas Reactors

15.4.1 Continuous-Flow Fixed-Dome Biogas Digester

Continuous-flow biogas digesters are mainly used for the treatment of animal manure and human waste in slurry form. If a reactor is fed with other feedstock, such as food wastes or vegetal biomass, homogeneous slurry must be prepared. A schematic diagram of this reactor is shown in Figure 15.1. The reactor is basically composed of three parts, the inlet, fermentation chamber, and outlet hydraulic chamber, and they are connected by inlet and outlet pipes. On top of the fermentation holder, the dome is equipped with a manhole cap filled with water to check for gas leaks. Biogas generated in the liquid phase is collected in the headspace, and variable pressure builds up. When slurry is fed, equal amounts of sludge displace into the hydraulic chamber (SLSI, 2006).

Available reactor sizes vary from 6 to 12 m^3, and their daily feedstock requirement and biogas generation potential are given in Table 15.2. A lack of mixing in the fermentation chamber is the major disadvantage. Since the reactor is built underground, expert masonry is required to prevent troubleshooting and reactor structure failures upon frequent use. If the water level of the ground is shallow, failure

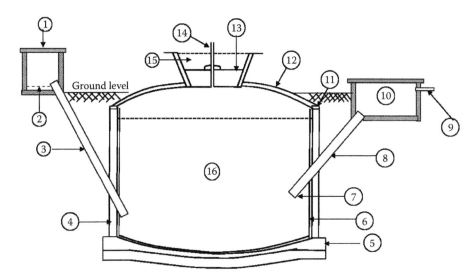

Figure 15.1 Schematic diagram of fixed-dome biogas reactor. 1, inlet; 2, sand trap; 3, inlet pipe; 4, brick wall; 5, lower ring beam; 6, gas tight plaster; 7, outlet opening; 8, outlet pipe; 9, overflow pipe; 10, hydraulic chamber (outlet); 11, upper ring beam; 12, dome; 13, movable cover; 14, gas pipe; 15, water seal; 16, fermentation chamber. (Adapted from SLSI [Sri Lanka Standard Institution], Code of practice for design and construction of biogas systems part 1—Domestic biogas systems, Sri Lanka Standard 1292:2006, SLSI, Colombo, 2006.)

TABLE 15.2   Biogas Production Potential from Animal Manure and Human Waste

Substrate	Daily Waste Production (kg/Animal)	% Dry Matter (DM)	Biogas Yield (m³/kg DM)	Biogas Yield (m³/Animal/Day)
Pig manure	2	17	3.6–4.8	1.43
Cow manure	8	16	0.2–0.3	0.32
Chicken manure	0.08	25	0.35–0.8	0.01
Human excrement	0.5	20	0.35–0.5	0.04

Source: Adapted from Bond, T., and Templeton, M.R., *Energy Sustain. Dev.*, 15, 347–354, 2011.

of the reactor can occur due to leakage of water into the fermentation chamber via cracks. The fermentation chamber of a continuous-flow biogas digester during the initial construction stage and after completion of construction of the dome is shown in Figure 15.2a and b.

In China and India, the growth of biogas technology happened using pig and cow manure as substrates. Cattle dung is considered an ideal substrate due to the presence of methanogenic bacteria in the stomachs of ruminants (Bond and Templeton, 2011). Thus, information on the amount of manure production by animals and their biogas generation potential is required when sizing these reactors and is shown in Tables 15.2 and 15.3.

(a)

(b)

Figure 15.2   (a) Initial stage of constructing the dome of a digester. (b) External view of the fermentation chamber of the continuous-flow biogas digester after the construction stage. (Adapted from Switch Asia project report by people in need Srilanka, 2014.)

TABLE 15.3   Daily Feed Requirement and Biogas Generation from Continuous-Flow Biogas Digester

Digester Volume (m³)	Amount of Biogas Production (m³/day)	Amount of Waste Required Daily (kg)	Approx. No. of Animals Required for Each Type		
			Cow/Buffalo	Poultry	Pig
6	1.5	25	2–4	140	11–12
8	2.0	50	4–6	300	22–23
10	2.5	75	6–8	–	33–34
12	3.0	100	8–10	–	44–45

Source: Adapted from SLSI (Sri Lanka Standard Institution), Code of practice for design and construction of biogas systems part 1—Domestic biogas systems, Sri Lanka Standard 1292:2006, SLSI, Colombo, 2006.

15.4.2 Floating-Drum Digester

The floating-drum digester was developed by KVIC in India in 1962. A schematic diagram of this digester is shown in Figure 15.3. This design consists of a movable inverted steel drum placed on a cylindrical-shaped digester. An inverted steel drum acts as a gas storage tank and moves up and down depending on the amount of accumulated gas at the top of the slurry. Floating-drum digesters produce biogas at a constant pressure with variable volume, and also from the position of the drum, the amount of biogas accumulated under the drum is easily detectable (Rajendran et al., 2012). Movement of the gas holder causes it to break scum formation and improve reactor performance. However, the acidic nature of the slurry causes the steel drum to corrode, requiring a coating of paint (Deepanraj et al., 2014).

15.4.2.1 Dry-Batch Digester
This is a version of the floating-drum reactor, but gas collection is conducted using an external water jacket via displacement of water. The feedstock to the dry-batch digester can be a mixture of straw and cow dung, food waste, market waste, aquatic

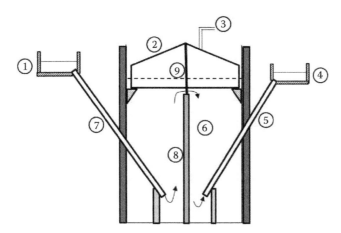

Figure 15.3 Schematic diagram of floating-drum biogas reactor. 1, mixing pit; 2, gas holder; 3, gas outlet; 4, outlet pit; 5, outlet pipe; 6, slurry; 7, inlet pipe; 8, partition wall; 9, central guide.

weed (e.g., water hyacinth), and other agricultural residues. This reactor was developed by the National Engineering Research and Development Centre (NERDC) of Sri Lanka in 1992. Since then, it has been used at the domestic and institutional levels. In this dry-batch process, feedstock is fed with a required inoculum, mostly cow manure, and the reactor is sealed. Water is also added at the start-up, to make a semisolid mixture. A schematic diagram of this reactor is shown in Figure 15.4.

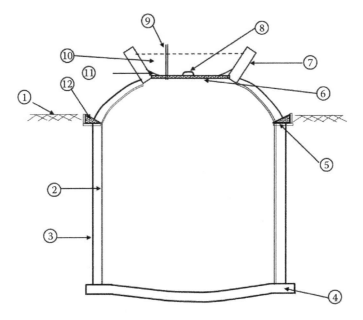

Figure 15.4 Schematic diagram of dry-batch reactor. 1, ground level; 2, gastight plaster; 3, brick work; 4, concrete bottom; 5, brick work; 6, concrete lid; 7, brick work; 8, manhole handle; 9, biogas outlet; 10, water seal; 11, clay sealing; 12, concrete.

The complete cycle time for degradation varies according to the type of substrates. For example, paddy straw may need about 6 months, while 3 months is needed for grass. This is a typical batch reactor process, and the gas production rate varies with time. Following start-up, until 2–3 weeks, no combustible biogas is produced, and it mostly contains carbon dioxide. When the gas production diminishes, the reactor must again be charged with fresh feedstock. Biogas generated is stored in inverted barrels floating in a water tank (Figure 15.5a and b). Because of this feature, constant-pressure gas output is available.

A 1-ton-capacity dry-batch digester is normally recommended for an average four-member family for the purposes of basic cooking and lighting needs. Although this is recommended for solid-state biomass feedstocks, it suffers from the following drawbacks: the conversion efficiency decreases due to a lack of mixing and leachate recirculation, and pretreatment such as disintegration is required when charging the reactor. Different stages of the start-up procedure are shown in Figures 15.6 and 15.7.

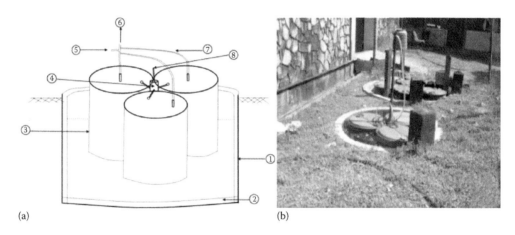

(a)   (b)

Figure 15.5 Gas collection system. (a) Schematic diagram of gas collection system. 1, brick work; 2, bottom concrete; 3, barrel; 4, barrel clamp; 5, gas inlet; 6, gas outlet; 7, connecting tube; 8, center G.I. pipe. (b) Picture of gas collection system.

Figure 15.6 Initial collection and preparation of feedstock (straw).

(a)   (b)

(c)   (d)

Figure 15.7   Different stages of start-up of dry-batch anaerobic digester. (a) Soaking straw in a liquid mixture of manure and water. (b) Mixture inside fermentation chamber. (c) Placing manhole, that is, concrete lid. (d) Concrete lid is sealed with clay to prevent gas leakage.

15.4.3 Plug-Flow Digesters

15.4.3.1 ASTRA Model
Continuous-flow fixed-dome-type anaerobic reactors have been designed to treat feedstock in slurry form and are not successful in treating biomass feedstocks, such as agricultural residues and market wastes. In order to ferment typical biomass residues in conventional slurry-based biogas plants, they need to be subjected to either physical or chemical treatment. When residues are converted into particle forms and fed into the reactor, they float on top of the liquid and retard the fermentation process. To avoid this drawback and incorporate pretreatment into the reactor itself, the plug-flow concept was adopted by Chanakya et al. (2004). Based on research and development studies conducted at the ASTRA Indian Institute of Science (IIsc) in Bangalore, a new plug-flow digester for the treatment of biomass feedstocks came into operation (Figures 15.8 and 15.9). When fresh biomass is fed into the reactor,

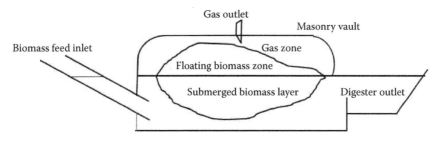

Figure 15.8   Schematic diagram of biomass-based plug-flow reactor.

(a)   (b)

(c)

Figure 15.9 (a) Plug-flow reactor (Indian model). (b) Modified Indian model with external hydraulic chamber at the construction stage (Sirilak Umaga). (c) Installed Sirilak Umaga.

at the front end acedogenesis begins after 3–4 days and volatile fatty acids (VFAs) generate. Because of slow movement, accumulation of VFAs minimizes and thereby diffuses, reducing inhibition. Digestate from this reactor can be further subjected to aerobic composting.

A plug-flow reactor is shown in Figure 15.9a. To recirculate leachate enriched with active methanogens, this reactor has been modified with an external recirculation loop and is shown in Figure 15.9b (Janathaksan, 2016). The Sirilak Umaga reactor following completion of construction is shown in Figure 15.9c.

15.4.3.2 Hybrid Plug-Flow Reactor

By retaining inherent Indian plug-flow reactor features, an additional external hydraulic chamber was added to enhance the recycling of digested slurry from the outlet of the reactor to the feeding tank. This enhances the recycling of active biomass into fresh feed, and thereby improves the methane yield. This model was further improved by adding an open tank on top of the reactor. Organic waste is fed to this liquid separator tank, and water is separately added to enhance digestion (Figures 15.10 and 15.11). Instead of water, leachate from the outlet tank can be recycled. The rotor has been fixed on top of the tank to remove light materials, for example, plastics and mesh to remove grit and nonbiodegradable materials, which mostly float on the

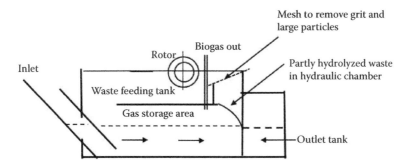

Figure 15.10   Schematic diagram of a hybrid plug-flow reactor. (Adapted from Janathaksan, Colombo, Sri Lanka.)

(a)   (b)

Figure 15.11   (a) Feeding source-separated organic materials into hydrolysis chamber. (b) Manual separation of nonbiodegradable materials.

liquid surface. This reactor technology has been implemented in a local municipality to treat 1 ton/day of source-separated municipal solid waste, which mainly consists of household food waste and market waste. Following 4 days' retention time to enhance hydrolysis, acidified feed is transferred to the bottom chamber, where anaerobic digestion occurs. The sizes of the first and second tanks are 30 and 50 m^3, respectively. The expected daily biogas generation is 88 m^3, and the sludge generated is used in composting. The biogas generated is used to run a 70 kW generator, and when it is not in operation, hot water is generated for on-site requirements of the workers.

15.4.4 Prefabricated Biogas Reactors

The three major low-rate reactors discussed in the above sections are built on-site using concrete and bricks. These reactors are also not movable. Construction of these

TABLE 15.4   Advantages and Disadvantages of Two Biogas Reactor Technologies

Digester Type	Advantages	Disadvantages
Fixed-dome digester	Low initial cost	Requires high technical skills for gastight construction
	Long, useful life span	Difficult to repair in case of leakage
	No moving or rusting parts involved	Requires heavy construction materials
	Compact basic design	Amount of gas produced is not immediately visible
	Less land required if built underground	
	Low maintenance	
Floating-drum digester	Simple and easy to understand operation	
	Visible stored gas volume	
	Constant gas pressure	
	Relatively easy construction	

Source: Adapted from Cheng, S. et al., *Renew. Sustain. Energy Rev.*, 34, 387–400, 2014.

digesters takes considerable time, and special skills are needed. Because of the inferior quality of construction and the acidic environment, reactor failure is possible. The advantages and disadvantages of fixed-dome and floating-drum reactors are given in Table 15.4. Thus, to overcome these drawbacks and adopt reactors under different geographical, topographical, and climate conditions, new bioreactor models are needed (Cheng et al., 2014). Prefabricated biogas reactors (PBDs) are strongly recommended under the following circumstances: where sites are located in remote areas and it is difficult to transport construction materials, where the ground water level is high, where it is difficult to construct with brick and concrete, where the rebuilding of residential areas affects conventional digesters, and where there is limited availability of a skilled workforce.

Prototypes of PBDs are available in fixed-dome, floating-drum, and plug-flow configurations. China dominates in the production of PBDs. According to the classifications made by the China Association of Rural Energy (CAREI), two PBD models are available: composite material digesters (CMDs) and bag digesters (BD). CMDs are fabricated from either fiber-reinforced plastic (FRP) or soft plastic. BDs are fabricated from hard plastic (Cheng et al., 2014).

15.4.4.1 Composite Material Digesters
Appropriate Rural Technology Institute (ARTI) developed a CMD for the treatment of food waste generated in households. This compact biogas digester resembles the floating-drum digester and is made using two cut-down high-density poly ethylene (HDPE) tanks. The larger tank acts as the container containing the waste material, while the smaller one is inverted and telescoped into the larger one (Voegeli et al., 2009). The small tank serves as the gas holder, while the larger tank serves as the fermentation chamber. A similar reactor technology, named "green gas reactor," was developed by a local entrepreneur in Sri Lanka to treat biodegradable organic fractions of kitchen waste. This reactor resembles a portable floating-drum-type biogas unit that requires

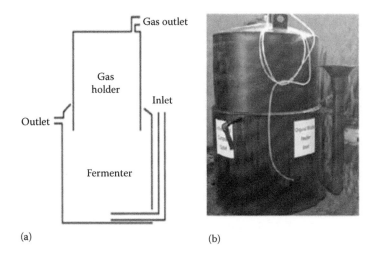

(a)   (b)

Figure 15.12   Green gas reactor: (a) schematic diagram and (b) picture of operating unit.

TABLE 15.5   Reactor Performance of Green Gas Reactor

Reactor Capacity (m³)	Kitchen Waste/ Day (kg)	Water Input/ Day (m³)	Biogas Production/ Day (kg)
0.5	4	0.008	0.16
1.0	8	0.015	0.4
5.0	30	0.060	2

Source: Adapted from Arpico, Arpico green gas unit, Arpico, Maharagama, Sri Lanka, http://www.rpcpolymers.com/greengass.php.

batch or semibatch feeding. A schematic diagram and picture of this reactor are shown in Figure 15.12. This reactor is available in three different sizes, varying from 0.5 to 5 m³, and the performances are given in Table 15.5. The major deficiency of this reactor is the lack of stirring, and the particle size should also be reduced before feeding.

BiogasPro Agama is a CMD that was developed in South Africa and is popular among countries such as Botswana, Namibia, Mozambique, and Tanzania (BiogasPro Agama, 2016). Although these PBDs are popular among rural and urban households, there are challenges for the dissemination of this technology, such as low quality, high capital investments, and lack of standards and after-sale service facilities.

15.4.4.2 Bag Digesters (Tubular Digesters)
The low-cost tubular digester design was first introduced by Taiwan and later disseminated among countries such as Vietnam, Barbados, Colombia, Kenya, Bolivia, Costa Rica, the Philippines, and Kenya. Construction of the tubular digester is not as complex as for CMDs. These digesters are mostly suitable for remote areas where it is difficult to transport or acquire conventional constructional materials. The tubular digester is made from a low-cost polyethylene tube, and both ends of the tubular film are bent around a 6-inch PVC drain pipe, which is then wound with a recycled tire tube (Cheng et al., 2014). According to the review conducted by Kinyua et al. (2016), the digester length can vary from 8 to 40 m, with a circumference of, 3.6–5 m.

The volume of the digester varies from 2.4 to 12 m^3, and the approximate liquid volume is 75% of the total volume. The digester is placed inside a trench to provide support for the weight of the reactor liquid. Inlet and outlet pipes are positioned at 45° to maintain equal influent and effluent flows. When there is no insulation, the biogas production rate of this digester is about 30% lower than that of the fixed-dome digester. Therefore, roof shelters are frequently used to regulate temperature inside the digester. Depending on the availability of funding, a biogas storage bag can be installed above the digester. For the treatment of swine and cow manure, this digester has been installed in the Monteverde region of Costa Rica, and situations with and without a gas storage bag are shown in Figure 15.13.

With the objective of scaling and commercializing biogas systems in Kenya, the IFAD and Biogas International distributed 500 FBSs among rural households (Sovacool et al., 2015). A FBS contains four parts: plastic input and output pipes, a reinforced plastic digester bag, plastic piping to transport biogas, and a plastic greenhouse covering. The major advantages of this system are its low cost, mobility, requirement of less manure for start-up, and shorter retention time. To start the FBS, 200–400 kg of fresh manure and an equal amount of water is required. When the reactor is stable, a daily feeding of 20 kg and 20 L of water will produce 700–1000 L of biogas, equivalent to 2–3 h of cooking. Similar biogas development was launched in Bolivia by the Energising Development Programme, from 2007 to 2012. The low-cost tubular reactor technology was introduced to three different climatic regions

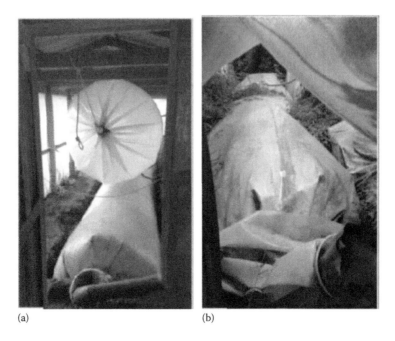

(a)   (b)

Figure 15.13 Tubular digester treating swine waste manure generated in Costa Rica: (a) with a gas bag (a) and (b) without a gas bag. (Adapted from Kinyua, M.N. et al., *Renew. Sustain. Energy Rev.*, 58, 896–910, 2016.)

within Bolivia: cold, tropical, and warm. In total, 747 digesters were installed under this program (Martí-Herrero et al., 2014). This tubular reactor is made from double tubular polyethylene, each with a thickness of 0.15–0.3 mm. Other technical features resemble the typical low-cost tubular technology. From the total number of reactors installed, 87% dairy cattle feedstock and 11.5% pig manure are used, and the rest is poultry and sheep manure. This digester is fed daily with 20 kg of manure and 60 L of water, producing 0.7–0.8 m^3 of biogas with 80 L of liquid biofertilizer, which is known as biol in Bolivia. Even in cold climates, these results are achieved by insulation of the digester by means of a greenhouse.

15.5 Case Study

15.5.1 Introduction to a Pilot Plant

Source-separated food waste collected from a university student canteen was used as the feedstock for the pilot plant. The major objective of establishing this pilot plant was to produce biomethane and use purified upgraded biomethane as a transport fuel in a three-wheeler, a vehicle used by common people in Sri Lanka. In addition, this pilot plant (see Figure 15.15) was used as a demonstration plant to conduct anaerobic codigestion using canteen food waste as the major substrate and locally available substrates, such as straw, water hyacinth, and Gliricidia leaves as cosubstrates (Figure 15.14).

15.5.2 Sizing of an Anaerobic Digester

To enhance the efficiency of the anaerobic conversion process, it was decided to use a high-rate reactor with mixing (Figure 15.15). The amount of daily methane requirement was fixed at 5 m^3. By assuming a typical methane composition of biogas (55% v/v) and by allowing a safety factor of 25%, the daily biogas requirement was 11.4 m^3. According to several studies conducted with food waste by Mata-Alvarez (2003), the biogas yield is 0.8 m^3/kg of total volatile solids (TVS) degraded. In order to characterize

(a)   (b)   (c)

Figure 15.14 Feedstocks for anaerobic codigestion plant: (a) Gliricidia leaves, (b) water hyacinth, and (c) canteen food waste.

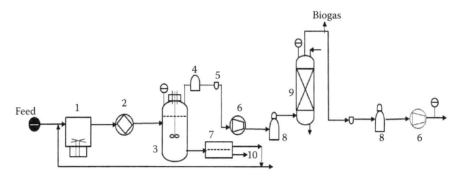

Figure 15.15 Schematic diagram of pilot-scale anaerobic digestion plant to treat canteen food waste and other cosubstrates. 1, food waste crusher; 2, submerged pump; 3, anaerobic reactor; 4, gas storage with water displacement; 5, moisture trap; 6, compressor; 7, sedimentation tank; 8, intermediate gas storage cylinders; 9, water scrubber; 10, sludge.

canteen food waste, Kularathne (2010) conducted physicochemical analyses and lab-scale batch experiments. The TS content and TVS content were 37% (w/w) and 24% (w/w), respectively. By dividing the daily biogas requirement by the biogas yield, the daily TVS was calculated as 14.2 kg and the daily food waste requirement was estimated to be 59 kg. Since the moisture content of the canteen food waste was 63%, the solid and water contents were calculated as 22 and 37 kg/day, respectively. By varying the TS contents at 5%, 7%, 10%, 12%, and 15%, lab-scale experiments were conducted with 500 mL bottles and the optimum solid content was found at 10%. The amount of water to be added to make a suspension was calculated as 161 kg/day, leading to a total mass flow to the reactor of 220 kg/day. Assuming a resulting suspension density of 1000 kg/m³ and assigning a hydraulic retention time of 20 days, the reactor bulk liquid volume was estimated at 4.4 m³. As a consequence, a 5 m³ biodigester was built considering 10% excess volume for headspace.

15.5.3 Operation of a Pilot Plant

Before start-up, the piping and reactor were tested for any gas leaks. This was conducted by pressurizing the reactor (Figure 15.16a) at 0.2 bar over pressure. According to the lab-scale experiments, Gliricidia leaves were identified as the best cosubstrate for anaerobic digestion, and a daily feed mixture was prepared according to Table 15.6.

A food waste mixture was crushed for 15 min using a high-speed crusher to reduce the particle size to less than 0.5 mm. This slurry, with a TS content of 10% (w/w), was then transferred into the tank (Figure 15.16c) equipped with a submersible sewage pump (Model V1500DF) to feed the anaerobic reactor semicontinuously at a flow rate of 22 m³/day for the 5 working days of the week. The reactor was operated under mesophilic conditions and mixed for 30 min/day using a vertical shaft equipped with three twisted peddle-type impellers rotating at 20 rpm.

The biogas generated in the digester was collected in an inverted barrel (Figure 15.16c) with water displacement and then passed through a moisture trap before

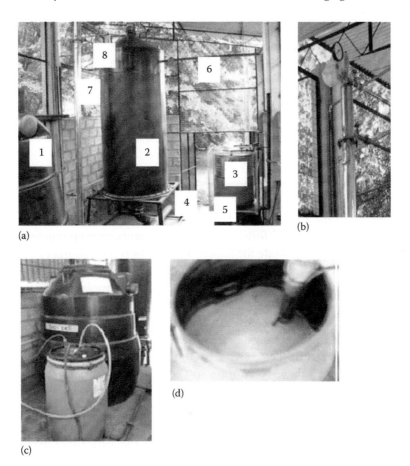

Figure 15.16 (a) Picture of the pilot plant. 1, sedimentation tank; 2, high-rate anaerobic reactor; 3, food waste crusher; 4, sludge drain pipe; 5, slurry tank; 6, input feed pipe; 7, scrubber; 8, biogas outlet pipe. (b) Water scrubber. (c) Biogas collection tank. (d) Slurry tank with submerged pump.

TABLE 15.6 Composition of Different Samples Used for Codigestion Experiments

Substrate	Sample 1	Sample 2	Sample 3	Sample 4
Food waste (%)	70	70	70	90
Water hyacinth (%)	20	–		
Gliricidia (%)	–	20		
Rice straw (%)	–	–	20	
Inoculum	10	10	10	10

being compressed and stored inside intermediate storage cylinders at a pressure of 10 bar. A moisture trap made from a sealed glass tube with a gas inlet and outlet was immersed in a cold water bath. The compressor used was a scuba-type compressor (GS 160 HONDA 5.5). A scrubber made with plastic ring-type packing was used to upgrade the biogas. Gas from the storage tank at a pressure of 10 bar entered into the scrubber (Figure 15.16b) from the bottom, and water at pH 7 was sprayed from the top of the column. This gas scrubbing was conducted batch-wise, and the methane composition was upgraded from 60% to 85%. The methane composition was measured using a potable methane analyzer (MT 28-2005). Upgraded biogas was again stored in high-pressure cylinders and further compressed before use in vehicle gas cylinders at a pressure of 20 bar. Digested slurry could be either drained out to a closed pit or transferred into a sedimentation tank, from where water is recycled into the crusher.

15.5.4 Optimization of Anaerobic Codigestion Process Using Lab Experiments

It was concluded that the instability of the pilot-scale reactor operation was due to a lack of nitrogen source, and therefore lab-scale experiments were conducted to determine the most efficient cosubstrate for anaerobic digestion (Dilnayana et al., 2010).

For these anaerobic codigestion experiments, 60 mL syringes were used as batch reactors containing 50 mL of sample (Figure 15.17). Four samples were investigated according to the different compositions given in Table 15.6.

Inoculum was obtained from an active upflow anaerobic sludge bed (UASB) reactor operating at a local wastewater treatment plant. All experiments were conducted at room temperature (i.e., 35°C ± 3°C). Cumulative biogas production was measured every 3 h for 5 days. According to the batch experiment results (Figure 15.18), the highest cumulative biogas production and initial biogas production rate were observed for sample 2, where gliricedia was used as the cosubstrate. The highest total chemical oxygen demand (COD) removal efficiency (Figure 15.19) was also obtained with this sample. Thus, it was concluded that the gliricedia leaves were the

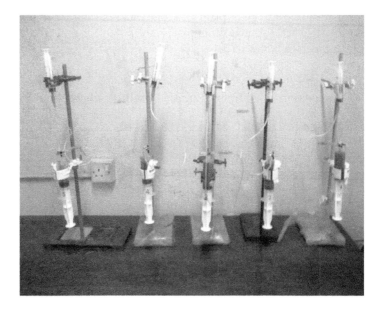

Figure 15.17 Picture of batch experiment.

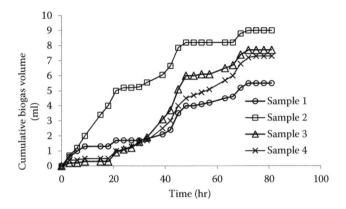

Figure 15.18   Cumulative biogas production from codigestion experiment.

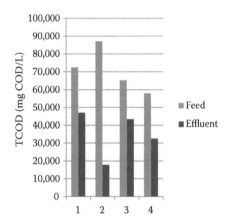

Figure 15.19   Influent (feed) and effluent total COD following batch experiment.

most relevant cosubstrate for anaerobic codigestion with canteen food waste on a larger scale.

15.6 Conclusion

According to the review and critical analysis of existing low-rate reactors for biomethane generation potential, there are pros and cons for each reactor mentioned. However, when modifications are done to improve process performance, significant conversion efficiency can be achieved. When it comes to social acceptance, such as a user-friendly feeding system, compact, highly efficient reactors; easy handling of sludge; and affordability are paramount. To improve the efficiency of low-rate reactors, a new high-rate, low-cost pilot-scale reactor was designed, fabricated, and commissioned at university premises. It was found that anaerobic codigestion improves conversion efficiency, and economically viable less capital-intensive systems can be developed locally and even upgraded to large-scale systems to solve solid waste problems in local municipalities. Implementation of bioethanol and biohydrogen as alternatives to biomethanation at the domestic and institutional levels in low-income countries is not economically and practically viable in the current context.

References

Arpico. Arpico green gas unit. Maharagama, Sri Lanka: Arpico. http://www.rpcpolymers .com/greengass.php (accessed October 5, 2016).

BiogasPro Agama. www.biogaspro.com (accessed on March 6, 2016).

Bond, T., and Templeton, M.R. 2011. History and future of domestic biogas plants in the developing world. *Energy for Sustainable Development* 15: 347–354.

Brenter, L., Peccia, J., and Zimmermann, J.B. 2010. Challenges in developing biohydrogen as a sustainable energy source. Implications for a research agenda. *Environmental Science and Technology* 44: 2243–2254.

Chanakya, H.N., Modak, J.M., and Rajabapaiah, P. 2004. Evolving biomass based biogas plants: The ASTRA experience. *Current Science* 87: 917–925.

Cheng, S., Li, Z., Mang, H., Huba, E., Gao, R., and Wang, X. 2014. Development and application of prefabricated biogas digesters in developing countries. *Renewable and Sustainable Energy Reviews* 34: 387–400.

De Alwis, A.A. 2012. A tool for sustainability: A case for biogas in Sri Lanka. *Journal of Tropical Forestry and Environment* 2(1): 1–9.

Deepanraj B., Sivasubramanian, V., and Jayaraj, S. 2014. Review paper: Biogas generation through anaerobic digestion process: An overview. *Research Journal of Chemistry and Environment* 18(5): 83–90.

Dilnayana, K.W.N., Kularathne, M.A.D.I.C., Rathnasiri, P.G., Joseph, P.G., and De Alwis, A.A.P. 2010. Optimization of an anaerobic co-digestion process and use of bio methane as a transport fuel. Presented at the National energy symposium by SriLanka Sustainable Energy Authority, Colombo, Srilanka, 4–8 August, 2010.

Gupta, S.K., Kumari, S., Reddy, K., and Bux, F. 2013. Trends in biohydrogen: Major challenges and state of the art developments. *Environmental Technology* 34(13–14): 1653–1670.

Janathaksan (GTE) Limited, Sri Lanka. http://janathakshan.com/index.php?lang=en (accessed on March 5, 2016).

Kamp, L.M., and Forn, E.B. 2016. Ethiopia's emerging domestic biogas sector: Current status, bottlenecks and drivers. *Renewable and Sustainable Energy Reviews* 60: 475–488.

Khan, E.U., and Martin, A.R. 2016. Review of biogas digester technology in rural Bangladesh. *Renewable and Sustainable Energy Reviews* 62: 247–259.

Kinyua, M.N., Rowse, L.E., and Ergas, S.J. 2016. Review of small-scale tubular anaerobic digesters treating livestock waste in the developing world. *Renewable and Sustainable Energy Reviews* 58: 896–910.

Kularathne, M.A.D.I.C. 2010. Development of a pilot scale biogas plant to utilize biomethane as a transport fuel. MSc dissertation, University of Moratuwa, Sri Lanka.

Levin, D.B., Carere, C.R., Cicek, N., and Sparling, R. 2009. Challenges for biohydrogen production via direct lignocellulose fermentation. *International Journal of Hydrogen Energy* 34: 7390–7403.

Lohan, S.K., Dixit, J., Kumar, R., Pandey, Y., Khan, J., Ishaq, M., Modasir, S., and Kumar, D. 2015. Biogas: A boon for sustainable energy development in India's cold climate. *Renewable and Sustainable Energy Reviews* 43: 95–101.

Martí-Herrero, J., Chipana, M., Cuevas, C., Paco, G., Serrano, V., Zymla, B., Heising, K., Sologuren, J., and Gamarra, A. 2014. Low cost tubular digesters as appropriate technology for widespread application: Results and lessons learned from Bolivia. *Renewable Energy* 71: 156–165.

Mata-Alvarez, J. 2003. *Biomethanation of Organic Fraction of Municipal Solid Wastes.* London: IWA Publishing.

Mungwe, J.N., and Colombo, E. 2014. Domestic biogas digesters in developing countries: Performance and selection of appropriate technologies for mass dissemination. Presented at the Global Humanitarian Technology Conference (GHTC), San Jose, CA, October 10–13.

NERDC (National Engineering Research and Development Centre of Sri Lanka). http:// nerdc.lk/ (accessed on January 4, 2016).

Rajendran, K., Aslanzadeh, S., and Taherzadeh, M.J. 2012. Household biogas digesters—A review. *Energies* 5: 2911–2942.

Rupf, G., Bahri, P.A., Boer, K., and McHenry, M.P. 2016. Broadening the potential of biogas in sub-Saharan Africa: An assessment of feasible technologies and feed stocks. *Renewable and Sustainable Energy Reviews* 61: 556–571.

SLSI (Sri Lanka Standard Institution). 2006. Code of practice for design and construction of biogas systems part 1—Domestic biogas systems. Sri Lanka Standard 1292:2006. Colombo: SLSI.

Sovacool, B.K., Kryman, M., and Smith, T. 2015. Scaling and commercializing mobile biogas systems in Kenya: A qualitative pilot study. *Renewable Energy* 76: 115–125.

Surendra, K.C., Takara, D., Hashimoto, A.G., and Khanal, S.K. 2014. Biogas as a sustainable energy source for developing countries: Opportunities and challenges. *Renewable and Sustainable Energy Reviews* 31: 846–859.

Voegeli, Y., Lohri, C., Kassenga, G., Baier, U., and Zurbrügg, C. 2009. Performance of the ARTI compact biogas plant for kitchen waste—Case study from Tanzania. Paper presented at the Twelfth International Waste Management and Landfill Symposium, Cagliari, Italy, October 5–9.

Index

Page numbers followed by f and t indicate figures and tables, respectively.

Printed and bound by CPI Group (UK) Ltd, Croydon, CR0 4YY

01/11/2024

01782603-0013